Pelagic Sediments: on Land and under the Sea

Pelagic Sediments: on Land and under the Sea

EDITED BY KENNETH J. HSÜ
AND HUGH C. JENKYNS

Proceedings of a symposium held at the Swiss Federal Institute of Technology, Zürich, 25–6 September 1973
Sponsored by the International Association of Sedimentologists and the European Geophysical Society

SPECIAL PUBLICATION NUMBER 1 OF THE
INTERNATIONAL ASSOCIATION OF SEDIMENTOLOGISTS
PUBLISHED BY BLACKWELL SCIENTIFIC PUBLICATIONS
OXFORD LONDON EDINBURGH MELBOURNE

Published by Blackwell Scientific Publications
Osney Mead, Oxford
85 Marylebone High Street, London W1
9 Forrest Road, Edinburgh
P.O. Box 9, North Balwyn, Victoria, Australia

ISBN 0 632 00167 4

First published 1974

Contents

Spec. Publs int. Ass. Sediment. (1974) **1,** 1–10

Pelagic Sediments: on Land and under the Sea

An introduction

HUGH C. JENKYNS *and* KENNETH J. HSÜ

Department of Geological Sciences, University of Durham, and Geologisches Institut, E.T.H., Zürich, Switzerland

A little over 100 years ago, on 21 December 1872, H.M.S. *Challenger* sailed from Portsmouth Harbour. By the time she returned to England, on 24 May 1876, she had founded the science of oceanography and ushered in a new era of geology. It is to C. Wyville Thomson and John Murray of the *Challenger's* scientific staff that we owe the pioneer studies of Recent oceanic deposits that decades of further work have merely embellished. A preliminary report on the nature of Recent pelagic sediments was published by Thomson in 1874(a); an account by Murray followed in 1876. These papers documented the pteropod, *Globigerina*, diatom and radiolarian oozes and the abyssal red clays. Both of these papers also refer to the first discovery of marine ferromanganese deposits, one catch of which was described by Thomson (1874a, p. 45), in flowing Victorian prose, as 'nearly a bushel of nodules, from the size of a walnut to an orange'.

Some 15 years prior to the departure of the *Challenger*, H.M.S. *Cyclops* had been despatched by the British Admiralty to ascertain the depth and nature of the sea bottom where it was proposed to lay the Atlantic telegraph cable. A sample of the mud obtained was sent to Thomas Henry Huxley for examination. In an appendix to the captain's report, published in 1858, Huxley described from the sediment some tiny concentrically layered calcareous objects that he termed 'coccoliths' (Huxley, 1868).

In 1861 Henry Clifton Sorby published a paper in which he mentioned the occurrence of bodies in the English Chalk that were 'identical with the objects described as Coccoliths by Professor Huxley' (Sorby, 1861, p. 193). The stage was thus set for a comparison between Recent *Globigerina*-bearing ooze and Cretaceous Chalk. Such graphic expressions as 'we are still living in the Cretaceous epoch' began to ornament the literature, although this particular remark incurred the wrath of Sir Charles Lyell and Sir Roderick Murchison (Thomson, 1874b, p. 470). In his classic paper of 1879 Sorby wrote (p. 78) that Chalk was 'very far from being identical with the *Globigerina*-ooze of our modern deep oceans' but was analogous to 'deep ocean mud comparatively free from volcanic and other mechanical mineral impurities . . .'. Although we would not now agree with the imputation of great depth inherent in Sorby's comparison of an ocean mud with the Chalk, his comment underlines recognition of the pelagic nature of this shelf-sea sediment. And by 1891 Cayeux had stressed that the presence of sponges and bivalves in this Cretaceous deposit and the local abundance of rounded quartz grains suggested more near-shore shallow-water environments.

Interpretation of Alpine rocks was, however, to prove more problematic.

Suess, writing in *Entstehung der Alpen* (1875), clearly felt that Alpine Mesozoic rocks were pelagic. However, he wrote before the *Challenger's* results were widely known and his conclusions may have been largely instinctive. He assumed (Suess, 1875, p. 98) that in the northern part of the Eastern Alps the Rhaetian rocks were the least pelagic and that overlying deposits had been laid down in progressively deeper water. This is perfectly in accord with modern interpretation (e.g. Garrison & Fischer, 1969). Suess also suggested (1875, pp. 98–99) that uninterrupted marine sedimentation across the Jurassic-Cretaceous boundary in the Alps signified pelagic conditions; and he contrasted this with the fresh-water Wealden environments that characterized north Germany and England at the same time. He furthermore attributed the formation of the great limestone deposits of the Alps ('die . . . Bildung der grossen Kalkablagerungen der Alpen') to a pelagic setting; although it is unclear to which formations he was referring. If he had waited a year or two he could have made more sophisticated interpretations on the basis of comparative sedimentology by utilizing the *Challenger's* results.

In 1877 Fuchs published a brilliant study on the origin of Aptychus Limestones. Aware of the differential solubility of calcite and aragonite, and inspired by the findings of the *Challenger* on carbonate dissolution in deep seas, Fuchs rejected the idea of post-mortal separation of aptychi from buoyant air-filled ammonite shells, and concluded that the preferential preservation of calcitic operculae resulted from selective dissolution of the aragonitic phragmocones.

In the following year Gümbel (1878) compared ferromanganese nodules from red pelagic limestones of the Lower Jurassic, Eastern Alps to those forming on the ocean floors. Fuchs, following up ideas expressed in his earlier (1877) paper produced, in 1883, a comprehensive treatise that posed the question: 'Welche Ablagerungen haben wir als Tiefseebildungen zu betrachten?'. In this account Fuchs over-reached somewhat in his enthusiasm to attribute a deep-sea setting to all rocks containing pelagic organisms. Nevertheless he was correct in interpreting the red ammonite limestones of the Alpine chain as pelagic deposits and, returning to the problem of the Aptychus Limestones (Fuchs, 1883, pp. 510–512), had no alternative but to suggest their formation in great depths since the *Challenger* had found that delicate aragonitic pteropods could survive down to 4000 m.

Neumayr published his *Erdgeschichte* in 1887 in which he reiterated some of the interpretations of his colleague Fuchs on Alpine Mesozoic rocks. Realizing that the red ammonite limestones of the Trias (Hallstatt facies) and Jurassic owed their pigment to iron-rich argillaceous matter, Neumayr (1887, pp. 364, 366) suggested that they had formed in depths between the present accumulation level of *Globigerina* ooze and red deep-sea clay. Neumayr also considered that the ferromanganese nodules in these ammonite limestones signified great depths. He was thus at pains to stress the similarity of Alpine rocks to Recent oceanic deposits, and (Neumayr, 1887, p. 364) wrote the following: 'Murray, der die Challenger-Expedition mitgemacht und die bei dieser Gelegenheit gesammelten Meeresgrundproben untersucht hat, fand eine Probe jener alpinen Vorkommnisse den Tiefseevorkomnissen unter allen ihm bekannten Gesteinen am nächsten stehend.' Murray's seal of approval was thus stamped on the deep-water interpretation of Alpine pelagic rocks. This, however, was not to last.

Recognition of deep-sea deposits on land was not limited to the Alpine-Mediterranean area. Some Recent *Globigerina* and pteropod limestones from the Solomon Islands described in 1885 by Guppy (in a paper communicated by John Murray), radiolarian cherts from the Ordovician of southern Scotland described by Hinde (1890), certain Tertiary chalks and marls from Barbados (Harrison & Jukes-Browne, 1890), were also interpreted as oceanic sediments. Nicholson, Regius Professor of Natural History at the University of Aberdeen, summarizing these occurrences, wrote the following in his address on recent progress in palaeontology for the year 1890: 'An interesting point in connection with the Radiolarian hornstones of the Mesozoic period, and the chalk-like Radiolarian marls of Barbados and of various parts of Southern Europe and Northern Africa, is that these deposits seem to be clearly ancient representatives of the modern "Radiolarian ooze" of the deep sea. If this point be admitted . . . then we have in these old accumulations indubitable "deep sea deposits;" and it cannot be asserted that no deposits similar to the deep-sea oozes of the present day are to be recognised among the stratified rocks which compose the greater part of the earth's crust. This admission will necessarily have an important bearing upon the modern theory that the present continental areas have been in the main regions of elevation, and the existing oceans in the main areas of depression, since the beginning of the Cambrian period, if not from still earlier times.' (Nicholson, 1890, p. 56.) This last sentence of Nicholson's embodies a controversy; clearly the occurrence of true oceanic deposits on land was incompatible with the permanency of continents and ocean basins. And this particular dogma was to pervade late nineteenth and twentieth century geological thought. This conceptual strait-jacket seems finally to have led John Murray to reject the oceanic nature of pelagic rocks on land. Murray's weighty words in the *Challenger* report on deep-sea deposits, co-authored with the Belgian Renard, read as follows: 'With some doubtful exceptions, it has been impossible to recognise in the rocks of the continents formations identical with these pelagic deposits.' (Murray & Renard, 1891, p. 189.) The 'doubtful exceptions' included the Scottish Ordovician cherts and the Oceanic Deposits of Barbados* alluded to above.

Murray's influence was considerable. No less an authority than Johannes Walther (1897) expressed the view that pelagic sediments were not necessarily deep and oceanic; he even ascribed to a near-shore environment some of the deposits claimed as deep-sea sediments by Fuchs (1883). The inspiration for Walther's interpretations are betrayed by the following sentence: 'Der beste Kenner recenter Tiefsee-Ablagerungen, Dr. John Murray, liess sich von vielen Geologen solche Gesteine zusenden, die man für Tiefsee-Ablagerungen hielt, und konnte feststellen, dass unter diesen Proben mit Ausnahme des Kalkes von Malta kein Sediment sei, das mit recenten Tiefsee-Ablagerungen übereinstimmt.' (Walther, 1897, p. 237.)

Murray's views found many adherents amongst the Anglo-American geological fraternity. Nevertheless certain European workers persisted in the belief that oceanic deposits could be found in ancient mountain chains. Most important among these was Steinmann who in 1905 (p. 50) wrote the following: 'Im Gegensatz zu Murray . . .

* There seems little doubt, in fact, that these rocks *are* truly oceanic (Lohmann, 1973) and their equivalents have been sampled by Deep Sea Drilling in the western Atlantic (Bader *et al.*, 1970).

habe ich mit Hinde . . . u.a. stets die Ansicht vertreten, dass die reinen, kalkfreien Radiolariengesteine, wie sie uns in der Form gleichartiger Massen von mehr oder minder erheblicher Mächtigkeit aus mesozoischen und paläozoischen Formationen bekannt sind, echte Tiefseeabsätze darstellen, denen die gleiche geologische Bedeutung zuzuerkennen ist wie dem Radiolarienschlamm der heutigen Tiefsee.' He then goes on to state (p. 57): 'Die ophiolithischen Massengesteine sind in ihrem Auftreten an die Tiefseezone gebunden . . .' and later adds (p. 59): 'Wir können uns wohl vorstellen, dass unter den grossen Meerestiefen sich magmatische Massen von extremer Basizität ansammeln und dass bei der Auffaltung der abyssischen Regionen diese Massen mit aufsteigen und zur Injektion gelangen . . .'. Already implicit in this paper is the concept of ophiolites as ocean crust and radiolarites as ancient analogues of deep-sea siliceous oozes; with the advantage of hindsight it makes sobering reading. Yet clearly Steinmann was decades before his time; outside continental Europe his paper was largely ignored. Arnold Heim's (1924) classic publication on Alpine sedimentary facies suffered the same fate, as he himself sadly observed in a later work (Heim, 1958). A subsequent paper of Steinmann (1925), including some pioneering petrographic studies, also made little impression in extra-Alpine circles. Molengraaf's work (1909, 1915, 1922) on Mesozoic deep-sea deposits and ferromanganese nodules from Borneo, Timor and Rotti went generally unheeded; certainly the tectonic implications were never followed through. These early European workers waited in limbo until the more permissive framework of the new global tectonics earned them their respectability.

Perhaps at least some of the controversy was semantic in nature. A clear distinction between the words 'pelagic'=pertaining to the open sea, with no depth connotation, and 'oceanic' with implications of great depth and specific type of basement might have clarified the issue. In the Alpine-Mediterranean system the situation is complicated by the fact that Mesozoic pelagic sediments are floored by both continental and oceanic crust (Bernoulli & Jenkyns, 1974). And the Jurassic radiolarites, much discussed by Steinmann, occur in both continental-margin and true oceanic settings. Clearly the situation was ripe for confusion.

On 28 July 1968, the newly built drilling vessel *Glomar Challenger* sailed from Orange, Texas. This date marked the start of a new era in marine geology. Early drilling results (e.g. Peterson *et al.*, 1970; Maxwell *et al.*, 1970), persuaded all but a few diehards of the reality of a spreading ocean, of drifting and colliding continents. The classical Alpine concepts were exhumed: it was no longer unthinkable that oceanic sediments should be found in mountain chains. On the contrary, the zone of collision of two continental masses was the logical place to find slivers of oceanic crust and their sedimentary cover. Deep Sea Drilling results have thus given new impetus and new understanding to studies of Alpine pelagic rocks.

A further push to the bandwagon has been given by the recent discoveries of prolific oil fields under the North Sea where the Cretaceous chalk is a main petroleum reservoir.

Although some preliminary studies of core material have been already published in the Initial Reports of the Deep Sea Drilling Project, it was felt that much detailed information and final interpretations remained in the files of individual workers. And land-locked geologists studying ancient pelagic sediments could clearly benefit from close co-operation with sea-borne specialists. Accordingly, it was proposed that the International Association of Sedimentologists hold a special symposium on pelagic sediments and that the papers be published as the first of a series of special publications. Both of these proposals were approved by the council. The symposium was held in Zürich during 25–28 September 1973, and all but a few of the oral presentations are included in this volume.

The first paper, by Berger and Winterer, is an analytical treatise on the interrelationships between plate tectonics and pelagic sedimentation. An underlying assumption is the rule of thumb discovered during the first *Challenger* expedition that carbonate dissolution closely governs the nature of deep-sea sediments. As an oceanic plate moves laterally away and vertically down from a ridge crest the sediments deposited upon it should change from calcareous to non-calcareous as the calcite compensation depth is crossed. This idealized model is, however, complicated by horizontal displacement of plates across belts of differing organic productivity. Synthesizing all these variables, Berger and Winterer have produced a survey of *Plate stratigraphy and the fluctuating carbonate line*. They demonstrate the great temporal variation in the calcite compensation depth.

Studies on dissolution facies in Recent oceanic facies have been tolerably common (e.g. Hay, 1970; Hsü & Andrews, 1970; Berger, 1972; Roth & Thierstein, 1972). However, in pelagic sediments on land, diagenesis has greatly obscured the evidence of penecontemporaneous solution of micro- and nannofossils; only the corrosion of megafossils can be readily documented. Carrying the study of Garrison & Fischer (1969) a step further, Schlager has investigated the *Preservation of cephalopod skeletons and carbonate dissolution on ancient Tethyan sea floors*. He describes the preferential solution of aragonitic phragmocones over calcitic skeletal parts, and stresses that leaching of most aragonite took place after the fossil had been embedded in bottom sediment. By studying the mode of cephalopod preservation he proposes a hierarchy of dissolution facies.

The next two papers present case histories of Palaeozoic cephalopod-bearing pelagic limestones. Tucker deals with all aspects of the *Sedimentology of Palaeozoic pelagic limestones: the Devonian Griotte (Southern France) and Cephalopodenkalk (Germany)*. These facies, partly nodular, contain a fauna of cephalopods, thin-shelled bivalves, conodonts, styliolinids and ostracods; the rocks also bear witness to early lithification and solution, and they are stratigraphically condensed. Bandel, more specifically, describes *Deep-water limestones from the Devonian-Carboniferous of the Carnic Alps, Austria*. He distinguishes two broad facies groups, one containing only rare redeposited units, the other containing abundant turbidites derived from a shallow-water carbonate platform. Widespread traces of dissolution in the pelagic facies serve as the main criterion for deciphering palaeobathymetry.

The next four articles have chalk as their theme. Schlanger and Douglas, both veterans of the *Glomar Challenger*, have studied the *Pelagic ooze-chalk-limestone transition and its implications for marine stratigraphy*. They describe the diagenesis of calcareous oozes in an oceanic environment where meteoric ground water is necessarily excluded, and show that although lithification generally increases with age and depth of burial there are many departures from this ideal state. They introduce the concept of 'diagenetic potential' (the potential of a sediment to form a chalk or limestone) to explain this, such potential being dependent on the nature and abundance of certain diagenetically soluble calcareous components in the original bottom sediment, this in turn being governed by palaeo-oceanographic conditions.

Neugebauer, concerning himself with both oceanic and shelf-sea facies, has looked at *Some aspects of cementation in chalk*. His main thesis is that chalk remains soft and uncemented because marine magnesium-rich pore fluids, which are oversaturated with respect to low-magnesian calcite, inhibit pressure solution/precipitation until considerable overloads are reached. At some critical depth solution of low-magnesian calcite at grain contacts is possible, and its reprecipitation as a higher magnesian calcite elsewhere leads to progressive depletion of magnesium in the pore fluid. With this inhibiting factor removed pressure solution and cementation can proceed at an accelerated pace. There is some agreement here with the trends observed by Schlanger and Douglas in Deep Sea Drilling cores.

Scholle presents his ideas on *Diagenesis of Upper Cretaceous chalks from England, Northern Ireland, and the North Sea*. It is well known that the chalk in Great Britain differs radically in hardness; soft chalks in southern England become harder when traced northward into Yorkshire, being replaced by the completely lithified White Limestone in Northern Ireland. The hardness of the Irish Chalk has been variously attributed to presence of aragonite in the initial sediment and to metamorphism by overlying basalts (see Black, 1953; Hancock, 1961, 1963). Drawing on petrographic and isotopic data and scanning electron microscopy, Scholle proposes an increasing gradient of hydrothermal and meteoric recrystallization from the North Sea to Northern Ireland related to rifting in the North Atlantic. Compaction and lithification of Irish chalk are attributed to expulsion of original marine pore fluids (after the hypothesis of Neugebauer) and loading by basalts.

The final paper on chalk, co-authored by Håkansson, Bromley and Perch-Nielsen, is titled, *Maastrichtian chalk of north-west Europe—a pelagic shelf sediment*. They remind us that pelagic sediments can be formed in shallow epeiric environments, a fact which may, however, be discernible from the presence of coccolith floras that are not fully oceanic. They also discuss the genesis of flint in the light of studies on cherts from Deep Sea Drilling cores.

Nodular limestones constitute our next theme. Alpine geologists have long been intrigued by the Triassic and Jurassic red nodular limestones that have facies equivalents in the Devonian and Carboniferous of central Europe (described here by Tucker and Bandel). The Tethyan Jurassic facies are known locally as *Knollenkalk* and *Ammonitico Rosso*. Up till now no modern equivalent of these facies had been found. The paper of Müller and Fabricius, *Magnesian-calcite nodules in the Ionian deep sea: an actualistic model for the formation of some nodular limestones* may, however, have solved this age-old puzzle. These authors describe nodules of centimetre scale that apparently form at or just below the sediment-water interface, their cement being derived from Mediterranean sea water. They suggest that the Tethyan nodular limestones were formed in a setting similar to the present-day Mediterranean.

Working independently, Jenkyns has also suggested a diagenetic process for nodule formation; his paper is titled: *Origin of red nodular limestones (Ammonitico Rosso, Knollenkalke) in the Mediterranean Jurassic: a diagenetic model*. He presents a series of petrographic criteria that suggest the nodules were formed by a solution-precipitation process rather than by irregular dissolution of a cemented calcareous sea bottom (a process advocated by Bandel for Palaeozoic pelagic nodular limestones). The cement for the nodules is assumed to derive from very fine-grained low-magnesian calcite and aragonite dissolved from within the sediment. An origin of the cement as a direct precipitate from sea water is not favoured since partial dissolution of aragonitic fossils on the sea floor (as documented by Schlager) suggests undersaturation of the bottom waters with respect to this polymorph of calcium carbonate.

Silica constitutes the next topic. Calvert presents a comprehensive survey of *Deposition and diagenesis of silica in marine sediments*. He notes that silica accumulates in marine sediments below waters of high biological productivity and in areas where volcanic products are widespread. He describes the solution-precipitation processes that turn opaline silica into chert. He furthermore notes that the concentration of dissolved silicon in interstitial waters of marine sediments is considerably higher than at the ocean bottom and there must therefore be a flux of silicon into the overlying waters. It is not unreasonable to suppose that this flux might be retarded when a siliceous ooze is suddenly buried by a turbidite or an ash bed. Could this explain the relationship between local silicification and redeposited horizons? Such a relationship may be observed in Deep Sea Drilling cores (e.g. Beall & Fischer, 1969) and on land (e.g. Garrison, 1967).

Results of Deep Sea Drilling have shown that only a small fraction of the siliceous oozes under the ocean bottom has been converted to chert; the rest is soft and unconsolidated. The exact mechanism of chert formation is still being worked out. Wise and Weaver, documenting the *Chertification of oceanic sediments*, illustrate the transition from opaline silica through disordered cristobalite lepispheres to quartz. They hold the 'maturation' theory of oceanic chert formation, that is, that the mineralogy of a siliceous rock is predominantly a function of its age. Wise and Weaver also stress the biological origin of most chert, even when recognizable siliceous remains are scanty. In describing the *Petrography and diagenesis of deep-sea cherts from the central Atlantic*, von Rad and Rösch also stress the 'maturation' theory. They note that the end product of quartz contains very few recognizable siliceous fossils yet its precursors are clearly biogenic. In contrast to Wise and Weaver and von Rad and Rösch, Lancelot proposes that the host sediments control the mineralogy of silica phases. His abstract is titled, *Formation of deep-sea chert: role of the sedimentary environment*. According to Lancelot, who worked with samples from the central Pacific, chert in clay-rich sediments comprises cristobalite while quartz is a primary precipitate in calcareous oozes, chalks and limestones. His complete account is published in the Initial Reports of the Deep Sea Drilling Project (Lancelot, 1973).

The next paper deals with cherts on land. Studying some Mesozoic radiolarites in the Othris Mountains of Greece, Nisbet and Price noticed that certain of the chert beds were graded and contained distinct intervals which were either structureless or parallel- and cross-laminated. The clay content of the radiolarites resembles the submarine weathering products of basalt; the tectonic position of the radiolarites suggests that they were originally deposited near the foot of a continental margin abutting a newly formed ocean. The title of their paper is thus, *Siliceous turbidites: bedded cherts as redeposited ocean ridge-derived sediments*.

Garrison gives us a general survey of *Radiolarian cherts, pelagic limestones, and igneous rocks in eugeosynclinal assemblages*. He stresses the effects that temporal changes in the carbonate compensation depth can have upon facies, and documents the biogenic nature of most sediments that lie upon or are interbedded with igneous rocks. He notes particularly the types of igneous rocks which underlie pelagic sediments and the nature of the contact. Reviewing occurrences from the Alps, Mediterranean and circum-Pacific regions, he presents models for the depositional setting of these sediments.

Boström discusses the *Origin and fate of ferromanganoan active ridge sediments.* These deposits are apparently formed through the expulsion of deep-seated magmatic fluids. Most of these sediments, having been formed on oceanic crust, will eventually be subducted or metamorphosed; ancient examples may, however, be sought in association with ophiolite complexes. Relevant in this connection are the *Pelagic sediments in the Cretaceous and Tertiary history of the Troodos massif, Cyprus* described by Robertson and Hudson. The basal member of the sedimentary series overlying the oceanic pillow lavas is constituted by the so-called *umbers* which are ancient analogues of active ridge sediments. The umbers pass up into radiolarian cherts which in turn give way to chalks. The general stratigraphic arrangement, albeit of differing age, resembles that in the Ligurian Apennines, as reviewed by Garrison.

Finally, Wendt demonstrates the occurrence of *Encrusting organisms in deep-sea manganese nodules.* The organisms comprise sessile arenaceous Foraminifera and may be found in depths as great as 5 km. Other organisms such as serpulids, corals, bryozoans and sponges commonly occur on nodules from seamounts. The sessile Foraminifera grow relatively rapidly, which suggests that nodules may, at certain periods, grow much faster than the average rates calculated from radiochemical analyses.

In conclusion it is clear that, with reference to pelagic sediments, the Recent is no skeleton key to the past. At the present time pelagic sedimentation is virtually confined to ocean basins. Clearly this was not the case during deposition of the Cretaceous chalks and those Tethyan Mesozoic pelagic facies which were laid down on continental basement. Such a setting, of course, gave them a favourable preservational bias. But even the true oceanic sediments that overlie ophiolites do not entirely correspond with what has been revealed by Deep Sea Drilling. The ascending stratigraphic sequence of pillow lavas, radiolarites, chalks (Cyprus, Ligurian Apennines) is not what one would expect to have formed on a spreading ridge moving below the calcite compensation depth. Before we rush to illustrate the close similarity between oceanic sediments on land and those under the sea perhaps we should look more critically at the differences.

REFERENCES

BADER, R.G. *et al.* (1970) *Initial Reports of the Deep Sea Drilling Project,* Vol. IV, pp. 753. U.S. Government Printing Office, Washington.

BEALL, A.O., JR & FISCHER, A.G. (1969) Sedimentology. In: *Initial Reports of the Deep Sea Drilling Project,* Vol. I (M. Ewing *et al.*), pp. 521–593. U.S. Government Printing Office, Washington.

BERGER, W.H. (1972) Dissolution facies and age-depth constancy. *Nature, Lond.* **236,** 392–395.

BERNOULLI, D. & JENKYNS, H.C. (1974) Alpine, Mediterranean and central Atlantic Mesozoic facies in relation to the early evolution of the Tethys. In: *Modern and Ancient Geosynclinal Sedimentation* (Ed. by R. H. Dott, Jr and R. H. Shaver), *Spec. Publs Soc. econ. Paleont. Miner., Tulsa,* **19,** 129–160.

BLACK, M. (1953) The constitution of the Chalk. *Proc. geol. Soc. Lond.* no. **1499,** 81–86.

CAYEUX, L. (1891) La Craie du nord de la France et la boue à Globigérines. *Annls Soc. geol. N.* **19,** 95–102.

FUCHS, T. (1877) Über die Entstehung der Aptychenkalke. *Sber. Akad. Wiss. Wien Math. Nat. Klasse,* Abt. I, **76,** 329–334.

FUCHS, T. (1883) Welche Ablagerungen haben wir als Tiefseebildungen zu betrachten? *Neues Jb. Miner. Geol. Paläont,* Beil. Bd. **2,** 487–584.

GARRISON, R.E. (1967) Pelagic limestones of the Oberalm Beds (Upper Jurassic-Lower Cretaceous), Austrian Alps. *Bull. Can. Petrol. Geol.* **15,** 21–49.

GARRISON, R.E. & FISCHER, A.G. (1969) Deep-water limestones of the Alpine Jurassic. In: *Depositional Environments in Carbonate Rocks, a Symposium* (Ed. by G. M. Friedman). *Spec. Publs Soc. econ. Paleont. Miner., Tulsa,* **14,** 20–56.

GÜMBEL, W. (1878) Ueber die im stillen Ocean auf dem Meeresgrunde vorkommenden Manganknollen. *Sber. bayer. Akad. Wiss.* **8,** 189–209.

GUPPY, H.B. (1885) Observations on the Recent calcareous formations of the Solomon group made during 1882–84. *Trans. Roy. Soc. Edin.* **32,** 545–581.

HANCOCK, J.M. (1961) The Cretaceous System in Northern Ireland. *Q. Jl geol. Soc. Lond.* **117,** 11–36.

HANCOCK, J.M. (1963). The hardness of the Irish Chalk. *Ir. Nat. J.* **14,** 157–164.

HARRISON, J.B. & JUKES-BROWNE, A.J. (1890) *The geology of Barbados* (being an explanation of the geological map of Barbados prepared by the same authors), pp. 64. Published by authority of the Barbadian legislature.

HAY, W.W. (1970) Calcium carbonate compensation. In: *Initial Reports of the Deep Sea Drilling Project,* Vol. IV (R. G. Bader *et al.*), pp. 672–673. U.S. Government Printing Office, Washington.

HEIM, A. (1924) Über submarine Denudation und chemische Sedimente. *Geol. Rdsch.* **15,** 1–47.

HEIM, A. (1958) Oceanic sedimentation and submarine discontinuities. *Eclog. geol. Helv.* **51,** 642–649.

HINDE, G.J. (1890) Notes on Radiolaria from the Lower Palaeozoic rocks (Llandeilo-Caradoc) of the south of Scotland. *Ann. Mag. nat. Hist.*, ser. 5, **6,** 40–59.

HSÜ, K.J. & ANDREWS, J.E. (1970) History of South Atlantic basin. In: *Initial Reports of the Deep Sea Drilling Project,* Vol. III (A.E. Maxwell *et al.*), pp. 464–471. U.S. Government Printing Office, Washington.

HUXLEY, T.H. (1868) On some organisms living at great depths in the North Atlantic Ocean. *Q. Jl microsc. Sci.* **8,** N.S., 203–212.

LANCELOT, Y. (1973) Chert and silica diagenesis in sediments from the central Pacific. In: *Initial Reports of the Deep Sea Drilling Project,* Vol. XVII (E. L. Winterer, J. I. Ewing *et al.*), pp. 377–405. U.S. Government Printing Office, Washington.

LOHMANN, G.P. (1973) Stratigraphy and sedimentation of deep-sea Oceanic Formation on Barbados, West Indies. *Bull. Am. Ass. Petrol. Geol.* **57,** 791. (Abstract.)

MAXWELL, A.E. *et al.* (1970) *Initial Reports of the Deep Sea Drilling Project,* Vol. III, pp. 806. U.S. Government Printing Office, Washington.

MOLENGRAAF, G.A.F. (1909) On oceanic deep-sea deposits of Central Borneo. *Proc. Sect. Sci. K. ned. Akad. Wet.* **12,** 141–147

MOLENGRAAF, G.A.F. (1915) On the occurrence of nodules of manganese in Mesozoic deep-sea deposits from Borneo, Timor, and Rotti, their significance and mode of formation. *Proc. Sect. Sci. K. ned. Akad. Wet.* **18,** 415–430.

MOLENGRAAF, G.A.F. (1922) On manganese nodules in Mesozoic deep-sea deposits of Dutch Timor. *Proc. Sect. Sci. K. ned. Akad. Wet.* **23,** 997–1012.

MURRAY, J. (1876) Preliminary report on specimens of the sea bottom. *Proc. R. Soc.* **24,** 471–532.

MURRAY, J. & RENARD, A.F. (1891) Report on deep-sea deposits based on specimens collected during the voyage of H.M.S. 'Challenger' in the years 1873–1876. In: *'Challenger' Reports,* pp. 525. H.M.S.O., Edinburgh.

NEUMAYR, M. (1887) *Erdgeschichte. Erster Band, Allgemeine Geologie,* pp. 653. Bibliographisches Institut, Leipzig.

NICHOLSON, H.A. (1890) Address on recent progress in palaeontology as regards invertebrate animals. *Trans. Edin. geol. Soc.* **6,** 53–69.

PETERSON, M.N.A. *et al.* (1970) *Initial Reports of the Deep Sea Drilling Project,* Vol. II, pp. 501. U.S. Government Printing Office, Washington.

ROTH, P.H. & THIERSTEIN, H. (1972) Calcareous nannoplankton: leg 14 of the Deep Sea Drilling Project. In: *Initial Reports of the Deep Sea Drilling Project*, Vol. XIV (D. E. Hayes, A. C. Pimm *et al.*), pp. 421–485. U.S. Government Printing Office, Washington.

SORBY, H.C. (1861) On the organic origin of the so-called 'Crystalloids' of the Chalk. *Ann. Mag. nat. hist.* ser. 3, **8,** 193–200.

SORBY, H.C. (1879) Anniversary address of the President (on the structure and origin of limestones). *Proc. geol. Soc.* **35,** 56–95.

STEINMANN, G. (1905) Geologische Beobachtungen in den Alpen. II. Die Schardtsche Überfaltungstheorie und die geologische Bedeutung der Tiefseeabsätze und der ophiolithischen Massengesteine. *Ber. naturf. Ges. Freiburg,* **16,** 18–67.

STEINMANN, G. (1925) Gibt es fossile Tiefseeablagerungen von erdgeschichtliche Bedeutung? *Geol. Rdsch.* **16,** 435–468.

SUESS, E. (1875) *Entstehung der Alpen,* pp. 168. W. Braumüller, Vienna.

THOMSON, C.W. (1874a) Preliminary notes on the nature of sea-bottom procured by the soundings of H.M.S. 'Challenger' during her cruise in the 'Southern Sea' in the early part of the year 1874. *Proc. R. Soc.* **23,** 32–49.

THOMSON, C.W. (1874b) *The Depths of the Sea,* pp. 527. Macmillan, London.

WALTHER, J. (1897) Ueber Lebensweise fossiler Meeresthiere. *Z. dt. geol. Ges.* **49,** 209–273.

Spec. Publs int. Ass. Sediment. (1974) **1,** 11–48

Plate stratigraphy and the fluctuating carbonate line

W. H. BERGER *and* E. L. WINTERER

Scripps Institution of Oceanography, University of California, San Diego, La Jolla, California 92037, *U.S.A.*

ABSTRACT

The most conspicuous feature of deep-sea sedimentation is the carbonate line, the facies boundary between calcareous ooze and pelagic clay. In any one facies regime, the carbonate line tends to follow depth contours, but on a global scale its depth range is considerable, about two kilometres. The topography of the surface described by the carbonate line is controlled by regional differences in carbonate supply, dissolution and, to some extent, redeposition. Supply and dissolution processes lead to geochemical fractionation, whereby carbonate is separated from other oceanic sediments. We identify basin-basin fractionation, bathymetric, latitudinal trophic, abyssal and diversity-related fractionation.

The palaeogeography and palaeobathymetry of carbonate deposits can be reconstructed using principles from plate tectonics and stratigraphy ('plate stratigraphy'), where information is available from drill holes. The reconstructions demonstrate that the carbonate line fluctuated in some areas during certain times, and stayed nearly constant in others. Arguments from plate tectonics provide clues on the variability of the fractionation processes which control the position of the carbonate line.

INTRODUCTION

The surface of Earth has several first order quasi-linear features which are controlled largely by elevation, such as the coast line, the snow line, the tree line, and the carbonate line, that is, the facies boundary between carbonate-rich and carbonate-poor sediments. Just as the migrating tree line and snow line tell the story of varying precipitation and temperature on land, the changing topography of the carbonate line contains information on the chemical climate of the sea and hence on the planetary environment. Oceanic environments are shaped by plate tectonics and its ramifications. To the extent that plate tectonics are responsible for sedimentation patterns through time and to the extent that plate-tectonic principles are applicable to deciphering the stratigraphic record, the science of stratigraphy becomes 'plate stratigraphy'.

In this essay we explore a number of relationships between deep-sea sedimentation and plate tectonics, focusing on the carbonate line and the regional and global controls which it reflects. We begin by discussing the present pattern of the carbonate line and then marshal evidence to show that this pattern varied through time, using palinspastic reconstructions derived from plate stratigraphy.

THE CONCEPTS 'CARBONATE LINE' AND 'COMPENSATION SURFACE'

Definitions

About one half of the deep ocean floor is covered by calcareous ooze, the other, deeper half, by sediment with only a few percent of calcium carbonate (Sverdrup, Johnson & Fleming, 1942, p. 977, Table 106). The boundary between the two facies regimes roughly follows depth contours over large areas (Fig. 1), owing to increased carbonate dissolution rates at depth (Murray, in Murray & Renard, 1891, pp. 277–279). The boundary has been called a compensation depth for calcium carbonate, generally abbreviated CCD, because rate of supply and rate of dissolution are approximately compensated at this level (Bramlette, 1961). Unfortunately, this term has occasionally given rise to a misconception that the level is a depth marker which, once established, is valid for all or most of the deep ocean. As both Arrhenius (1952) and Bramlette (1961) emphasized, this is not so. Instead, the surface defined by all local compensation levels (here called the carbonate compensation surface, abbreviated CCS) can display considerable topography near the equator in the Pacific and elsewhere (see Fig. 1). Arrhenius (1952, p. 190) introduced the term 'carbonate compensation line' to denote the facies boundary on the present sea floor; we use 'carbonate line' with a view to the snow line analogy (Peterson, 1966). The carbonate line, then, is the intersection between the carbonate compensation surface (CCS) and the sea-floor topography. We retain 'CCD' as a general undifferentiated term for both carbonate line and CCS.

Applicability

Like the snow line, the carbonate line is a statistical concept: from afar, the facies boundary appears quite well defined and seems to follow depth contours. On closer inspection, however, the line proves less distinct and merely denotes a probability limit: a few hundred metres above the line a surface sediment sample will probably be rich in carbonate; a similar distance below the line it will have none, or only a few percent, of carbonate. The sharpness of the carbonate line may be objectively defined, using the appropriate probabilities for finding carbonate, stated with respect to a given deviation from the line, say, 200 m above or below. This sharpness is a function of the relief of the compensation surface, of the relief of the bottom and associated redeposition processes, and of rate distributions of supply and removal of carbonate particles.

The zone which separates well-preserved from poorly preserved calcareous assemblages, the lysocline, is interpreted as denoting a marked increase in dissolution rate over a short depth interval at this level (Berger, 1971). In places, this level is well defined, as in the western trough of the South Atlantic, where it coincides with the boundary between Antarctic Bottom Water and North Atlantic Deep Water, and here the carbonate line also is well defined. A similar relationship between sharpness of lysocline and of carbonate line holds for pelagic areas with relatively low rates of supply, as on the East Pacific Rise, well south of the equator. The carbonate line is close to the lysocline here, providing for a distinct demarcation between carbonate and clay regime. Conversely, where the lysocline is indistinct and well separated from the carbonate line, as in the equatorial Pacific, the boundary between calcareous facies

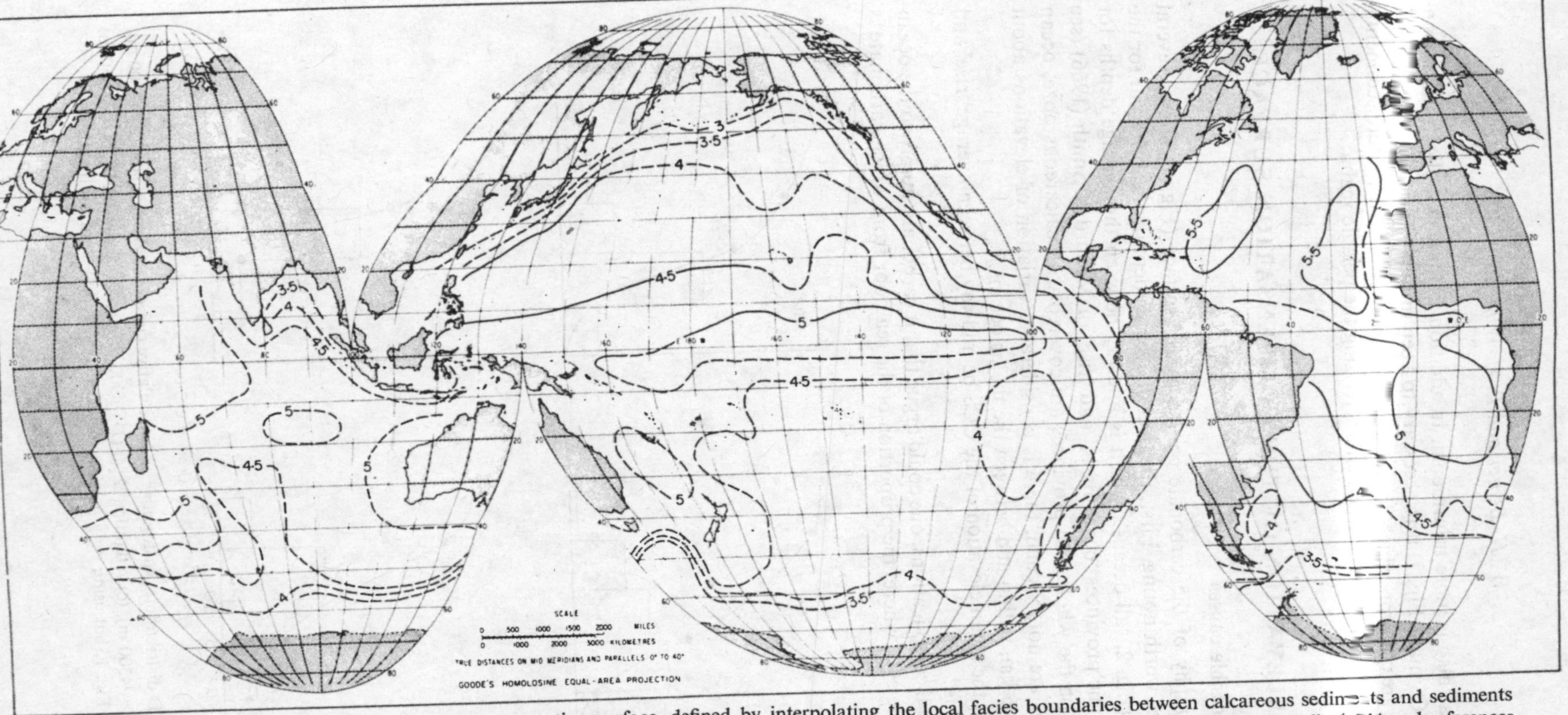

Fig. 1. Topography of the calcium carbonate compensation surface, defined by interpolating the local facies boundaries between calcareous sediments and sediments with no or only a few percent of carbonate. Numbers on contours denote km below sea surface. Sources: Murray & Renard (1891); Revelle (1944 and references therein); Olausson (1960a, b, c); Ericson *et al.* (1961); a great number of less comprehensive studies cited in Smith, Dygas & Chave (1968); carbonate maps by Russian scientists (Emelyanov and associates in Atlantic, Lisitzin and Petelin in Pacific, see Lisitzin, 1972); additional information collected in the World Ocean Sediment Data Bank, SIO and retrieved with the help of Jane Z. Frazer. Elevations were determined by finding the most probable compensation level for each 10° square. ——, More than twenty control samples per 10° square; – – –, less than twenty control samples.

and pelagic clay also is more transitional. In addition to these complications, in considering the CCS map (Fig. 1) it is necessary to keep in mind that both lysocline and carbonate line are concepts which do not work well near continents, where sedimentation processes are rather unlike those characterizing the pelagic realm.

MEAN ELEVATION OF THE COMPENSATION SURFACE

Position and possible causes

The topography of the carbonate compensation surface (Fig. 1) shows several general features worth noting. First, the overall average depth is near 4·6 km for the pelagic realm (Fig. 2, 'all oceans') which is midway between the average depths for the physiographic provinces 'ocean basin' and 'rise' of Menard & Smith (1966) (see Fig. 3). Thus, on the whole, rise provinces are covered with calcareous ooze, ocean basin provinces are not. Second, there is considerable variation of elevations about this mean of 4·6 km, within and between the major oceans.

Why should the mean elevation of the CCS be midway between average 'rise' and 'basin' elevations?

Neglecting all complications, one could argue that if carbonate supply to the ocean floor is the same everywhere, the proportion of the sea floor from which carbonate is

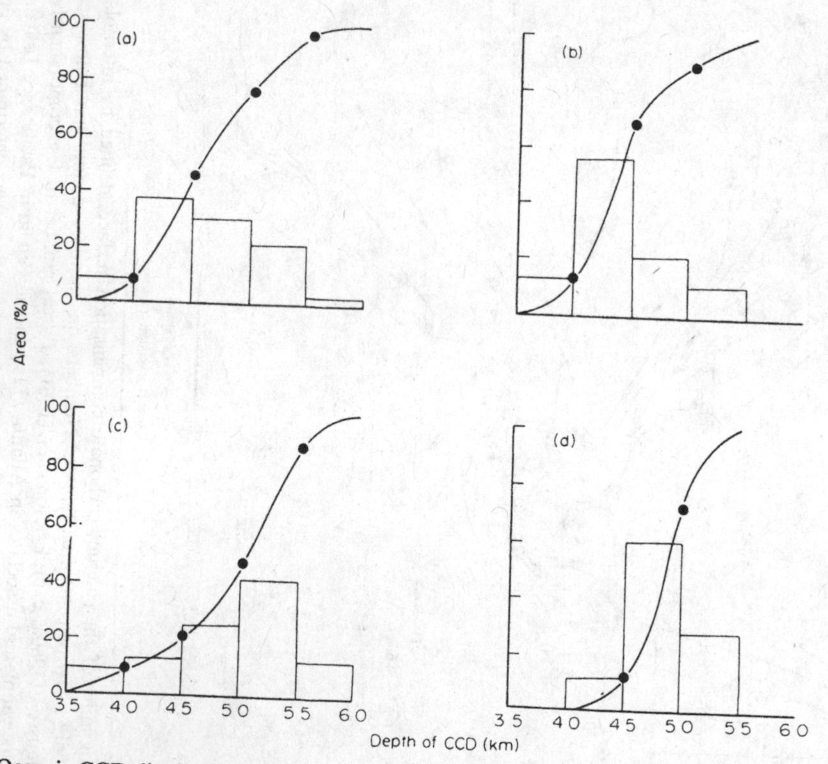

Fig. 2. Oceanic CCD distributions, histograms and cumulative curves in (a) all oceans ($\overline{D}$=4600 m), (b) Pacific (57%) ($\overline{D}$=4300 m), (c) Atlantic (22%) ($\overline{D}$=4900 m) and (d) Indian (21%) ($\overline{D}$=4850 m) Oceans. Data from Fig. 1, but using only portions of oceans >800 km from continents.

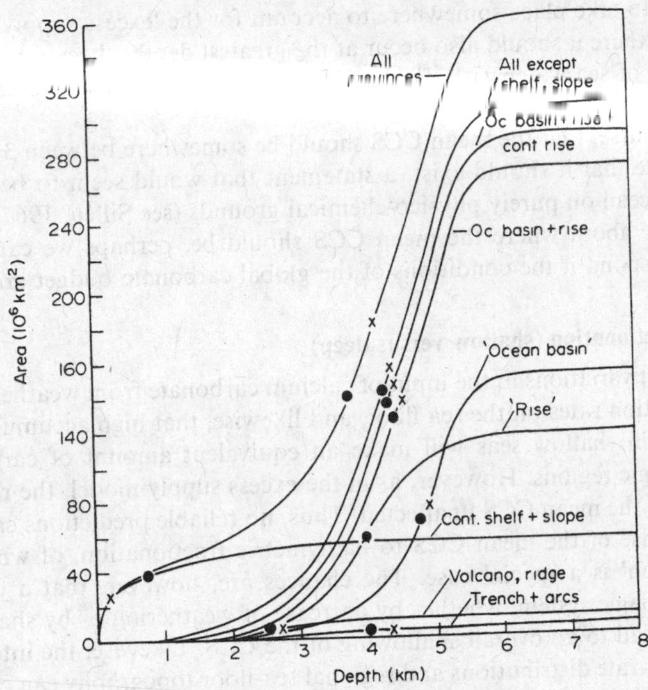

Fig. 3. Hypsometric curves for various physiographic provinces of the world's oceans (after Menard & Smith, 1966, but including unpublished data compiled by Menard & Smith). ×, Median depth for province; ●, mean depth for province.

removed is the same as the proportion of carbonate supply that is dissolved (Broecker, 1971, p. 249). The proportion of supply to be dissolved corresponds to the 'excess' delivery by organisms. Since shell-forming organisms care not about the input of carbonate to the system, but only about growth-controlling factors, they precipitate carbonate in excess of input, an excess that is redissolved under steady-state conditions (Berger, 1970).

Within this 'linear' model, then, the CCS adjusts to a level where the proportion of non-calcareous sea floor corresponds to the excess supply of carbonate by shell-forming organisms. High rates of oceanic mixing, high fertility, excess supply, and a relatively shallow overall CCS would be positively correlated. However, the linear model crucially depends on the assumption that there is no dissolution in areas above the CCS, an assumption that is demonstrably false for pelagic areas (Murray & Renard, 1891, p. 267; Schott, 1935; Arrhenius, 1952; Phleger, Parker & Peirson, 1953; Ruddiman & Heezen, 1967; Berger, 1968) as well as for the continental slopes up to very shallow depths (Berger, 1968, 1970, 1971; Berger & Soutar, 1970; Thiede, 1971, 1973). The other major assumption, the evenness of carbonate supply rates, also is grossly violated by the fertility variations in the productive surface waters.

There is, then, no simple answer to the question of why the mean depth of the CCS is where it is. Carbonate has to accumulate somewhere as long as it enters the ocean, and since rates in shallow water appear too low to accommodate the precipitation required for steady-state concentrations in sea water, the mean CCS must be deeper than 3 km to intersect a significant part of the deep ocean floor (Fig. 3). Also, dissolution has to take place somewhere to account for the 'excess supply'. If dissolution occurs anywhere it should also occur at the greatest depths, because of increasing undersaturation of sea water with depth. The lower limit for a mean CCS, therefore, is in the vicinity of 5·5 km.

To say, of course, that the mean CCS should be somewhere between 3 and 5·5 km is merely to state that it should exist, a statement that would seem to be reasonable for any mixed ocean on purely physico-chemical grounds (see Sillén, 1967). Although we can say little about where the mean CCS should be, perhaps we can anticipate how it would respond if the conditions of the global carbonate budget are varied.

Bathymetric fractionation (shallow versus deep)

It is clear that variations in the input of calcium carbonate from weathering should affect sedimentation rates on the sea floor, and likewise, that high accumulation rates on shelves and in shallow seas will make an equivalent amount of carbonate unavailable in pelagic regions. However, as in the excess supply model, the rates can be adjusted to leave the mean CCS unaffected. Thus, no reliable predictions can be made about the response of the mean CCS to bathymetric fractionation, of which 'basin-shelf fractionation' is a special case. The chances are, however, that a decrease in input into the pelagic system, whether by decrease of weathering or by shallow-water extraction, will lead to an overall shallowing of the CCS. Likewise, the interaction of carbonate supply-rate distributions and regional sea-floor topography can conceivably lead to changes in bathymetric fractionation within the pelagic realm itself, with important implications for the mean position of the CCS. For example, at the present time the North Atlantic has a shallow deep-sea floor, providing a huge area for carbonate accumulation. If for some reason the North Atlantic sea floor would receive carbonate at a rate similar to that in the eastern equatorial Pacific, this region could act as an even greater sink for the oceanic carbonate output than it does already.

The major controls on the global balance between deep- and shallow-water accumulation of calcium carbonate can be summarized as follows.

(A) Changes in rate of $CaCO_3$ accumulation on shelves.
   (1) Changes in total area and/or capacity of shelf seas due to transgression and regression.
      (a) Local subsidence or uplift.
      (b) Opening or closing of seaways creating or destroying continental margins and 'carbonate platforms'.
      (c) Worldwide changes in sea level.
   (2) No change in total area of shelf seas.
      (a) Changes in water temperature or fertility over shelves, due to climatic or oceanographic changes.
      (b) Changes in nature of carbonate-fixing shelf organisms through evolution.
   (3) Changes in position of shelves with respect to world climatic belts, through continental drift.

(B) Changes in rate of $CaCO_3$ accumulation on oceanic shallows (mainly rise crests and continental slopes).
   (1) Evolutionary changes in carbonate-fixing planktonic organisms (biogeography, relative abundance of solution-susceptible and solution-resistant forms, skeletal mineralogy, shell morphology).

(2) Changes of proportion of ocean floor in shallow part of rise crests.
(3) Changes in relationship between sea floor hypsography and calcite-saturation patterns in the water column (saturation gradients, oxygen minimum effects, etc.).
(4) Changes in fertility gradients across continental slopes.

In the following sections we will take up some of these mechanisms, with emphasis on the ones directly controlled by plate tectonics.

Transgressions and regressions

Today's shelf seas, from the shoreline to a depth of 200 m, have an area of about 27 million km², roughly 7·5% of the total area of the world ocean (Menard & Smith, 1966). A slightly larger area—about 40 million km²—lies between sea level and +200 m (Kossinna, 1921). It has been suggested (Eicher, 1969) that the middle Cretaceous transgression produced seas about 500 m deep over parts of the North American Continent. Hallam (1963) suggests a minimum global sea level change of 100 m and Fairbridge (1961) proposes an overall change of 200 m. A transgression of a few hundred metres also accords with the results of Ronov (1968) on a worldwide scale. Within limits of – 200 to +200 m, the average change in area of shelf seas, for each 1 m change in sea level, is about 150,000 km², neglecting isostatic effects, or about 200,000 km² including isostasy (any rise or fall of sea level over the shelves is increased somewhat because of the weight of water causing isostatic adjustment).

Maximum sustained regional carbonate accumulation rates on shelves appear to be about the same as maximum oceanic rates; that is, in the range of 20–40 m/million years, judging from the thickness and age spans of Cretaceous and Cenozoic shelf limestone sections in various parts of the world (Brinkmann, 1960; Kummel, 1970). Gross short-term carbonate production rates (see, for example, Chave, Smith & Roy, 1972) are probably at least ten times this great in tropical seas, and we suspect therefore that regional subsidence rates are the limiting control on shelf accumulation rates.

To get an estimate of the possible effects of a transgression today on the world mean of the CCS, compare the area of carbonate accumulation in the ocean basins—about 130×10^6 cm² (Sverdrup *et al.*, 1942)—with the area of shelf seas where carbonates could accumulate. A rise of 100 m would increase the shelf area by about 20×10^6 km². Shift of carbonate deposition from the deep sea to the shelf would shift the CCS from its present mean of 4600 m (Fig. 2a) to about 4300 m (Fig. 3, ocean basin and rise curve), to account for a decrease of the carbonate area from 130 to 110×10^6 km². Under the assumptions of the 'linear' model, then, with the additional proviso that carbonate accumulates everywhere at maximum sustained rates seen in the geological record, a 1 m change in sea level produces about three times that change in the CCS, within the present configuration of hypsography. For shelf accumulation rates less than the observed maximum, the effect of sea-level change on CCS change is correspondingly reduced. We suggest that changes in sea level and in CCS level are parallel and comparable in magnitude, as a working hypothesis. We shall return to this question when considering the evidence for CCS fluctuations in the past.

Changes in oceanographic conditions on shelf seas

The amount of carbonate accumulation on shelves depends partly on the marine climates prevailing over shelves. Conditions in humid temperate and cool regions

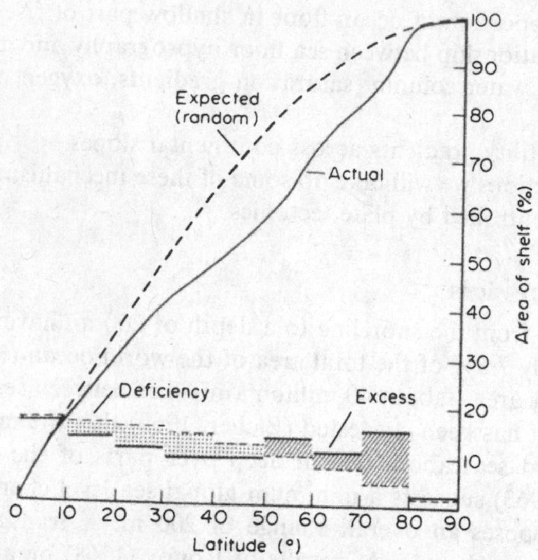

Fig. 4. Shelf area *v.* latitude, histograms and cumulative curves. Actual shelves shown as solid lines; 'expected' distribution (shelf area proportional to area of sphere at same latitude) shown as broken lines. Excess over expected, hatched shading; deficiency, stipples. Data from Kossinna (1921).

tend to inhibit carbonate accumulation in shelf seas (Seibold, 1970). Today's shelves are located in higher latitudes than in a 'random' world (Fig. 4): only 40% of today's shelves lie between 0 and 30°. In Early Tertiary and Mesozoic times, the large northern continents were farther south than they are today, and more equable and warm climates prevailed over the globe. Thus, a Mesozoic transgression would have a potentially much greater effect on raising the global CCD than an equal transgression in late Cenozoic times.

Evolutionary changes in shell properties

We have already cited evidence that the CCD is a rate-governed phenomenon, that is, that dissolution occurs above the regional carbonate line, but at a rate less than that of the supply. It is clear, therefore, that the carbonate line can be depressed by supplying increasingly corrosion-resistant material.

Imagine, for example, that there are no calcite-fixing pelagic organisms, only aragonitic pteropods. All calcite would be deposited by shallow-water organisms and if their fixation rate accounts for the total influx to the ocean, no carbonate would be left for the deep sea, the pteropod shells, which are of the more soluble aragonite, being dissolved.

Conversely, imagine that a pelagic organism evolves which produces enough optical calcite spheres similar to those of Peterson (1966), so as to use much of the total influx of carbonate to the ocean. The CCS would deepen, but only for the spheres. Because the spheres extract so much carbonate and refuse to recycle it at a rate commensurate with oceanic mixing rates, undersaturation would increase and the compensation level for all other shells would rise.

At present, we do have an analogous if less extreme situation: the most resistant Foraminifera dissolve at least fifty times more slowly than the least resistant ones

(Berger & Soutar, 1970). The compensation level for the more delicate shells coincides by definition with the 'lysocline'. At the lysocline, the more easily dissolved particles disappear; at the CCD, the most resistant have time to dissolve. The CCD is thus sensitive to the evolutionary changes in the microstructure of foraminiferal tests and of coccolithophorid plates, and to the changing biogeographic patterns of these organisms. Skeletal materials, especially in fertile and in shallow waters, are subject to post-depositional attack by various organisms such as deposit feeders and boring algae and fungi. These activities reduce the resistance of shells, making possible increased dissolution rates. Since evolutionary effects are probably very long term, we will not consider them further here, except to note that the spectrum of relative solubilities shown by modern calcareous plankton, which makes possible the discrimination of solution levels—calcite-CD, lysocline, aragonite-CD—is a product of evolution. The present-day distributional patterns of oceanic carbonate sediments is at the same time more complex than it was in the past, and more rich in information.

Changes in hypsography and sea-floor spreading

We have seen that the mean CCS elevation is between the mean depths for the provinces 'ocean basin' and 'rise' (Figs 2a, 3). Changes in the global rate of sea-floor spreading (Larson & Pitman, 1972) would change the shape of spreading ridges with two important consequences for the mean position of the CCS. First, a change in volume of the spreading ridges would lead to changes in the total volume of the ocean basins, and hence to transgression and regression over the continental shelves (Hallam, 1963; Menard, 1964; Frerichs, 1970; Pessagno, 1972; Flemming & Roberts, 1973; Hays & Pitman, 1973; Rona, 1973). Second, the depth level separating calcareous from non-calcareous sea floor would have to change if the ratio between the two facies domains were to stay constant.

To make quantitative estimates of these effects, we need to understand just how the hypsometric curve for the oceans would be changed by altering spreading rates. When spreading rates increase, consumption rates must increase correspondingly, and it is the combined effect of these two processes that determines changes in the hypsometric curve. To visualize these effects we need a model.

The age-depth curve for spreading ridges, based on data by Sclater, Anderson & Bell (1971) (Fig. 5), provides one element in the model, but we need a design for consumption also. A guide to this design is the crustal age pattern in today's ocean (Fig. 6). From inspecting the map it is obvious that there is a large proportion of young ocean floor and only small remnants of old floor. We would expect, therefore, that the younger (and therefore shallower) the floor, the greater is its chance to be consumed by a trench, if trenches develop independently of the age structure. Likewise, if trenches are indeed random slices across the oceanic crust, a regular decrease of the abundance of ages should result. The 'observations' (in places greatly extended by interpolations and extrapolations, especially in the Indian and Antarctic Oceans) show this decrease (Fig. 7), although the age from 20 to 40 million years appears under-represented (prematurely consumed?) while the range from 60 to 80 million years is somewhat too abundant (escaped 'normal' consumption?). The uncertainties in the crustal age map prevents us from attaching much significance to the detailed form of the age histogram, although we believe that its general form accurately reflects present knowledge. The greater reliability of the younger ages suggests that the

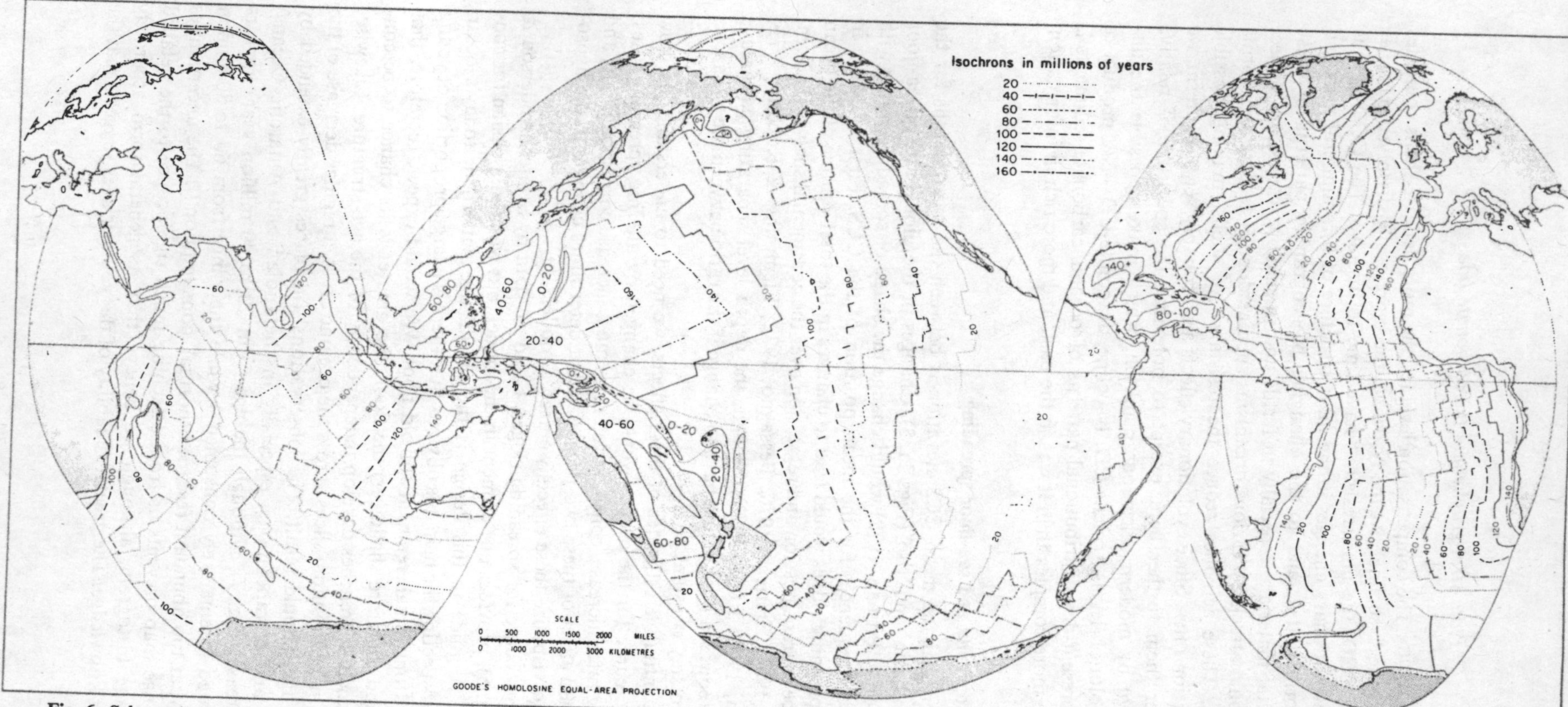

Fig. 6. Schematic crustal age map. Ages based on data from magnetic anomalies (Le Pichon, 1968; Heirtzler *et al.*, 1968; Vogt & Ostenso, 1970; Laughton, 1971; Pitman & Talwani, 1972; Fisher, Sclater & McKenzie, 1971; McKenzie & Sclater, 1971; Weissel & Hayes, 1971; Molnar *et al.*, in press; Herron, 1972; Larson & Chase, 1972; Hayes & Pitman, 1970; Atwater & Menard, 1970) and Deep Sea Drilling Project data (Initial Reports, Legs 1–14, 16–19; Geotimes for other legs) very liberally interpreted and extrapolated especially in the Indian Ocean.

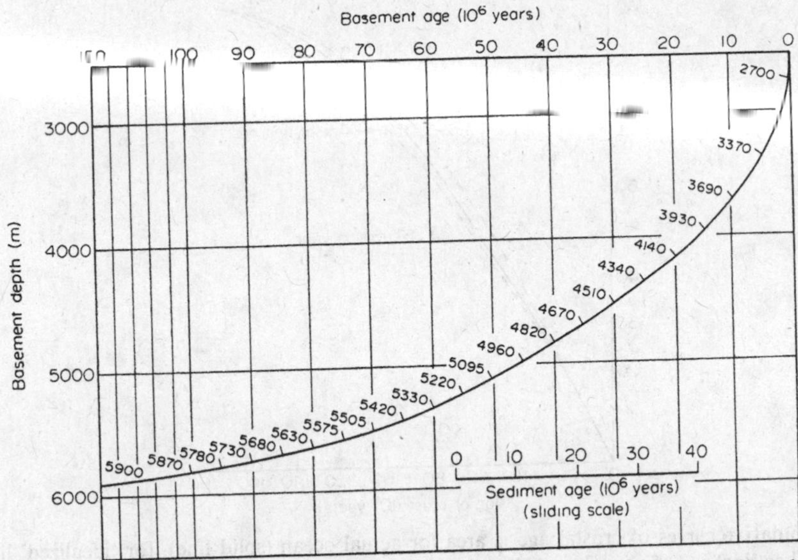

Fig. 5. Generalized age-depth relationship for actively spreading sea floor, based on data by Sclater, Anderson, Bell (1971). Ridge crest arbitrarily fixed at the average elevation of 2700 m. Extrapolated beyond 80 million years on the basis of Deep Sea Drilling Project data (in Berger, 1972).

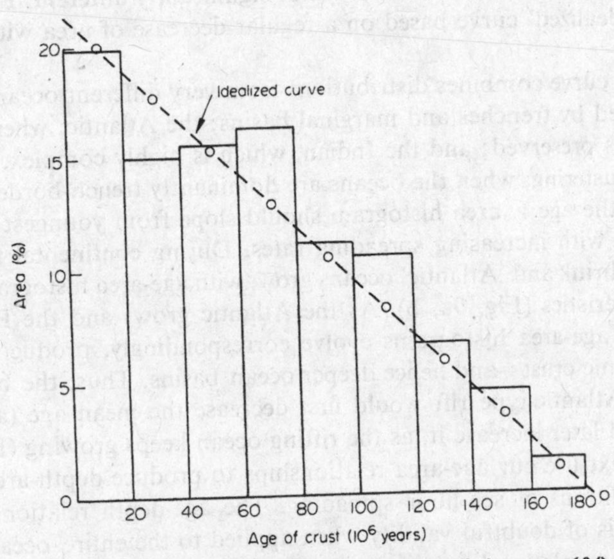

Fig. 7. Histogram of oceanic crustal age (in 20 million years units) *v.* area (%) for today's ocean, derived from measuring areas on crustal age map (Fig. 6). The 'idealized' line simply averages out the steps in the histogram.

apparent deficiency of 20–40 million years crust may be real (Indian Ocean spreading halt), but the apparent excess of 60–80 million years crust may be just an artefact.

The general impression that the trenches are more or less indiscriminate in their consumption is borne out by inspection of the ages of crust at trenches today. A cumulative curve of age *v.* length of trench (Fig. 8, 'age at trenches') is very similar to the age *v.* area curve. The median age of crust is about 60 million years, and the median

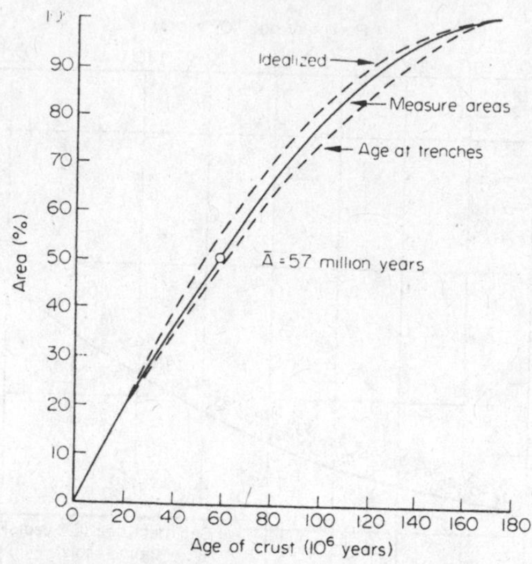

Fig. 8. Cumulative curves of crustal age *v.* area for actual ocean (solid line), for 'idealized' line of Fig. 20 (broken line) and of crustal age *v.* length (%) for crust being now consumed at today's trenches (from map in Fig. 6).

age of crust being consumed at trenches is not significantly different. Both curves are similar to the 'idealized' curve based on a regular decrease of area with age (dashed line in Fig. 7).

The age-area curve combines distributions from very different oceans: the Pacific, nearly surrounded by trenches and marginal basins; the Atlantic, where virtually all crust produced is preserved; and the Indian, which is highly complex. During times of continental clustering, when the oceans are dominantly trench-bordered, that is of a 'Pacific' type, the age *v.* area histogram should slope from youngest to oldest, the slope steepening with increasing spreading rates. During continental fragmentation 'Pacific' oceans shrink and 'Atlantic' oceans grow, with age-area histograms of entirely different characteristics (Fig. 9a, b). As the Atlantic grows and the Pacific shrinks (Fig. 9c–g), the age-area histograms evolve correspondingly, producing ever older ages for the oceanic crust—and hence deeper ocean basins. Thus, the break-up of a Pangaea by an Atlantic-type rift would first decrease the mean age (and depth) of oceanic crust and later increase it, as the rifting ocean keeps growing (Fig. 10).

We should next use our age-area relationships to produce depth-area histograms for various conditions of sea-floor spreading. The age-depth relationship given in Fig. 5, however, is of doubtful validity when applied to the entire ocean floor. That there is no absolute relationship between depth and age was, of course, recognized by Sclater *et al.* (1971) who stressed differences between ridge crest and associated flank depths. To illustrate the discrepancy between the age-depth curve (Fig. 5) and the actual world, we have constructed a synthetic hypsometric curve, from age *v.* area (Fig. 7) and age *v.* depth (Fig. 5). A comparison of the deduced and the actual (Menard & Smith, 1966, ocean basin + rise) hypsometric curves for the ocean basins (less trenches, volcanoes, continental rises, and aseismic ridges) is shown in Fig. 11. The deduced is deeper than actual, and the discrepancy is not removed even when using the high accumulation rates for pelagic sediments in today's equatorial Pacific at various depths (Berger, 1973a), and correcting for isostatic loading.

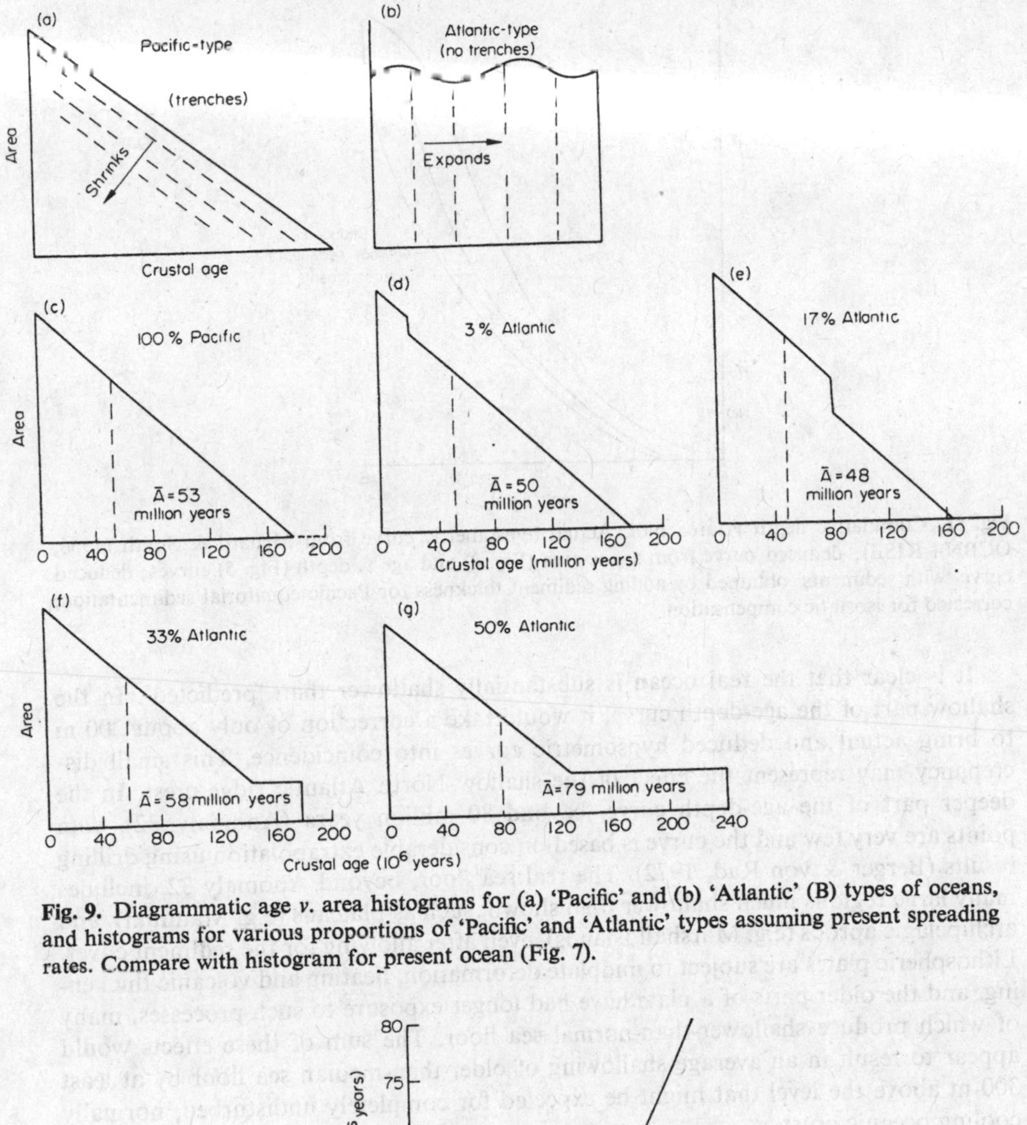

Fig. 9. Diagrammatic age *v.* area histograms for (a) 'Pacific' and (b) 'Atlantic' (B) types of oceans, and histograms for various proportions of 'Pacific' and 'Atlantic' types assuming present spreading rates. Compare with histogram for present ocean (Fig. 7).

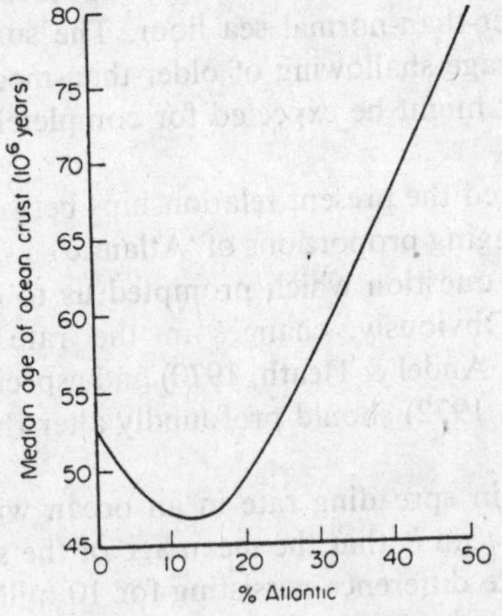

Fig. 10. Mean age of oceanic crust as a function of percentage of 'Atlantic'-type ocean assuming present spreading rates.

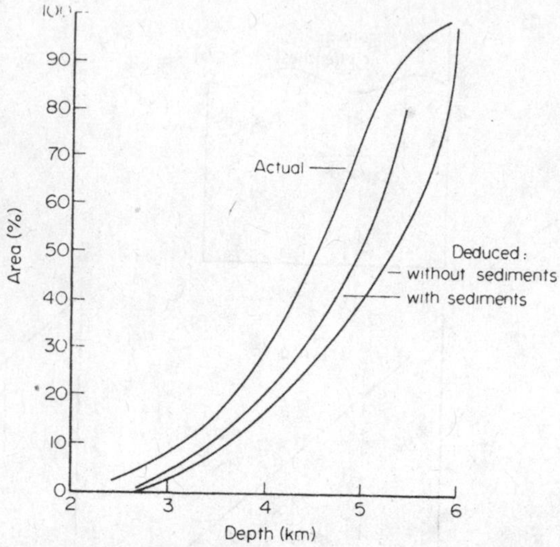

Fig. 11. Cumulative depth *v.* area plot. Actual hypsometric curve from Menard & Smith (1966, OCBN+RISE); deduced curve from age *v.* area (Fig. 8) and age *v.* depth (Fig. 5) curves; deduced curve 'with sediments' obtained by adding sediment thickness for Pacific equatorial sedimentation, corrected for isostatic compensation.

It is clear that the real ocean is substantially shallower than 'predicted'. In the shallow part of the age-depth curve, it would take a correction of only about 200 m to bring actual and deduced hypsometric curves into coincidence. This small discrepancy may represent the effect of the shallow North Atlantic ridge crest. In the deeper part of the age-depth curve, beyond 80 million years (Anomaly 32), data points are very few and the curve is based on considerable extrapolation using drilling results (Berger & von Rad, 1972). The real sea floor, beyond Anomaly 32, includes many large regions much shallower than shown, such as plateaus (e.g. Manihiki), and archipelagic aprons (e.g. Marshall Islands), even after allowing for the sediment cover. Lithospheric plates are subject to midplate deformation, heating and volcanic thickening, and the older parts of a plate have had longer exposure to such processes, many of which produce shallower-than-normal sea floor. The sum of these effects would appear to result in an average shallowing of older-than-median sea floor by at least 300 m above the level that might be expected for completely undisturbed, normally cooling oceanic crust.

So far we have considered the present relationships between age, depth, and area, as well as the effects of changing proportions of 'Atlantic'- *v.* 'Pacific'-type oceans. We now return to the original question which prompted us to examine the make-up of the hypsographic curve. Obviously, changes in the rate of sea-floor spreading (Ewing & Ewing, 1967; van Andel & Heath, 1970) and especially simultaneous global changes (Larson & Pitman, 1972) should profoundly alter the hypsometry of the sea floor.

The effect of a change in spreading rate in an ocean with a crustal age *v.* area pattern similar to today's ocean is that the mean age of the sea floor shifts about 0·8 million years for a 10% rate difference persisting for 10 million years (Fig. 12). The results are nearly the same, whether the 'ideal' curve or the actual curve for age *v.* area is used. The corresponding change in mean depth (Fig. 5) is 20 m for a 10% rate change, after 10 million years at the new rate. The effects are non-linear for very large rate changes or for very long time periods. Some idea of the very long-term effects can be seen by calculating mean depths for mature 'Pacific'-type oceans, with crust produced at various spreading rates (Fig. 13). Spreading rates for the past 20 million years have averaged very close to 3 km² of new crust per year, since about 60×10^6 km² of crust are 0–20 million years old (Fig. 6).

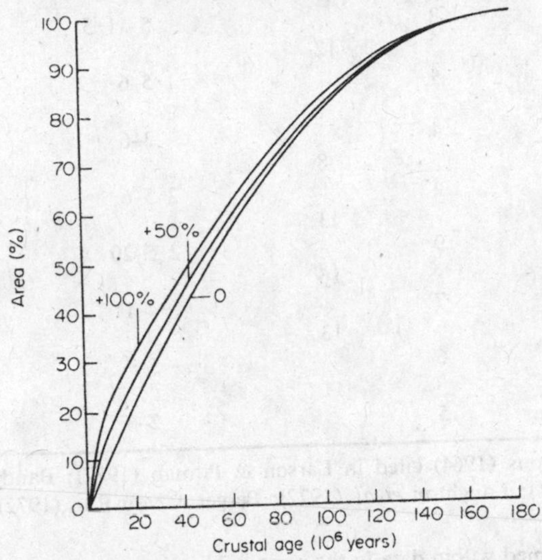

Fig. 12. Crustal age *v.* area (cumulative %) for 'ideal' ocean, and after a 10-million-year period in which spreading rates were 50% and 100% faster than today's rates. Median crustal ages are: ideal, 52 million years; 50% increase, 48 million years; 100% increase, 44 million years.

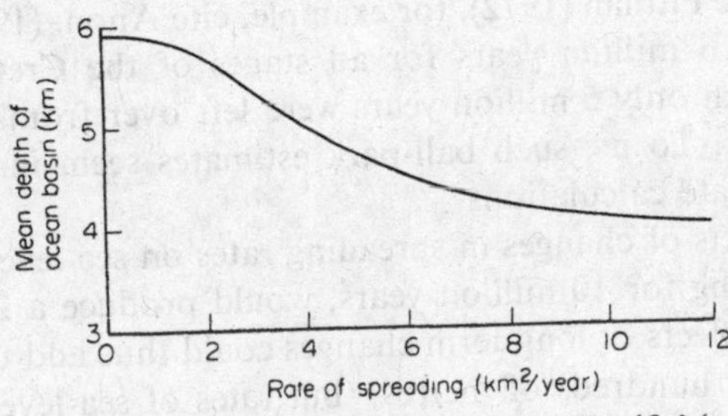

Fig. 13. Mean depth *v.* spreading rate (in km²/year) for mature 'Pacific' (trench-surrounded) oceans. Rate for the past 20 million years, about 3 km²/year.

Very great departures from the present rate have been proposed by Hays & Pitman (1973, Table 3); e.g., 5·6 km²/year for the period 50–85 million years ago, and 12·6 km²/year for the period 85–110 million years ago. The highest global rates are deduced mainly from long extrapolation of short segments of old ridge crests in the Pacific. To put these estimates and the ones by Larson & Pitman (1972) into perspective, we have collected age estimates from six recent publications, the oldest of which is the paper (Anon., 1964) cited in Larson & Pitman (1972). There is considerable range in the estimates for the duration of Cretaceous stages (Table 1). Since these durations form the denominator in the spreading-rate calculations, rate estimates can be made

Table 1. Duration of Cretaceous stages, single and as pairs, as estimated in six recent publications*

	Average		Range	
	Single	Pairs†	Single	Pairs†
Maastrichtian	7		5–9	
		15		11–19·5
Campanian	8		5–11·5	
		12		9–13
Santonian	4		1·5–6	
		8		4·5–12
Coniacian	4		3–6	
		8		6–12
Turonian	4		2·5–6	
		13		8–22·5
Cenomanian	9		2·5–20	
		16		6·5–30
Albian	7		4–11	
		13		9–17
Aptian	6		5–7	
		11		9–12
Barremian	5		2–6	

* Anonymous (1964) cited in Larson & Pitman (1972); Bandy (1967); Lambert (1971); Laughton *et al.* (1972); Berger & von Rad (1972); Gilluly (1973).

† Pairs formed within data in the same publications.

to differ by factors of two or three simply by using different sources for the time ranges. At the same time, rate changes calculated in such a fashion would automatically apply to all oceans. Larson & Pitman (1972), for example, cite Anon. (1964), who estimated a duration of exactly 6 million years for all stages of the Cretaceous, except the Maastrichtian for which only 5 million years were left over from the assumed duration of the Cretaceous. To us, such ball-park estimates seem inappropriate for the purpose of spreading-rate calculations.

To sum up the effects of changes in spreading rates on sea level, a 10% change in spreading rate, persisting for 10 million years, would produce a 20 m change in sea level. The cumulative effects of long-term changes could thus add up to produce large excursions of sea level, hundreds of metres—but rates of sea-level change would be slow—probably around 10m/million years. According to our previous estimate of the basin-shelf effect, a change in the CCS of roughly the same magnitude might be expected. It is difficult to apply these estimates to the distant past; e.g. the Mesozoic. First, only fragments of the old ridge systems are preserved, and entire Mesozoic plates may have been consumed leaving hardly a trace. Equally discouraging, the dating of Mesozoic oceanic crust is as yet uncertain. The biostratigraphic ages of key Mesozoic anomalies (32, M-1, M-22) are not known to within a stage, and even if they were, the tie to the radiometric scale is still rather tenuous (Table 1).

Changes in hypsography from other causes

Overall shallowing or deepening of the sea floor—and hence sea-level changes—attributable to the building of volcanic seamounts is minor. Seamounts occupy only about 3% of the ocean floor today (Fig. 3), and represent a volume of about $1{\cdot}5 \times 10^7$ km^3 (assuming 4500 m as a mean height). If they can erupt anywhere with equal probability, then their areal density pattern is determined by crustal age patterns. An estimate of the average rate of volcanism over the past 160 million years, obtained by dividing the total volume by the mean age of the crust (60 million years), is $0{\cdot}25 \times 10^6$ km^3/million years. At these rates, cessation of *all* seamount building for a million years would result in a regression of less than 1 m.

Non-uniform growth and decay of 'mantle convection plumes' (Morgan, 1972) or asthenospheric 'bumps' (Menard, 1973), or changes in distributions of these features beneath oceans and continents, may cause small sea-level fluctuations. A typical bump measures about 1000 km across and 0·5 km high, or about $0{\cdot}5 \times 10^6$ km^3 in volume. Addition or subtraction of such a feature on the sea floor would change sea level by about 1·5 m. The postulated life span of these features is 10^7–10^8 years, and it is thus unlikely that they have been a major factor in sea-level changes.

A change in sea level from changing global average rates of sedimentation also appears negligible. A doubling of the sedimentation rate for 1 million years would induce a transgression of about 5 m.

Changes in the total area of continent caused by subduction of one continent beneath another, can cause some effect on global sea level. If the Tibetan Plateau represents a 'doubled' continental crust, then 3×10^6 km^2 of crust has been added to the oceans at the expense of the continents, for constant total area. This would cause a sea-level fall of about 40 m. The rate lies in the range 1–3 m/million years.

Effects of glaciation and of desiccation of isolated basins

There are at least two ways of producing large changes in sea level—and hence hypsometry—which are unrelated to displacement by crustal motions: glaciation and evaporation of water from isolated basins.

During the Pleistocene, sea-level oscillations between about −130 m and perhaps +20 m resulted largely from the waxing and waning of northern continental ice sheets (Flint, 1971). The potential effect of these sea-level changes on the mean CCD was probably diminished by relatively low global accumulation rates of carbonate on the present shelves, a disproportionate share of which is in high latitudes (Fig. 4). The pronounced CCD (and lysocline) fluctuations in Pacific equatorial regions known from the Pleistocene (Arrhenius, 1952) continue back through the Miocene (Tracey *et al.*, 1971), when glacially induced sea-level changes were small. There is evidence for glaciation in Antarctica as long ago as Early Tertiary (Hayes, Pimm, *et al.*, 1973), and the change in sea level caused by growth and wasting of glaciers in this region alone would amount to about 60 m (Flint, 1971).

Desiccation of an ocean basin isolated from the world ocean would cause a transgression elsewhere. The present day Mediterranean, with a volume of $3{\cdot}8 \times 10^6$ km^3 (Menard & Smith, 1966), contains 0·28% of the volume of the world ocean. Consequently, emptying of the present basin, similar to the drying proposed for the Late Miocene (Ryan, Hsü *et al.*, 1973) would raise the sea level of the remaining world ocean by about 10 m. Larger effects require larger basins. Thus, the Early Cretaceous South Atlantic may have been nearly isolated during the time when South America and Africa were still juxtaposed across the Guinea Fracture Zone. Evaporation of this basin, about five times as large as the Mediterranean, would have provoked a 50 m transgression elsewhere.

It is a special feature of both glaciation and evaporation that the associated rise and fall of sea level, and hence of the CCD, can happen very quickly, in a few thousand or tens of thousands of years, in contrast to the much slower rates resulting from tectonic changes in the shape of the hypsometric curve. Likely effects of a fast transgression on carbonate distribution include creation of clear-water shelves favourable for carbonate accumulation, and reduced clastic dilution of pelagic carbonates on the continental slope and on turbidite plains.

RELIEF OF THE COMPENSATION SURFACE BETWEEN AND WITHIN BASINS

Range and possible causes

The existence of a calcite compensation level is contingent upon the fact that dissolution rates in the ocean increase with depth, mainly in response to increasing pressure and decreasing temperature. The depth of the mean position of the CCS is largely a matter of mass balance, bathymetric fractionation, and hypsography. The relief of the CCS topography reflects a host of additional controlling factors virtually all of which are poorly studied and understood. Rather than joining the discussion on the relationship of the CCD to saturation level as calculated by thermodynamics (a discussion which has contributed little to geological understanding so far), we inspect the CCS map (Fig. 1) for general patterns in relief and offer suggestions of how these patterns might be produced.

The total depth range of the pelagic carbonate line is roughly 2 km, that is, the factors responsible for the topography are comparable to the effects of pressure and temperature difference corresponding to a 2 km water column at bathyal and abyssal depths. Considerable variations in supply rates of carbonate and in dissolution rates within and between ocean basins are indicated, therefore.

Differences in supply arise from:

(1) Overall fertility variations, which are a function of nutrient supply and available sunlight;

(2) the amount of preservable calcareous shells fixed per organic production unit;

(3) the range of resistance to dissolution of these shells;

(4) the efficiency of transfer to the ocean floor.

Differences in dissolution depend on:

(5) The amount and quality of organic matter that reaches the sea floor in addition to the calcareous shells;

(6) the benthonic activity that this food supply engenders;

(7) the concentration of dissolved carbon dioxide in the water directly overlying the sediment;

(8) the rate of burial;

(9) various aspects of interstitial-water chemistry.

In the following sections we will discuss some of the oceanographic and geological processes that produce these differences.

Inter-ocean fractionation (basin versus basin)

Inter-ocean or basin-basin fractionation (Berger, 1970) differs from bathymetric fractionation in that pressure, temperature and light gradients play a negligible role in producing favourable and unfavourable environments for carbonate deposition, while chemical gradients are established between communicating ocean basins such that one basin is characterized by 'deep-water', the other by 'shallow-water' aspects. Thus, the North Atlantic tends to collect 'young' surface-type waters, depleted in nutrients, relatively warm, oxygen-rich and carbon dioxide-poor. It is characterized by carbonate sedimentation and has the greatest CCS depths in the open ocean (Fig. 2). Conversely, the North Pacific has relatively shallow compensation levels; its deep waters are 'old', relatively cold, and rich in carbon dioxide. The differences in water are maintained by fractionation processes resulting from biological activity and from water exchange patterns between the basins.

Latitudinal (high *v.* low), trophic (fertile *v.* barren), and diversity-related fractionation

Tropical shallow seas are both warm and well lit and accumulate carbonate, while there is a tendency to exclude carbonate from seas in mid and high latitudes. Precipation patterns play an important role in this respect, since brackish seas favour dissolution while hypersaline ones favour precipitation of calcium carbonate (compare Baltic Sea with Persian Gulf; see Seibold, 1970).

In the deep sea also there is a latitudinal fractionation effect: poleward of the polar fronts coccolithophorids are virtually absent (Hasle, 1960; McIntyre, Ruddiman & Jantzen, 1972). Foraminifera do persist, however (Bé, 1960, 1969). The reason that the CCS rises toward high latitudes (Fig. 1), therefore, is not simply an effect of decreased shell production, but is largely due to increased dissolution (Parker & Berger, 1971), owing to cold corrosive bottom waters, current action and benthonic activity. Between the polar fronts latitudinal effects, if any, appear of minor importance (Fig. 14) except for the equatorial belt itself.

A striking feature of the CCS map (Fig. 1) is the rapid rise of the compensation line toward the continents. Since productivity is generally high around continents because of upwelling and mixing processes delivering nutrients to surface waters, the rise of the CCS must be entirely due to greatly intensified dissolution, despite the increased burial rates provided by terrigenous material. Where the CCS rise is not obvious, as in the North Atlantic, it is because there is no intersection of the CCS with the topography. A similar rapid increase in dissolution rates is, however, indicated from foraminiferal evidence (Berger, 1968, 1970; Parker & Berger, 1971; Thiede, 1973).

We suggest that two factors contribute significantly to this continental slope dissolution effect: (1) the intense development of CO_2 from a rich supply of organic matter, which is responsible for a considerable lowering of the pH, and (2) the greatly enhanced activity of benthonic organisms which digest and damage the particles, as well as stir and mix them, preventing the interstitial waters from becoming saturated with calcite. Where combustion and mixing is prevented by oxygen depletion, dissolution of calcite near the water-sediment interface virtually ceases (Berger & Soutar, 1970).

If high fertility engenders dissolution of calcite to the degree indicated, how is it that the equatorial high fertility zone in the Pacific is clearly characterized by a carbonate belt (Arrhenius, 1952), that is, by a greatly depressed CCS (Fig. 1)? This paradox is indeed puzzling and has no ready explanation. We submit the following thoughts for consideration.

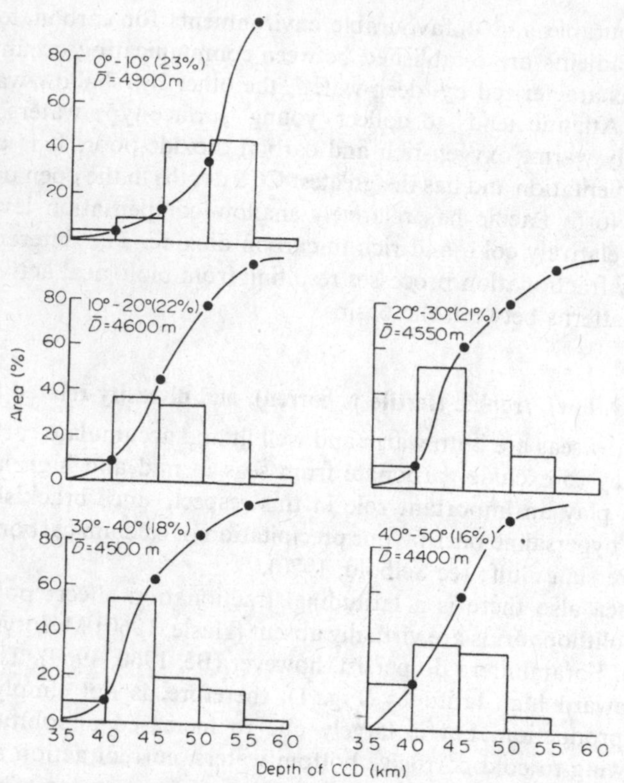

Fig. 14. Depth distribution of the carbonate line by latitudinal belts (percentage of total area of belt in parentheses).

(1) The bulk of the carbonate belt consists of residual assemblages from which the more soluble shells have been removed (Berger, 1971, 1973b). To a considerable, but undetermined degree, therefore, the belt exists because there is a relatively high proportion of resistant shells within the already high supply of carbonate. In the eastern equatorial Pacific, for example, the resistant Foraminifera *Globorotalia menardii, G. tumida, Globoquadrina dutertrei,* and *Pulleniatina obliquiloculata* can provide one-third of a well-preserved assemblage (Parker & Berger, 1971). Among the coccolithophorids the resistant forms *Cyclococcolithina leptopora* and *Geophyrocapsa* are very abundant (P. H. Roth, personal communication).

Thus, to a certain extent, the equatorial belt is a product of diversity-related fractionation, with the carbonate being shifted to the place where there is a high supply of very resistant shells. Around continents, the pelagic carbonate supply is much more easily dissolved, since by far the dominant portion of the supply consists of *Globigerina* and *Emiliania huxleyi.*

(2) The modes of transfer of shell material and of organic matter from surface waters to the sea floor differ greatly in quality and efficiency between peri-continental and oceanic regimes. The food chains in the pelagic equatorial area are expected to be rather different from those around the continents, with profound effects on the mode of transfer, the condition in which shells (especially coccoliths) are delivered and especially the amount and kind of organic matter finding its way to the floor. We

suggest that the initial calcite-to-organics ratio in surficial sediments of the equatorial belt is considerably greater than around the continents, so that destructive effects of [illegible] development and benthonic activity are less along the equator than along the continental slope. Also, the seasonal and quasi-seasonal boom and bust productivity around the continents leaves much organic matter incompletely digested. This material is then easily oxidized on the sea floor, whereas the organic matter delivered from more stable environments should be well sifted by digestion and hence rather refractory.

(3) Deep bottom currents tend to be intensified along the continental slope, where they follow elevation contours (Defant, 1961, Figs. 320, 322; Heezen, Hollister & Ruddiman, 1966). Likewise internal waves and tidal currents may interact with sediments on the slope to a greater degree than with those in the deep ocean. We think it likely that such bottom-current intensification contributes significantly to the more intense dissolution on the continental slope.

An additional general feature of the CCS map may be mentioned, although it is much less striking than the inter-ocean contrast, the peri-continental effect, or the equatorial belt. There appears to be a tendency of the CCS topography to mirror the deep-sea bottom topography in a much subdued fashion. We suggest redeposition effects as a likely explanation for this phenomenon: to some degree, carbonate is removed from topographic highs and carried to the deep plains.

Besides these general patterns, there are more local ones related to the flow of Antarctic Bottom Water. In the western South Atlantic, Antarctic Bottom Water enters through a passage at the Rio Grande Rise, opened through sea-floor spreading, but is barred from the eastern South Atlantic by the Walvis Ridge. Thus, the pattern of the flow of corrosive bottom water is of major importance, producing shifts in carbonate deposition by 'abyssal fractionation'. A similar effect is seen south of Tasmania (Conolly & Payne, 1972; Watkins & Kennett, 1972).

THE CONCEPT 'PLATE STRATIGRAPHY'

Definition

In the geological world view introduced by Hess (1962), oceanic crust continuously accretes at the mid-ocean ridges, accumulates sediment as it moves away from its origin, and finally disappears within the island-arc system. Mid-ocean ridges, being the surface manifestation of the ascending limbs of mantle convection current, are relatively hot. The ridges are elevated with respect to the older sea floor, therefore, owing to thermal expansion as well as phase changes. Continents drift passively with the horizontally moving upper mantle. Hess' concepts, amplified and modified by subsequent studies (Vine & Matthews, 1963; Wilson, 1963, 1965; McKenzie & Parker, 1967; Morgan, 1968; LePichon, 1968; Isacks, Oliver & Sykes, 1968) form the core of the theory of plate tectonics.

We have already applied the concepts of this theory in attempting to understand the factors that control the position and topography of the calcite compensation surface. We feel that even the qualitative and speculative understanding thus gained should be helpful in interpreting the stratigraphic record, to which we now turn our attention.

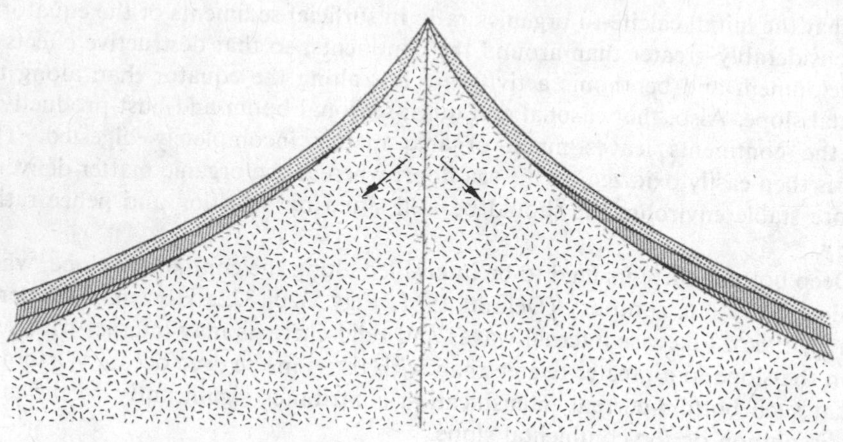

Fig. 15. Quasi-transgressive stratigraphic sequence on a spreading and subsiding ridge flank—the first principle of plate stratigraphy. (According to Hess, 1962.)

For deep-sea stratigraphy, plate tectonics implies sedimentation on horizontally moving plates which are slowly subsiding as they cool. The principles and operations appropriate to the treatment and interpretation of the resulting facies relationships (Berger, 1972; Winterer, 1971, 1973) constitute the concept of 'plate stratigraphy' (Berger, 1973a). Hess (1962) recognized the first of these principles, that of onlapping chrono-stratigraphic units onto a spreading ridge, simulating a transgressive sequence (Fig. 15). Other stratigraphic relationships can be readily deduced from simple models.

Sedimentological corollaries to Hess' model of sea-floor spreading and subsidence

The depth of the calcite compensation surface is largely between 4 and 5 km (Fig. 1), and the average elevation of mid-ocean ridges is between 2·5 and 3 km. Thus, the upper flanks of mid-ocean ridges accumulate carbonate, while clay is deposited further down. The sediment in contact with newly created basaltic sea floor, therefore, is carbonate, and it is later covered by clay as it subsides below the compensation level. The resulting record shows basalt overlain by carbonate, in turn overlain by clay in the area below the carbonate line (Fig. 16a). The subsurface boundary between carbonate and clay ('carbonate-clay boundary', abbreviated CCB) constitutes the fossil trace of the ancient carbonate line, whose age is that of the associated sediment. This implication of sea-floor spreading was verified by the early legs of the Deep Sea Drilling Project (Peterson *et al.*, 1970, Leg 2; Hsü & Andrews, 1970, Leg 3; Benson, Gerard & Hay, 1970, Leg 4; Fischer *et al.*, 1970, Leg 6), as well as by subsequent legs. We can make a reasonable estimate of the expected thickness of carbonate overlying the basalt. The sinking sea floor traverses a depth difference of somewhat less than 2 km on the average, assuming the present topography of mid-ocean ridges and of the compensation surface. The time needed to subside through this interval is approximately 30–35 million years (Menard, 1969; Sclater, Anderson & Bell, 1971). Sedimentation rates of carbonate on ridge flanks vary greatly, but are generally between 4 and 20 m per million years at the present time, although the central part of the ridge apparently collects very little sediment (Ewing & Ewing, 1967). Allowing for this effect, whatever its cause, we can expect between 120 m and 600 m of carbonate, or near 250 m on the (geometric) average. This rough estimate is somewhat high when compared with the thickness of carbonate in Atlantic drill holes thought to have reached basement (Fig. 17). The present CCD, therefore, appears unusually deep or carbonate sedimentation rates are unusually high compared with conditions in Tertiary and Late Cretaceous time, or both.

Several sites show a sequence from basal carbonate to clay to upper carbonate (Fig. 17, Sites 10, 20, 19, 15). Such a sequence can be produced by a fluctuating CCD, provided that the site intersected the palaeo-CCD and did not stay above the fluctuation zone throughout its existence (Fig. 16b). Under the same assumption of a fluctuating CCD, the carbonate stratigraphy of a non-spreading (non-subsiding) ridge would be quite different from that of a spreading (subsiding) one (Fig. 18). Thus, the interpretation of the carbonate record is intimately tied to the kinds of vertical motions that are to be postulated.

The effects of horizontal plate motions are best illustrated in the equatorial Pacific. The carbonate belt (Fig. 1) extends as a long narrow tongue from the East Pacific Rise in a westward direction, until the depth of the ocean is well in excess of 5 km. The belt

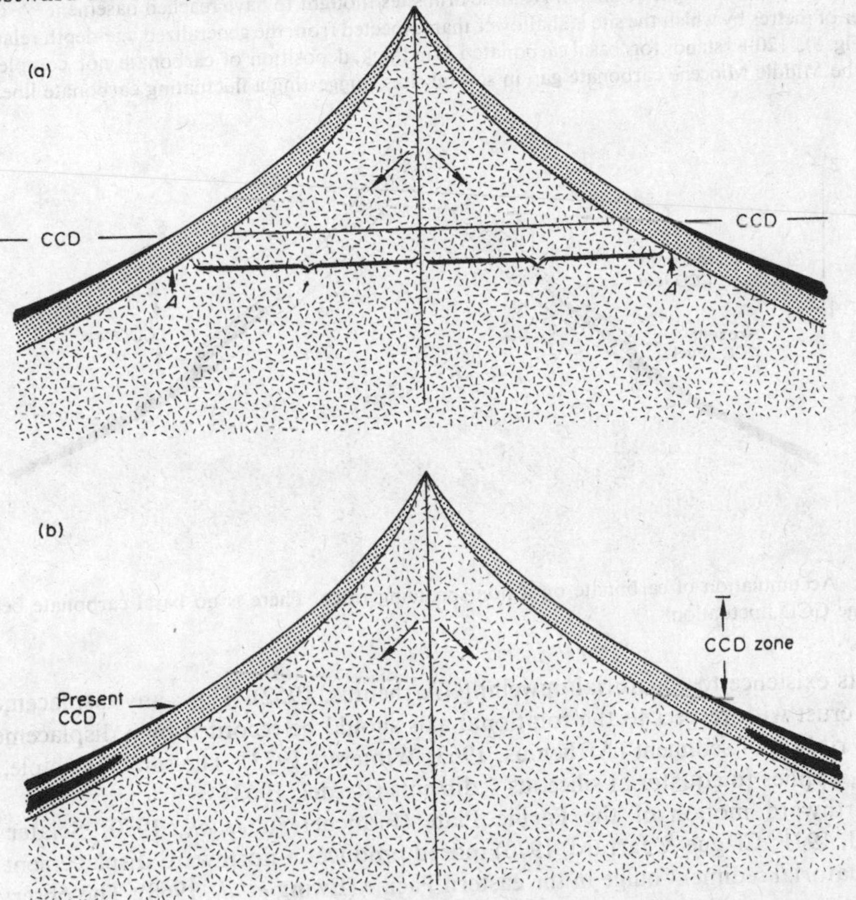

Fig. 16. Carbonate deposition on spreading and subsiding ridge flanks for (a) stationary and (b) fluctuating CCD levels. Stipples, calcareous ooze; black, clay; *t*, distance from ridge crest corresponding to the time *t* during which carbonate accumulates; *A*, final thickness of carbonate, corresponding to time *t*; CCD zone: range of fluctuation of local CCD level.

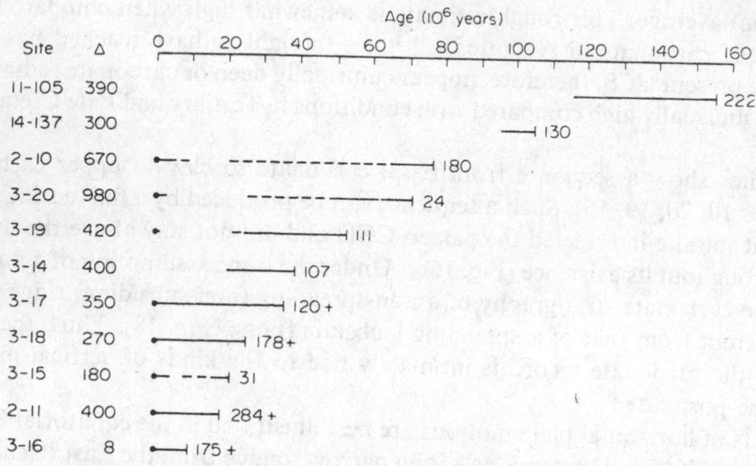

Fig. 17. Time interval of carbonate deposition (horizontal lines) and total thickness of basal carbonate (number to the right of line) in Atlantic drill sites thought to have reached basement. △ is the number of metres by which the site is shallower than expected from the generalized age-depth relationship (Fig. 5). '120+' stands for 'basal carbonate 120 m thick, deposition of carbonate not completed'. Note the Middle Miocene carbonate gap in several sites, suggesting a fluctuating carbonate line.

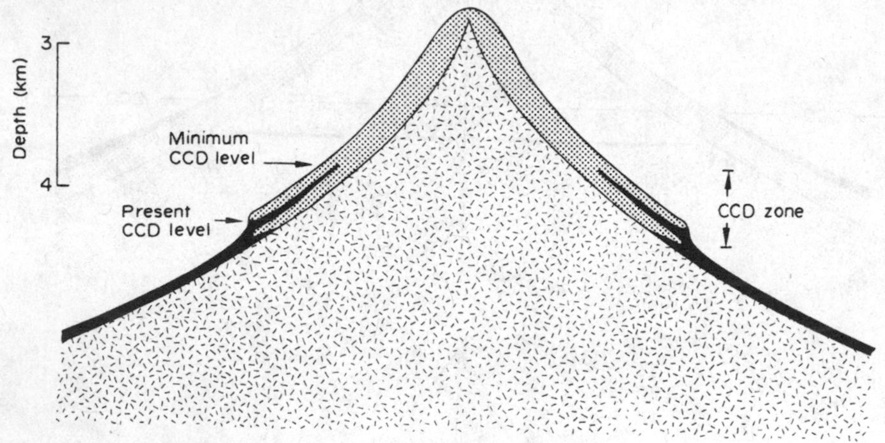

Fig. 18. Accumulation of carbonate on a non-spreading ridge. There is no basal carbonate below the range CCD fluctuation.

owes its existence to equatorial upwelling (Arrhenius, 1952). Thus, any displacement of the crust with respect to the rotational axis should be recorded by a displacement of this ribbon of sediment (Arrhenius, 1963; Winterer, 1973). The same principle, of course, applies to equatorial siliceous sediments (Riedel, 1963).

Northward motion of the Pacific Plate (Francheteau *et al.*, 1970; Sclater & Jarrard, 1971; Clague & Jarrard, 1973) indeed produces a northward displacement of the equatorial sediment bulge in the eastern Pacific (Ewing *et al.*, 1968). The observed shift is an average displacement for acoustically transparent sediments of various ages back to Eocene time. At each geographical location on this bulge, sediments deposited north of the equator will overlie equatorial ones, which may in turn overlie sediments accumulated south of the equator (Fig. 19). Between any two sites north of the equator, the ages of the ancient equatorial sediments will differ, unless the two points lie along the track of the palaeo-equator.

Given the age and latitude of ancient equatorial sediments, the northward component of the plate motion in degrees per million years can be determined (Winterer, 1973; van Andel & Heath, 1973; Berger, 1973a). This is the most striking instance for the general principle that facies properties tied to latitudinal zonation make it possible to determine latitudinal migration of the sea floor by their progressive displacement with time.

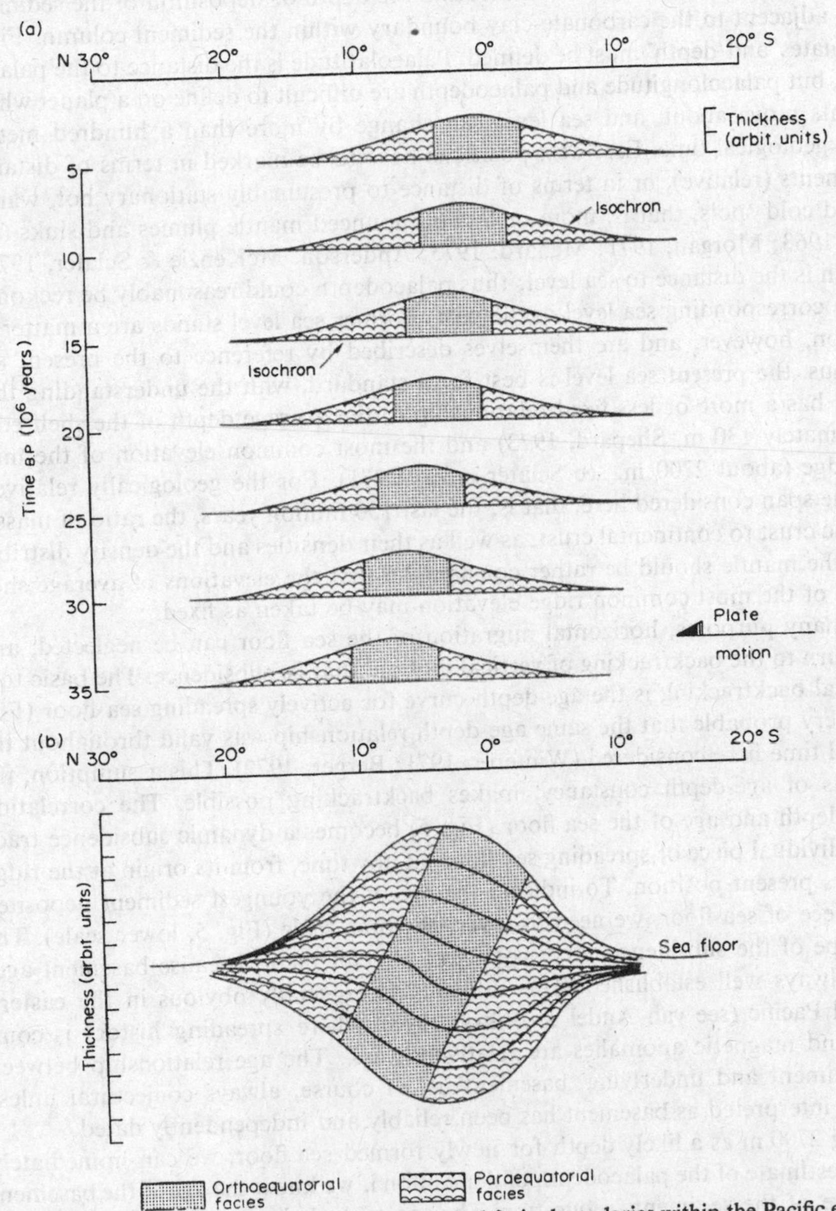

Fig. 19. Schematic model of generating isochrons and facies boundaries within the Pacific equatorial bulge of pelagic sediments. Orthoequatorial: carbonate-silica cycles. Paraequatorial: siliceous ooze.

The sediment bodies that result from subjecting the compensation surface (Fig. 1) to both subsidence and migration become difficult to visualize, especially when the surface is allowed to change its topography through time. For this reason, it is advantageous to view sedimentary facies distributions within age-palaeodepth matrices or palaeodepth–palaeolatitude matrices.

Vertical backtracking

To determine the palaeobathymetry of the carbonate line from Deep Sea Drilling data, we need to know the co-ordinates and the depth of deposition of the sediment samples adjacent to the carbonate-clay boundary within the sediment column. First, 'co-ordinates' and 'depth' must be defined. Palaeolatitude is the distance to the palaeo-equator, but palaeolongitude and palaeodepth are difficult to define on a planet where continents move about, and sea level can change by more than a hundred metres through geological time. East-west positions have to be marked in terms of distance to continents (relative), or in terms of distance to presumably stationary hot, warm, cool, and cold spots, that is, more or less pronounced mantle plumes and sinks (see Wilson, 1963; Morgan, 1971; Menard, 1973; Anderson, McKenzie & Sclater, 1973).

Depth is the distance to sea level; thus palaeodepth could reasonably be reckoned from the corresponding sea level in the past. Former sea level stands are a matter of contention, however, and are themselves described by reference to the present sea level. Thus, the present sea level is best for a standard, with the understanding that this level has a more or less fixed relationship to the average depth of the shelf edge (approximately 130 m, Shepard, 1973) and the most common elevation of the mid-ocean ridge (about 2700 m, see Sclater *et al.*, 1971). For the geologically relatively short time span considered here, that is, the last 150 million years, the ratio of masses of oceanic crust to continental crust, as well as their densities and the density distributions in the mantle should be rather constant, so that the elevations of average shelf edge and of the most common ridge elevation may be taken as fixed.

For many purposes, horizontal migration of the sea floor can be neglected, and we first turn to the backtracking of vertical motion, that is, subsidence. The basic tool for vertical backtracking is the age-depth curve for actively spreading sea floor (Fig. 5). It is very probable that the same age-depth relationship was valid throughout the geological time here considered (Winterer, 1971; Berger, 1972). This assumption, the hypothesis of age-depth constancy, makes backtracking possible. The correlation between depth and age of the sea floor (Fig. 5) becomes a dynamic subsidence track of any individual piece of spreading sea floor, at any time, from its origin at the ridge crest to its present position. To indicate the age of the youngest sediment deposited on this piece of sea floor, we need to invert the time scale (Fig. 5, lower scale). The exact shape of the subsidence curve is subject to refinement, because basement ages are not always well established. This difficulty is especially obvious in the eastern equatorial Pacific (see van Andel & Bukry, 1973), where spreading history is complicated and magnetic anomalies are relatively weak. The age relationship between oldest sediment and underlying 'basement' is, of course, always conjectural unless the basalt interpreted as basement has been reliably and independently dated.

Taking 2700 m as a likely depth for newly formed sea floor, we can immediately obtain an estimate of the palaeodepth of deposition if we know the age of the basement and the age of the sediment in question. The age of the basement defines the starting point (A) on the idealized subsidence curve, whereas the age of the sediment indicates the distance we have to back up on the track to find the point B at which this sediment was being deposited (Fig. 20; basement age=60 million years; sediment age=40 million years; A=5330 m: B=4140 m). This method is rather crude, because it neglects that a particular piece of sea floor need not follow the absolute palaeodepth values of the subsidence track, even though it may subside parallel to the curve, and also neglects that the basement depth should change through isostatic adjustment to the sediment load. To correct for this omission, we need to use site depth and depth-in-hole. We now relocate the curve parallel to itself by the vertical distance between actual site depth at C and the expected depth at A. Now the subsidence curve goes through the actual site depth at C, and we can find a preliminary palaeodepth at D as before. To compensate for isostatic loading, we go downward from point D by one half the depth-in-hole of the sediment in question. Thus, we obtain point P. For example, for a stack of sediment 400 m thick and for Z=670 m, the final palaeodepth is 3670 m (Fig. 20). We assume, obviously, that as the sea floor subsided from P through time, sedimentation kept making the sea floor somewhat shallower than it would have been without sediments. Thus, the sea floor ended up at C instead of below this point. This upbuilding, however, proceeds only at one half the rate of sediment accumulation, because the sediment simultaneously depresses the crust isostatically under its increasing weight. The correction factor of one half may be justified as follows.

The density of the upper mantle sima is approximately 3·3 g/cm³, that of the sediment between about 1·6 g/cm³ for unconsolidated ooze and 2·4 g/cm³ for highly compacted material (Holmes, 1965, p. 30; Beall & Fischer, 1969; Gealy, 1971). Replacing the water overlying the sea floor by sediment adds roughly 1 g/cm³. This corresponds to 1 kg/cm² for each 10 m of load, a pressure that is equivalent to a column

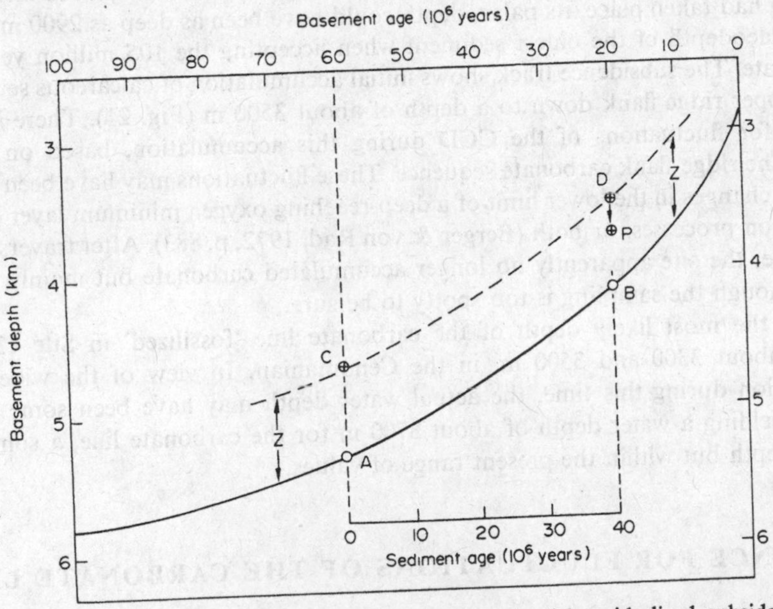

Fig. 20. Palaeodepth determination by vertical backtracking parallel to idealized subsidence track. A and B: present depth and palaeodepth (40 million years age) on idealized curve; C: actual site and D: analogue to B on parallel curve; Z: distance between A and C; P: final palaeodepth after correction for isostatic loading (see text).

of mantle rock 3 m in height. As the crust is isostatically depressed, an additional load is provided by the increase in water depth. Thirty per cent of the displaced sima balances increased water pressure, rather than the sediment load, so that a subsidence of 4·3 m becomes necessary for isostatic adjustment of a 10 m sediment load. Forty-three per cent is close enough to one half for the present purposes.

We consider the palaeodepth estimate resulting from the backtracking procedure to be the best guess that can be made on the basis of information on basement age, sediment age, present depth of site and depth-in-hole of the sediment in question. Further refinement may be possible by taking into account any anomalous vertical motions of the sea floor produced by its riding over 'asthenospheric bumps' or holes as indicated by gravity anomalies (Menard, 1973). When dealing with pelagic sediments in Alpine-type mountain chains on land, the palaeodepth estimate based on the idealized age-depth curve may in cases be the only facies-independent estimate. The time available for subsidence from an original ridge crest depth of about 2700 m would be the age difference between the sediment in question and the basal sediment or basalt. We now turn to carbonate-line reconstruction, using actual drill sites. A simple example is provided by Site 137, in Leg 14 of the Deep Sea Drilling Project.

Palaeodepth reconstruction of the carbonate line, Site 137

Site 137, drilled during Leg 14 (Hayes, Pimm *et al.*, 1972), is about 105 million years old (Berger & von Rad, 1972, p. 877) and is at 5361 m water depth at the present time. Sediment thickness is 397 m. The amount of subsidence for 105 million years is estimated at 3160 m (Fig. 5). The site started, therefore, at 5361 m minus 3160 m plus 199 m ($\frac{1}{2} \times 397$ m), that is at a palaeodepth of about 2400 m (Fig. 21). If the basalt was emplaced at the ridge crest, its palaeodepth is 2400 m, if it was emplaced after some spreading had taken place, its palaeodepth could have been as deep as 2900 m, which is the palaeodepth of the oldest sediment when accepting the 105 million years site age estimate. The subsidence track shows initial accumulation of calcareous sediments on the upper ridge flank down to a depth of about 3500 m (Fig. 21). There is some evidence for fluctuations of the CCD during this accumulation, based on clayey layers in the ridge-flank carbonate sequence. These fluctuations may have been associated with changes in the lower limit of a deep-reaching oxygen minimum layer or with redeposition processes, or both (Berger & von Rad, 1972, p. 883). After traversing the CCD zone, the site apparently no longer accumulated carbonate but mainly barren clays, although the sampling is too spotty to be sure.

Thus, the most likely depth of the carbonate line 'fossilized' in Site 137 was between about 3300 and 3500 m, in the Cenomanian. In view of the widespread transgression during this time, the actual water depth may have been some 300 m greater, yielding a water depth of about 3700 m for the carbonate line, a somewhat shallow depth but within the present range of values.

EVIDENCE FOR FLUCTUATIONS OF THE CARBONATE LINE

Quaternary fluctuations

We have already mentioned the carbonate cycles of the equatorial Pacific which were first described by Arrhenius (1952). These cycles are largely due to dissolution

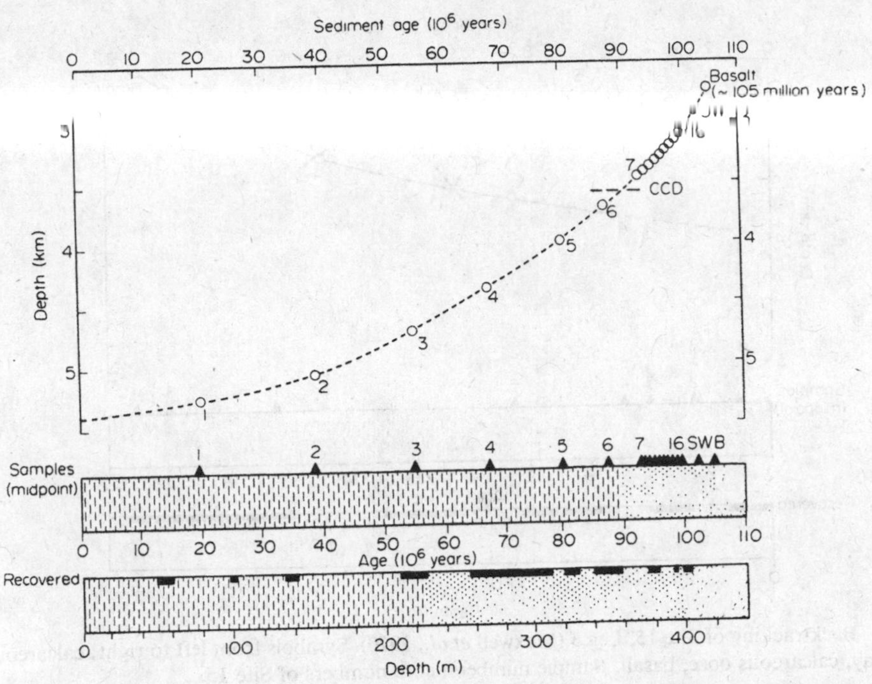

Fig. 21. Backtracking of Site 137, Leg 14 (Hayes, Pimm *et al.*, 1972). Symbols from left to right: clay, calcareous ooze, basalt. Sample numbers: core number of Site 137.

pulses, with greatly increased corrosion of calcareous tests during the stages characterized by low carbonate content and interpreted as interglacials (Olausson, 1965; Hays *et al.*, 1969; Broecker, 1971; Berger, 1973b). Thus the carbonate compensation surface was shallower during interglacials, although we cannot say by how much. Similar dissolution cycles exist in the central Atlantic (Berger, 1968; Ruddiman, 1971). During glacial times, there may have been sufficient suppression of $CaCO_3$ production in high latitudes to raise the carbonate line in high latitudes and allow a corresponding deepening in low latitudes on a world-wide scale. A shelf effect also may have been active, as discussed in the section on bathymetric fractionation. In any case, there is evidence from the Quaternary record that a 'warm ocean' is characterized by increased dissolution of carbonate in tropics and subtropics, while a 'cold ocean' has a relatively deeper carbonate line in the same area.

Tertiary fluctuations

We have seen that the carbonate stratigraphy of the Atlantic suggests alternations of carbonate and non-carbonate deposition (Fig. 17) which implies CCD fluctuations whether subsidence is invoked or not (Figs 16b, 18). Using the backtracking procedure, we can now attempt to quantify these Tertiary fluctuations, in order to relate them to the fractionation mechanisms discussed earlier.

A striking example for a fluctuating CCD is provided by Site 15, in Leg 3 of the Deep Sea Drilling Project (Fig. 22). At this site, a relatively thin clayey sediment layer representing a considerable time span is sandwiched between two carbonate units. The interpretation is that the carbonate line was near 3250 m about 15 million years

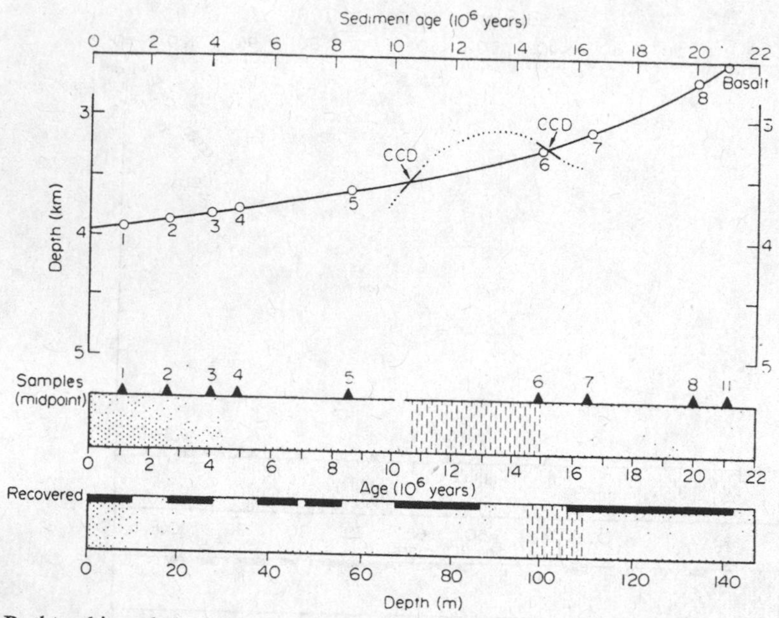

Fig. 22. Backtracking of Site 15, Leg 3 (Maxwell *et al.*, 1970). Symbols from left to right: calcareous ooze, clay, calcareous ooze, basalt. Sample numbers: core numbers of Site 15.

ago and subsequently stayed at or above this depth, leading to dissolution of carbonate on the subsiding sea floor. Later, about 10–11 million years ago, the carbonate line came down to 3500 m and kept falling, staying below the subsiding sea floor. Thus, carbonate accumulated on the bottom despite the increasing depth of deposition.

When analysing the available drilling data for the Atlantic in the same fashion, it is found that North and South Atlantic compensation surfaces appear to fluctuate together, although their absolute palaeodepths may differ by several hundred metres. A highly generalized CCD curve for the central Atlantic summarizes this information (Fig. 23, dashed line). In view of the Pleistocene warm-ocean versus cold-ocean contrast in carbonate deposition, it is interesting that the shape of the central Atlantic CCD curve resembles Tertiary temperature curves derived from oxygen isotope measurements (Devereux, 1967; Douglas & Savin, 1971). A correspondence between high O^{18} palaeotemperatures and low carbonate values in Tertiary deep-sea sediments has previously been noted (Moore, 1972, Fig. 9).

What is the relationship between the CCD fluctuations in the tropical Pacific and those in the Atlantic? To answer this question we must first reconstruct the CCS of the equatorial Pacific. Since the CCS here has great relief (Fig. 1), we can no longer ignore horizontal motion. Comparing palaeodepths of CCD's from different ages, some from their palaeoequator and some from more distant areas could easily introduce spurious 'fluctuations'. To overcome this difficulty, latitude must be introduced in addition to depth and time. In the appropriate diagram, subsidence tracks form a concave roof-like surface (Fig. 24). If uniform latitudinal motion is added during subsidence, the path will proceed from A to E, through B, C, and D, in the manner indicated. If the rate of migration is known, each degree latitude corresponds to a certain period of time, so that the original subsidence track (Fig. 5) can be used again, with both time units and latitudinal units on the *x*-axis.

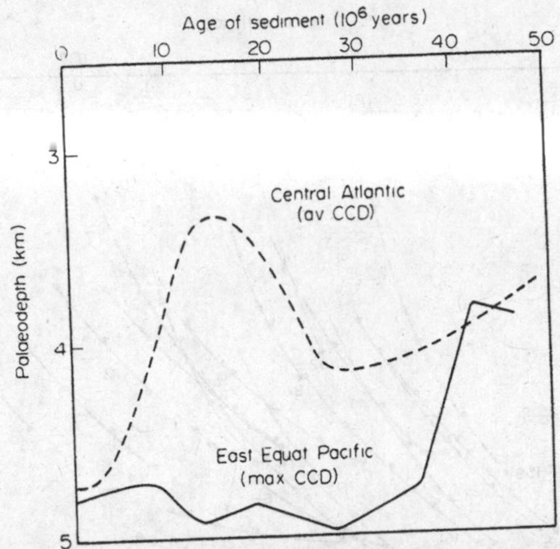

Fig. 23. Fluctuations of the carbonate line during Tertiary time, from backtracking of carbonate-clay boundaries in Deep Sea Drilling holes (Atlantic: adapted from Berger & von Rad, 1972; Pacific: Berger, 1973a).

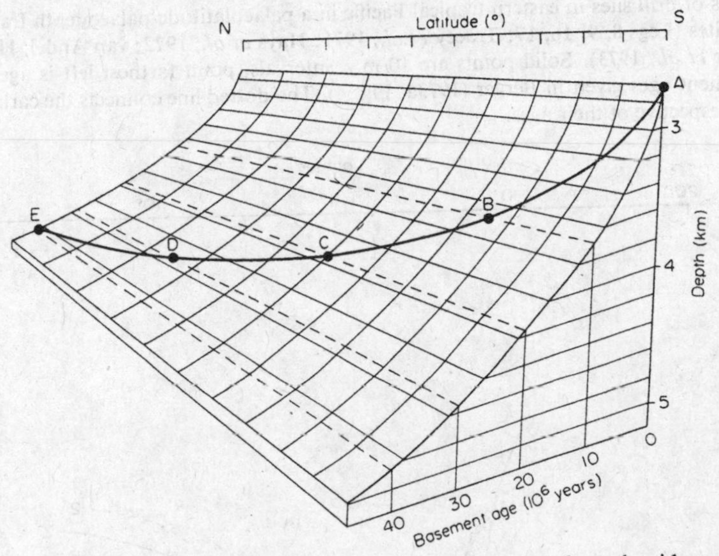

Fig. 24. Path of a piece of sea floor generated at a ridge, moving away from the ridge and subsiding, as well as moving northward at a uniform rate. Positions A, B, C, D, E correspond to ages 0, 10, 20, 30 and 40 million years.

In the eastern tropical Pacific the sea floor has been moving north by about 10° in the last 40 million years (Winterer, 1973; van Andel & Heath, 1973; Berger, 1973a). Thus, every 2·5° correspond to 10 million years and the paths of the various sites drilled by DSDP again appear as subsidence tracks in an age-depth frame, where age and palaeolatitude both use the *x*-axis (Fig. 25). The CCD crossings include a wide time range, the connecting line, therefore, has no physical meaning if the CCS fluctuated through time. Surprisingly, such fluctuations, although present, are not very pronounced, so that an average CCS profile can be given for post-Eocene time (Fig. 26, Curve 2) which compares well with the profile for surface sediments (Fig. 26,

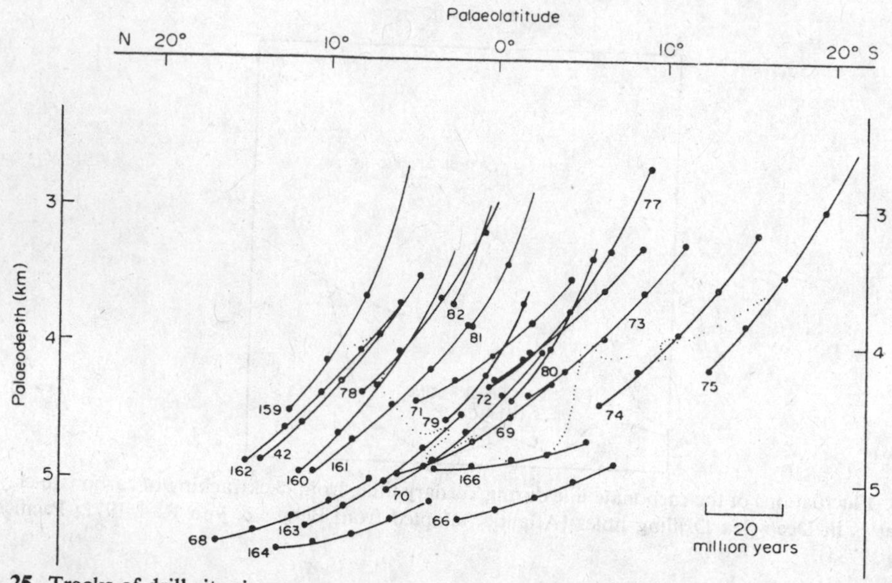

Fig. 25. Tracks of drill sites in eastern tropical Pacific in a palaeolatitude-palaeodepth frame. Numbers are drill sites (Legs 8, 9, 16, 17; Tracey *et al.*, 1971; Hays *et al.*, 1972; van Andel, Heath *et al.*, 1973; Winterer *et al.*, 1973). Solid points are 10 m.y. intervals, point farthest left is age 0. Tracks based on basement ages given in Berger (1973a, Fig. 4). The dotted line connects the carbonate-clay boundaries, irrespective of their ages.

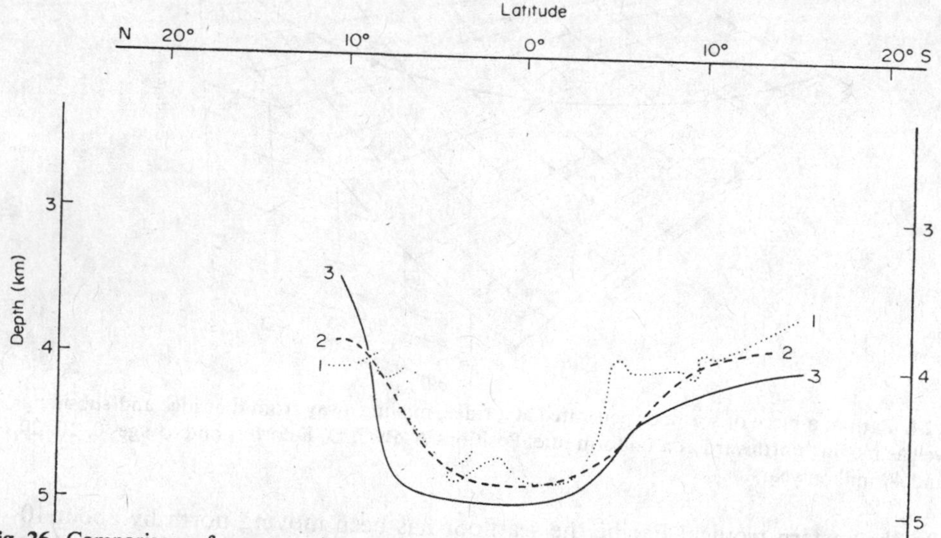

Fig. 26. Comparison of cross sections through the Tertiary and the present CCS in the eastern tropical Pacific, between 100°W and 150°W. Line 1: clay-carbonate boundary, irrespective of age, same as in Fig. 25. Line 2: generalized post-Eocene CCS profile (Berger, 1973a, Fig. 11). Line 3: present-day CCS profile.

Curve 3). The apparent stability of the Pacific equatorial CCS would seem to imply a feedback mechanism between supply from surface waters and dissolution rates at depth, which provides a certain measure of independence from productivity fluctuations and changes in bottom-water circulation. That such environmental changes did take place may be inferred from the distinct variation in rates and composition of equatorial sedimentation during post-Eocene time.

When connecting the maximum CCS palaeodepths for the various geological time spans from Quaternary to Mid-Eocene, a curve results which shows but slight variation since the Eocene, but marks a sudden lowering of the CCS at the end of the Eocene (Fig. 23). This lowering is reflected in the sharp contrast between carbonate-poor Eocene and carbonate-rich Oligocene sediments outcropping over large parts of the tropical North Pacific (Riedel & Funnell, 1964; Heath, 1969).

SUMMARY AND CONCLUSIONS

We have attempted to clarify the following points.

(1) The 'carbonate compensation depth' or carbonate line can be statistically defined and is a useful concept.

(2) The carbonate distribution in the world ocean defines a 'carbonate compensation surface', which is by no means flat, so that there is no justification for extrapolating a compensation 'depth' from one region to another.

(3) The mean depth of the compensation surface results from complex interaction of many factors. While it may be possible to identify some of the terms which govern changes of the mean depth, an integration of these terms does not seem possible at this time.

(4) The relief of the carbonate compensation depth is considerable, about 2 km. General features of the topography are readily apparent and make it possible to identify factors influencing carbonate distribution, such as inter-ocean fractionation, latitudinal, abyssal, trophic and diversity-related fractionation.

(5) Applying the principles of plate tectonics to the stratigraphic record of the deep sea makes it possible to reconstruct the fluctuations of 'fossil' carbonate lines, within the framework of plate stratigraphy.

(6) The range of fluctuations of the carbonate line through the Tertiary in any one place is comparable to the relief in topography of the present compensation surface.

(7) The reasons for the fluctuations are not known. There seems to be a tendency for warm climates having a relatively carbonate-poor deep-sea floor in mid and low latitudes and for cold oceans having a good deep-sea carbonate record. Flooding (warm climate) and exposure (cold climate) of shelves (cf. Frerichs, 1970) as well as latitudinal shifts in carbonate production appear promising as possible causes.

(8) Owing to the large number of factors affecting CCS positions, fluctuations of the carbonate line can be interpreted in many ways, and additional equations are needed to solve for so many variables. These equations will come from the kind of geoeconomy advocated by Kuenen (1950), and extended into the past as a type of geoeconomic stratigraphy built on the concepts of plate tectonics.

In striving to understand the development of the pelagic compensation surface as a result of geological history and evolution, it will be increasingly necessary to turn to the record on land. The evolutionary emergence of calcareous nannoplankton and of planktonic Foraminifera made possible a redistribution of carbonate sedimentation from shelf environments to the ocean floor. The prevalence of Mesozoic radiolarian cherts as the oldest sediments resting on tholeiitic pillow basalts at numerous localities in the Alpine-Mediterranean or in the circum-Pacific belt (Garrison, 1974; Robertson

& Hudson, 1974), contrasts markedly with the more typical sequence in younger pelagic successions, where carbonate rocks lie on the pillow basalts. Originally, then, a shallow pelagic carbonate compensation surface may have been enforced by rarity of calcareous pelagic material or by shell mineralogy unfavourable for deep-ocean preservation (aragonite), or by the action of large-scale shallow-water carbonate sinks (shelves, platforms), or a combination of these factors.

ACKNOWLEDGMENTS

We gratefully acknowledge assistance by Stuart M. Smith and Jane Z. Frazer, with retrieval of hypsographic and sedimentary data respectively. The research was supported by the National Science Foundation (GA 35451 and GA 36697) and by the Office of Naval Research (N00014-69-A-0200-6006).

REFERENCES

ANDERSON, R.N., MCKENZIE, D.P. & SCLATER, J.G. (1973) Gravity, bathymetry, and convection in earth. *Earth Planet. Sci. Letts*, **18**, 391–407.

ANON. (1964) Geological Society Phanerozoic time-scale. *Q. J. geol. Soc. Lond.* **120**, supp., 260–262.

ARRHENIUS, G. (1952) Sediment cores from the East Pacific. *Rep. Swed. deep Sea Exped.* (1947–1948), Parts 1–4, **5**, 1–288.

ARRHENIUS, G. (1963) Pelagic sediments. In: *The Sea* (Ed. by M. N. Hill), Vol. 3, pp. 655–727. Interscience, New York.

ATWATER, T. & MENARD, H.W. (1970) Magnetic lineations in the Northeast Pacific. *Earth Planet. Sci. Lett.* **7**, 445–450.

BANDY, O.L. (1967) Cretaceous planktonic foraminiferal zonation. *Micropaleontology*, **13**, 1–31.

BÉ, A.W.H. (1960) Some observations on arctic planktonic Foraminifera. *Contr. Cushman Fdn foramin. Res.* **11**, 64–68.

BÉ, A.W.H. (1969) Planktonic Foraminifera. In: *Distribution of Selected Groups of Marine Invertebrates in Waters South of 35°S Latitude. Antarctic Map Folio. Ser., Am. geogr. Soc.* **11**, 9–12.

BEALL, A.O. & FISCHER, A.G. (1969) Sedimentology. In: *Initial Reports of the Deep Sea Drilling Project*, Vol. I (M. Ewing *et al.*), pp. 521–593. U.S. Government Printing Office, Washington.

BENSON, W.E., GERARD, R.D. & HAY, W.W. (1970) Summary and conclusions. In: *Initial Reports of the Deep Sea Drilling Project*, Vol. IV (R. G. Bader *et al.*), pp. 659–673. U.S. Government Printing Office, Washington.

BERGER, W.H. (1968) Planktonic Foraminifera: selective solution and paleoclimatic interpretation. *Deep Sea Res.* **15**, 31–43.

BERGER, W.H. (1970) Biogenous deep-sea sediments: fractionation by deep-sea circulation. *Bull. geol. Soc. Am.* **81**, 1385–1402.

BERGER, W.H. (1971) Sedimentation of planktonic Foraminifera. *Mar. Geol.* **11**, 325–358.

BERGER, W.H. (1972) Deep-sea carbonates: dissolution facies and age-depth constancy. *Nature, Lond.* **236**, 392–395.

BERGER, W.H. (1973a) Cenozoic sedimentation in the eastern tropical Pacific. *Bull. geol. Soc. Am.* **84**, 1941–1954.

BERGER, W.H. (1973b) Deep-sea carbonates: Pleistocene dissolution cycles. *J. Foraminer. Res.* **3**, 187–195.

BERGER, W.H. & SOUTAR, A. (1970) Preservation of plankton shells in an anaerobic basic off California. *Bull. geol. Soc. Am.* **81**, 275–282.

BERGER, W.H. & VON RAD, U. (1972) Cretaceous and Cenozoic sediments from the Atlantic Ocean. In: *Initial Reports of the Deep Sea Drilling Project*, Vol. XIV (D. E. Hayes and A. C. Pimm *et al.*), pp. 787–954. U.S. Government Printing Office, Washington.

BRAMLETTE, M.N. (1961) Pelagic sediments. In: *Oceanography* (Ed. by M. Sears). *Publs. Am. Ass. Advmt Sci.* **67,** 345–366.

BRINKMANN, R. (1960) (Trans. J. E. Sanders) *Geologic Evolution of Europe*, pp. 161. Hafner, New York.

[illegible], W.S. (1971) Calcite accumulation rates and glacial to interglacial changes in oceanic mixing. In: *The Late Cenozoic Glacial Ages* (Ed. by K. K. Turekian), pp. 239–265. Yale University Press.

CHAVE, K.E., SMITH, S.V. & ROY, K.J. (1972) Carbonate production by coral reefs. *Mar. Geol.* **12,** 123–140.

CLAGUE, D.A. & JARRARD, R.D. (1973) Tertiary Pacific plate motion deduced from the Hawaiian-Emperor chain. *Bull. geol. Soc. Am.* **84,** 1135–1154.

CONOLLY, J.R. & PAYNE, R.R. (1972) Sedimentary patterns within a continent–mid-ocean ridge–continent profile: Indian Ocean South of Australia. *Antarctic Res. Series, Am. Geophys. Union,* **19,** 295–315.

DEFANT, A. (1961) *Physical Oceanography,* pp. 729. Pergamon Press, Oxford.

DEVEREUX, I. (1967) Oxygen isotope paleotemperatures on New Zealand Tertiary fossils. *N. Z. J. Sci.* **10,** 988–1011.

DOUGLAS, R.G. & SAVIN, S.M. (1971) Isotopic analyses of planktonic Foraminifera from the Cenozoic of the Northwest Pacific, Leg 6. In: *Initial Reports of the Deep Sea Drilling Project,* Vol. VI (A. G. Fischer *et al.*), pp. 1123–1127. U.S. Government Printing Office, Washington.

EICHER, D.L. (1969) Paleobathymetry of Cretaceous Greenhorn sea in eastern Colorado. *Bull. Am. Ass. Petrol. Geol.* **53,** 1075–1090.

ERICSON, D.B., EWING, M., WOLLIN, G. & HEEZEN, B.C. (1961) Atlantic deep-sea sediment cores. *Bull. geol. Soc. Am.* **72,** 193–286.

EWING, J. & EWING, M. (1967) Sediment distribution on the mid-ocean ridges with respect to spreading of the sea floor. *Science,* **156,** 1591–1592.

EWING, J., EWING, M., AITKEN, T. & LUDWIG, W.J. (1968) North Pacific sediment layers measured by seismic profiling. *Geophys. Monogr.* **12,** 147–173.

FAIRBRIDGE, R.W. (1961) Eustatic changes in sea level. In: *Physics and Chemistry of the Earth* (Ed. by L. H. Ahrens), Vol. 4, pp. 99–185. Pergamon, Oxford.

FISCHER, A.G., HEEZEN, B.C., BOYCE, R.E., BUKRY, D., DOUGLAS, R.G., GARRISON, R.E., KLING, S.A., KRASHENINNIKOV, V., LISITZIN, A.P. & PIMM, A.C. (1970) Geological history of the western North Pacific. *Science,* **168,** 1210–1214.

FISHER, R.L., SCLATER, J.G. & MCKENZIE, D.P. (1971) Evolution of the Central Indian Ridge, Western Indian Ocean. *Bull. geol. Soc. Am.* **82,** 553–562.

FLEMMING, N.C. & ROBERTS, D.G. (1973) Tectono-eustatic changes in sea level and seafloor spreading. *Nature, Lond.* **243,** 19–22.

FLINT, R.F. (1971) *Glacial and Quaternary Geology,* pp. 892. John Wiley, New York.

FRANCHETEAU, J., HARRISON, C.G.A., SCLATER, J.G. & RICHARDS, M.L. (1970) Magnetization of Pacific seamounts: a preliminary polar curve for the northeastern Pacific. *J. Geophys. Res.* **75,** 2035–2061.

FRERICHS, W.E. (1970) Paleobathymetry, paleotemperature and tectonism. *Bull. geol. Soc. Am.* **81,** 3445–3452.

GARRISON, R.E. (1974) Radiolarian cherts, pelagic limestones, and igneous rocks in eugeosynclinal assemblages. In: *Pelagic Sediments: on Land and under the Sea* (Ed. by K. J. Hsü and H. C. Jenkyns). *Spec. Publs int. Ass. Sediment.* **1,** 367–399.

GEALY, E.L. (1971) Saturated bulk density, grain density and porosity of sediment cores from the western equatorial Pacific: Leg 7, *Glomar Challenger*. In: *Initial Reports of the Deep Sea Drilling Project,* Vol. VII (E. L. Winterer *et al.*), pp. 1081–1104. U.S. Government Printing Office, Washington.

GILLULY, J. (1973) Steady plate motion and episodic orogeny and magmatism. *Bull geol. Soc. Am.* **84,** 499–514.

HALLAM, A. (1963) Major epeirogenic and eustatic changes since the Cretaceous, and their possible relationship to crustal structure. *Am. J. Sci.* **261,** 397–423.

HASLE, G.R. (1960) Plankton coccolithophorids from the subantarctic and equatorial Pacific. *Nytt Mag. Bot.* **8,** 77–88.

HAYES, D.E., FRAKES, L.A., BARRETT, P., BURNS, D.A., CHEN, P., FORD, A.B., KANEPS, A.G., KEMP, E.M., MCCOLLUM, D.W., PIPER, D.J.W., WALL, R.E. & WEBB, P.N. (1973) Leg 28 *Deep sea drilling in the Southern Ocean. Geotimes,* **18** (6), 19–24.

HAYES, D.E., PIMM, A.C. *et al.* (1972) *Initial Reports of the Deep Sea Drilling Project,* Vol. **XIV**, pp. 975. U.S. Government Printing Office, Washington.

HAYES, D.E. & PITMAN, W.C. III (1970) Magnetic lineations in the North Pacific. In: Geological investigations of the North Pacific (Ed. by J. D. Hays). *Mem. geol. Soc. Am.* **126,** 291–314.

HAYS, J.D. & PITMAN, W.C. III (1973) Lithospheric plate motion, sea level changes and climatic and ecological consequences. *Nature, Lond.* **246,** 18–22.

HAYS, J.D., SAITO, T., OPKYKE, N.D. & BURCKLE, L.H. (1969) Pliocene-Pleistocene sediments of the Equatorial Pacific: their paleomagnetic, biostratigraphic, and climatic record. *Bull. geol. Soc. Am.* **80,** 1481–1514.

HAYS, J.D. *et al.* (1972) *Initial Reports of the Deep Sea Drilling Project,* Vol. **IX,** pp. 1205. U.S. Government Printing Office, Washington.

HEATH, G.R. (1969) Carbonate sedimentation in the abyssal Equatorial Pacific during the past 50 million years. *Bull. geol. Soc. Am.* **80,** 689–694.

HEEZEN, B.C., HOLLISTER, C.D. & RUDDIMAN, W.F. (1966) Shaping of the continental rise by deep geostrophic contour currents. *Science,* **152,** 502–508.

HEIRTZLER, J.R., DICKSON, G.O., HERRON, E.J., PITMAN, W.C. & LEPICHON, X. (1968) Marine magnetic anomalies, geomagnetic field reversals, and the motions of the ocean floor and continents. *J. Geophys. Res.* **73,** 2119–2136.

HERRON, E.M. (1972) Sea-floor spreading and the Cenozoic history of the East-Central Pacific. *Bull. geol. Soc. Am.* **83,** 1671–1692.

HESS, H.H. (1962) History of ocean basins. In: *Petrologic Studies: A Volume to Honor A. F. Buddington* (Ed. by A. E. J. Engel *et al.*), pp. 599–620. Geological Society of America, New York.

HOLMES, A. (1965) *Principles of Physical Geology,* pp. 1288. Ronald Press, New York.

HSÜ, K.J. & ANDREWS, J.E. (1970) History of South Atlantic Basin. In: *Initial Reports of the Deep Sea Drilling Project,* Vol. III (A. E. Maxwell *et al.*), pp. 464–467. U.S. Government Printing Office, Washington.

ISACKS, B., OLIVER, J. & SYKES, L.R. (1968) Seismology and the new global tectonics. *J. Geophys. Res.* **73,** 5855–5899.

KOSSINNA, E. (1921) Die Tiefen des Weltmeers. *Veröff. Inst. Meeresk. Geogr. naturw.* **9,** 70 pp.

KUENEN, PH.H. (1950) *Marine Geology,* pp. 568. John Wiley, N.Y.

KUMMEL, B. (1970) *History of the Earth,* 2nd edn, pp. 707. Freeman, San Francisco.

LAMBERT, R. (1971) The pre-Pleistocene Phanerozoic time scale—a review. *Spec. Publs geol. Soc. Lond.* **5,** 9–31.

LARSON, R.L. & CHASE, C.G. (1972) Late Mesozoic Evolution of the Western Pacific Ocean. *Bull. geol. Soc. Am.* **83,** 3627–3644.

LARSON, R.L. & PITMAN, W.C. (1972) World-wide correlation of Mesozoic magnetic anomalies and its implications. *Bull. geol. Soc. Am.* **83,** 3645–3662.

LAUGHTON, A.S. (1971) South Labrador Sea and the evolution of the North Atlantic. *Nature, Lond.* **232,** 612–617.

LAUGHTON, A.S., BERGGREN, W.A., BENSON, R.N., DAVIES, T.A., FRANZ, U., MUSICH, L.F., PERCH-NIELSEN, K., RUFFMAN, A.S., VAN HINTE, J.E. & WHITMARSH, R.B. (1972) Explanatory Notes. In: *Initial Reports of the Deep Sea Drilling Project,* Vol. XII (A. S. Laughton, W. A. Berggren *et al.*) pp. 9–30. U.S. Government Printing Office, Washington.

LE PICHON, X. (1968) Sea-floor spreading and continental drift. *J. Geophys. Res.* **73,** 3661–3697.

LISITZIN, A.P. (1972) Sedimentation in the World Ocean (Ed. by K. S. Rodolfo). *Spec. Publs Soc. econ. Paleont. Miner., Tulsa,* **17,** 1–218.

MAXWELL, A.E. *et al.* (1970) *Initial Reports of the Deep Sea Drilling Project,* Vol. III, pp. 806. U.S. Government Printing Office, Washington.

MCINTYRE, A., RUDDIMAN, W.F. & JANTZEN, R. (1972) Southward penetrations of the North Atlantic Polar Front: faunal and floral evidence of large-scale surface water mass movements over the last 225,000 years. *Deep Sea Res.* **19,** 61–77.

MCKENZIE, D.P. & PARKER, R.L. (1967) The North Pacific: an example of tectonics on a sphere. *Nature, Lond.* **216,** 1276–1280.

MCKENZIE, D.P. & SCLATER, J.G. (1971) The evolution of the Indian Ocean since the late Cretaceous. *Geophys. J.* **25,** 437–528.

MENARD, H.W. (1964) *Marine Geology of the Pacific*, pp. 271. McGraw-Hill, New York.
MENARD, H.W. (1969) Elevation and subsidence of oceanic crust. *Earth Planet. Sci. Letts*, **6**, 275–284.
MENARD, H.W. (1973) Depth anomalies and the bobbing motion of drifting islands. *J. Geophys.Res.* **78**, 5128–5137.
MENARD, H.W. & SMITH, S.M. (1966) Hypsometry of ocean basin provinces. *J. Geophys. Res.* **71**, 4305–4325.
MOLNAR, P., ATWATER, T., MAMMERICKX, J. & SMITH, S.M. (in press) Magnetic anomalies, bathymetry and tectonic evolution of the South Pacific since the late Cretaceous. *Bull. geol. Soc. Am.*
MOORE, T.C., JR, (1972) DSDP: successes, failures, proposals. *Geotimes*, **17**, (7), 27–31.
MORGAN, W.J. (1968) Rises, trenches, great faults and crustal blocks. *J. Geophys. Res.* **73**, 1959–1982.
MORGAN, W.J. (1971) Convection plumes in the lower mantle. *Nature, Lond.* **230**, 42–43.
MORGAN, W.J. (1972) Deep mantle convection plumes and plate motions. *Bull. Am. Ass. Petrol. Geol.* **56**, 203–213.
MURRAY, J. & RENARD, A.F. (1891) Report on deep-sea deposits based on the specimens collected during the voyage of H.M.S. *Challenger* in the years 1872 to 1876. In: *'Challenger Reports'*, pp. 525. H.M.S.O., Edinburgh.
OLAUSSON, E. (1960a) Sediment cores from the West Pacific. *Rep. Swed. deep Sea Exped.* 1947–1948, **6**, 161–214.
OLAUSSON, E. (1960b) Sediment cores from the Indian Ocean. *Rep. Swed. deep Sea Exped.* 1947–1948, **9**, 53–88.
OLAUSSON, E. (1960c) Sediment cores from the North Atlantic Ocean. *Rep. Swed. deep Sea Exped.* 1947–1948, **7**, 227–286.
OLAUSSON, E. (1965) Evidence of climatic changes in North Atlantic deep-sea cores, with remarks on isotopic paleotemperature analysis. *Prog. Oceanogr.* **3**, 221–252.
PARKER, F.L. & BERGER, W.H. (1971) Faunal and solution patterns of planktonic Foraminifera in surface sediments of the South Pacific. *Deep Sea Res.* **18**, 73–107.
PESSAGNO, E.A. (1972) Pulsations, interpulsations and sea-floor spreading. *Mem. geol. Soc. Am.* **132**, 67–73.
PETERSON, M.N.A. (1966) Calcite: rates of dissolution in a vertical profile in the central Pacific. *Science*, **154**, 1542–1544.
PETERSON, M.N.A., EDGAR, N.T., VON DER BORCH, C.C. & REX, R.W. (1970) Cruise leg summary and discussion. In: *Initial Reports of the Deep Sea Drilling Project*, Vol. II (M. N. A. Peterson *et al.*), pp. 413–427. U.S. Government Printing Office, Washington.
PHLEGER, F.B., PARKER, F.L. & PEIRSON, J.F. (1953) North Atlantic foraminifera. *Rep. Swed. deep Sea Exped.* 1947–1948, **7**, 122 pp.
PITMAN, W.C. & TALWANI, M. (1972) Sea-floor spreading in the North Atlantic. *Bull. geol. Soc. Am.* **83**, 619–646.
REVELLE, R.R. (1944) Marine bottom samples collected in the Pacific Ocean by the *Carnegie* on its seventh cruise. *Publs Carnegie Inst.* **556**, part 1, 180 pp.
RIEDEL, W.R. (1963) The preserved record: paleontology of pelagic sediments. In: *The Sea* (Ed. by M. H. Hill), Vol. 3, pp. 866–887. Interscience, New York.
RIEDEL, W.R. & FUNNELL, B.M. (1964) Tertiary sediment cores and microfossils from the Pacific Ocean floor. *Q. J. geol. Soc. Lond.* **120**, 305–368.
ROBERTSON, A.H.F. & HUDSON, J.D. (1974) Pelagic sediments in the Cretaceous and Tertiary history of the Troodos massif, Cyprus. In: *Pelagic sediments: on Land and under the Sea* (Ed. by K. J. Hsü and H. C. Jenkyns). *Spec. Publs int. Ass. Sediment.* **1**, 403–436.
RONA, P.A. (1973) Relations between rates of sediment accumulation on continental shelves, sea-floor spreading, and eustacy inferred from the central North Atlantic. *Bull. geol. Soc. Am.* **84**, 2851–2872.
RONOV, A.B. (1968) Probable changes in the composition of sea water during the course of geologic time. *Sedimentology*, **10**, 25–43.
RUDDIMAN, W.F. (1971) Pleistocene sedimentation in the Equatorial Atlantic: stratigraphy and faunal paleoclimatology. *Bull. geol. Soc. Am.* **82**, 283–302.
RUDDIMAN, W.F. & HEEZEN, B.C. (1967) Differential solution of planktonic Foraminifera. *Deep Sea Res.* **14**, 801–808.
RYAN, W.B.F., HSÜ, K.J. *et al.* (1973) *Initial Reports of the Deep Sea Drilling Project*, Vol. XIII, Parts 1 and 2, pp. 1447. U.S. Government Printing Office, Washington.

SCHOTT, W. (1935) Die Foraminiferen in dem äquatorialen Teil des Atlantischen Ozeans. *Wiss. Ergebn. dt atlant. Exped. 'Meteor'* 1925–1927. Bd. III (3) Sect. B, 43–134.
SCLATER, J.G. & JARRARD, R.D. (1971) Preliminary paleomagnetic results. In: *Initial Reports of the Deep Sea Drilling Project,* Vol. VII (E. L. Winterer *et al.*), pp. 1227–1234. U.S. Government Printing Office, Washington.
SCLATER, J.G., ANDERSON, R.N. & BELL, M.L. (1971) Elevation of ridges and evolution of the central eastern Pacific. *J. Geophys. Res.* **76,** 7888–7915.
SEIBOLD, E. (1970) Nebenmeere im humiden und ariden Klimabereich. *Geol. Rdsch.* **60,** 73–105.
SHEPARD, F.P. (1973) *Submarine Geology,* 3rd edn, pp. 517. Harper and Row, New York.
SILLÉN, L.G. (1967) The ocean as a chemical system. *Science,* **156,** 1189–1197.
SMITH, S.V., DYGAS, J.A. & CHAVE, K.E. (1968) Distribution of calcium carbonate in pelagic sediments. *Mar. Geol.* **6,** 391–400.
SVERDRUP, H.U., JOHNSON, M.W. & FLEMING, R.H. (1942) *The Oceans, their Physics, Chemistry and General Biology,* pp. 1087. Prentice Hall, Englewood Cliffs, N.J.
THIEDE, J. (1971) Planktonische Foraminiferen in Sedimenten vom ibero-marokkanischen Kontinentalrand. *Meteor-Forsch. Ergebn.* **7,** 15–102.
THIEDE, J. (1973) Planktonic Foraminifera in hemipelagic sediments: shell preservation off Portugal and Morocco. *Bull. geol. Soc. Am.* **84,** 2749–2754.
TRACEY, J.I. Jr *et al.* (1971) *Initial Reports of the Deep Sea Drilling Project,* Vol. VIII, pp. 1037. U.S. Government Printing Office, Washington.
VAN ANDEL, TJ.H. & BUKRY, D. (1973) Basement ages and basement depths in the eastern equatorial Pacific from Deep Sea Drilling Project, Legs 5, 8, 9 and 16. *Bull. geol. Soc. Am.* **84,** 2361–2370.
VAN ANDEL, TJ.H. & HEATH, G.R. (1970) Tectonics of the Mid-Atlantic Ridge, 6–8° south latitude. *Mar. Geophys. Res.* **1,** 5–36.
VAN ANDEL, TJ.H. & HEATH, G.R. (1973) Geological results of Leg 16: the central equatorial Pacific Rise. In: *Initial Reports of the Deep Sea Drilling Project,* Vol. XVI (Tj. van Andel and G. R. Heath *et al.*), pp. 937–949. U.S. Government Printing Office, Washington.
VAN ANDEL, TJ.H., HEATH, G.R. *et al.* (1973) *Initial Reports of the Deep Sea Drilling Project,* Vol. XVI, pp. 949. U.S. Government Printing Office, Washington.
VINE, F.J. & MATTHEWS, D.H. (1963) Magnetic anomalies over ocean ridges. *Nature, Lond.* **199,** 947–949.
VOGT, P.R. & OSTENSO, N.A. (1970) Magnetic and gravity profiles across the Alpha Cordillera and their relation to Arctic sea-floor spreading. *J. Geophys. Res.* **75,** 4925–4937.
WATKINS, N.D. & KENNETT, J.P. (1972) Regional sedimentary disconformities and Upper Cenozoic changes in bottom water velocities between Australasia and Antarctica. *Antarctic Res. Series, Am. Geophys. Union,* **19,** 273–293.
WEISSEL, J. & HAYES, D.E. (1971) Asymmetric seafloor spreading south of Australia. *Nature, Lond.* **231,** 518–521.
WILSON, J.T. (1963) Evidence from islands on the spreading of ocean floors. *Nature, Lond.* **197,** 536–538.
WILSON, J.T. (1965) A new class of faults and their bearing upon continental drift. *Nature, Lond.* **207,** 343–347.
WINTERER, E.L. (1971) History of the Pacific Ocean Basin *Abstr. geol. Soc. Am.* (with Programs), **7,** 754.
WINTERER, E.L. (1973) Sedimentary facies and plate tectonics of Equatorial Pacific. *Bull. Am. ass. Petrol. Geol.* **57,** 265–282.
WINTERER, E.L., EWING, J.I. *et al.* (1973) *Initial Reports of the Deep Sea Drilling Project,* Vol. XVII, pp. 930. U.S. Government Printing Office, Washington.

Spec. Publs int. Ass. Sediment. (1974) **1**, 49–70

Preservation of cephalopod skeletons and carbonate dissolution on ancient Tethyan sea floors

WOLFGANG SCHLAGER

Shell Research, Volmerlaan 6, Rijswijk (ZH), The Netherlands

ABSTRACT

Cephalopods often hold the documents of carbonate dissolution in ancient pelagic environments because of their large size and the possession of aragonitic and calcitic skeletal parts. In Tethyan cephalopod limestones the majority of the aragonitic phragmocones were leached after being embedded in sediment. The moulds were later filled with cement and/or a geopetal sediment, now present as a crystal mosaic or fossiliferous micrite. The fabric resembles that of freshwater leaching in shallow-water carbonates and must have formed in a semi-lithified sediment, close to or at the sediment surface and during halts in deposition. Calcite appears to have been practically stable in this environment. In aptychus beds and radiolarites most of the phragmocones were dissolved before they could be embedded in sediment, only the calcitic aptychi being preserved. Mere transport separation of phragmocones and aptychi is rejected on several grounds.

On a large scale, the intensity of dissolution in the various facies parallels the depth of deposition assumed on circumstantial evidence. Thus, undersaturation of the sea water was probably the main reason for the differential dissolution patterns. Oxidation of the sediment and rate of deposition were less important.

The way ammonites are preserved in the cephalopod limestones presents some arguments against extensive dissolution of sediment in this facies. The 'Subsolution' features are more readily explained by an interplay of differential cementation, winnowing of loose sediment, and organic corrosion during periods of non-deposition.

ZUSAMMENFASSUNG

Begünstigt durch ihre Grösse und den Besitz aragonitischer und calcitischer Skelett-Elemente, bewahren Cephalopoden oft die Spuren der Karbonat-Lösung fossiler, pelagischer Environments. In den Cephalopodenkalken der Tethys fiel ein grosser Teil der aragonitischen Phragmocone nach der Einbettung der Lösung zum Opfer. Die Hohlformen wurden später mit Karbonat-Zement und/oder einem Geopetalsediment (jetzt als feinkörniger Pflastercalcit oder Mikrit vorliegend) verfüllt. Das dabei entstandene Gefüge gleicht dem einer Auslaugung durch Süsswasser in Flachwasser-Karbonaten. Es muss während Omissions-Perioden, in einem schwach lithifizierten Sediment und nahe der Sediment-Oberfläche entstanden sein. Calcit war in diesem Environment fast stabil. In Aptychenschichten und Radiolariten wurden die meisten der Phragmocone bereits vor der Einbettung aufgelöst, nur die calcitischen Aptychen wurden eingebettet. Blosse Frachtsonderung als Ursache der Trennung von Phragmoconen und Aptychen wird aus verschiedenen Gründen abgelehnt.

Im grossen gesehen verändert sich die Lösungs-Intensität gleichsinnig mit der, aus der Gesamtsituation abgeleiteten Wassertiefe der Ablagerungs-Raumes. Die unterschiedlichen Lösungs-Gefüge spiegeln daher vor allem tiefenbedingte Änderungen der Karbonat-Untersättigung des Bodenwassers wider. Oxydationsgrad der Sedimente und Sedimentationsrate waren demgegenüber von geringerer Bedeutung.

Abgesehen von der selektiven Zerstörung der Aragonit-Schalen war die anorganische Karbonat-Lösung im Environment der Cephalopodenkalke gering. Der Grossteil der sogenannten Subsolutions-Erscheinungen dieser Gesteine wird daher auf Auswaschung und organische Korrosion zurückgeführt.

INTRODUCTION

The progressive dissolution of carbonate with increasing water depth in Recent oceans has become a well-documented and generally accepted fact. Apart from chemical analyses of sea water the carbonate content of the sediments and differing solution resistance of micro-organisms have proved to be the most efficient tools in these investigations (Berger, 1970; Lisitzin, 1972). Application of these techniques to the geological past is severely hampered, however, as sediments alter during diagenesis and micro-organisms evolve and change in time. In consequence, there is considerable disagreement in the interpretation of the environment and bathymetry of ancient pelagic sediments, such as the cephalopod limestones, radiolarites and aptychus beds of the Alpine-Mediterranean orogenic belt.

Starting from the observations of Garrison & Fischer (1969) and earlier authors, I have attempted to evaluate the impact of carbonate dissolution on ancient sea floors from the preservation of cephalopod shells. Compared to micro-organisms, they have the drawback of being somewhat rarer, though they are fairly common in many Tethyan pelagic sediments. Their advantages, on the other hand, are obvious: the skeletons are so large that the rock fabric in many cases documents early diagenetic processes, even after severe diagenetic alterations; and very often one specimen originally contained both aragonitic and calcitic skeletal elements.

The way in which cephalopods are dealt with here differs considerably from normal systematic descriptions and this has influenced my terminology to some degree. It has been necessary to draw distinctions which are irrelevant for a systematic treatment and *vice versa*. I have used 'cephalopod skeleton' for the hard parts taken collectively, 'phragmocone' for the chambered test (including the living chamber, if present), 'shell' for the walls of the phragmocone, 'aptychi' for the pair of valves commonly interpreted as opercula of the living chamber, 'anaptychus' for the single valve of Palaeozoic and Lower Jurassic ammonites (the latter having been reinterpreted as rhyncholites by Lehmann, 1970). 'Rhyncholite' stands for nautiloid beaks. The distinction between 'phragmocone' and 'shell' is necessary in the context of the leaching processes described on the following pages. For instance, a shell mould means the void formerly occupied by the wall of the phragmocone only, and not by the whole phragmocone including the volume of the chambers.

With regard to the original mineralogy of cephalopod skeletons it should be noted that calcite is the only $CaCO_3$-modification present in this material now and that there is no direct proof of the former aragonitic nature of the phragmocones and the former calcitic nature of aptychi and rhyncholites. However, there seems to be a general agreement on this statement amongst students of cephalopods and my own observations in various marls of the Alpine Mesozoic have confirmed these assumptions.

The results presented here are based on three weeks field work in the Northern and Southern Limestone Alps and on material kindly provided by colleagues from other parts of the Alpine-Mediterranean belt. In total, some 250 samples have been studied in slabs and thin sections. They have been obtained from the following formations.

Northern Limestone Alps
- Upper Triassic Hallstatt Limestone (Krystyn, Schäffer & Schlager, 1971)
- Lower Jurassic Adnet Limestone (Garrison & Fischer, 1969; Jurgan, 1969)
- Middle Jurassic Klaus Limestone (Krystyn, 1972)
- Upper Jurassic Ruhpolding Radiolarite (Garrison & Fischer, 1969)
- Upper Jurassic to Lower Cretaceous Oberalm and Schrambach Beds (Garrison & Fischer, 1969)

Southern Limestone Alps
- Middle-Upper Jurassic Ammonitico Rosso (Sturani, 1964, 1971)
- Upper Jurassic Radiolarite (Bernoulli, 1972)
- Upper Jurassic to Lower Cretaceous Maiolica (Bernoulli, 1972)

Central Apennines
- Middle-Upper Jurassic Ammonitico Rosso + other pelagic Limestone (Bernoulli, 1972)

Western Sicily
- Jurassic cephalopod Limestone (Wendt, 1969a; Jenkyns, 1970)

The above formations can be attributed to four facies.

(a) *Crinoidal limestones* (Hierlatz Lst., Ammonitico Rosso p.p.). This facies occurs in lenses between shallow-water carbonates and the overlying cephalopod limestones and comprises grain- or packstones, consisting of disintegrated crinoids, with locally abundant brachiopods and some cephalopods. Jenkyns (1971a, p. 472) has interpreted these bioclastic sands as 'sand-waves formed on current-swept pelagic seamounts' in some tens of metres of water depth. Redeposited material of this type is found in tectonic fissures and as (partly turbiditic) intercalations in other pelagic deposits, e.g. the cephalopod limestones (Bernoulli & Jenkyns, 1970).

(b) *Cephalopod limestones* (Ammonitico Rosso p.p., Adnet Lst., Hallstatt Lst., Klaus Lst.). This facies comprises fine-grained, highly oxidized, predominantly red (or pale) skeletal wackestones with abundant ammonite phragmocones and some aptychi, and a smaller number of nautiloids and belemnites. Sedimentation was slow, sometimes continuous, but more often interrupted by frequent periods of erosion and non-deposition causing stratigraphic condensation, submarine erosion surfaces and hardgrounds. Depth of deposition is thought to have ranged from a few hundreds to several thousands of metres (Garrison & Fischer, 1969; Wendt, 1970, p. 445; Jenkyns, 1971b; Bernoulli, 1972, p. 813).

(c) *Aptychus beds* (Biancone, Maiolica, Oberalm and Schrambach Beds). This term stands for light-coloured, coccolith-rich, chert-bearing limestones, that are all characterized by the relative abundance of aptychi and the scarcity of other mega-fossils.

They are almost unanimously interpreted as analogues of the nannoplankton ooze in modern oceans (Garrison & Fischer, 1969, p. 35; Bernoulli, 1972, p. 813).

(d) *Radiolarites* (Ruhpolding Radiolarites and others). These are highly oxidized calcareous radiolarian muds, diagenetically altered to well-bedded cherts, siliceous marls or limestones. Depth of deposition was probably close to the calcite compensation level (Garrison & Fischer, 1969, p. 47).

Jurassic sequences in the Alpine-Mediterranean region are characterized by a gradual deepening of the depositional environment, with shallow-water carbonates being overlain by various pelagic deposits. In this case, a rather constant succession from shallow-water carbonates through crinoidal limestones, to cephalopod limestones, and finally to radiolarites and aptychus beds can be observed. The shallower part of this sequence, i.e. the combination shallow-water carbonates–crinoidal limestone–cephalopod limestone is also frequently found in the Middle or Upper Triassic (Bystricky, 1972, p. 300; Kubanek, 1969, p. 12a).

PRESERVATION OF CEPHALOPOD REMAINS

Cephalopod and crinoidal limestones

Red cephalopod limestones of the Hallstatt, Adnet or Ammonitico Rosso type are the most prominent ammonite-bearing rocks of the Tethyan Mesozoic. A variety of corrosion phenomena, attributed partly to carbonate dissolution, submarine erosion and boring activity, have been described from them.

Starting from the often incomplete preservation of cephalopod phragmocones, several authors (e.g. Garrison & Fischer, 1969 and Jenkyns, 1971b, Fig. 3) have concluded that dissolution of aragonite must have taken place. I can but confirm this view. Generally, the cephalopod limestones have suffered extensive leaching of aragonite and the patterns very much resemble those of selective vadose leaching in shallow-water carbonates (see Figs 3–6; as opposed to Figs 1, 2 without leaching). The shells were removed from between the sediment, sometimes also from between a drusy cement in the chambers. The moulds created by this process are a perfect and undeformed model of the shell. They were later filled by either a thin druse and/or a blocky cement or a fine-grained crystal mosaic, closely resembling the 'vadose silt' of Dunham (1969) or by a younger generation of muddy sediment. In this latter case the mould must have been open to the sediment-water interface (Figs 3, 4). Moulds filled with silt and blocky cement presumably remained covered by sediment (Figs 5, 6). The origin of the crystal sediment presents a problem which is discussed below.

The degree of dissolution, and the percentage of specimen affected by it, vary from case to case and seem to be influenced by the distance from omission surfaces. In some beds, practically all phragmocones were leached and the shells replaced by a sediment fill. In other rocks, such as parts of the Hallstatt Limestone of the Eastern Alps, or certain pelagic limestones from Sicily, the formation of iron-manganese oxide coatings or sediment deposition was fast enough to prevent extensive aragonite leaching. Most of the so-protected shell material stabilized by *in situ* transformation of aragonite to calcite (Figs 1, 2). Ammonites in sediment-filled fissures ('neptunian dykes') were apparently also protected from dissolution and underwent *in situ* transformation to calcite.

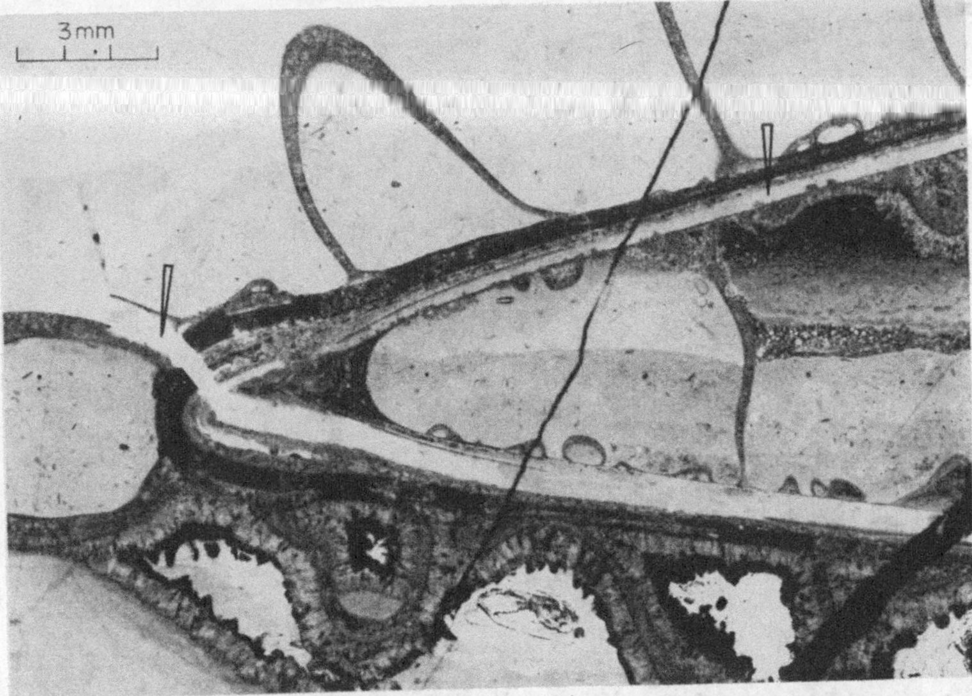

Fig. 1. Inner whorl of a *Pinacoceras* sp. without dissolution effects. Preservation of original fine laminar shell structure indicates *in situ* transformation of aragonite to calcite. Note incipient spalling of outer whorl along two sediment-filled fractures (arrows). Upper Triassic Hallstatt Limestone, Sommeraukogel near Hallstatt, Austria. Thin section, negative print.

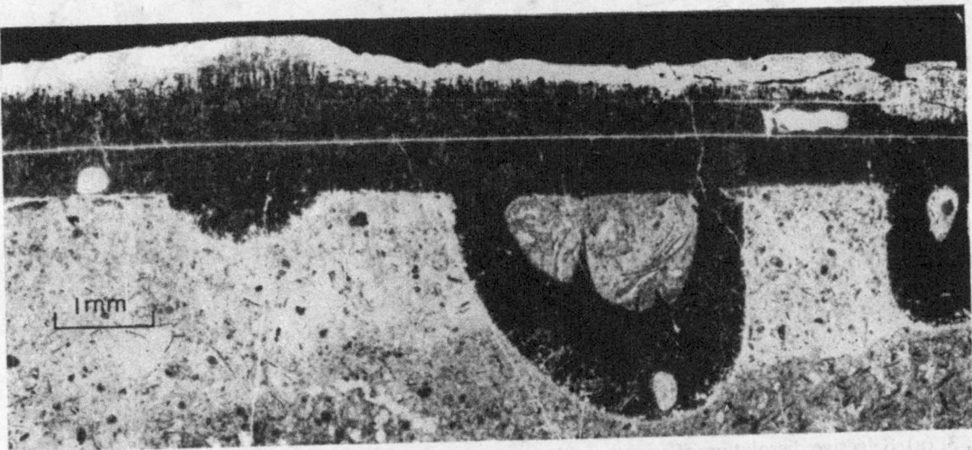

Fig. 2. Ammonite shell with original shell structure still recognizable through lamination and orientation of calcite prisms perpendicular to it. Only outermost layer has been dissolved, the rest was protected by iron-manganese oxide coating (white) and underwent *in situ* transformation of aragonite to calcite. Middle Jurassic cephalopod limestone, Monte Inici, Sicily. Thin section, negative print (sample courtesy of H. C. Jenkyns).

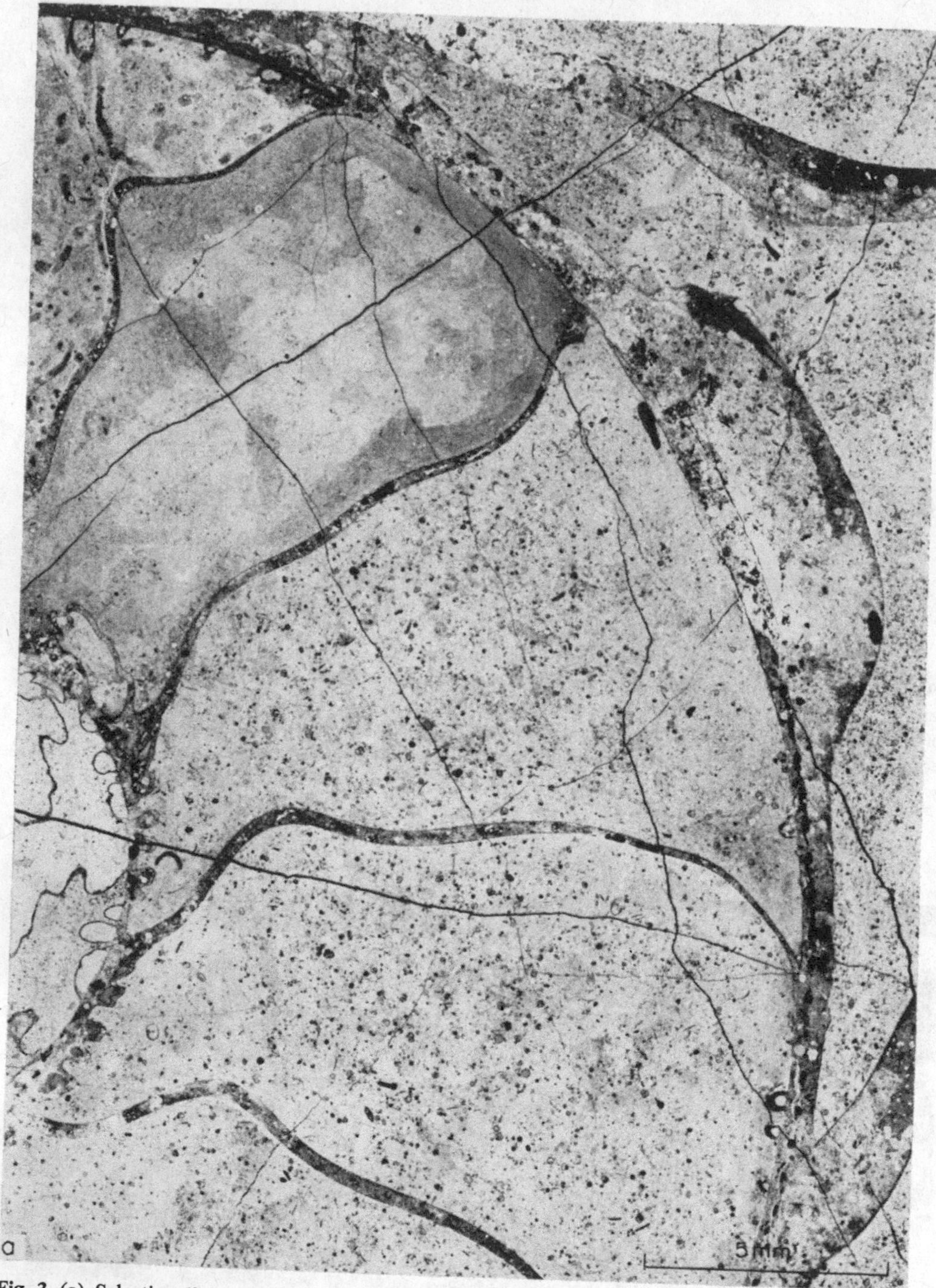

Fig. 3. (a) Selective dissolution of an aragonitic ammonite shell below a hardground. Except for the innermost whorl, all shell material has been removed and the undeformed mould is filled with normal bottom sediment. Upper Jurassic Ammonitico Rosso, La Stua, Dolomites, Italy. Acetate peel, negative print.

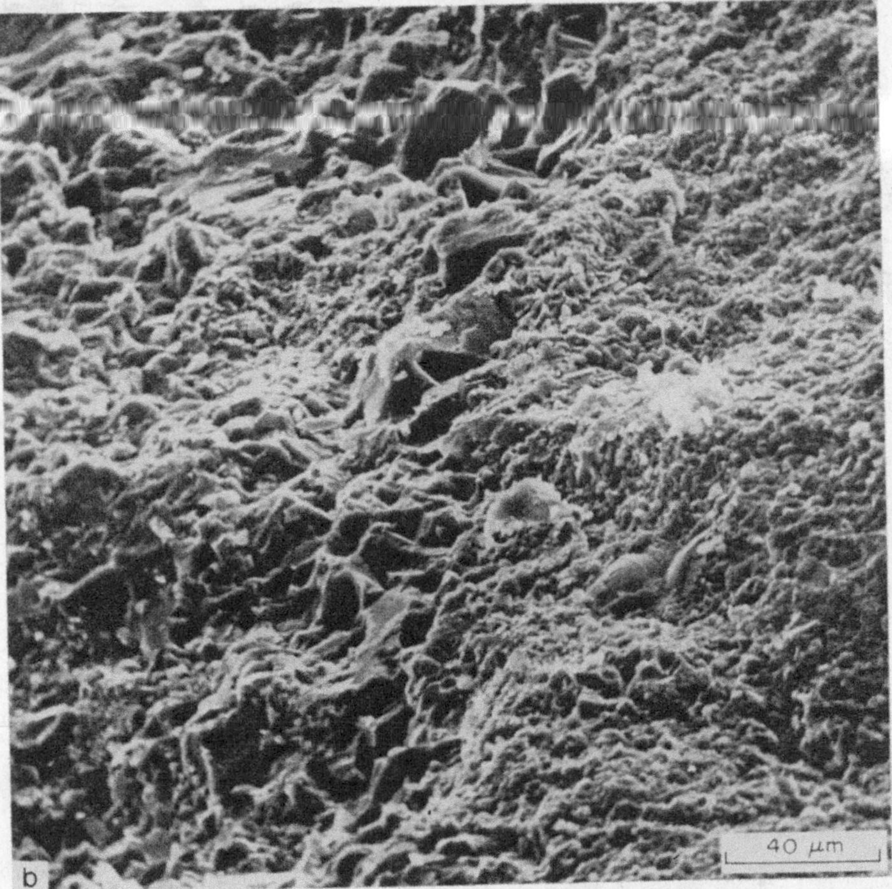

Fig. 3. (b) A zone of blocky or drusy cement separates embedding sediment (right) from the infill of the mould (left). Same specimen as in Fig. 3 (a). Scanning electron micrograph by Centr. Lab. TNO Delft. × 550.

Outside the Tethyan realm, I have observed selective leaching, with subsequent fill by bottom sediment, on cephalopods from a phosphatic-glauconitic hardground in a deep-marine setting of the Bassin de Beausset (Cenomanian, SE France). Similar patterns seem to occur in the Cretaceous Chalk hardgrounds of England (Bromley in Bathurst, 1971, p. 412).

The section in Fig. 8 yields some information about the onset and the increase of aragonite dissolution in the above-mentioned deepening sequences. It ranges from shallow-water oolites to a pelagic tintinnid mudstone. Leaching phenomena have been observed on the phragmocones throughout the cephalopod facies and, in one case, also within the crinoidal limestone. However, dissolution effects become more prominent upwards in the sequence. This may be illustrated by comparing the ammonite beds at the transition from crinoidal to cephalopod limestone with those near the overlying tintinnid limestone. The basal bed is shown in Fig. 7. Some rounded ammonite fragments are seen, together with obviously redeposited phragmocones that have been intensively bored. Their partially missing outer wall may have been

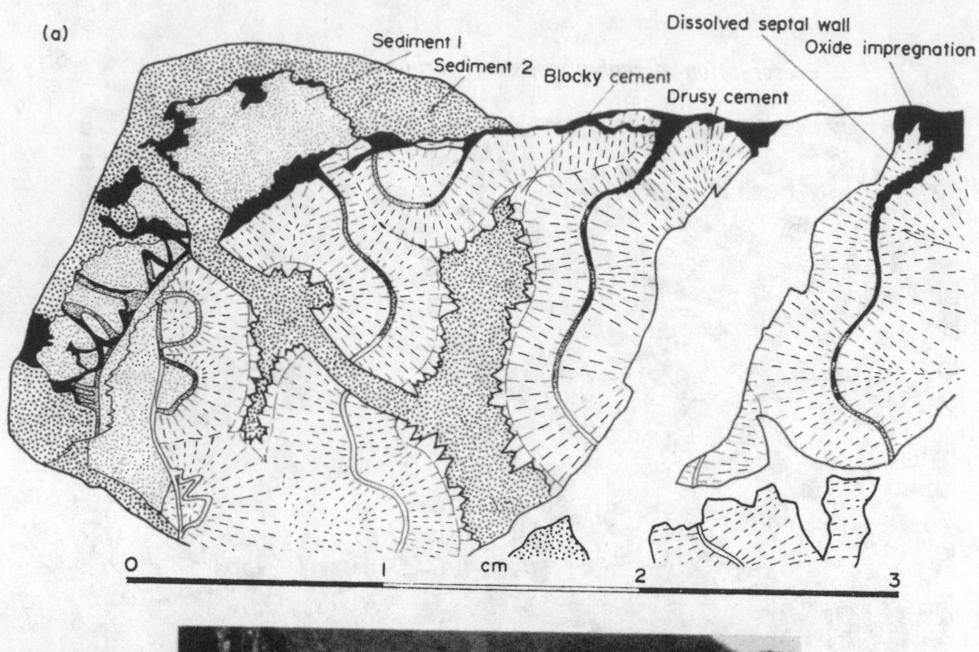

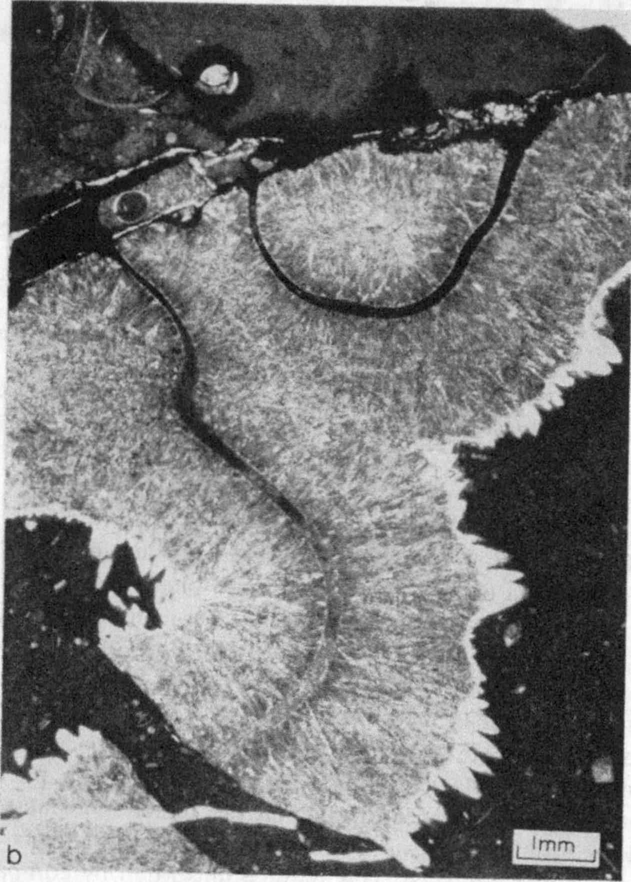

Fig. 4. (a) Ammonite phragmocone showing selective dissolution of aragonite from between a non-aragonitic cement. Starting from a hardground on top of the fossil, the walls were largely dissolved and the mould filled by muddy sediment and iron-manganese oxide. The (presumably high-magnesian calcitic) druse in the chambers was only marginally affected by dissolution. To the left two generations of embedding sediment, separated by ferromanganese oxide-coated corrosion surface. Sediment 2 penetrates along fractures into chamber cavities. Upper Triassic Hallstatt Limestone, Feuerkogel, Austria. Drawn from thin section. (b) Detail of Fig. 4 (a). Thin-section, positive print.

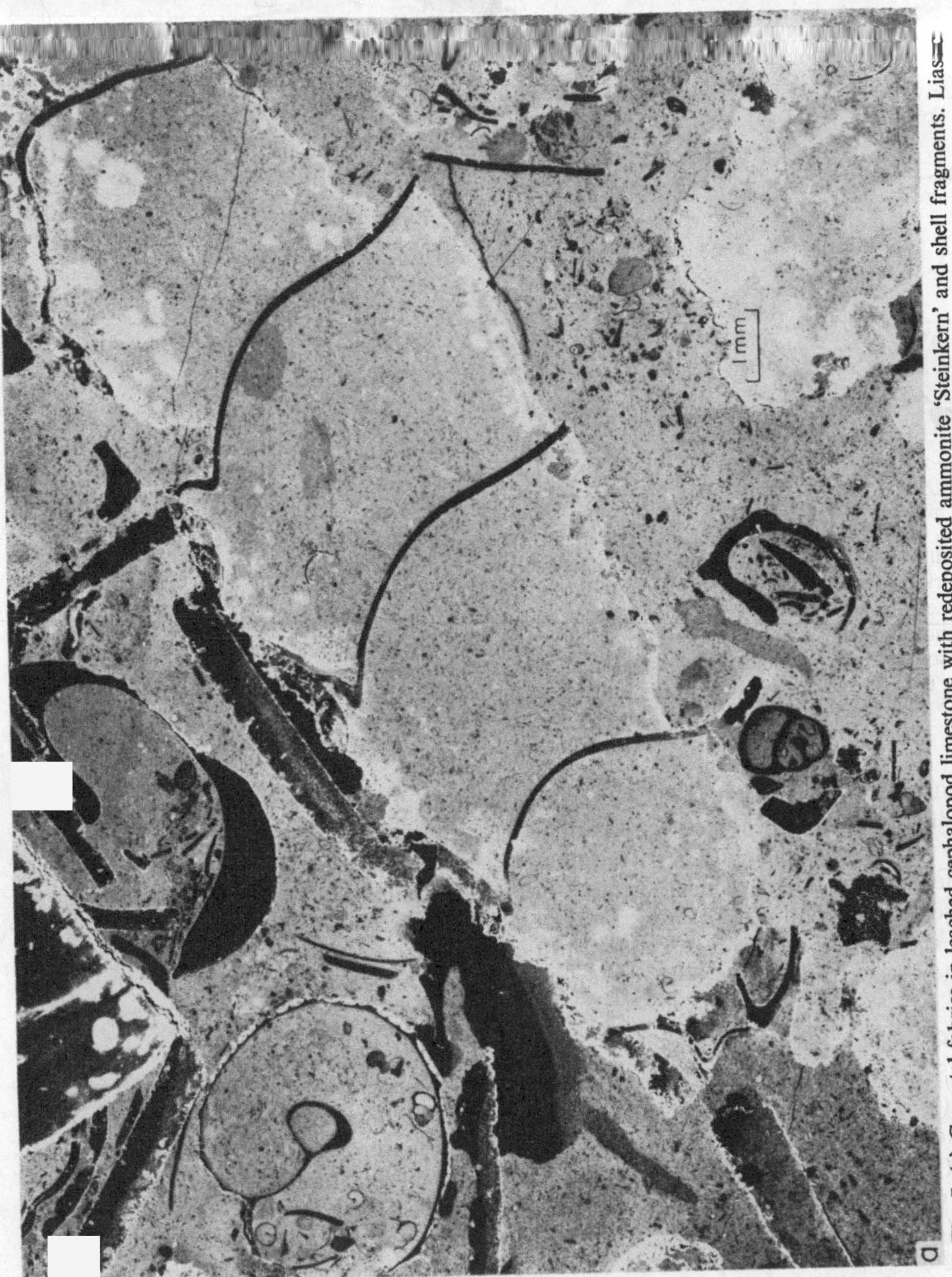

Fig. 5. (a) Geopetal fabrics in leached cephalopod limestone with redeposited ammonite 'Steinkern' and shell fragments. Lias; Adnet Limestone, Schneibstein near Salzburg. Thin section, negative print (sample courtesy of H. Jurgan).

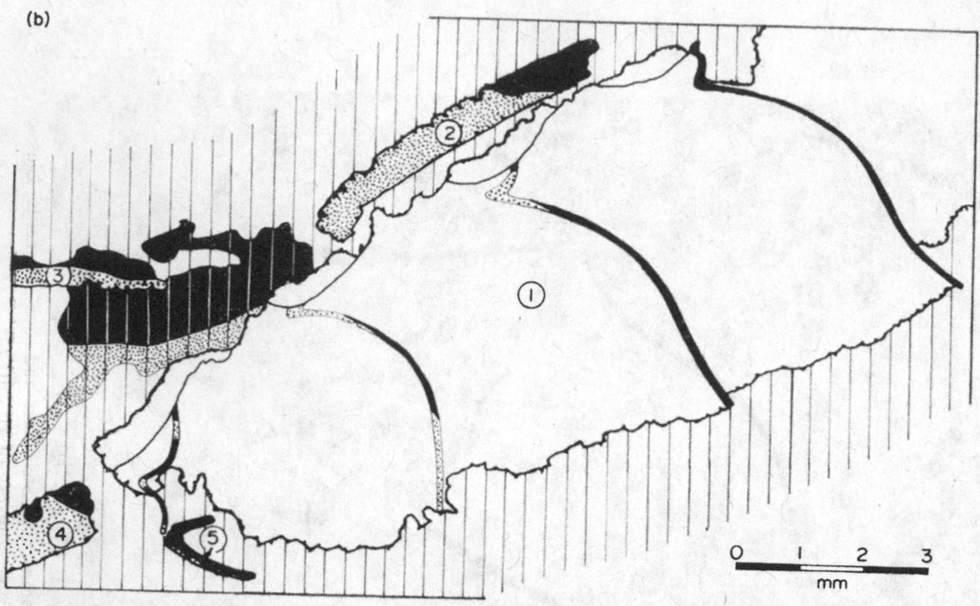

Fig. 5. (b) Line drawing from thin section 5(a) showing redeposited ammonite 'Steinkern' (1) and shell fragments (2–5). Embedding muddy sediment around ammonites is dashed, cement in geopetal fabrics is shown in black, the internal sediment (crystal mosaic) dotted. Cavity fills are oriented the same way up in shell moulds and in the nearby cavity in the embedding matrix, and septal walls stand out of the corroded surface of the 'Steinkern'. This suggests that aragonite of the shells was dissolved after embedding of the fragments 1–4 and not earlier. Corrosion of the fragments must be due to boring and winnowing of sediment rather than inorganic dissolution.

spalled off, rather than leached away, because the remaining shell shows internal laminations, documenting at least partial *in situ* transformation of the aragonite. Only the first signs of dissolution of the septal walls can be observed in some places. This holds true for other specimens in the same beds as well. A fully leached shell was not observed in this horizon. Fig. 3, on the other hand, is typical for the higher part of the cephalopod beds in this section. Most of the phragmocone shells have been completely leached and replaced by bottom sediment or blocky cement.

Discussion

Several conclusions can be drawn directly from the above observations.

(1) The preservational characteristics are such that the shells must have been removed by chemical dissolution rather than by boring or etching organisms.

(2) Dissolution was highly selective, removing only the aragonitic shells and leaving muddy sediment, echinoderms and other calcitic fossils intact.

(3) Aragonite was the most soluble carbonate mineral present in these sediments. Magnesian calcite, as represented by echinoderm fragments, was either less soluble or had stabilized to calcite before chemical dissolution could seriously affect it (for an example of rapid transformation from magnesian calcite to calcite in the deep sea, see Milliman, 1966; Milliman & Müller, 1973, p. 41). At least one carbonate modification, presumably calcite, was stable, or quasi-stable, in this same environment. It

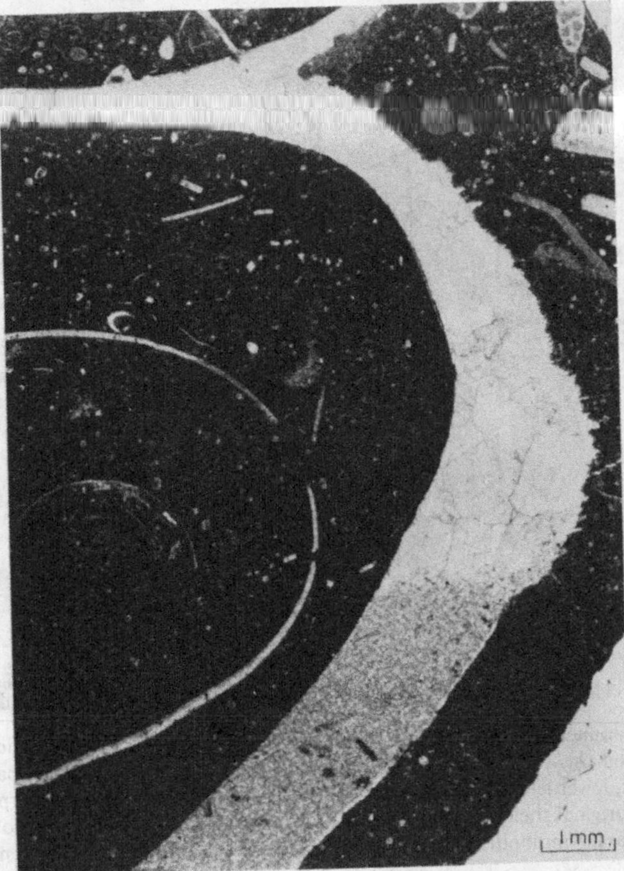

Fig. 6. Ammonite shell mould filled with geopetal sediment (crystal fragments) and blocky cement. (Stylolitic contact between cement and embedding sediment in top right corner.) Tithonian cephalopod limestone, Furlo, Apennines, thin section, positive print (sample courtesy of D. Bernoulli).

formed the solution-resistant cement shown in Fig. 4 and the thin druse that was precipitated in moulds where aragonite had been previously dissolved (Fig. 3b).

(4) Cephalopod dissolution must have taken place within a firm, if not lithified, well-permeable sediment, close to the sediment-water interface. These conditions are required to explain the lack of compaction of the moulds, the efficiency of the leaching and the presence of bottom sediment. Such conditions are best fulfilled near hard-grounds and other omission surfaces.

(5) Sea water is the most plausible leaching agent in this environment as carbonate saturation of the pore fluids in pelagic sediments tends to increase with increasing age and sediment cover (e.g. Presley & Kaplan, 1972, p. 1011 on Deep Sea Drilling material). A secondary source of CO_2 (like breakdown of organic matter by bacteria in reducing environments) is not likely in highly oxidized sediments.

(6) The onset of aragonite dissolution seems to coincide with the facies of crinoidal sands in the sense of Jenkyns (1971a).

Although I believe the above conclusions to be fairly well established they pose as many new questions as they answer. The problems are best illustrated if one confronts

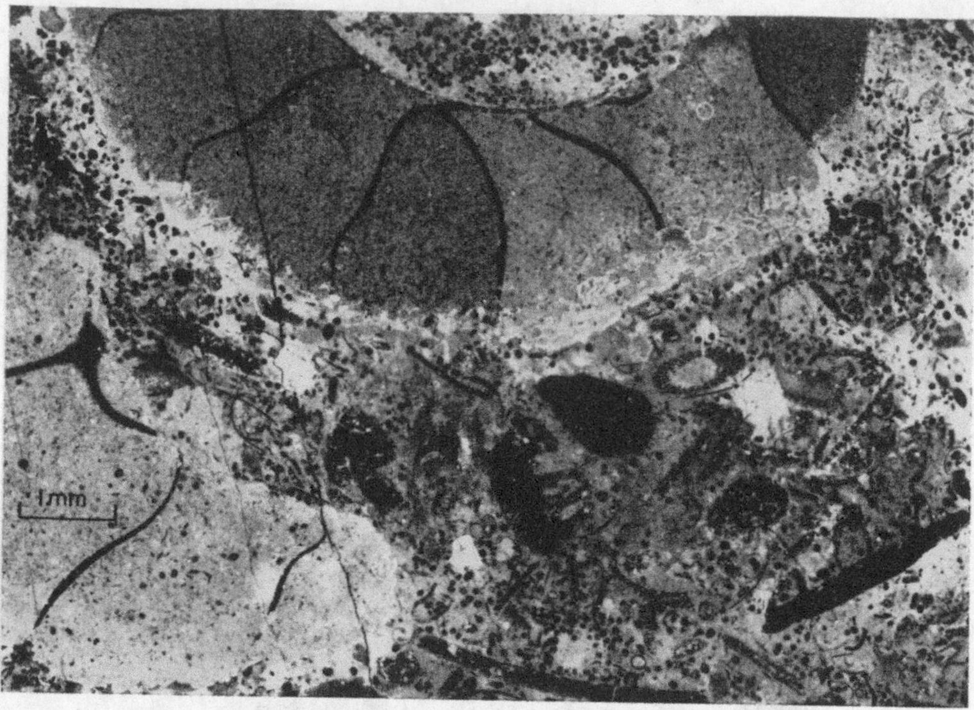

Fig. 7. Exhumed ammonite clasts and worn fragments of chamber walls in a transition zone from crinoidal to cephalopod Lst. Only slight aragonite dissolution is documented by the partial removal of septal walls of clast, lower left. Shell fragments in lower right were not dissolved despite transport. In upper 'Steinkern', original shell lamination partially preserved. Dark spheres are probably sponge spicules. Liassic (?), basal Ammonitico Rosso, La Stua, Dolomites, Italy. Thin section, negative print.

the observations presented here with Dunham's (1969) description of vadose diagenesis in subaerially exposed carbonates. The similarities are obvious. The most striking parallel is a diagenetic sediment, now represented by a crystal mosaic of sand to silt size. It is texturally almost identical with Dunham's 'vadose silt', similarly showing inclined bedding and complete fill of voids—all characteristics that have been taken as evidence for deposition from streaming water. It is generally less well sorted and coarser than typical vadose silt. Another one of the few differences lies in the depth penetration of the leaching phenomena. The leaching in the cephalopod limestones seems to be more strictly bound to omission surfaces and to have less vertical penetration than its vadose counterpart (a distinction which equally applies to submarine compared to subaerial cementation). Flushing with undersaturated water is, therefore, required in a near-surface layer of pelagic sediment to leach the aragonite shells and deposit the internal sediment in the moulds. Episodic, strong bottom currents, possibly tsunami waves as suggested by Lucas (1966), may have accounted for this. Apart from the diagenetic fabrics described here, such currents are also indicated by various features of the hardground faunas in the cephalopod limestones, such as current imbrication of phragmocones (Wendt, 1973), turning over of phragmocones up to 30 cm diameter (Wendt, 1969b, p. 224) and winnowing of the smaller forms from hardgrounds and their deposition in fissures (Krystyn *et al.*, 1971, p. 301). In the vadose setting, Dunham (1969) assumes the crystal silt to form by an interplay of erosion, which

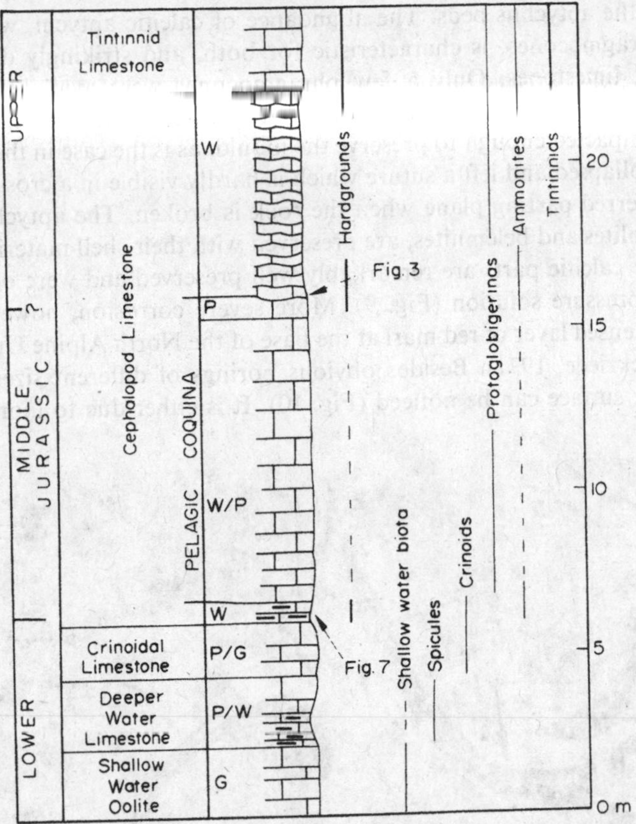

Fig. 8. A typical deepening sequence of the Alpine Jurassic (La Stua, north of Cortina d'Ampezzo, Italy).

would produce the sediment, chemical dissolution, which would take away the fines, and transport separation, which would leave behind the coarse material. Only silt is sufficiently coarse to resist dissolution and sufficiently fine to be carried along by the streaming pore water. Long transport distances are necessary to make these processes efficient. This is not the case for the pelagic leaching with only shallow penetration in the cephalopod limestones. Here, the sediment of crystal fragments could represent a residue that formed within the mould from parts of the shell already converted to low-magnesian calcite. It is also possible that the equigranular crystal mosaic in the moulds was not deposited as a crystal sand or silt at all, but formed from pelagic mud by aggrading recrystallization during diagenesis. Fabrics of this kind have been observed in submarine hard layers of shallow-water environments (Bathurst, 1971, p. 373). Their appearance is very similar to that of vadose silts.

Radiolarites, aptychus beds

The two formations, although clearly representing different facies, mainly display the same preservation of cephalopods, apart from the generally greater abundance of these fossils in the aptychus beds. The abundance of calcitic aptychi, which far outnumber the phragmocones, is characteristic for both, and strikingly different from the cephalopod limestones. Only a few phragmocones were available to me; the ammonite shells were leached after burial, but the sediment was apparently not cemented or compacted enough to preserve the mould, as is the case in the cephalopod limestones. It collapsed and left a suture which is hardly visible in a cross-section, but serves as a preferred parting plane when the rock is broken. The aptychi, as well as the rare rhyncholites and belemnites, are preserved with their shell material. With few exceptions these calcitic parts are remarkably well preserved and were only corroded by borers and pressure solution (Fig. 9). More severe corrosion, however, was observed in a condensed layer of red marl at the base of the North Alpine Upper Jurassic radiolarite (Huckriede, 1971). Besides obvious borings of different sizes, an overall corrosion of the surface can be noticed (Fig. 10). It is either due to inorganic calcite

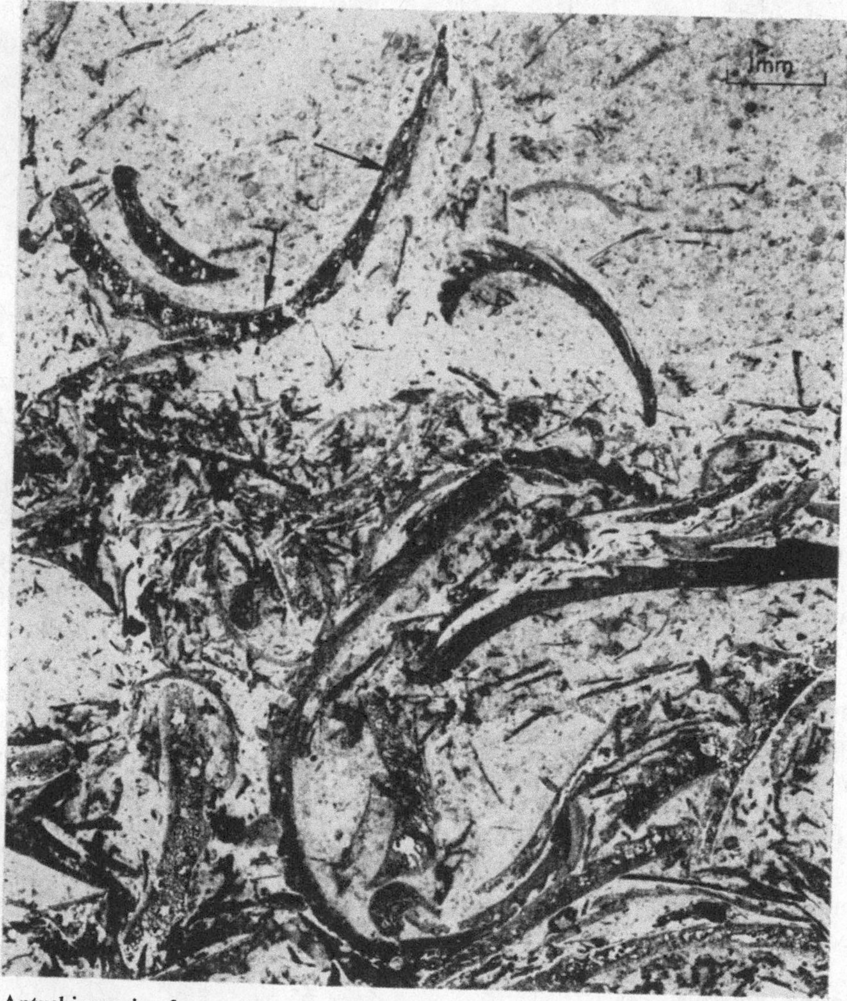

Fig. 9. Aptychi coquina from red aptychus beds. Oxide impregnated borings (arrow) and occasional stylolites are the only traces of corrosion on the calcitic valves. Upper Jurassic, Glasenbach-Klamm near Salzburg, Austria. Thin section, negative print.

Fig. 10. Rhyncholite from a condensed layer at the base of a radiolarite. Corrosion by boring is obvious (arrows), but additional inorganic dissolution at the surface of the calcitic fragment seems likely. Upper Jurassic, Tauglboden near Salzburg, Austria. Thin section, positive print.

dissolution or to etching by micro-organisms. Garrison & Fischer (1969, p. 34) also report corrosion of aptychi in Upper Jurassic radiolarites, though their figured example is from a weathered rock surface and thus not very convincing.

Discussion

The rare phragmocones embedded in the sediment of the aptychus beds were apparently dissolved in a similar way to their counterparts in the cephalopod limestones, the only major difference being the lesser degree of sediment lithification at the time of leaching. The main distinction to be drawn between radiolarites and aptychus beds on one side and cephalopod limestones on the other lies in the general deficit of phragmocones compared to the number of aptychi in the same beds. Consequently, any further discussion on the preservation of cephalopods in these facies can be reduced to the following two questions.

Why are there so many more aptychi in the (Upper Jurassic to Cretaceous) aptychus beds and radiolarites and so few in the underlying (mostly Lower to Middle Jurassic) cephalopod limestones?

What caused the marked deficit of phragmocones in these Upper Jurassic to Cretaceous rocks?

The first question refers to a world-wide phenomenon not specific to our facies, and probably caused by the appearance of more ammonite species with calcitic aptychi (Trauth, 1927).

The various aspects of the second question have recently been discussed by Garrison & Fischer (1969, p. 20, 34, 35, 41). They arrived at the conclusion that aragonite dissolution, rather than transport separation, was responsible for the dominance of aptychi over phragmocones. Transport only caused local accumulation of aptychi coquinas (which were taken as an argument for a shallow depositional depth by Zapfe, 1963). I fully support Garrison's and Fischer's view and would like to point out a few more supporting arguments.

In the Alpine aptychus beds, coquinas are the exception rather than the rule. Most aptychi occur scattered over bedding planes and double valves are rather frequent. Thus water agitation was apparently very low and the aptychi were not transported after decay of the fleshy parts enclosing the two valves.

The facies of the aptychus beds is amongst the most extensive in the Tethyan Mesozoic. Its extension has to be measured in hundreds, if not thousands of kilometres (Trümpy, 1971, p. 305). Transport separation as a rule cannot be held to account for such figures. The separation of gastropod tests and opercula, an example frequently quoted in this context, has been reported to amount to tens of kilometres (Trusheim, 1931).

The phenomenon of the aptychi surplus seems only to occur in pelagic sediments. In coeval deposits of shallow epeiric seas, such as the Solnhofen limestone, phragmocones and aptychi are found in the same beds with the former normally outnumbering the latter.

The anaptychi of the German Devonian present another test case, in which the differences in transportability are much the same as those between Jurassic phragmocones and aptychi; the chemical composition, however, is different since anaptychi are chitinous objects belonging to aragonite phragmocones. Consequently, they sometimes occur in great numbers in reducing environments such as the Büdesheimer Schichten (Matern, 1931, p. 164; Clausen, 1968, p. 21), and are extremely rare in the contemporaneous oxidized sediments, such as the cephalopod limestones. In both sediments, however, they are found together with and often within the phragmocones. Clearly, in this case, chemical dissolution, rather than transport sorting, controls the distribution of the anaptychi and this interpretation has been widely accepted (e.g. Müller, 1963, p. 39).

From all this, one has to conclude that most of the aragonitic phragmocones were dissolved before burial, only a few were embedded and subsequently dissolved within the sediment to yield 'Steinkerns'. This illustrates a higher intensity of dissolution than in the cephalopod limestones, where, in spite of lower rates of deposition, a large number of cephalopod shells escaped dissolution at the sea floor and became embedded in the sediment.

Between aptychus beds and radiolarites, I found no principal difference with regard to preservation of cephalopods. The smaller number of aptychi and the almost complete absence of phragmocones in the radiolarites, may indicate a higher dissolution intensity, but environmental differences can also be held responsible. The above-mentioned condensed horizon with calcite corrosion cannot be taken as being representative of the radiolarite facies in general, because its fossil content is abnormal, its rate of deposition extremely low and its chemical composition altered by an admixture of volcanic material.

FACTORS INFLUENCING THE DISSOLUTION PATTERNS

To analyse the causes for the differential dissolution patterns of cephalopod shells, let us turn briefly to the interplay of pelagic sedimentation and carbonate dissolution in the Recent. In modern oceans, the general increase of carbonate dissolution with water depth is reflected in the consecutive disappearance of aragonitic petropods, calcitic globigerinids and calcitic coccoliths (e.g. Lisitzin, 1972, p. 144ff; for a subdivision of the calcite dissolution facies, see Hsü & Andrews, in Maxwell *et al.*, 1970). This view has been confirmed and more precisely documented by suspending carbonate matter at various depths in sea water and measuring its dissolution (Peterson, 1966; Berger, 1970). These experiments allowed definition of several levels of pronounced increase of dissolution effects on carbonate. One of them marks the onset of carbonate dissolution (generally within the thermocline), another (the lysocline) marks the beginning of strong dissolution effects on calcite, several hundred metres above the compensation depth. Berger's results also show a wide depth range below the thermocline (120–3800 m at this particular site) in which aragonite is almost exclusively dissolved, whilst calcite remains practically unaffected.

Besides the increase of undersaturation with water depth, there is good evidence for the impact of other factors on the rate of carbonate dissolution: Berger (1970, Fig. 2) shows the delay of dissolution caused by organic coatings on carbonate particles. Rapid removal of these coatings in oxidizing environments accelerates dissolution. Last, but not least, the time of exposure of a carbonate particle plays an important role. Once it has reached the bottom, this is largely controlled by the rate of sedimentation and bioturbation. The combined effects of these factors can locally overrule the depth control, as has been demonstrated by Berger & Soutar (1970) for a setting of swells (with dissolution facies) and basins (without dissolution).

The relevant question for us is: 'Which of the above mentioned factors has controlled the dissolution of cephalopod remains in the Tethyan Mesozoic?' *Oxidation* of the sediments shows little variation throughout the sequences. Apart from the aptychus beds, only highly oxidized, mostly red sediments have been considered here. The characteristics of the aptychus facies, as far as I have referred to them, appear in much the same way in radiolarites as well as in marls (e.g. Fig. 9, or the Italian Rosso ad Aptici or the Deep Sea Drilling site 105 in the western central Atlantic, Bernoulli, 1972; Renz, 1972). A higher degree of oxidation does not seem to fundamentally change the picture.

Table 1. Some sedimentation rates of Tethyan pelagic sediments (in Bubnoff units, i.e. mm/1000 years)

	Cephalopod limestone	Aptychus beds	Radiolarites
Garrison & Fischer (1969) (Northern Limestone Alps)	0·6–1	17–51	3–4
Bernoulli (1972) (Apennines)	(< 1)–2·5–6·5	8–10	3–9*
Wendt (1971) (Northern Limestone Alps)	0·5–1·5		
Own estimate (Northern Limestone Alps)		50†	

* Own calculations from Bernoulli's published datings and sections.
† Includes shallow-water turbidites.

Overall sedimentation rates (and thus duration of exposure of the skeletons to sea water) are similar for cephalopod limestones and radiolarites and higher for aptychus beds (see Table 1). Crinoidal limestones are too thin to yield reliable average values. On a small scale, there was probably more variation in cephalopod and crinoidal limestones, because of their 'pulsating' sedimentation with depositional periods being interrupted by intervals of non-deposition. This has indeed influenced the preservation of cephalopods since selective dissolution is preferentially bound to periods of non-deposition.

However, neither degree of oxidation nor rates of deposition display the same trend as we have observed with dissolution effects on cephalopods. What does coincide with this trend is the general deepening tendency of the sequence if we accept the environmental interpretations of Garrison & Fischer (1969), Bernoulli (1972) and others. Thus, it seems to me, that an increase of undersaturation of sea water is the most suitable explanation for the dissolution patterns of cephalopod shells.

Under this assumption, cephalopod limestones, aptychus beds and parts of the radiolarites would have to be correlated with the zone of almost exclusive aragonite dissolution in modern oceans and possibly some of the radiolarites with the zone of severe calcite dissolution below the present-day lysocline. It is, however, difficult to assess absolute values for the ancient water depths even if one accepts this correlation. Firstly, because there is increasing evidence for major fluctuations in the dissolution levels in the past, and secondly because cephalopods and plankton shells, used as a correlation tool in the Recent, differ in size by several orders of magnitude.

REMARKS ON SUBSOLUTION FEATURES IN TETHYAN LIMESTONES

Since Hollmann's (1964) paper, a group of corrosion phenomena, associated with hardgrounds and other omission surfaces, have been commonly summarized under the heading 'Subsolution', the term being originally defined by Heim (1958, p. 643) as submarine dissolution of aragonite, calcite, dolomite or gypsum. It was soon recognized that other processes, such as organic boring, etching and winnowing of sediment, have contributed considerably to the corrosion phenomena in Tethyan pelagic sediments (e.g. Wendt, 1970, p. 445). Some of the above observations on cephalopod phragmocones tend to reduce the importance of inorganic dissolution in the 'Subsolution' facies even further. This is so for the following reasons.

(a) We have seen that most of the aragonitic ammonite shells were first embedded in muddy sediment and subsequently removed by inorganic dissolution.

(b) This dissolution worked selectively and clearly demonstrates a higher solubility of the aragonitic shell over the embedding mud (Fig. 3). The latter was as resistant to dissolution as the non-aragonitic skeletal fragments and cements. This suggests that the mud—at least at this stage of diagenesis—contained little or no aragonite.

(c) The aragonite leaching does not explain the typical 'Subsolution' features for which the fragments 1–5 in Fig. 5b can be taken as examples. They must have formed by a corrosion process operating at the sea floor because of their intimate association with borings and sessile benthos.

(d) This corrosion at the sea floor does not seem to have preferentially attacked the aragonite shells. On the contrary, in some cases, like the 'Steinkern' shown in Fig. 5, the shell stands out of the strongly bored sediment and must have been more resistant to corrosion than the embedding mud.

(e) Consequently, formation of 'Subsolution' surfaces at the sea floor and selective aragonite leaching must be of different origin. Assuming that leaching by undersaturated sea water has removed the aragonite (see section 'Cephalopod and crinoidal limestones'), the 'Subsolution' phenomena must be caused by other processes, such as organic boring, etching and winnowing of loose sediment. Their impact was strong at the sediment surface, where they could compensate or even overcompensate for the dissolution by sea water. Within the sediment, inorganic leaching became dominant.

This hypothesis on the formation of 'Subsolution' surfaces does not explain the removal of the outer shell of phragmocones, which must have preceded the corrosion of the sediment fill. Here, spalling comes to mind as a possible explanation. Most of the ammonite 'Subsolution' relicts are exhumed fragments and the ammonite shells tend to break at the contact of outer and septal walls when filled with sediment. The initial stage of this process is illustrated in Fig. 1. (It should be noted, however, that this applies only to a very special type of 'Steinkern'. All the others, which are enclosed by a mould filled with sediment, must have lost their walls by aragonite leaching within the sediment.)

All this points to the same conclusions reached by Bromley (1965) and Bathurst (1971, p. 399) in their analysis of the chalk hardgrounds in N.W. Europe. They have made a strong case for the importance of both winnowing and organic corrosion and have shown that the hardgrounds, amongst others, go through an initial stage of cementation in patches, which latter coalesce to form a hard surface layer. Winnowing, during the incipient stage, creates a rugged surface and occasional isolated clasts by removing loose sediment from between the cemented patches. Boring and organic corrosion do the rest. This same process, applied to the Tethyan hardgrounds, would explain the karst-like relief of many of them (Wendt, 1969b, p. 223), as well as the numerous 'Subsolution' relicts, the size of which is often incompatible with other evidence on the capacity of current transport (Krystyn, 1972, p. 204). The importance of organic corrosion is illustrated by the almost ubiquitous network of thallophyte borings on Tethyan hardgrounds and their fossils.

CONCLUSIONS

In Tethyan Mesozoic limestones, the impact of carbonate dissolution on aragonitic and calcitic parts of cephalopods has given rise to several different fabric patterns. The main types, in order of increasing dissolution effects, are as follows.

(a) Phragmocones were not dissolved at all but embedded in muddy sediment and, later, transformed into calcite without passing a void stage. This is common, firstly, in cephalopod limestones without omission surfaces and/or relatively high rates of deposition; secondly, in hardground sequences near the transition from shallow-water carbonates to cephalopod limestones; thirdly, in neptunian dykes.

(b) Phragmocones were selectively leached from a lithified or firmly compacted sediment at or very close to the sediment surface, with the undeformed mould being filled with cement, redeposited crystal fragments of sand to silt size, or bottom sediment. This preservation is typical for the majority of cephalopod limestones but is also occasionally found in the (more permeable) crinoidal grainstones.

(c) Most phragmocones were dissolved before being embedded in sediment, a few thereafter, yielding deformed 'Steinkerns'. The calcitic remains (aptychi, rhyncholites) were not affected by dissolution and are therefore disproportionately abundant in the sediments. This mode of preservation is well known from, and typically developed in, the aptychus beds and most of the radiolarites.

(d) All aragonite is dissolved prior to burial and calcitic parts are corroded. This situation was only found in a condensed layer at the base of a North Alpine radiolarite.

In all cases, the dissolved skeletons were either in direct contact with, or not far from, sea water. Distribution of the dissolution patterns over the various depositional facies is such that the increasing effects of dissolution roughly parallel the deepening tendency of the sequences. In the material studied here, it seems that the preservation of cephalopod skeletons reflects mainly the increase of undersaturation of carbonate in sea water. Other factors, such as oxidation of the sediment, rate of deposition and speed of lithification, have only modified this picture.

The way ammonites are preserved in the cephalopod limestones presents some arguments against extensive dissolution of sediment in this facies. The 'Subsolution' features are more readily explained by an interplay of differential cementation, winnowing of loose sediment, and organic corrosion in periods of non-deposition.

ACKNOWLEDGMENTS

C. W. Wagner (Shell Research, Rijswijk) and A. Ford (S.I.P.M., Rijswijk) kindly went through an earlier draft of the manuscript. D. Bernoulli (Basel), H. C. Jenkyns (Durham), H. Jurgan (Berlin) and L. Krystyn (Wien) provided material for comparison from various parts of the Tethyan Mesozoic. J. Wendt (Tübingen) made available the proofs of his publication on East-Mediterranean Hallstatt limestones. All this help is most gratefully acknowledged. I am furthermore indebted to the Koninklijke/Shell Exploratie en Produktie Laboratorium and Shell Research B.V. for permission to publish this study.

REFERENCES

BATHURST, R.G.C. (1971) *Carbonate Sediments and their Diagenesis*, pp. 620 Elsevier, Amsterdam.

BERGER, W.H. (1970) Planktonic Foraminifera: selective solution and the lysocline. *Mar. Geol.* **8,** 111–138.

BERGER, W.H. & SOUTAR, A. (1970) Preservation of plankton shells in an anaerobic basic off California. *Bull. geol. Soc. Am.* **81,** 275–282.

BERNOULLI, D. (1972) North Atlantic and Mediterranean Mesozoic facies: a comparison. In: *Initial Reports of the Deep Sea Drilling Project,* Vol. XI (C. D. Hollister, J. I. Ewing *et al.*), pp. 801–872. U.S. Government Printing Office, Washington.

BERNOULLI, D. & JENKYNS, H.C. (1970) A Jurassic basin: the Glasenbach Gorge, Salzburg, Austria. *Verh. geol. Bundesanst., Wien,* **1970,** 504–531.

BROMLEY, R.G. (1965) *Studies in the lithology and conditions of sedimentation of the chalk rock and comparable horizons,* pp. 355. Unpublished thesis, University of London.

BYSTRICKY, J. (1972) Faziesverteilung der mittleren und oberen Trias in den Westkarpaten. *Mitt. Ges. Geol. Bergbaustud.* **21,** 289–310.

CLAUSEN, C.D. (1968) Oberdevonische Cephalopoden aus dem Rheinischen Schiefergebirge. I. *Palaeontographica,* **128a,** 1–86.

DUNHAM, R.J. (1969) Early vadose silt in Townsend Mound (reef), New Mexico. In: *Depositional Environments in Carbonate Rocks* (Ed. by G. M. Friedman). *Spec. Publs Soc. econ. Paleont. Miner., Tulsa,* **14,** 139–181.

GARRISON, R.E. & FISCHER, A.G. (1969) Deep-water limestones and radiolarites of the Alpine Jurassic. In: *Depositional Environments in Carbonate Rocks* (Ed. by G. M. Friedman). *Spec. Publs Soc. econ. Paleont. Miner., Tulsa,* **14,** 20–55.

HEIM, A. (1958) Oceanic sedimentation and submarine discontinuities. *Eclog. geol. Helv.* **51,** 642–649.

HOLLMANN, R. (1964) Subsolutions-Fragmente. (Zur Biostratinomie der Ammonoidea im Malm des Monte Baldo (Malm; Norditalien). *Neues Jb. Geol. Paläont., Abh.* **119,** 22–82.

HUCKRIEDE, R. (1971) Rhyncholithen-Anreicherung (Oxfordium) an der Basis des Älteren Radiolarits der Salzburger Kalkalpen. *Geologica Palaeont.* **5,** 131–147.

JENKYNS, H.C. (1970) The Jurassic of Western Sicily. In: *Geology and History of Sicily* (Ed. by W. Alvarez and K. H. A. Gohrbandt), pp. 245–254. Petroleum Exploration Society of Libya, Tripoli.

JENKYNS, H.C. (1971a) Speculations on the genesis of crinoidal limestones in the Tethyan Jurassic. *Geol. Rdsch.* **60,** 471–488.

JENKYNS, H.C. (1971b) The genesis of condensed sequences in the Tethyan Jurassic. *Lethaia,* **4,** 327–352.

JURGAN, H. (1969) Sedimentologie des Lias der Berchtesgadener Kalkalpen. *Geol. Rdsch.* **58,** 464–501.

KRYSTYN, L. (1972) Die Oberbajocium- und Bathonium-Ammoniten der Klaus-Schichten des Steinbruches Neumühle bei Wien (Österreich). In: *Ehrenberg-Festschrift* (Ed. by F. Bachmayer and H. Zapfe), pp. 195–310. Österr. Paläont Ges. Wien.

KRYSTYN, L., SCHAFFER, G. & SCHLAGER, W. (1971) Über die Fossil-Lagerstätten in den triadischen Hallstätter Kalken der Ostalpen. *Neues Jb. Geol. Paläont. Abh.* **137,** 284–304.

KUBANEK, F. (1969) *Sedimentologie des alpinen Muschelkalks (Mitteltrias) am Kalkalpensudrand zwischen Kufstein (Tirol) und Saalfelden (Salzburg),* pp. 202. Thesis Techn. Universität, Berlin.

LEHMANN, U. (1970) Lias-Anaptychen als Kieferelemente (Ammonoidea). *Paläont. Z.* **44,** 25–31.

LISITZIN, A.P. (1972) *Sedimentation in the World Ocean* (Ed. by K. S. Rodolfo). *Spec. Publs Soc. econ. Paleont. Miner., Tulsa,* **17,** pp. 218.

LUCAS, G. (1966) Fonds durcis, lacunes sous-marines, séries condensées et ondes marines seismiques ou 'Tsunami'. *C. r. hebd. séanc. Acad. Sci., Paris,* **262,** 2141–2144.

MATERN, H. (1931) Oberdevonische Anaptychen in situ und über die Erhaltung von Chitin-Substanzen. *Senckenbergiana,* **13,** 160–167.

MAXWELL, A.E. *et al.* (1970) *Initial Reports of the Deep Sea Drilling Project,* Vol. III, pp. 806. U.S. Government Printing Office, Washington.

MILLIMAN, J.D. (1966) Submarine lithification of carbonate sediments. *Science,* **153,** 994–997.

MILLIMAN, J.D. & MÜLLER, J. (1973) Precipitation of magnesian calcite in the deep-sea sediments of the Eastern Mediterranean Sea. *Sedimentology,* **20,** 29–45.

MÜLLER, A.H. (1963) *Lehrbuch der Paläozoologie* I, 2nd edn, pp. 387. Gustav Fischer, Jena.

PETERSON, M.N.A. (1966) Calcite: rates of dissolution in a vertical profile in the central Pacific. *Science,* **154,** 1542–1544.

PRESLEY, B.J. & KAPLAN, I.R. (1972) Interstitial water chemistry: Deep Sea Drilling Project, leg 11. In: *Initial Reports of the Deep Sea Drilling Project,* Vol. XI (C. D. Hollister, J. I. Ewing *et al.*), pp. 1009–1019. U.S. Government Printing Office, Washington.

RENZ, O. (1972) Aptychi (Ammonoidea) from the Upper Jurassic and Lower Cretaceous of the Western North Atlantic (Site 105, leg 11 DSDP). In: *Initial Reports of the Deep Sea Drilling Project,* Vol. XI (C. D. Hollister, J. I. Ewing *et al.*), pp. 607–629. U.S. Government Printing Office, Washington.

STURANI, C. (1964) La successione delle faune ad ammoniti nelle formazioni mediogiurassiche delle Prealpi venete occidentali. *Memorie Ist. geol. miner. Univ. Padova,* **24,** 1–63.

STURANI, C. (1971) Ammonites and stratigraphy of the 'Posidonia alpina' beds of the Venetian Alps (Middle Jurassic, mainly Bajocian). *Memorie Ist. geol. miner. Univ. Padova,* **28,** 1–190.

TRAUTH, F. (1927) Aptychenstudien I. *Annln naturh. Mus. Wien,* **41,** 171–259.

TRUSHEIM, F. (1931) Schneckendeckel-Ablagerungen in interglazialen Brack- und Süsswasserbildungen am Boden der Nordsee. *Senckenbergiana,* **13,** 153–159.

TRÜMPY, R. (1971) Stratigraphy in mountain belts. *Q. Jl geol. Soc. Lond.* **126,** 293–318.

WENDT, J. (1969a) Die stratigraphisch-paläogeographische Entwicklung des Jura in Westsizilien. *Geol. Rdsch.* **58,** 735–755.

WENDT, J. (1969b) Stratigraphie und Paläogeographie des Roten Jurakalks im Sonnwendgebirge (Tirol, Österreich). *Neues Jb. Geol. Paläont. Abh.* **132,** 219–238.

WENDT, J. (1970) Stratigraphische Kondensation in triadischen und jurassischen Cephalopodenkalken der Tethys. *Neues Jb. Geol. Paläont. Mh.* **1970,** 433–448.

WENDT, J. (1973) Cephalopod accumulations in the Middle Triassic Hallstatt-Limestone of Jugoslavia and Greece. *Neues Jb. Geol. Paläont. Mh.* **1973,** 624–640.

ZAPFE, H. (1963) Aptychen-Lumachellen. *Annln naturh. Mus. Wien,* **66,** 261–266.

Spec. Publs int. Ass. Sediment. (1974) 1, 71–92

Sedimentology of Palaeozoic pelagic limestones: the Devonian Griotte (Southern France) and Cephalopodenkalk (Germany)

MAURICE E. TUCKER

Department of Geology, University College, Cardiff

ABSTRACT

Pelagic limestones, widely developed during the Devonian (Griotte and Cephalopodenkalk) of the Variscan Geosyncline, are mainly biomicrosparites dominated by a fauna of cephalopods, thin-shelled bivalves, conodonts, styliolinids and ostracods. The main features of these limstones are: (a) a dominantly pelagic fauna; (b) evidence of early lithification (hardgrounds, sheet cracks, neptunian dykes and lack of compaction); (c) local submarine solution; (d) extensive development of pressure-solution planes (stylolites and solution stringers or flasers); and (e) the general condensed nature of their sequences relative to contemporaneous facies. The Griotte is generally red and contains ferromanganese nodules and encrustations around skeletal debris. The Cephalopodenkalk is chiefly grey and contains pyrite. Much of the skeletal debris is bored and there is evidence of extensive bioturbation contributing (with diagenesis) towards the generation of a homogeneous microsparitic mosaic. Faunal and stratigraphic evidence suggest deposition at depths of several tens to hundreds of metres.

INTRODUCTION

True pelagic sediments contain more than 30% biogenic pelagic material (Shepard, 1963). In modern oozes, the biogenic part comprises planktonic and nektonic organic remains, mainly of the calcareous foraminiferids, pteropods and coccoliths, and the siliceous radiolarians and diatoms. Ancient equivalents are chalks and radiolarian cherts. Hemipelagic constituents are present in variable proportions and consist of continent-derived material (normally of very fine grain) which has been transported by rivers, density currents, nepheloid layers and wind. Chemically precipitated deposits such as manganese nodules and pavements, and phosphates, may also occur in the pelagic environment.

Pelagic sedimentation need not involve great depths. Indeed, on some continental shelves pelagic oozes are accumulating from depths of 50 m (e.g. the Yucatan Shelf, Logan *et al.*, 1969) and upon seamounts in oceanic areas from depths of 200 m (Fischer & Garrison, 1967). One of the main factors in the formation of pelagic oozes is the paucity of silt and clay, either through lack of input from nearby continental areas or through deposition upon topographic highs where terrigenous material is channelled into neighbouring deeper-water areas. The compensation depths for aragonite and calcite (averaging about 3500 and 4500 m respectively in present-day oceans) affect the maximum depths of accumulation of calcareous oozes. However, with slow sedimentation rates and long exposure of calcareous material on the sea floor, dissolution of carbonate can be expected at much shallower depths (as for example is recorded from depths of a few hundred metres off Barbados, Garrison & Fischer, 1969).

Modern pelagic carbonates may accumulate upon both oceanic and continental crust, but those deposited on continental crust have a greater preservation potential. In the former case, pelagic sediments are laid down directly upon the ocean floor or, more commonly, upon submarine rises formed by mid-ocean ridges and submerged volcanoes and reefs (seamounts and guyots). Through the process of sea-floor spreading, however, these sediments will be eventually carried down subduction zones, scraped off descending oceanic lithosphere to be preserved in suture zones, or (at best) associated with obducted blocks of oceanic crust. Pelagic carbonates are accumulating upon modern continental slopes and shelves in many parts of the world. They overlie paralic and shelf sands, platform and reef carbonates. The pelagic sediments are commonly part of a transgressive sequence prograding shorewards over shallow-water sediments (e.g. the West Florida and Yucatan Shelves) resulting from the eustatic rise in sea level over the past 20,000 years. It is likely that most fossil pelagic sediments were deposited on continental shelves and slopes.

One other site of accumulation for pelagic sediments is in marginal seas behind island arcs. In this situation, upfaulted basement ridges and volcanic extrusions occur which can support biogenic pelagic sediments (and reefs) while terrigenous material is deposited in neighbouring troughs (e.g. the East China Sea, Wageman, Hilde & Emery, 1970, and behind the Tonga-Kermadec island arc system, Karig, 1970).

PALAEOZOIC PELAGIC SEDIMENTS

Pelagic carbonates only appear to assume importance from the Devonian. During the Devonian, pelagic limestones became widespread and are recorded from the Rheinisches Schiefergebirge, Harz Mountains, Thuringia, Sudetic Mountains, Bohemian Massif, Carnic Alps (see Bandel, 1974), the Bosporus region, Montagne Noire, Pyrenees, Cantabrians, North Africa and S.W. England. These rocks, generally fine-grained cephalopod limestones, are known as the Griotte in southern Europe and Cephalopodenkalk in Germany. They were deposited at various times and in various situations (described below) but were particularly well-developed during the Late Devonian. It is the sedimentological features of these limestones which are the subject of this paper. Upper Silurian limestones containing cephalopods (Orthocerenkalk) in the Carnic Alps (von Gaertner, 1934; Bandel, 1972, 1974) may represent the earliest occurrence of this type of pelagic facies.

Sediments occupying an equivalent position in Lower Palaeozoic geosynclinal successions are hemipelagic black shales with a fauna of graptolites, orthocones and bivalves (the graptolitic facies of Holland, 1971, and others). From the Appalachians, there are records of Cambrian and Ordovician fine-grained limestones (which may have a pelagic origin) occurring in basinal environments. They may be interbedded with turbidites, as in the Cloridorme Formation (Enos, 1969). These and other possible pelagic limestones, such as the Conestooga Limestone (Rodgers, 1968), have commonly suffered extensive neomorphism and structural deformation making an analysis of their original texture difficult or impossible. Some basinal limestones, cited as possibly of pelagic origin (e.g. the Cambrian West Castleton Formation, Sanders & Friedman, 1967), contain shallow-water fossils and are probably partly turbidites derived from nearby shallow-water carbonate platforms (Bird & Theokritoff, 1967).

Carboniferous and Permian pelagic limestones are restricted in Central Europe since, during this time, the Variscan geosynclinal troughs were being filled up with clastic sediments, deformed and uplifted (Hercynian orogeny) and then subjected to subaerial erosion. Devonian pelagic carbonate sedimentation, though, did continue into the Early Carboniferous in some parts of Central Europe, as, for instance, in the Carnic Alps (Bandel, 1974). However, in the Cantabrians of N.W. Spain, extensive Viséan pelagic limestones do occur (the Alba Griotte, Kanis, 1956; Wagner, 1963; Frets, 1965). This is a red nodular limestone rich in goniatites and conodonts, identical to the Devonian Griotte of the Pyrenees and Montagne Noire. The Alba Griotte is extensively developed over both the Leonide and Asturian facies areas as part of a transgressive blanket deposit, overlying shallow-water Upper Devonian quartzites.

Some Lower Palaeozoic limestones, such as the Cambrian red griotte of the Lancara Formation in the Cantabrians (Meer Mohr & Schreuder, 1967; Zamarreño, 1972) resemble later pelagic limestones in lithology, but contain a rich benthos including ribbed brachiopods and trilobites (P. Wallace, personal communication, 1973). The Lower Ordovician *Orthoceras* Limestones of South Sweden are also reminiscent of later pelagic limestones (being rich in cephalopods and conodonts, stratigraphically condensed and containing evidence of early lithification, Lindström, 1963), but yield a similar benthonic fauna.

There have been marked changes in the faunal composition of pelagic carbonates with time. It is considered (Tucker, 1973a; Tucker & Kendall, 1973, and this paper) that the Devonian pelagic limestones were originally composed of skeletal material (chiefly molluscan and ostracod). During the Jurassic, coccolithophorids replaced thin-shelled bivalves and ostracods as the main component of calcareous pelagic oozes. Scanning electron microscope studies have shown that coccoliths are the main constituents of the red pelagic limestones of the Tethyan Mesozoic (e.g. Garrison & Fischer, 1969; Jenkyns, 1971; Bernoulli, 1972). Planktonic Foraminifera became an important constituent of pelagic carbonates from the Cretaceous onwards (Riedel, 1963) and in modern seas coccolith and pteropod oozes can be subordinate to globigerinid oozes (Rodgers, 1957). Ancient analogues of pteropod oozes are found in the Devonian where cricoconarids (particularly the styliolinids) were locally sufficiently abundant to form limestones (Tucker & Kendall, 1973).

DEVONIAN PELAGIC LIMESTONES

The stratigraphy, context and facies of the Griotte and Cephalopodenkalk are first described, and then their fauna, sedimentology and diagenesis.

The Griotte of the Montagne Noire

The Montagne Noire (Fig. 1) is chiefly an area of unmetamorphosed, but folded and thrusted, Palaeozoic strata situated to the south of the Massif Central, in the

Départments of Aude and l'Herault, Southern France. The Upper Devonian is developed in two main lithofacies (Gèze, 1949); (a) the *Griotte* (Figs 2–4)—a pelagic limestone sequence cropping out in the regions of Caunes Minervois, Mont Peyroux and Cabrières (Fig. 1) and (b) the *Calcschiste*—a more shaley, probably deeper-water facies, occurring to the north-north-west of the Griotte, in the regions of St Pons, Faugères and Nord Minervois. Boyer (1964) and Boyer *et al* (1968) divided the Upper Devonian Griotte into three formations: the Infragriotte, Vrai Griotte and Supragriotte. The Vrai Griotte (to which the term 'griotte' was originally applied) is a red, argillaceous nodular limestone, commonly weathering to give a brecciated appearance (Fig. 4). It is generally 10–15 m thick, rich in ammonoids, and of early Famennian age. The Supragriotte above (30–40 m thick) of Famennian to earliest Carboniferous age, consists mostly of grey or buff-coloured biomicrosparitic limestones with some red horizons. The Infragriotte (20–40 m thick), of Frasnian age, is more variable along the length of the Montagne Noire, than the Vrai Griotte or Supragriotte. Consisting generally of red and grey fine-grained limestones and nodular limestones, the Infragriotte contains breccias at Caunes Minervois. At Coumiac near Mont Peyroux cherts and siliceous shales form the lowest Frasnian. At Cabrières, cherts, shales and limestones rich in *Tentaculites*, of the lower Frasnian, are overlain by the Calcaires à Galettes (3 m thick), a sequence of black bituminous limestones which Boyer *et al.* (1968) correlate with the Kellwasserkalk of the Harz Mountains (Buggisch, 1972).

The Upper Devonian Griotte succeeds Givetian shallow-water massive limestones containing corals, stromatoporoids and brachiopods (the Calcaire à Polypiers Silicieux). In the region of Caunes Minervois pelagic limestones overlie, and also occur contemporaneously with, a Givetian reef complex (Boyer, 1964).

Sedimentary facies

Six lithofacies may be distinguished in the Montagne Noire Griotte.

(a) *Red and grey pelagic limestones* (Fig. 2). Fine-grained limestones, locally nodular, containing a pelagic fauna mostly of microscopic organisms (thin-shelled bivalves, ostracods, styliolinids and conodonts) but locally with many ammonoids. Both red (through finely disseminated haematite) and grey (pyritic) limestones occur. Structures present include current laminations, burrow fills, sediment discontinuities and, less commonly, sheet cracks. The limestones are evenly bedded (beds up to 1 m thick) and contain numerous pressure-solution planes (mostly horizontal stylolites). Limestones of this type are very uniform and dominate the Supragriotte of Caunes Minervois and Mont Peyroux (where they are mostly grey) and occur also in the Infragriotte.

(b) *Pelagic limestones with ferromanganese encrustations.* Sediments as (a) above, but containing ferromanganese nodules and encrustations around lithoclasts, skeletal debris and upon hardground surfaces (Tucker, 1973b). Limestones of this type are mostly red, but ferromanganese nodules do occur in grey or buff-coloured sediments. Sedimentary structures (laminations, hardgrounds and sheet cracks) are more common. Limestones with ferromanganese encrustations are most common in the Infragriotte—as at Mont Peyroux—but constitute much of the Supragriotte at Combe d'Izarne, Cabrières.

(c) *Intraformational breccias.* Beds composed of pelagic limestone clasts, from a few millimetres to 2 m in diameter, embedded in a red argillaceous matrix. Crude

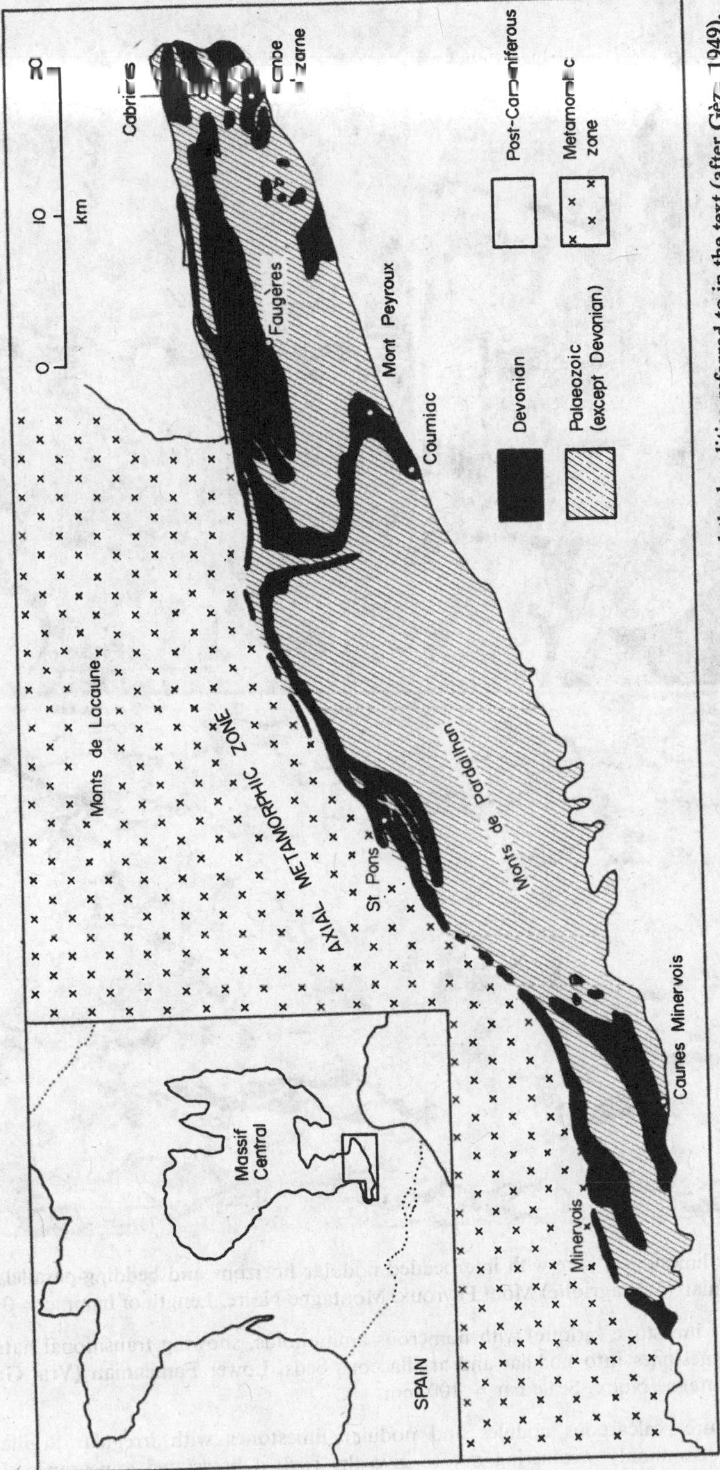

Fig 1. Sketch-map of the Montagne Noire, showing the location of the Devonian strata and the localities referred to in the text (after Gèze, 1949).

Fig. 2. Red pelagic limestones, here with interbedded nodular horizons and bedding-parallel stylolites. Lower Famennian (Supragriotte) Mont Peyroux, Montagne Noire. Length of hammer = 0·35 m.

Fig. 3. Red nodular limestone (griotte) with numerous ammonoids, showing transitional nature of more continuous limestones into nodular and argillaceous beds. Lower Famennian (Vrai Griotte) Mont Peyroux, Montagne Noire. Scale bar = 100 mm.

Fig. 4. Griotte texture—calcareous nodules and nodular limestones with irregular argillaceous partings—the latter commonly weathering out to give the rock a brecciated appearance. Lower Famennian (Vrai Griotte) Mont Peyroux, Montagne Noire. Scale bar = 50 mm.

grading and imbrication may be present, but the breccias are mostly structureless. They are probably submarine slump deposits or fault-scarp breccias. The breccias mostly occur within the Frasnian Infragriotte, and at Caunes Minervois 20 m of breccia are present.

(d) *Red nodular limestones* (Figs 3, 4). Red nodular limestones separated by thin, irregular, shaley partings—often looking brecciated (the griotte texture). The fauna is similar to (a) and (b) above but is commonly enriched in ammonoids which form the centres to nodules. Sedimentary structures, apart from burrow fills, are rare. The origin of the griotte texture is considered later. This facies is developed over much of the Montagne Noire in the lower Famennian (Vrai Griotte), but also occurs at other horizons.

(e) *Shale interbeds*. Thin argillaceous beds, mostly less than 5 cm thick but reaching up to 15 cm, interbedded with the pelagic limestones. Fossils are less common in the shales than in the adjacent limestones.

(f) *Cherts and siliceous shales*. Finely laminated grey cherts and siliceous shales, occurring in the lowest part of the Infragriotte at Coumiac and Cabrières, also succeeding the Griotte in the Lower Carboniferous (the Schistes and Lydiennes, Gèze, 1949).

Cephalopodenkalk of the Rhenohercynian Geosyncline

In the Rheinisches Schiefergebirge and Harz Mountains pelagic limestones are developed from the Lower Devonian till the lower part of the Carboniferous, the most extensive development being in the Upper Devonian. Schmidt (1925) recognized two main facies in the Upper Devonian; (a) the *Schwellen* facies, consisting of pelagic limestones locally rich in ammonoids (stratigraphically termed Cephalopodenkalk), which were considered to have been deposited upon submarine rises or Schwellen (Rabien, 1956; Tucker, 1973a); and (b) the *Becken* facies, of silty shales locally rich in ostracods (Cypridinenschiefer), deposited in neighbouring basinal areas.

The Rhenohercynian Geosyncline (Fig. 5) was divided into a northern and southern basin by a mid-geosynclinal rise (Meischner, 1968, 1971), an upfaulted basement block. This rise was divided into subsidiary ridges and troughs, and upon two of these ridges, pelagic limestones accumulated from the Early Devonian. Within the basins adjacent to the mid-geosynclinal rise, and along the rise itself, extensional igneous activity during the Givetian and early Frasnian gave rise to volcanic ridges and swells, upon some of which (e.g. the Hauptgrünsteinzug) pelagic carbonates accumulated. Reef limestones developed upon other volcanic ridges (e.g. at Langenaubach) and on the northern continental shelf from the Givetian (e.g. the reefs of Attendorn, Brilon and Arnsberg, Krebs, 1968). Pelagic limestones also accumulated on the flanks of, and are intercalated with, these reefs (Krebs, 1972).

During the later Frasnian, subsidence affected much of the Rhenohercynian Geosyncline and led to the development of cephalopod limestones over the former reef and shallow-water carbonate areas. From the late Frasnian through the Famennian, pelagic limestones continued to form over the submerged reefs and volcanic ridges, and upon subsidiary ridges of the mid-geosynclinal rise, while ostracod shales and locally turbidites were deposited in the basinal areas.

The limestones of the Schwellen facies are generally monotonously uniform sequences of grey biomicrosparites with numerous pressure-solution planes (lithologically termed Flaserkalk), occurring in beds from a few centimetres to a metre in thickness and separated by thin shale seams. Red Cephalopodenkalk does occur,

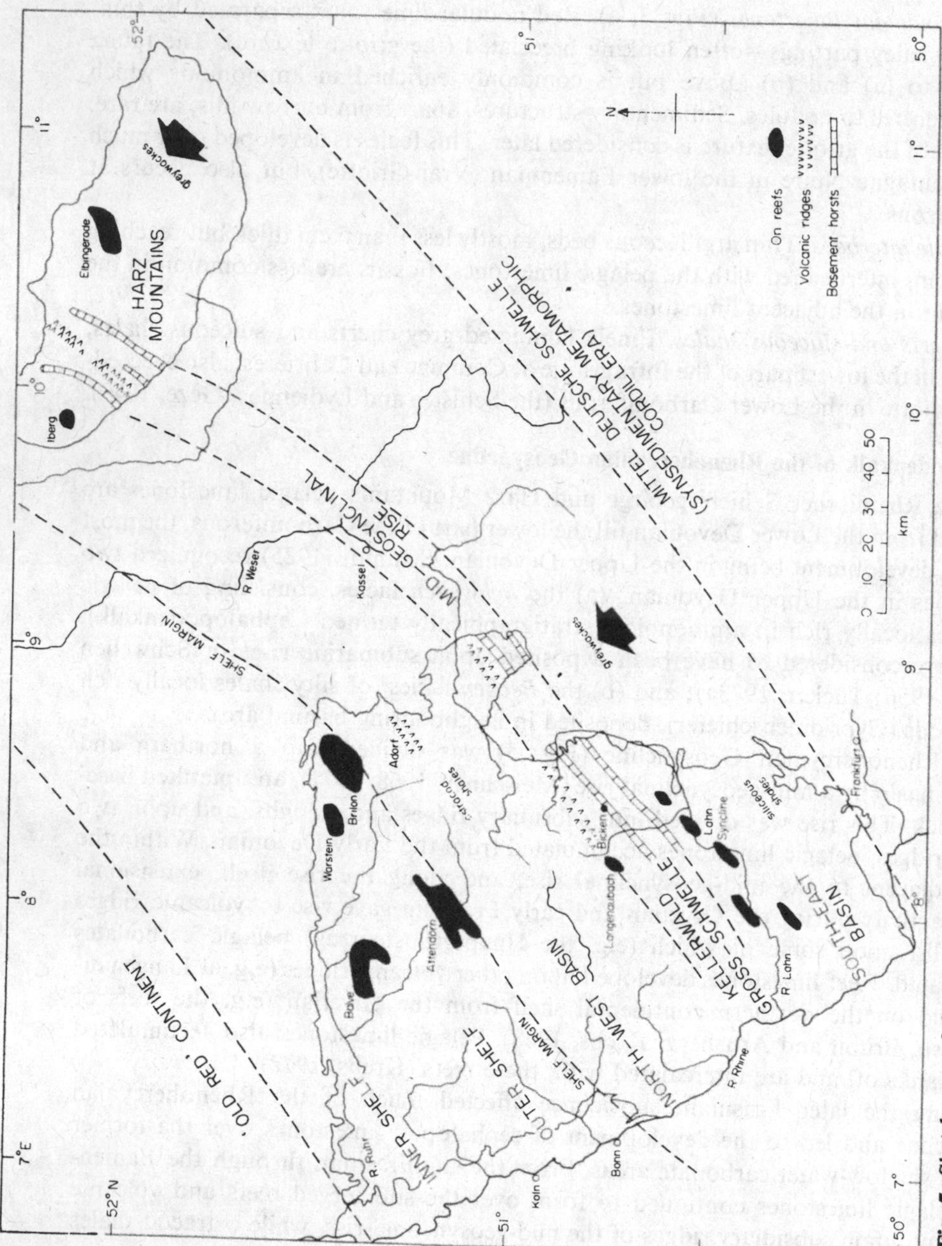

Fig. 5. Sketch-map of the Rhenohercynian Geosyncline showing the occurrence of pelagic limestones (Cephalopodenkalk) and the palaeogeography at the end of the Frasnian, Late Devonian. The pelagic limestones are shown occurring above volcanic swells, reefs and shallow-water carbonate areas and upfaulted basement ridges; it should be noted, however, that those reefs developed within the basins were themselves formed on top of volcanic extrusions.

chiefly in the lower Famennian (*Platyclymenia* Stufe), and also where the limestones were deposited upon volcanic ridges and hydrothermal haematitic enrichment took place (Bottke, 1965).

For much of the Cephalopodenkalk, sedimentary structures, apart from bioturbation and lamination, are rare. However, at some localities, such as Benner Steinbruch, Bicken and Adorf am Martenberg, both in the Rheinisches Schiefergebirge, sheet cracks, hardgrounds and neptunian dykes are common.

In many cases the pelagic limestones can be shown to pass laterally into deeper-water basinal shales via a slope facies of nodular limestones and shales with nodules (also pelagic), which are commonly slump-folded or reworked into breccias (Tucker, 1973a). Cephalopod limestones occurring in a basinal environment, and not restricted to submarine highs, have been reported from the southern Rheinisches Schiefergebirge by Meyer & Bandel (personal communication). The thickness of the Cephalopodenkalk is generally about 20–30 m for the Upper Devonian (a rate of sedimentation, not allowing for pressure solution, of several mm per 1000 years), considerably condensed relative to the contemporaneous ostracod shales (about 300 m thick).

Fauna of Devonian pelagic limestones

The Griotte and Cephalopodenkalk are often extremely rich in fossils, mainly pelagic organisms, which, apart from the cephalopods, are mostly of microscopic size. Ammonoids are unbroken and occur at all angles to the bedding. Bivalves, especially thin-shelled types, are abundant and may form coquinas. The shells, mostly without ornamentation and occurring disarticulated, are generally less than 2 mm in diameter, with a valve thickness of 10 μm. Larger bivalves, with shell diameters up to 20 mm, belong to the genera *Buchiola* and *Cardiola*. Small gastropods, less than 2 mm in height, and acrotretid brachiopods of similar size also occur. The small size and thin nature of the molluscan shells is consistent with a nektonic or nektoplanktonic mode of life.

Ostracods, of both smooth and ornamented form, mostly belong to the Entomozoacea (Rabien, 1956) considered to be equivalent to the modern Halocypriden (active planktonic forms). Cricoconarids (mainly styliolinids) were an important constituent of pelagic limestones until the end of the Frasnian. They commonly form laminae and in other cases are the sole components of the limestones. Conodonts are ubiquitous throughout Devonian pelagic sediments.

Arenaceous Foraminifera, of the genus *Tolypammina,* encrust hard substrates within the limestones (shells, conodonts, lithoclasts and hardground surfaces). *Tolypammina* is also associated with ferromanganese nodules (Tucker, 1973b), and in the Cephalopodenkalk has been found intergrown with a red alga, forming nodules (Tucker & Kendall, 1973, Fig. 3).

Other faunal elements present include fragments of bryozoan colonies and trilobites (mainly of the blind phacopid group), calcispheres and (very rarely) solitary corals of the *Syringaxon* type. Isolated crinoid ossicles are found scattered throughout much of the Griotte and Cephalopodenkalk. Crinoid ossicles, the larger bivalves and goniatites are commonly bored.

Sedimentary structures

The sedimentology and diagenesis of the Cephalopodenkalk in Germany have been described in an earlier paper (Tucker, 1973a) and consequently only a résumé will be given here, with emphasis placed upon the Griotte.

Bioturbation

Within the Griotte and Cephalopodenkalk there is commonly much evidence of bioturbation, and simple burrows, up to 15 mm in diameter, weather out on bedding surfaces. Burrows in the Vrai Griotte occur in both the limestone nodules and in the surrounding argillaceous matrix (Fig. 6) where the burrows are more calcareous. Burrows in limestones containing many ferromanganese nodules may themselves be encrusted (Fig. 7) with ferromanganese oxides and Foraminifera, indicating a sedimentary break before filling of the burrows. Small geopetal structures (1–10 mm in diameter) filled with calcisiltite internal sediment and sparite druse, are interpreted as burrow fills. Concentrations of skeletal material (especially cricoconarids) into elongate patches, circular in cross-section (diameter less than 10 mm) are probably also burrow fills. Bandel (1972) has noted similar structures in Cephalopodenkalk from the Carnic Alps. The structureless (non-laminated) appearance of many of the limestone beds may in part be the result of extensive bioturbation.

Current structures

Current structures are mainly laminae, several mm thick, visible through their variable skeletal content. The laminae may be discontinuous or vaguely defined probably through the activity of burrowing organisms. Concentrations of skeletal debris commonly form lenses a few centimetres thick which can be traced over several metres.

Hardgrounds

Hardground surfaces from the Cephalopodenkalk, described by Tucker (1973a), are of two types. (1) Flat surfaces (penetrated by microscopic borings) truncating

Fig. 6. Bedding plane with simple burrow structures in both nodular limestone and the surrounding more argillaceous matrix. Lower Famennian (Vrai Griotte) Coumiac, Montagne Noire. Scale bar = 25 mm.

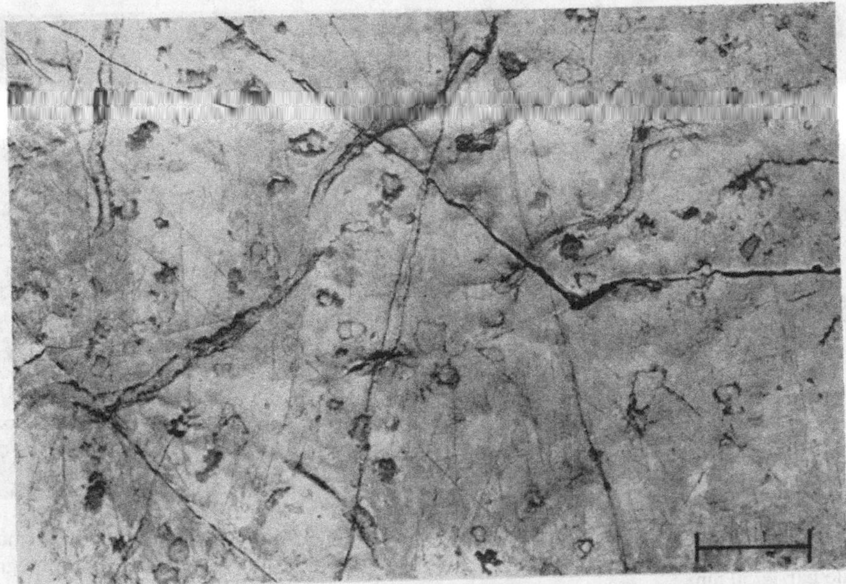

Fig. 7. Bedding-plane surface with many ferromanganese nodules and meandering burrow structures. Some of the latter are also encrusted with ferromanganese oxide, indicating a pause before they were filled with sediment. Lower Frasnian (Infragriotte) Mont Peyroux, Montagne Noire. Scale bar = 50 mm.

fossils and cavity-filling cements and (2) cryptohardgrounds—irregular surfaces encrusted by arenaceous Foraminifera which are only detectable in thin section. In the first case, the surfaces are interpreted as having formed by corrasion after early lithification of the sediment. It is likely, however, that submarine solution played a part in the development of the irregularity and relief of the cryptohardgrounds.

Corrasional hardground surfaces occur in the Infragriotte at Mont Peyroux (Kendall & Tucker, 1971, Fig. 1) truncating ferromanganese-encrusted bivalves, crinoids and fills of sheet cracks. Hardground surfaces encrusted with Foraminifera and ferromanganese oxide (Figs 8, 9) also occur in the Infragriotte. The Foraminifera form 'microreefs', which may consist of several generations (Fig. 10). Succeeding sediments are banked around the microreefs, which rise up to 20 mm above the hardground surfaces. The latter, when traced laterally, commonly pass into a haematitic pressure-solution plane. Foraminiferal microreefs identical to these have been described by Wendt (1969) in pelagic limestones from the Alpine Trias.

Within limestones of the Griotte, there are commonly distinct bedding changes, normally marked by a colour change from light to dark red, red to buff or grey. These sediment discontinuities can be traced for up to 50 m across the outcrop and have an irregular, sometimes undulating form, with a relief up to several centimetres (Figs 9, 11). In many cases, lithoclasts the colour of the underlying sediment, up to 15 mm in diameter, mostly sub-angular and sometimes coated with ferromanganese oxide, occur within the sediment above the discontinuity surface. The sediment immediately above is commonly slightly coarser and, in Frasnian examples, richer in styliolinids than that below. Bioturbation may be evident in the sediment below the discontinuity. These surfaces do not appear to be encrusted with Foraminifera and are rarely coated with ferromanganese oxides. Some thick beds of the Infragriotte are made up

Fig. 8. Hardground surface (arrowed) encrusted with ferromanganese oxide and upon which small foraminiferal microreefs have developed. Lower Frasnian (Infragriotte) Mont Peyroux, Montagne Noire. Scale bar = 50 mm.

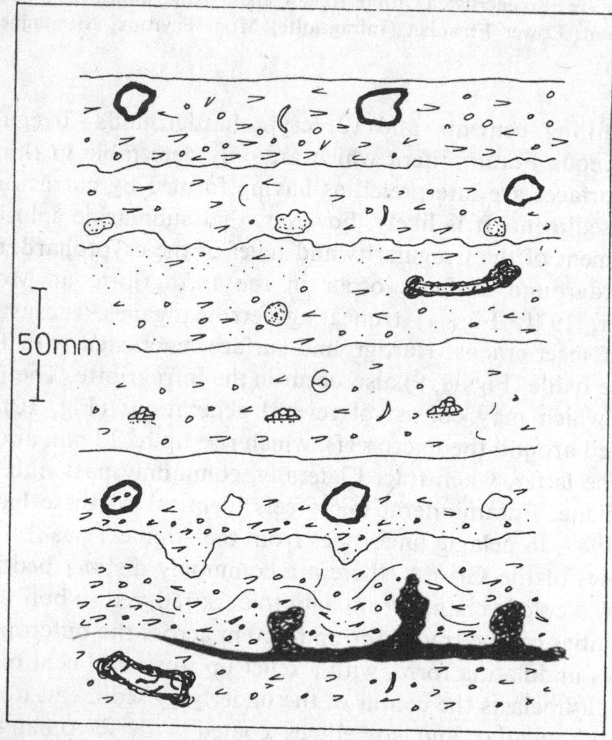

Fig. 9. Field-sketch showing various sediment discontinuities in a 0·30 m section of red pelagic limestones. The discontinuities are marked by irregular surfaces, colour changes, grain-size changes and lithoclasts. A hardground surface encrusted with ferromanganese oxide and foraminiferal microreefs passes laterally into a pressure-solution seam. Also shown are ferromanganese nodules and encrustations around bivalves, styliolinids and small cavity structures probably formed by bio-turbation. Lower Frasnian (Infragriotte) Mont Peyroux, Montagne Noire.

Fig. 10. Sketch from thin sections of foraminiferal microreefs encrusting a hardground surface. Ferromanganese oxides associated with the Foraminifera commonly develop colloform structures, replacing the surrounding styliolinid-rich sediment. Lower Frasnian (Infragriotte) Mont Peyroux, Montagne Noire.

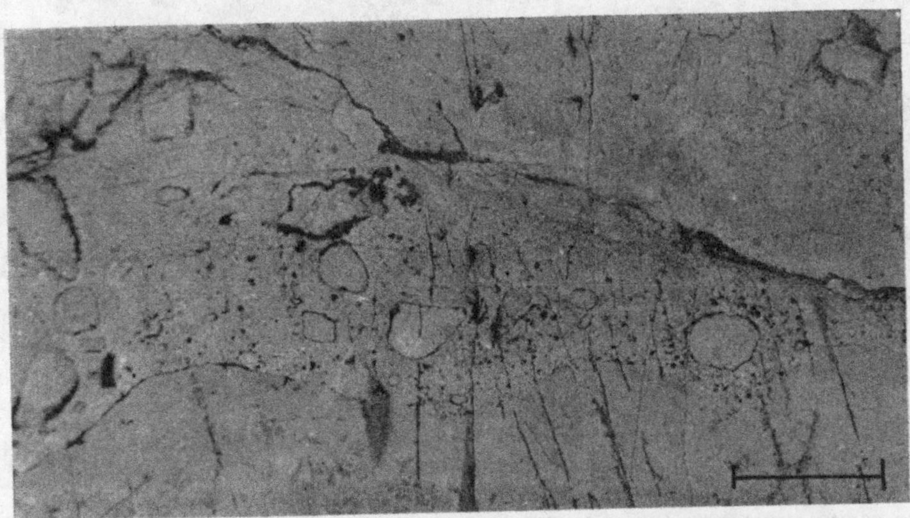

Fig. 11. Sediment discontinuity consisting of an irregular surface separating dark red carbonate sediment below, from lighter, slightly coarser, sediment above. Lithoclasts of the underlying sediment occur above the discontinuity. Frasnian (Infragriotte) Coumiac, Montagne Noire. Scale bar = 10 mm.

of many units, each with a distinct but irregular base and containing lithoclasts (e.g. Fig. 9).

Other bedding-parallel colour changes in the Griotte do occur without distinct boundaries, when lithoclasts are generally absent. Lateral variations in colour of the Griotte from red to grey also occur, but are probably secondary since there appears to be no change in the sediment itself.

Bedding-parallel colour changes in the Griotte represent changes in the physico-chemical environment, during and/or after sedimentation. The irregular but persistent nature of the discontinuities and the presence of lithoclasts suggest some degree of synsedimentary lithification, followed by submarine solution of the sediment. The discontinuities probably represent a short lull in sedimentation.

Sheet cracks and neptunian dykes

Bedding-parallel sheet cracks (Fig. 12) occur in both the Cephalopodenkalk (Tucker, 1973a) and Griotte, but only within compact limestones, never within nodular or shaley limestones such as the Vrai Griotte. The cracks are generally parallel-sided, up to 50 mm in height and may extend laterally for 3 m. Those in the Infragriotte tend to occur along particular horizons. Several sheet cracks may occur immediately above each other (Fig. 12) and be interconnected.

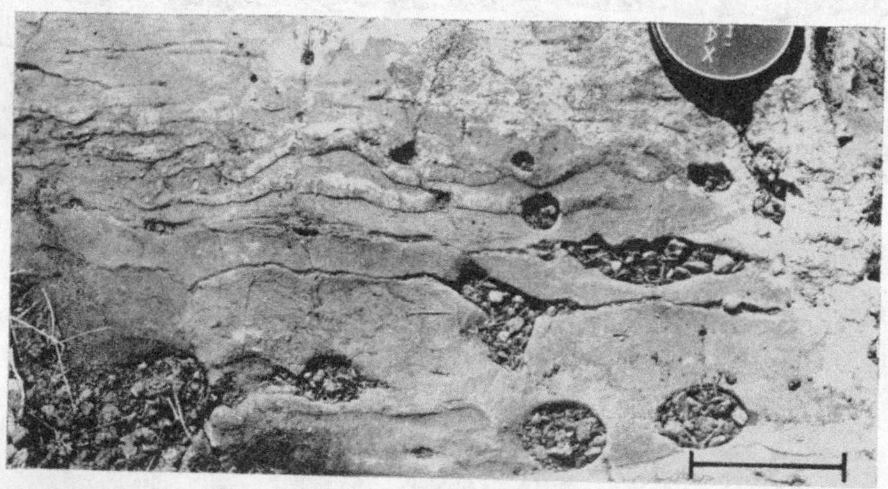

Fig. 12. Several sheet cracks in red pelagic limestone. These cavity structures, filled by internal sediment and fibrous calcite, occur along a particular horizon which can be traced laterally for 50 m. Lower Frasnian (Infragriotte) Mont Peyroux, Montagne Noire. Scale bar = 50 mm.

Smaller cavity structures, with irregular tops, up to several centimetres long and 20 mm high, occur within some limestones of the Griotte. In some cases they appear to be concentrated above more argillaceous parts of the bed.

The cavity structures are partly filled by internal sediment (commonly several periods of fill) which is often a red calcisiltite. Fibrous calcite, following and inter-bedded with the internal sediment, fills most of the cavities. Remaining cavities are occluded by equant drusy calcite, although in some cases a coarse haematitic silt was introduced after the formation of the fibrous calcite.

Neptunian dykes in the Cephalopodenkalk (Tucker, 1973a) vary from 1 to 15 cm in width and penetrate down at least 50 cm. They are filled with many saucer-shaped lenses of internal sediment, commonly with laminae of styliolinids, and contain clasts of the host sediment. Dykes are rare in the Griotte.

Processes invoked for the formation of sheet cracks and neptunian dykes include shear failure upon sedimentary slopes, bioturbation, desiccation, karstification, tectonic fracturing and localized sediment dewatering. Shear failure and the slight movement of a lithified sediment mass upon insignificant sedimentary slopes is the most likely mechanism for the bedding-parallel sheet cracks in the Griotte and Cephalopodenkalk. Smaller cavity structures with irregular tops may have formed by localized sediment dewatering (a process suggested by Heckel, 1972, for *Stromatactis* s.s.). The dykes probably formed through tectonic fracturing of previously lithified limestones (as Wendt (1971) suggested for similar dykes in Alpine Jurassic pelagic limestones).

Ferromanganese encrustations and lithoclasts

Ferromanganese nodules (up to 50 mm in diameter) are common in the Griotte (Tucker, 1973b) and take the form of encrustations around limestone clasts and skeletal debris. Some limestone beds, such as in the Infragriotte of Mont Peyroux or the Famennian Griotte of Combe d'Izarne, contain numerous ferromanganese-encrusted limestone clasts (Fig. 13). The clasts, subangular and of similar lithology to the surrounding sediment, probably formed through sea-floor solution of indurated sediment (the 'Subsolution' of Hollmann, 1964). Subangular limestone clasts without ferromanganese encrustations occur throughout the Griotte, but tend to be associated with the sediment discontinuities previously described.

Fig. 13. Limestone containing numerous ferromanganese encrustations around lithoclasts and bivalves. Some encrusted lithoclasts contain earlier ferromanganese crusts indicating reworking or two phases of 'Subsolution'. Famennian (Supragriotte) Combe d'Izarne, Montagne Noire. Scale bar = 50 mm.

Ferromanganese nodules and lithoclasts are less common in the Cephalopodenkalk, and were only encountered in association with hardgrounds and neptunian dykes. It has been suggested (Tucker, 1973b) that a high organic content of the original sediment may have been a factor in limiting the precipitation of ferromanganese oxides.

Diagenesis

Cementation

There is abundant evidence in the Griotte and Cephalopodenkalk for early diagenetic or synsedimentary lithification. The presence of hardground and discontinuity surfaces, sheet cracks, neptunian dykes, and lithoclasts all suggest a prior lithification of the sediment, probably through precipitation of a micritic cement. In addition to this, there are no early compactional features and all fossils in the limestones are full-bodied.

Coarse calcite cements partly fill sheet cracks, neptunian dykes and geopetal voids produced by cephalopods and bivalves. Two cement generations (radiaxial fibrous calcite, followed by equant druse) are present, as is found in many limestones. The radiaxial fibrous calcite is interpreted as a replacement after an early diagenetic acicular cement (Kendall & Tucker, 1973). Fibrous calcite also occurs as envelopes around styliolinids, where replacement has again taken place (Tucker & Kendall, 1973).

Shell dissolution

The shells of bivalves and ammonoids are mostly preserved as sparite pseudomorphs. Apart from loss of shell structure, the existence of a void stage is also shown by fragments of the ferromanganese coating contained within the calcite of the shell wall (Fig. 14) (analogous to broken micritic envelopes, Bathurst, 1964). Evidence of early shell dissolution is provided by single bivalve shells containing calcisiltite, Foraminifera and minute bivalves, arranged as a geopetal internal sediment and overlain by sparite (Fig. 15). In some goniatite fills there is no evidence of the septa or outer shell although internal sediments are present and were clearly deposited upon the septa (Fig. 16). In this case lithification of the internal sediment before shell dissolution is necessary to explain the apparently 'floating' boat-shaped lenses of internal sediment now seen.

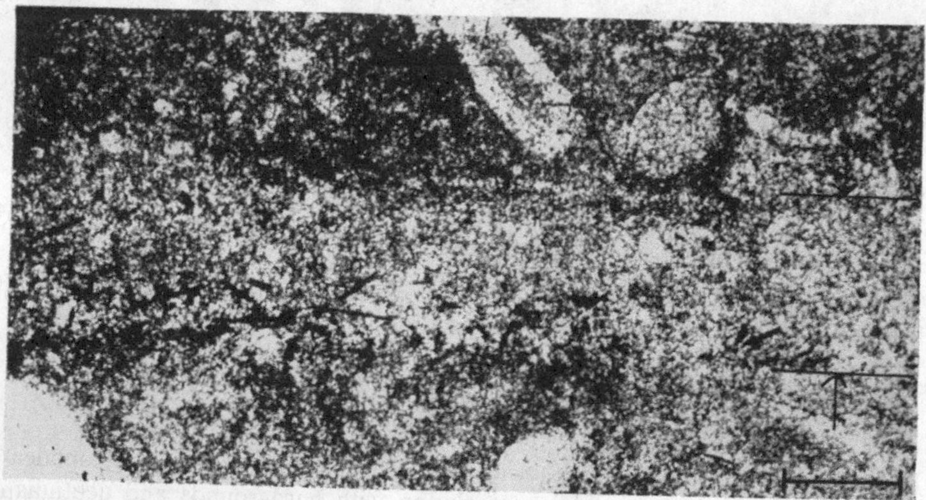

Fig. 14. Photomicrograph of bivalve (edge of shell arrowed) now preserved as microsparite, containing fragments of a former ferromanganese coating. The shell itself is now barely discernible from the matrix. Lower Frasnian (Infragriotte) Mont Peyroux, Montagne Noire. Thin section; scale bar = 0·2 mm.

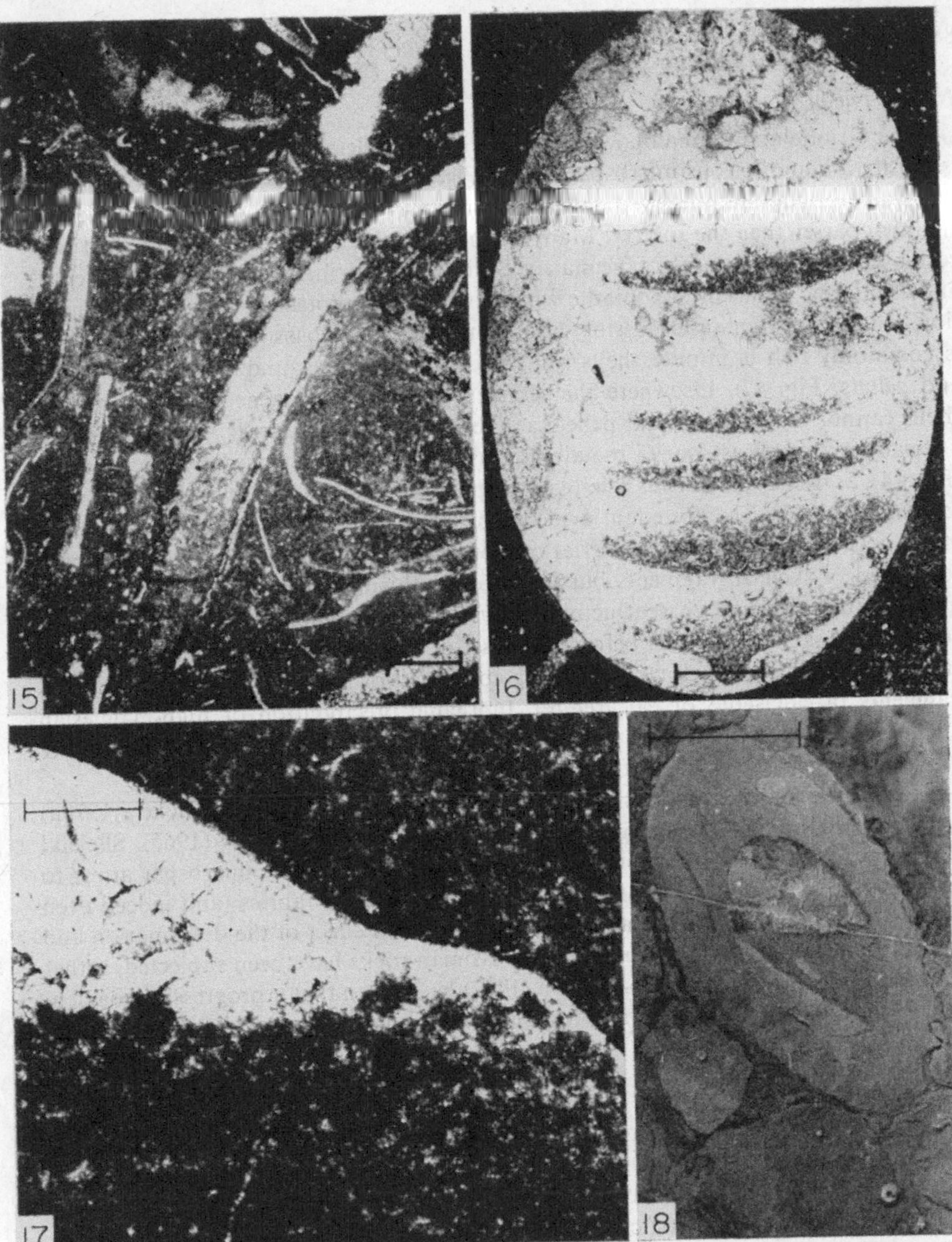

Fig. 15. Two single bivalve shells (centre and upper left) partly filled by internal sediment containing minute bivalved shells and Foraminifera, overlain by sparite druse—the whole constituting a geopetal structure and indicating early shell dissolution. Famennian (Supragriotte) Combe d'Izarne, Montagne Noire. Thin section; scale bar = 1 mm.

Fig. 16. Central part of a goniatite showing several lenses of internal sediment (with minute bivalved shells) which accumulated upon the septa. There is now no evidence of the septa. Famennian (Supragriotte) Coumiac, Montagne Noire. Thin section; scale bar = 1 mm.

Fig. 17. Pellets preserved below a bivalve geopetal structure. Upper Famennian (Supragriotte) Mont Peyroux, Montagne Noire. Thin section; scale bar = 0·5 mm.

Fig. 18. Goniatite from nodular limestone showing two periods of internal sedimentation, with some rotation of the goniatite in between. Lower Famennian (Vrai Griotte) Coumiac, Montagne Noire. Scale bar = 10 mm.

Neomorphism

The Griotte and Cephalopodenkalk are rich in skeletal debris, in a matrix which is mostly a homogeneous microsparite (e.g. Fig. 15). The finer skeletal grains, mostly of bivalve origin, are preserved as clear equigranular microspar of a similar size or slightly coarser than the matrix. Many of these bivalve fragments are only just discernible from the microspar of the matrix. Sediment occurring beneath shell 'umbrellas' and as internal sediment in goniatite whorls, where it is protected from any compaction and bioturbation (and possibly also from pore waters passing through the rock), is commonly rich in minute shells (microsparitized bivalves and ostracods, Fig. 16, and pellets, Fig. 17). Elsewhere the sediment appears homogeneous, these minute shells cannot be discerned and pellets are very rare indeed. Larger molluscan shells normally preserved as sparite pseudomorphs may also be replaced by microsparite (Fig. 14), as well as the fibrous calcite envelopes developed around styliolinids (Tucker & Kendall, 1973). The abundant evidence of microsparitization suggests that much of the matrix, where skeletal material cannot be discerned, may originally have been composed of skeletal fragments but that neomorphism has reduced them to microspar. Bioturbation and current activity may have contributed to mechanical breakdown of the skeletal material before the onset of diagenesis. The sediment then may originally have been much coarser, being composed of silt- or even sand-sized shell fragments (a calcisiltite or calcarenite). Inorganic precipitation of high-magnesian calcite, as reported from deep-water carbonates in the Mediterranean (Milliman & Müller, 1973), may also have contributed to the sediment.

Coarser neomorphic fabrics occur locally in the limestones (see Tucker, 1973a) and resemble the porphyroid aggrading neomorphic mosaic of Folk (1965). Skeletal fragments preserved as inclusions within these patches of coarse microspar attest to their replacement origin. The development of these coarser fabrics (and indeed even the microsparitization of skeletal material) could be an effect of the deformation and low-grade metamorphism to which these Devonian rocks have been subjected, rather than of late diagenetic processes. Brown (1972) has shown that a progressive coarsening of micritic limestones occurs in response to increasing low-grade metamorphism.

Origin of the griotte texture

The main feature of the Vrai Griotte is its nodularity (Figs 3, 4) and irregular argillaceous seams. The origin of the similar Alba Griotte in the Cantabrians has been much discussed (e.g. Kanis, 1956; Wagner, 1963; Frets, 1965) with theories mostly involving diagenetic or tectonic processes, although Nagtegaal (in Frets, 1965) considered synsedimentary solution an important factor. Similar nodular limestones occur in the Jurassic Adnet Limestone (Garrison & Fischer, 1969) and Ammonitico Rosso (Hollmann, 1962; Jenkyns, 1974).

An early diagenetic lithification of the nodules in the griotte is shown by the full-bodied nature of the ammonoids. Compaction after lithification, but before filling of voids by sparite druse, has resulted in cracking of some nodules.

There is generally a higher skeletal content within the nodules, than within the surrounding argillaceous limestone where shells are flattened or contorted. Where not affected by pressure solution, there is a gradual transition from the nodules to the surrounding clay and from limestone beds up into discrete nodules. Where nodules have formed around goniatites, there is commonly a sharp boundary with the shale, and the outer shell of the goniatites is absent. In many cases the outer shells of the goniatites have been replaced by haematite, which is concentrated in the argillaceous partings.

Small limestone lithoclasts (interpreted as solution relicts) do occur within nodules and nodular limestones. However, there is no evidence of large-scale 'Subsolution' (as envisaged by Hollmann (1964) for the similar Ammonitico Rosso) which could have produced the whole griotte texture.

Geopetal structures within goniatites do not vary more than 10° from each other, also suggesting that 'Subsolution' and reworking of the nodules has not occurred. Some rotation of goniatites is indicated by two distinct internal sediment generations occurring at different angles within the same goniatite shell (Fig. 18); but this probably took place on the sea floor.

Breccia horizons in the Griotte can be distinguished from the nodular griotte by their fabric and variable clast lithology.

Four processes have affected the Vrai Griotte: lithification, dissolution, compaction and pressure solution (the last two processes can have occurred both during diagenesis and tectonic deformation). Two main stages are probably involved in the formation of the griotte texture. (1) An early lithification of the carbonate sediment, with cementation taking place preferentially around and within ammonoids (this may have involved the migration of carbonate towards the ammonoids, leaving the surrounding sediment depleted in carbonate) and (2) compaction (and pressure solution) through overburden and tectonic pressure resulting in movement and squeezing of the argillaceous sediment between the indurated nodules (see discussion in Jenkyns, 1974). Tectonic pressures result in reorientation of nodules along the direction of cleavage. The transitional nature of limestone nodules into continuous limestone beds, and nodules into the argillaceous matrix, suggest it is the clay content and its later effects which determine the texture of the sediment.

Pressure solution

Pressure solution has considerably affected the Griotte and Cephalopodenkalk and three main stages can be recognized: two diagenetic phases of pressure solution, mostly producing stylolites and separated by a phase of vein development, and a third phase, coincident with the development of fracture cleavage, producing numerous solution stringers or flasers. The latter, truncating and displacing all earlier structures, are the sites of limestone solution.

The environment of deposition of the Griotte and Cephalopodenkalk

The stratigraphic position of the Cephalopodenkalk and Griotte (commonly following reef limestones) and the dominantly pelagic fauna indicate deposition at depths greater than a few tens of metres. Although the effects of submarine solution can be seen in these Devonian pelagic limestones, they are not pervasive and it is most unlikely that deposition took place below the calcite compensation depth (about 4500 m). Only in a few instances can more precise limits be placed upon the limestones, as for example when foraminiferal/algal nodules are present suggesting depths less than 200 m (Tucker & Kendall, 1973). Encrusting Foraminifera alone cannot be used as a bathymetric indicator since these are now known to occur down to depths of 5 km or more, commonly within manganese nodules (Wendt, 1974).

Of the sedimentary structures present in these deeper-water limestones, none are truly diagnostic of their depositional environment. Hardgrounds, sheet cracks, neptunian dykes and lithoclasts can occur in shallow-water and reef limestones, and manganese nodules can develop in freshwater. It is the pelagic aspect of the fauna which is the main criterion. The sedimentary structures are a consequence of extremely slow rates of sedimentation, early lithification and sea-floor solution; processes characteristic of the pelagic environment, but which can take place elsewhere. Since these processes operate at a wide range of depths, and pelagic faunas are by their very nature unrestricted, the determination of the depth of deposition relies on palaeogeographic, tectonic and stratigraphic considerations, or, if present, the occurrence of some benthos of limited bathymetry.

The sediments accumulated under conditions of slow deposition generally in two situations. (a) Upon submarine rises, formed by submerged reefs, volcanic extrusions and fault-blocks where terrigenous sediment was channelled into adjacent basinal areas (as in the Rhenohercynian Geosyncline) and (b) upon submerged platform and shelf areas during a general lull in terrigenous sedimentation (as in the Montagne Noire and Cantabrians). Both the Griotte and Cephalopodenkalk occur within stratigraphic sequences beginning with Lower Devonian shallow-water clastics, which were clearly deposited upon a continental crust of Lower Palaeozoic and Precambrian strata. For much of the Devonian, the Rhenohercynian Geosyncline was an area of tensional tectonic activity with the development of synsedimentary faults, fault-bounded troughs and volcanic ridges. An analogous modern situation is that of marginal basins situated behind island arcs in the Pacific, where extensional tectonic activity has led to the development of fault-bounded ridges and basins, accompanied by the extrusion of volcanics (e.g. Karig, 1970). There is at present, however, no evidence for the former existence of Devonian oceanic crust to the south of the Rhenohercynian Geosyncline (Krebs & Wachendorf, 1973). With the Griotte of the Montagne Noire, the situation is more analogous to a stable continental shelf (e.g. the Yucatan Shelf, Logan *et al.*, 1969) where there is no volcanic activity and only limited synsedimentary faulting.

ACKNOWLEDGMENTS

I am grateful to Dr Roland Goldring for advice and encouragement during this work, which constitutes part of a Ph.D. thesis at Reading University. Thanks are also extended to Professor Dieter Meischner and research students at Göttingen, Professor Wolfgang Krebs and Dr Alan Kendall for discussions and comments on Cephalopodenkalk. I acknowledge with thanks receipt of a Reading University Postgraduate Studentship (1968–71), expenses towards further fieldwork (1973) from University College, Cardiff, and a grant from the Royal Society to present this paper at the First European Geophysical Society meeting in Zürich (September 1973).

REFERENCES

BANDEL, K. (1972) Palökologie und Paläogeographie im Devon und Unterkarbon der Zentralen Karnischen Alpen. *Palaeontographica,* Abt. A, **141,** 1–117.

BANDEL, K. (1974) Deep-water limestones from the Devonian-Carboniferous of the Carnic Alps, Austria. In: *Pelagic Sediments: on Land and under the Sea* (Ed. by K. J. Hsü and H. C. Jenkyns). *Spec. Publs int. Ass. Sediment.* **1,** 93–115.

BATHURST, R.G.C. (1964) The replacement of aragonite by calcite in the molluscan shell wall. In: *Approaches to Paleoecology* (Ed. by J. Imbrie and N. Newell), pp. 357–376. John Wiley, New York.

BERNOULLI, D. (1972) North Atlantic and Mediterranean Mesozoic facies: a comparison. In: Initial Reports of the Deep Sea Drilling Project, Vol. XI (C. D. Hollister, J. I. Ewing *et al.*), pp. 801–871. U.S. Government Printing Office, Washington.

BIRD, J.M. & THEOKRITOFF, G. (1967) Mode of occurrence of fossils in the Taconic allochthon (abs). *Spec. Pap. geol. Soc. Am.* **101,** 248.

BOTTKE, H. (1965) Die exhalativ-sedimentären devonischen Roteisensteinlagerstätten des Ostsauerlandes. *Geol. Jb. Beih.* **63,** 1–147.

BOYER, F. (1964) Observations stratigraphiques et structurales sur le Dévonien de la région de Caunes-Minervois. *Bull. Carte géol. Fr.* **60,** 106–122.

BOYER, F., KRYLATOV, S., LE FEVRE, J. & STOPPEL, D. (1968) Le Dévonien Supérieur et la Limite Dévono-Carbonifère en Montagne Noire (France). *Bull. Cent. Rech. Pau* (S.N.P.A.), **2,** 5–33.

BROWN, P.R. (1972) Incipient metamorphic fabrics in some mud-supported carbonate rocks. *J. sedim. Petrol.* **42,** 841–847.

BUGGISCH, W. (1972) Zur Geologie und Geochemie der Kellwasserkalke und ihrer begleitenden Sedimente (Unteres Oberdevon). *Abh. hess. Landesamt. Bodenforsch.* **62,** 1–68.

ENOS, P. (1969) Cloridorme Formation, Middle Ordovician flysch northern Gaspé Peninsula, Quebec. *Spec. Pap. geol. Soc. Am.* **177,** 1–66.

FISCHER, A.G. & GARRISON, R.E. (1967) Carbonate lithification on the sea floor. *J. Geol.* **75,** 488–496.

FOLK, R.L. (1965) Some aspects of recrystallization in ancient limestones. In: *Dolomitization and Limestone Diagenesis* (Ed. by L. C. Pray and R. C. Murray). *Spec. Publs Soc. Econ. Paleont. Miner., Tulsa,* **13,** 14–48.

FRETS, D.C. (1965) The geology of the southern part of the Pisuerga Basin and the adjacent area of Santibanez de Resoba, Palencia, Spain. *Leid. geol. Meded.* **31,** 113–162

GARRISON, R.E. & FISCHER, A.G. (1969) Deep-water limestones and radiolarites of the Alpine Jurassic. In: *Depositional Environments in Carbonate Rocks, a Symposium* (Ed. by G. M. Friedman). *Spec. Publs Soc. Econ. Paleont. Miner., Tulsa,* **14,** 20–56.

GÈZE, B. (1949) Étude géologique de la Montagne Noire et des Cévennes meridionales. *Mém. Soc. géol. Fr.* **29,** 1–62.

HECKEL, P.H. (1972) Possible inorganic origin for *Stromatactis* in calcilutite mounds in the Tully Limestone, Devonian of New York. *J. sedim Petrol.* **42,** 7–18.

HOLLAND, C.H. (1971) Silurian faunal provinces? In: *Faunal Provinces in Space and Time* (Ed. by F. A. Middlemiss, P. F. Rawson and G. Newall). *Spec. Issue Geol. J.* **4,** 61–76. Seel House Press, Liverpool.

HOLLMANN, R. (1962) Über Subsolution und die "Knollenkalke" des Calcare Ammonitico Rosso Superiore im Monte Baldo (Malm; Norditalien). *Neues Jb. Geol. Paläont. Mh.* **1962,** 163–179.

HOLLMANN, R. (1964) Subsolutions-Fragmente (Zur Biostratinomie der Ammonoidea im Malm des Monte Baldo/Norditalien). *Neues Jb. Geol. Paläont. Abh.* **119,** 22–82.

JENKYNS, H.C. (1971) The genesis of condensed sequences in the Tethyan Jurassic. *Lethaia,* **4,** 327–352.

JENKYNS, H.C. (1974) Origin of red nodular limestones (Ammonitico Rosso, Knollenkalke) in the Mediterranean Jurassic: a diagenetic model. In: *Pelagic Sediments: on Land and under the Sea* (Ed. by K. J. Hsü and H. C. Jenkyns). *Spec. Publs int. Ass. Sediment.* **1,** 249–271.

KANIS, J. (1956) Geology of the Eastern Zone of the Sierra Del Brezo (Palencia, Spain). *Leid. geol. Meded.* **21,** 377–445.

KARIG, D.E. (1970) Ridges and basins of the Tonga-Kermadec Island Arc System. *J. geophys. Res.* **75,** 239–254.

KENDALL, A.C. & TUCKER, M.E. (1971) Radiaxial fibrous calcite as a replacement after syn-sedimentary cement. *Nature, Phys. Sci.* **232,** 162–263.

KENDALL, A.C. & TUCKER, M.E. (1973) Radiaxial fibrous calcite: a replacement after acicular carbonate. *Sedimentology,* **20,** 365–389.

KREBS, W. (1968) Reef development in the Devonian of the eastern Rhenish Slate Mountains, Germany. In: *International Symposium on the Devonian System* (Ed. by D. H. Oswald) II, pp. 295–306. Alberta Soc. Petrol. Geologists, Calgary.

KREBS, W. (1972) Facies and development of the Meggen Reef (Devonian, West Germany). *Geol. Rdsch.* **61**, 647–671.

KREBS, W. & WACHENDORF, H. (1973) Proterozoic-Paleozoic geosynclinal and orogenic evolution of Central Europe. *Bull. geol. Soc. Am.* **84**, 2611–2630.

LINDSTRÖM, M. (1963) Sedimentary folds and the development of limestone in an early Ordovician Sea. *Sedimentology,* **2**, 243–292.

LOGAN, B.W., HARDING, J.L., AHR, W.M., WILLIAMS, J.D. & SNEAD, R.G. (1969) Carbonate sediments and reefs, Yucatan Shelf, Mexico. *Mem. Am. Ass. Petrol. Geol.* **11,** 1–198.

MEER MOHR, C.G. VAN DER, & SCHREUDER, GAN. H. (1967) On the petrography of the Lancara Formation from the Sierra De La Filera (Spain). *Leid. geol. Meded.* **38,** 185–189.

MEISCHNER, D. (1968) Stratigraphische Gliederung des Kellerwaldes. *Notizbl. hess. Landesamt. Bodenforsch. Wiesbaden,* **96,** 18–30.

MEISCHNER, D. (1971) Clastic sedimentation in the Variscan Geosyncline east of the River Rhine. In: *Sedimentology of parts of Central Europe* (Ed. by G. Müller), pp. 9–43. *Guidebook,* VIII *Int. Sedim. Congr.*, Kramer, Frankfurt.

MILLIMAN, J.D. & MÜLLER, J. (1973) Precipitation and lithification of magnesian calcite in deep-sea sediments of the eastern Mediterranean Sea. *Sedimentology,* **20,** 29–45.

RABIEN, A. (1956) Zur Stratigraphie und Fazies des Obderdevons in der Waldecker Hauptmulde. *Abh. hess. Landesamt. Bodenforsch.* **16,** 1–83.

RIEDEL, W.R. (1963) The preserved record: paleontology of pelagic sediments. In: *The Sea* (Ed. by M. N. Hill), Vol. 3, pp. 866–887. Interscience, New York.

RODGERS, J. (1957) The distribution of marine carbonate sediments: a review. In: *Regional Aspects of Carbonate Deposition* (Ed. by R. J. Le Blanc and J. G. Breeding). *Spec. Publs Soc. Econ. Paleont. Miner., Tulsa,* **5,** 2–14.

RODGERS, J. (1968) The eastern edge of the North American Continent during the Cambrian and Early Ordovician. In: *Studies of Appalachian Geology, Northern and Maritime* (Ed. by E. Zen), pp. 141–149. Wiley-Interscience, New York.

SANDERS, J.E. & FRIEDMAN, G.M. (1967) Origin and occurrence of limestones. In: *Carbonate Rocks, Origin, occurrence and classification* (Ed. by C. V. Chilinger, H. J. Bissell and R. W. Fairbridge). *Developments in Sedimentology,* **9A,** 169–265. Elsevier, Amsterdam.

SCHMIDT, H. (1925) Schwellen- und Beckenfazies im ostrheinischen Paläozoikum. *Z. dt. geol. Ges.* **77,** 226–234.

SHEPARD, F.P. (1963) *Submarine Geology*, 2nd edn, pp. 557. Harper & Row, New York.

TUCKER, M.E. (1973a) Sedimentology and diagenesis of Devonian pelagic limestones (Cephalopodenkalk) and associated sediments of the Rhenohercynian Geosyncline, West Germany. *Neues Jb. Geol. Paläont. Abh.* **142,** 320–350.

TUCKER, M.E. (1973b) Ferromanganese nodules from the Devonian of the Montagne Noire (S. France) and West Germany. *Geol. Rdsch.* **62,** 138–153.

TUCKER, M.E. & KENDALL, A.C. (1973) The diagenesis and low-grade metamorphism of Devonian styliolinid-rich pelagic carbonates from West Germany: possible analogues of recent pteropod oozes. *J. sedim. Petrol.* **43,** 672–687.

VON GAERTNER, H.R. (1934) Die Eingliederung des ostalpinen Palaozoikums. *Z. dt. geol. Ges.* **86,** 241–265.

WAGEMAN, J.M., HILDE, T.W.C. & EMERY, K.O. (1970) Structural framework of East China Sea and Yellow Sea. *Bull. am. Ass. Petrol. Geol.* **54,** 1611–1643.

WAGNER, R.H. (1963) A general account of the Palaeozoic Rocks between the Rivers Porma and Bernesga (Léon, N.W. Spain). *Boln Inst. geol. min. Esp.* **74,** 1–159.

WENDT, J. (1969) Foraminiferen-'Riffe' im karnischen Hallstätter Kalk des Feuerkogels (Steiermark, Österreich). *Paläont. Z.* **43,** 177–193.

WENDT, J. (1971) Genese und fauna submariner sedimentärer Spaltenfullungen im mediterranen Jura. *Palaeontographica, Abt. A,* **136,** 122–192.

WENDT, J. (1974) Encrusting organisms in deep-sea manganese nodules. In: *Pelagic Sediments: on Land and under the Sea* (Ed. by K. J. Hsü and H. C. Jenkyns). *Spec. Publs int. Ass. Sediment.* **1,** 437–447.

ZAMARREÑO, I. (1972) Las Litofacies carbonatadas del Cambrico de la Zona Cantabrica (N.W. Espana) y su Distribucion Paleogeografica. *Trab. Geol.* **5,** 1–118.

Spec. Publs int. Ass. Sediment. (1974) **1,** 93–115

Deep-water limestones from the Devonian-Carboniferous of the Carnic Alps, Austria

KLAUS BANDEL

Institut für Paläontologie, Rhein. Friedrich-Wilhelms-Universität, Bonn, West Germany

ABSTRACT

The central Carnic Alps contain continuous sedimentary carbonate sequences that extend through the Devonian and lowermost part of the Carboniferous. Some sediments were deposited near mean sea level; other cephalopod-bearing facies were laid down in deeper water. Red mottled cephalopod limestones are continuous into the Upper Devonian. Grey cephalopod limestones, containing derived shallow-water calcarenites, occur in the Devonian and grade upwards into turbidite-free cephalopod limestones of Early Carboniferous age. These deep-water limestones are characterized by evidence of early lithification and solution, and by the presence of a characteristic fauna. Interpretation of their depositional depth is based on present-day compensation depths for aragonite and calcite: the original sediments apparently resembled modern pelagic oozes. Sedimentological and palaeontological criteria indicate a progressive deepening of the basinal areas accompanied by continuous growth of platform carbonates at the basin edge until Late Devonian time when the shallow-water areas subsided.

INTRODUCTION

The Carnic Alps contain essentially continuous sedimentary sequences extending through the whole Devonian into the Lower Carboniferous. The sections discussed in this paper occur in the central Carnic Alps, to the east and west of the Plöcken Pass of the Bundesland Kärnten in South Central Austria and small parts of the province Udine of Northern Italy (Fig. 1). Detailed descriptions of localities, lithologies and microfacies are given in Bandel (1972).

The sections on which this study is based are parts of an allochthonous structural unit which was subjected to tectonic movements in the Hercynian and Alpine orogenies. Broad palaeogeographic reconstructions can therefore only be made with certain reservations. Limestones are present throughout, and clastics are notably absent. From lowermost Devonian to the lower Upper Devonian (Frasnian) a shallow-water facies can be distinguished from a deeper-water facies. Turbidites, composed of material derived from these shallow-water areas, occur within the deeper-water facies.

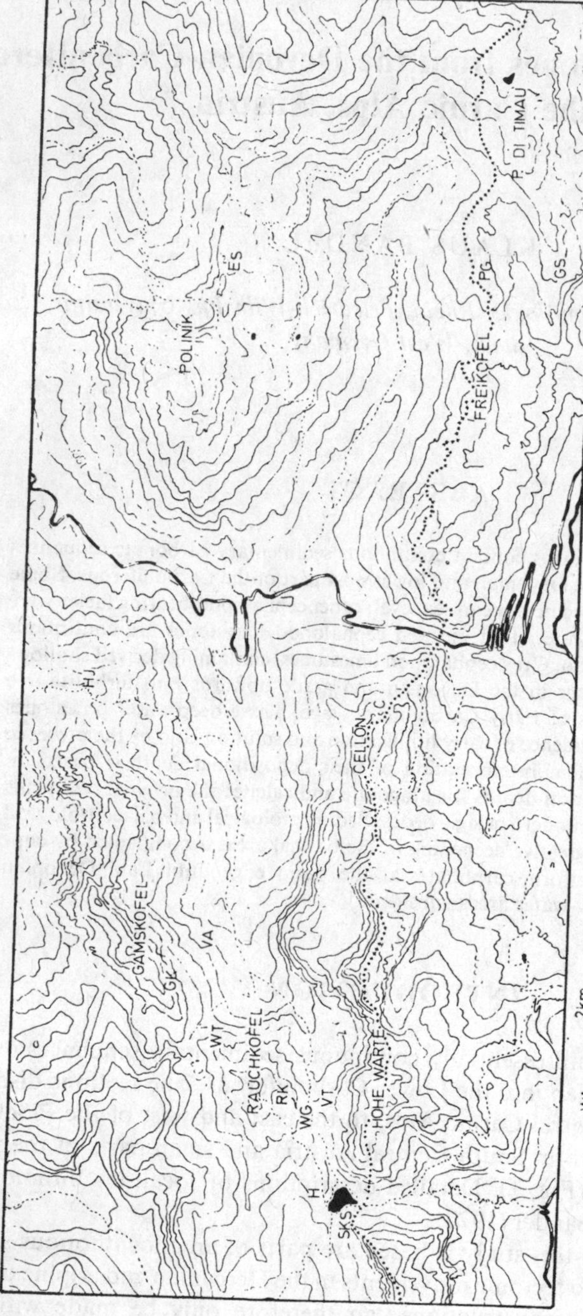

Fig. 1. Map of the area around the Plöcken Pass in the Central Carnic Alps. The location of each section is noted with its initial letters: SKS, Seekopfsockel; H, Hütte; WG, Wolayer Gletscher; VT, Valentin-Törl; RK, Rauchkofelboden; (Wolayer Lake sections): WT, Woderner Törl; CV, Cresta Verda; ES, Elferspitz; PG, Pal Grande; GS, Gamspitz.

Facies interpretations (Bandel, 1969, 1972) of the tectonically isolated 'Schuppen' can establish a palaeogeographic picture restricted to the central Carnic Alps. The deposits of deep-water origin are of two types, (a) pelagic limestones and (b) pelagic limestones with interbedded calcareous turbidites (redeposited or allodapic limestones of Meischner, 1964).

PELAGIC LIMESTONES WITH RARE REDEPOSITED BEDS

Description

Pelagic limestones associated with rare redeposited beds crop out at the localities of Seekopfsockel; Hütte; Wolayer-Gletscher; Valentintörl; Rauchkofelboden (Bandel, 1972). Stratigraphically, these Devonian pelagic limestones can be divided into five facies.

Fig. 2. Deep-water limestones mostly without redeposited beds at their outcrop at the Seekopf-Sockel. The Italian-Austrian border is situated at the small hut above the Wolayer Lake. The basal massive limestone bed is of pre-Devonian age. The tectonic unit is under- and overlain by Carboniferous shales and sandstones. The tectonic unit seen above contains mainly Lower Devonian shallow-water limestones.

Grauer Orthocerenkalk (grey limestone with abundant orthocone cephalopods)

The Silurian-Devonian boundary is situated within or at the base of well-bedded grey limestones characterized by the presence of numerous orthocone cephalopods which show a constant preferred orientation (pointed ends towards the south-south-west). The limestones are biomicrites with 5–60% biogenic material (mostly ostracod, tentaculitids, and Radiolaria). Although the sediment has been strongly bioturbated the original fine-stratification is preserved. Skeletal remains often show borings and evidence of bioturbation. Burrows are preserved in their original shape and have not been compacted (Fig. 12).

'*Roter Flaser- und Knollenkalk*' (red nodular limestone with and without shale interbeds)

The basal grey limestones slowly grade into red limestones. In the lower part of the sequence the red limestones consist of very regular nodular layers and thin continuous limestone beds with irregular upper and lower surfaces separated by red shales. The micritic nodules and beds are generally pink to light red and make up the bulk of the rock while the mudstone matrix is dark red. This matrix may include layers of abundant crinoid fragments, some small tabulate coral colonies, cephalopods and small angular micrite pebbles. In upper parts of the sequence shaly interbeds are absent but thick shaly partings follow bedding planes, while stylolites running at an angle to bedding planes show highly irregular indented sutures that contain only little insoluble residue. The limestones are biomicrites containing up to 30% biogenic material [mostly tentaculitids, Radiolaria, and ostracods (Fig. 11)]. The red colour of the limestones is caused by low concentrations of a haematite stain. A characteristic feature of these red limestones is the presence of rounded to angular intraclasts (up to 2 cm in diameter) which have a slightly lower insoluble residue than the surrounding limestone matrix. The boundaries of the clasts are generally perfectly sharp, mostly enriched with ferruginous minerals and commonly truncate fossils (Figs 4, 5). Stylolites may or may not form part of these surfaces. Many of these nodules are compound. Some of the clasts are coated with ferromanganese crusts similar to those described by Tucker (1973a, b, 1974) from Upper Devonian sediments of the Montagne Noire. Commonly these ferromanganese crusts show the development of colloform structures. Lattices of echinoderm skeletons, chambers of calcareous Foraminifera, intraseptal voids in tabulate corals and cephalopods, the interior of styliolinid shells, the early whorls of gastropods and borings in skeletal fragments are commonly filled with ferruginous material (Figs 6, 8). Skeletal fragments may also have a light external coating but massive concentric crusts with mushroom-like projections may also occur. Bioturbation has usually destroyed all internal stratification in these limestones but abundant geopetal structures in the fillings of burrows and the inside of fossil shells are present (Fig. 12). Only large skeletal fragments of cephalopods, continuous pieces of crinoid stems, and coral colonies are preserved parallel to the original bedding plane or in growth position. All burrow structures are preserved without any sign of compaction.

Some burrow systems are filled with many concentric layers of tentaculids, oriented with their long axes parallel to the length of the burrow. The inner cavity of many burrows is still preserved and now filled with sparry calcite. Burrows showing bends or sections parallel to bedding are often partly filled with finely laminated pelletal micrites.

'*Grauer Styliolinen-Flaserkalk*' (grey limestone with abundant tentaculitids)

In the uppermost Emsian the red limestones change their colour to grey. Ferruginous crusts around clasts and skeletal fragments are present but in the form of pyrite, barely visible in thin section and polished sample. (The crusts are best examined after heating the rock for an hour when the pyrite is oxidized.) Another difference from the red limestones is the occurrence of at least eight graded biosparites (up to 12 cm thick) in the Eifelian. These thin graded beds are strongly bioturbated and their internal structure has been destroyed. The thicker beds show grading with large skeletal remains and clasts in the lower part and a pelsparite in the upper part (Fig. 14).

'*Kalk mit phosphatischen Knollen*' (limestone with phosphatic nodules)

The grey Eifelian styliolinid limestones are overlain by thin beds of Late Devonian age, some of which are composed entirely of tentaculitid shells (Figs 9, 10). Givetian limestones are absent. Directly above the tentaculitid biosparites there occurs a bed with numerous large phosphatic clasts comprising up to 20% of the limestone. The angular phosphatic nodules, up to 10 cm long and 2 cm high, are composed of more than 90% phosphate, the remainder being calcitic skeletal fragments.

Each clast has a layered structure with a dark inner part and lighter outer part. Skeletal fragments in this bed are commonly coated with phosphate, arranged concentrically around small particles and upon the upper surfaces of large fragments. Sometimes such coated skeletal fragments are incorporated into large crusts but never seem to comprise its centre. As well as the inorganic phosphate, a large amount of biogenic phosphate is also present (conodonts, other phosphatic problematics, Arthodira plates, inarticulate brachiopod valves, fish teeth and scales).

'*Goniatiten-Flaserkalk*'

Of all macroscopic fossil remains goniatites are the most common in these limestones. The biomicrites contain up to 50% biogenic debris, mostly radiolarian. Ferruginous coatings and ferromanganese encrustations upon skeletal remains, as well as phosphatic crusts, occur at various horizons. Intraclasts are generally absent but irregular solution surfaces parallel to bedding planes are abundant and mostly coated with ferruginous crusts.

The extinction of pelagic tentaculitids that occurred in the late Frasnian is represented by a particular horizon (1 cm of sediment).

Discussion

'Grauer Orthocerenkalk' is of very variable thickness, at one section beginning at the Silurian-Devonian boundary (Seekopfsockel), elsewhere beginning in the Silurian (Hütte; Rauchkofelboden). Limestones of various compositions and ages are overlain by these grey pelagic sediments. There is no evidence of terrestrial conditions at this disconformity, suggesting that the stratigraphic break was not due to emersion. It is believed that submarine erosion and non-deposition were responsible for these stratigraphic breaks. Areas of erosion and non-deposition are not uncommon in present-day oceans, and strong currents have been recorded from all oceanic depths (Heezen, 1959). The Blake Plateau, for instance, situated at depths between 800 and 1400 m off the south-east coast of the USA is, for this reason, largely free of Recent and Pleistocene sediments.

Organic remains found in the Devonian limestones mainly belong to animals that swam in the open sea. Very stable depositional conditions are indicated by the fact that within different layers of these limestones the shells of orthocone cephalopods show an orientation of their pointed ends toward south-south-west. The same orientation of uncoiled cephalopod shells was measured at different sections on different bedding planes of Lower Devonian limestones of the overlying facies at various stratigraphical positions. These bedding planes belong to the same tectonic unit ('Schuppe') of the Rauchkofelboden-Valentingletscher sections. The extreme uniformity of orientation indicates considerable water depth and a continuity of conditions over many millions of years. These conditions are much more typical of deep-water environments than of those found in shallow water on shelves. Here the traces of larger differences in current directions should be expected in deposits spanning such a long time. The orientation of these large shells also provides evidence for current action within the depositional areas of these and overlying limestones. The lack of compaction in the grey limestones is proved by the preservation of burrow structures and thin fossil shells in their original shape. These suggest very early cementation for the grey limestones. Transitions with overlying facies are gradational. Red nodular limestones (Roter Flaser- und Knollenkalk) occurring in the Lower and Middle Devonian of the Carnic Alps are also developed in the Devonian of the Rheinisches Schiefergebirge, Frankenwald and Thuringia (Germany), and the Montagne Noire, France (Tucker, 1974). Similar lithologies occur in the Alpine Mesozoic (the Jurassic Adnet and Ammonitico Rosso facies and Triassic Hallstatt limestones). Hollmann's (1962) interpretation of the Upper Jurassic Calcare Ammonitico Rosso Superiore at Monte Baldo in North Italy, that is, that the facies formed primarily by submarine solution, can be applied to the Devonian red nodular limestones of the Carnic Alps. The nodules are thus interpreted as solution relicts of carbonate beds formed and subsequently destroyed on the sea floor. These nodules do not show any sign of concentric growth and can not be interpreted as concretions. They are discrete clast-like remnants of former beds which themselves were largely composed of angular solution clasts. The shale surrounding the clasts must at least partly be interpreted as an insoluble residue of limestone beds dissolved completely or to a great extent.

Fossil remains are enriched on some shale bedding planes and show strong evidence of dissolution, being relicts of beds that have been dissolved save for some small angular clasts. Nodular limestones grade into 'Flaserkalk' without a change in texture or composition. This means that here solution did not in general destroy limestone beds completely but left them mainly undissolved. Only undulating, millimetre-thick shale beds, occurring above hardground surfaces, may partly be diagenetically altered clays enriched by dissolution of calcite. The limestones are characterized by the presence of angular and rounded clasts, generally showing lighter colours than the surrounding matrix and which are themselves often compound. Clay mineral contents of these limestones (nodules included) are never higher than 6% if insoluble material enriched on styliolite seams is substracted. Cephalopod shells in general are poorly preserved due to dissolution, partly caused by pressure solution during stylolite formation. Remains of shells that have been redeposited are common. Fresh shells were apparently incorporated into carbonate oozes and filled with the soft mud. During non-deposition the lime-mud was indurated and subsequently dissolved. The fossil was thus exposed to solution which destroyed most of the shell but left the internal filling and septae to be preserved in newly deposited oozes.

There are now many records of lime-mud lithification in present-day oceans between depths of 200 and 3300 m (Bramlette, Faugn & Hurley, 1959; Milliman, 1967; Fischer & Garrison, 1967; Gevirtz & Friedman, 1966; Bartlett & Greggs, 1969; Milliman, Ross & Ku, 1969).

Many indurated Recent beds are commonly intensely bored by organisms and are associated with iron and manganese oxides which may either form a light coating over the surface of the lithified sediment or may take the form of nodules and encrustations. This shows that this type of cementation is active where deposition is virtually absent over a very long period of time. Very slow sedimentation and solution of calcium carbonate were the evidence on which Garrison & Fischer (1969) interpreted the Adnet limestone (Jurassic) as deposits formed in areas with depths up to 4100 m. Continuing sedimentation into progressively deeper water finally resulted in the deposition of radiolarites where calcium carbonate was dissolved (4500–5000 m). In the Devonian sections of the Carnic Alps radiolarian cherts occur in association with the red 'Flaserkalk' facies to the east of the sections discussed in this paper (i.e. at the Hohe Trieb, Schönlaub, 1969; at Findenig, Pölsler, 1969) and Monte Zermula.

The red colour of the limestone is due to disseminated haematite and is probably of diagenetic origin. It may have been derived from yellowish to light brown goethite that dehydrated rapidly during diagenesis into red haematite (Berner, 1969). Red colouration in the pelagic Devonian limestones of the Carnic Alps was interpreted by Brinkmann (1935) as typical for rapid subsidence and grey colours of the Kellerwand (shallow-water facies) were connected by him with low rates of subsidence. Normally traces of haematite in limestone are chemically difficult to detect, but in and on the rim of clasts as well as in and on skeletal fragments ferruginous material may be enriched (Figs 5, 6, 8). Hinze & Meischner (1968) described migration of iron-salts to the sedimentary surface of Recent sediments in the North Adriatic. pH and Eh conditions in the bottom muds, influenced by decomposing organic material and the oxygen consumption of burrowing organisms, cause migration of iron-salts towards the surface where oxidizing conditions prevail. A similar mechanism may have concentrated iron oxides on the outside of clasts and on solution surfaces in the limestone. Commonly, ferromanganese crusts on solution clasts show a development of colloform structures (Fig. 4) which were interpreted by Tucker (1973a, b, 1974) as diagenetic alterations of originally undulating crusts. Ferromanganese nodules, as described by Jenkyns (1970) from Jurassic beds, were not encountered, but Upper Devonian crusts from the Montagne Noire (Tucker, 1973a, b) are quite similar to those found in the pelagic limestones in the Carnic Alps from the lowermost Devonian into the Lower Carboniferous.

'Grauer Styliolinen-Flaserkalk' was deposited in similar conditions to the red limestones. The grey colour coincides with the occurrence of a few thin redeposited beds (see below).

After a period of non-deposition that lasted the whole Givetian and earliest Frasnian, thin layers consisting purely of tentaculitid shells formed locally (Figs 9, 10). The thin shells acted as nuclei around which acicular cement grew. Later on the acicular cement was transformed into large angular calcite crystals which contain ghosts of the original needle-like crystal growth. These layers resemble Recent cemented pteropod oozes (Milliman *et al.*, 1969; see also Tucker & Kendall, 1973).

The overlying bed with numerous phosphatic clasts and fossils formed under extremely slow rates of deposition. Phosphatic clasts and encrustations commonly follow hardgrounds recording long stratigraphical gaps (Jaanusson, 1961) or shorter ones (Bromley, 1967). In large nodules or crusts the basic layer grew on the surface of the sediment while the upper ones enlarged the crust in height and width. Material migrating within the sediment may well have concentrated phosphatic material around certain nuclei in concentric layers, but concentric coating of smaller fragments and one-sided coating of larger particles suggest precipitation of phosphate at the sediment-water interface. Crusts around skeletal fragments did not act as nuclei for the phosphatic growth of the larger clasts. Probably phosphatic material was delivered by up-welling carbon dioxide-rich waters.

At the Findenig, to the east of the sections considered here, 4 m of Givetian slump deposits occur (Pölsler, 1969), while at the Hohe Trieb, the Givetian is represented by 8 m of graded limestone within a sequence of radiolarites (Schönlaub, 1969). At the Poludnik, further east still, the Givetian is absent or very thin (Skala, 1969).

Sediments of the upper Frasnian are of particular interest and often contain lithologically curious layers (Buggisch, 1972). Buggisch studied the Kellwasserkalk in many parts of the Hercynian Geosyncline and found a very widespread occurrence of limestones rich in conodonts, tentaculitids, ostracods, goniatites, orthocone cephalopods, and fish remains (especially of pelagic Arthrodira which at that time were distributed world-wide). These fossils also characterize the 'Kalk mit phosphatischen Krusten' from the Carnic Alps. However, within the Kellwasser limestones, bioturbation is absent and there are no benthonic organisms, while in the Frasnian limestones of the Carnic Alps, corals, gastropods, bivalves, and inarticulate brachiopods are present.

With continuous sedimentation from the latest Frasnian into the late Famennian 'Goniatitenkalk', the record of deposition at these sections ends, but it is probable that pelagic sedimentation continued into the Early Carboniferous.

PELAGIC LIMESTONES WITH COMMON REDEPOSITED BEDS

Description

Devonian pelagic limestones associated with limestone turbidites occur to the west of the Plöckenpass (Woderner-Törl, Valentin-Alm, Cellon, Cresta di Colinetta) and to the east (Freikofel, Gamsspitz, Pal Grande, Pizzo di Timau, Elferspitz). Details are given in Bandel (1972).

'*Grauer Styliolinen-Flaserkalk mit turbiditischen Einlagerungen*' (Grey tentaculitid limestone with redeposited beds)

Grey pelagic limestones interleaved with graded, coarse-grained beds range from the Siegenian (Cellon) to the middle Eifelian (Woderner-Törl, Cellon, Freikofel). Gradations with 'roter Flaserkalk' are observable in the Lower Devonian of the Elferspitz. Gradations to 'grauer Orthocerenkalk' are present in the lower Emsian of the Woderner-Törl and the Siegenian of the Cellon.

The grey pelagic limestones are biomicrites containing between 10 and 30% biogenic material (mostly Radiolaria and tentaculitids). Apart from their colour the limestones are very similar to the 'roter Flaserkalk' and are particularly rich in intraclasts measuring up to 2 cm in diameter, giving the rock a vaguely nodular appearance. Iron sulphide coatings occur around clasts and skeletal fragments. Internal stratification has been completely destroyed by bioturbation. In upper Emsian and lower Eifelian deposits intraclasts are less common but irregular solution surfaces are found in great numbers. Authochtonous domed-shaped coral colonies, with diameters up to 20 cm, are found attached to the hardground surfaces.

Fig. 3. Deep-water limestones containing redeposited beds at their outcrop at the Cellon, seen from the Polinik.

The grey, 1–200 cm thick redeposited beds, rich in bitumen and poor in insoluble residue (below 1%, mostly consisting of authigenic quartz crystals), are distributed at random in these sections. They are composed of coarse-grained material, mostly echinoderm fragments, and exhibit good graded bedding. Other beds are composed of more diversified skeletal remains and may contain a basal layer of clasts, of the same type as those in the interbedded pelagic micrites. Clasts, large fragments of corals, stromatolitic algal remains, stromatoporoids, thick-shelled brachiopods, bryozoan colonies and other organic fragments disturb the grading at the base of a unit, while grading in upper parts is clearly visible. All these beds have been invaded by boring and burrowing organisms from above. Bioturbation penetrated to a maximum depth of 10 cm with increased intensity in the uppermost centimetres. Beds thinner than 10 cm are bioturbated to such an extent that they can only be detected with difficulty; those thinner than a few centimetres often only survived as coarse-grained burrow fills.

'*Lithoklastkalk*'

From the middle Eifelian to the late Frasnian only graded beds, one on top of the other, were deposited in these sections (Woderner-Törl, Cellon, Freikofel, Pizzo di Timau). Gradational beds from the lower basic tentaculitid limestones to the 'Lithoklastkalk' show higher density of graded beds with progressively fewer micrites between them. First biomicrites disappear, later pelmicrites and in some Givetian and lowermost Frasnian sections (Cellon, Freikofel) a climax is reached when even pelsparite disappears. From there on the tendency is reversed toward more fine-grained intermissions. In the late Frasnian a change in facies took place, accompanied by a change of colour from grey to red.

Skeletal fragments in the 'Lithoklastkalk' commonly have micritic envelopes. Composition of these beds is similar to that of the graded limestones interbedded with the tentaculitid limestones, apart from a higher content of pellets. The lithoclast limestones are bio-pelsparites with 10–50% pellets and 1–70% biogenic material of which 0–70% are echinoderms and 1–10% calcispheres. A characteristic feature of these limestones is the presence of large lithoclasts found at the base of almost every one of these 20–200 cm-thick graded beds. The basement for a bed is always formed by an irregular surface which is also the top of the graded bed below. Usually the lower part of a bed is rich in echinoderm (crinoid) fragments and large clasts which may be rounded (if smaller than 2 cm in diameter) or angular. The angular clasts are up to 6 cm thick, up to 50 cm wide and long, and never show any preferred orientation in the basal layers of these graded beds. Most clasts are bored parallel to the short sides. The borings may penetrate the whole thickness of a clast or continue only 1–2 cm into it. Fillings of these tunnels often show geopetal structures which are in accordance with stratification of the sequence. The internal structure of the clasts consists of stratified layers of pelmicrites or pelsparites parallel to the long sides and vertical to the thin edges. Except for bioturbation no deformation of stratification within clasts was noted. The lower part of each lithoclast bed is not graded because of the variable size of components, but higher up, echinoderm fragments grade into more pelletoidal sediment containing numerous calcispheres and Radiolaria. In the upper Eifelian some irregular bedding surfaces at the top of graded layers, and lithoclasts at the base of the next graded layer, are coated with phosphatic crusts and impregnations. Phosphatic material surrounds clasts, and clasts within clasts, as a thin rim; it forms many layered nodules, fills spaces within skeletal fragments, boreholes in clasts, and coats hardground surfaces (Fig. 15).

At the Woderner-Törl section spaces between lithoclasts at the base of a graded layer are filled with numerous valves of inarticulate brachiopods.

'*Bunter Radiolarienkalk*' (vari-coloured limestone with numerous Radiolaria)

In the upper Frasnian and the lowermost Famennian biomicrites with 5–10% biogenic material are interleaved with graded beds (mostly up to about 15 cm in thickness).

In the micrites, apart from characteristic radiolarians, numerous calcispheres, echinoderms, ostracods, small gastropods, cephalopods and trilobites are present. The limestone colours are variable (grey to light red and pink and brown) and clasts of these colours (and black also) occur at the base of the interbedded graded limestones. Biogenic material, clasts and hardgrounds commonly show ferruginous impregnations and may be encrusted with ferromanganese oxide. The graded pelsparites usually overlie an irregular surface with many sharp and rounded projections and tunnels leading down into the micrite (Fig. 14). Famennian pelagic micrites contain numerous strongly bioturbated pelsparite and pelmicrite beds, always less than 2 cm thick. Some of these limestones ('Flaserkalk mit *Stromatactis* Strukturen', Pal Grande) contain numerous bedding-parallel cavities (up to 2 cm high) filled with fibrous calcite that grew from all walls. Most of the cavities are shaped like flattened cushions or tubes. They have a smooth floor consisting of laminated micrite and an irregular roof. Often they form large bodies along bedding planes and resemble sheet cracks. However, even if they cover more than a square metre, they do not form a continuous cavity but rather a network of communicating tubes and cushions with pillars and walls between them. In smaller cavities a continuation of sediment-filled uncompacted burrow tubes with fibrous calcite-filled cavities can be seen. Commonly, vertical tunnels may lead from one oval cavity into another situated above, or horizontal tunnels may connect neighbouring cavities.

'*Goniatiten-Flaserkalk*'

Some beds of the micrites below this facies contain numerous goniatites (Cresta di Colinetta) and can be considered as gradational from 'Radiolarien-Flaserkalk' to the 'Goniatiten-Flaserkalk'. In the uppermost Famennian (Pal Grande) and the Lower Carboniferous (Cresta Verde, Colinetta di sopra) limestones characterized by goniatites are similar to those of the Fammenian from Wolayer See.

With deposits of limestone in this facies the carbonate sedimentation that characterized the pre-orogenic Palaeozoic history of the Carnic Alps came to an end.

In the goniatite limestones biogenic material, with a predominance of radiolarians, comprises up to 50% of the rock. In other respects the fauna is similar to that of the limestones below except for the presence of more benthonic organisms such as corals, Foraminifera and crinoids. Many biogenic fragments are encrusted with phosphate and in other layers with ferromanganese oxides. Clasts are rare, and graded biosparites are absent. Irregular bedding-plane surfaces are commonly encrusted and impregnated by ferruginous material. Adjacent skeletal remains may be strongly corroded or well preserved. Irregular hardground surfaces (relief up to 4 cm) are overlain by argillaceous layers. From such surfaces, tunnels with clay-coated walls extend vertically into the sediment.

Discussion

'Grauer Orthocerenkalk', similar to that of the Wolayer lake sections but interleaved with redeposited beds, is characteristic of the Lower Devonian up to the base of the 'Grauer Styliolinen-Flaserkalk'. The environment of deposition of the styliolinid limestone facies must have been very similar to that of the 'roter Flaserkalk' apart from the frequent incursions of turbidity currents.

The fine-grained texture of Liassic grey pelagic limestones was considered by Jurgan (1969) to indicate deposition in a quiet milieu without current activity. In the autumn of 1969, on a field trip in the northern Adriatic Sea, D. Meischner demonstrated to us that in Recent sediments originally coarse-grained carbonate material may be transformed into fine lime muds by the activity of dissolving and sediment-ingesting organisms. This mechanism of biological maceration of skeletal particles was active in the Devonian pelagic limestones of the Carnic Alps. Strong bioturbation, destroying an original fine stratification, is indicative of the existence of a rich endobenthonic fauna. Broken skeletal fragments and shells are bored to such an extent that only a thin outer skin has survived; this is a further indication of the action of biological grinding of skeletal remains to yield the fine-grained texture that characterizes these beds (Figs 6, 7).

Fabricius (1961) explained the difference in colour between Liassic red and grey faces in pelagic limestones as an effect of different sedimentation rates, grey colours produced under high rates of sedimentation, when complete decomposition of organic matter could not take place. The abundance of solution clasts suspended in the micritic matrix of limestones found between redeposited beds in the Lower Devonian sections of the Carnic Alps suggests an alternative interpretation. History of deposition started

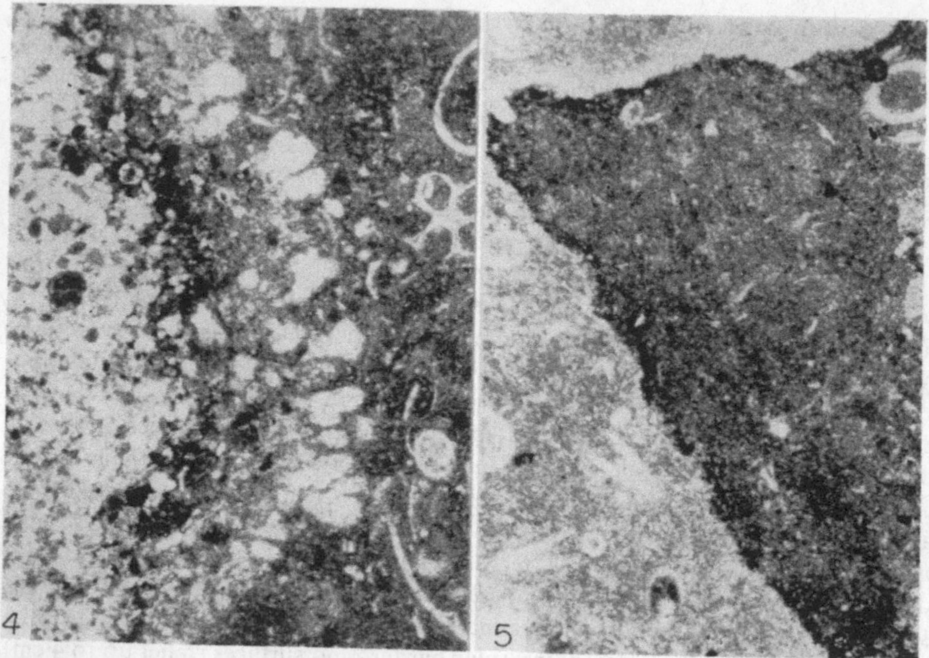

Fig. 4. Solution clast with club-like overgrowths. ×30. Thin section. (Frasnian of Hütte Section.)

Fig. 5. Solution clast with ferruginous coating. ×30. Thin section. (Emsian of Seekopfsockel Section.)

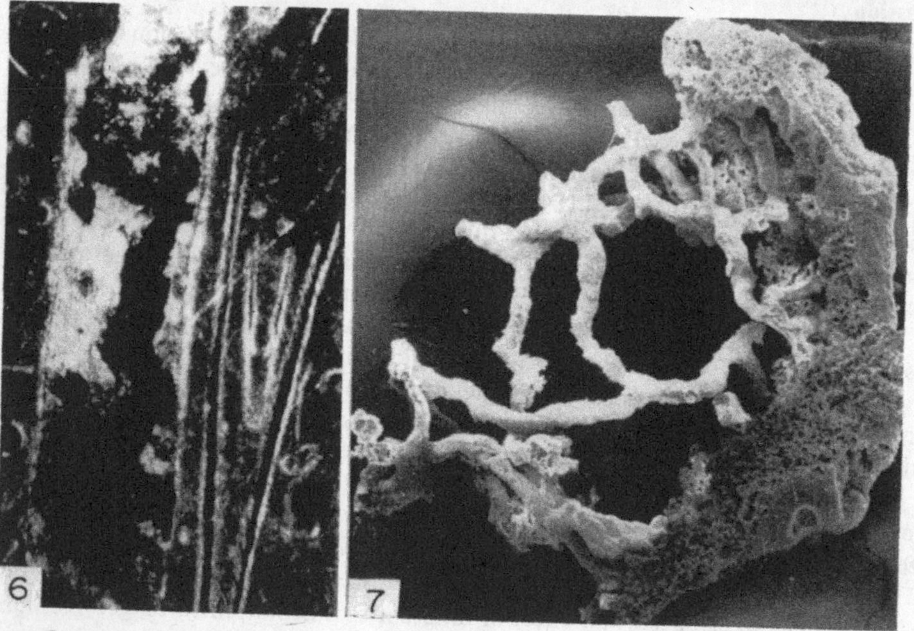

Fig. 6. Thick fragment of shell bored and filled with ferruginous material. ×30. Thin section. (Emsian of Wolayer Törl.)

Fig. 7. Dissolved crinoid ossicle with clay-filled borings. ×45. Gold-coated and photographed with the stereoscan electron microscope. (Lower Carboniferous of Cresta Verde.)

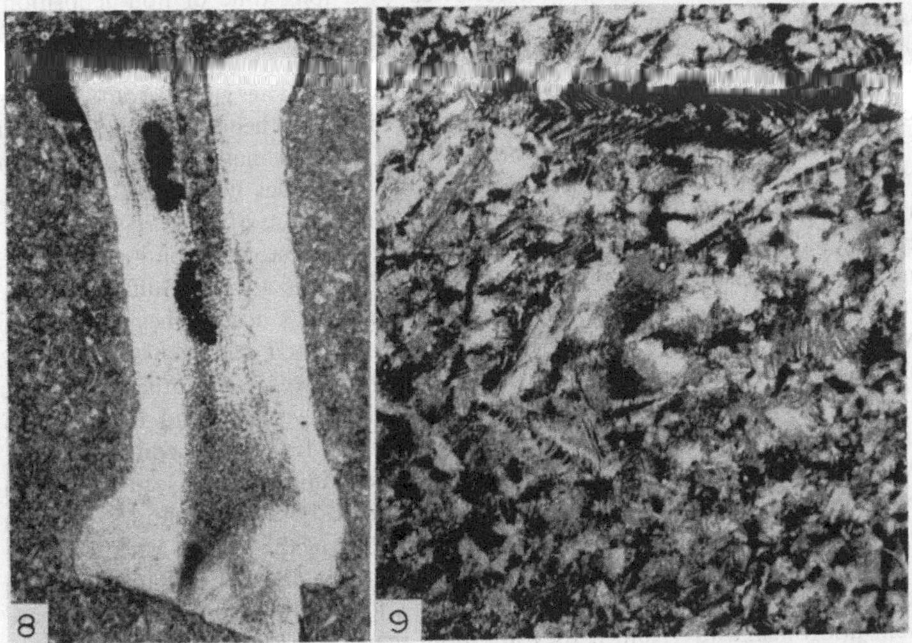

Fig. 8. Crinoid fragment with ferruginous lattice-fillings and ferromanganese crust on the outside. ×30. Thin section. (Eifelian, Hütte Section.)

Fig. 9. Pure tentaculitid biosparite consisting of the shells and their overgrowths. ×30. Thin section. (Frasnian, Wolayer Gletscher.)

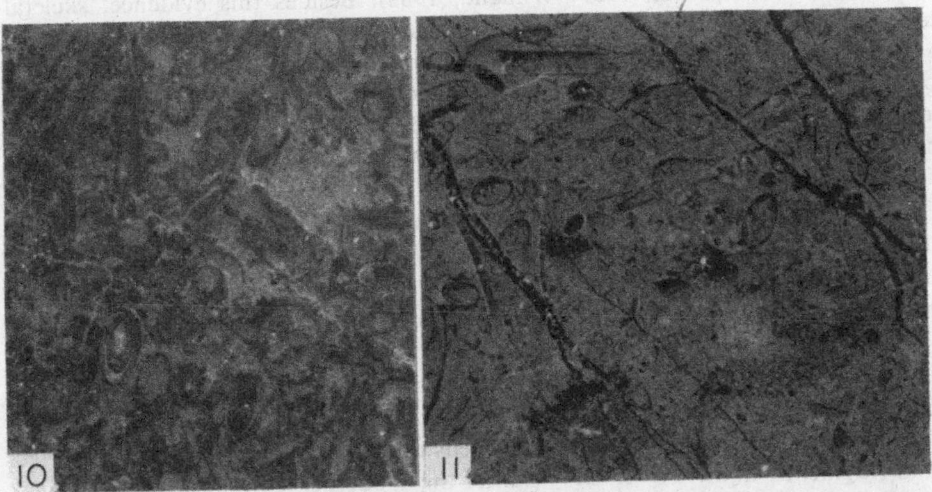

Fig. 10. Tentaculitid biosparite. Negative print from acetate peel. ×30. (Frasnian, Wolayer Gletscher.)

Fig. 11. Typical 'Styliolinen-Flaserkalk' biomicrite. ×30. Acetate peel. (Eifelian, Seekopfsockel Section).

here with sedimentation of lime ooze followed by a long time of non-deposition. Subsequently, bioturbation, induration and dissolution followed each other and left only solution clasts, preserved when new deposition began. This left enough time for oxidation and decomposition of all organic compounds. The grey colours of this facies must therefore be related to the influence of redeposited beds, and it is suggested that their deposition resulted in an enrichment of organic material to the bottom sediments. This would cause a transformation of iron-oxides into sulphides which, along with the bituminous material, provide the grey pigment of these rocks.

Recent sediments in many respects resembling the limestone turbidites from the Carnic Alps have been described by Davies (1968) from the abyssal Gulf of Mexico. In these deep-sea oozes 2–120 cm thick, light carbonate beds occur. They are graded, show an erosive base, often contain micritic pellets at their basal layers, are only bioturbated in the uppermost section, and are separated from each other by pelagic oozes. The thickness and composition of these beds is variable and they contain dominantly benthonic shallow-water remains mixed with few pelagic components.

All these characteristics can be seen in graded beds in Devonian sections. It can therefore be concluded that in their origin they parallel those beds found in modern deposits of the Gulf of Mexico. The skeletal fragments of the graded beds from the Gulf of Mexico derive from the shallow-water area of the Campeche Bank and were carried in suspension currents to abyssal depths. Close to their source area they are frequent and thick, away from it they are more scarce and thin. Turbidity currents move fast and far and cover distances up to 500 miles in the Gulf of Mexico.

The origin of skeletal material in the redeposited beds of the Carnic Alps is easily recognized. They are remains of benthonic organisms clearly of shallow-water origin as found in the same tectonic unit of the central Carnic Alps at the same stratigraphic position (Fig. 13). Indices of derivation from shallow-water environments are skeletal materials with micritic envelopes which suggest water depth no greater than 60 m (Swinchatt, 1969; Bathurst, 1967; Winland, 1968). Besides this evidence, skeletal remains of many organisms characteristic of reef, fore-reef and back-reef deposits can be determined from redeposited beds (e.g. crinoids, corals, algal crusts, oncolites, calcispheres, brachiopods and bryozoa). The composition of resedimented beds reflects the history of shallow-water deposition, where in Early Devonian times crinoid limestones were more common whereas in Middle Devonian times pelletoid sediments were more abundant (Bandel, 1969, 1972).

The erosive force of a turbidity current diminishes with the distance from the source area, a fact that is well documented in pelagic deposits from the Carnic Alps. Strong erosive forces were active at the deposition of the 'Lithoklastkalk' where the largest components are found at the base of graded beds. Strong currents must still have been active during the deposition of the Lower Devonian graded beds at the Cellon section because they are rather coarse-grained at the base with unsorted skeletal fragments and large lithoclasts. At the Lower Devonian Gamspitz section, however, lithoclasts are not as common and graded beds are mainly composed of well-sorted crinoid fragments. In the Lower Devonian of the Woderner-Törl section redeposited beds make up less of the rock column than in previously mentioned sections. Emplacement of Eifelian resedimented beds occurred furthest away from the source area at the Wolayer Lake sections. Here the admixture of pelagic components is prominent but currents were still strong enough to pick up solution clasts and skeletal fragments from the sea bottom and incorporate them into the basal part of resedimented beds.

Fig. 12. Burrow filled at the bottom with fine-grained sediment and at the top with calcite spar. ×30. Thin section. (Emsian, Cellon Section.)

Fig. 13. Redeposited bed whose basal part shows mainly echinoderm remains with their overgrowths forming most of the sparite matrix. ×30. Thin section. (Eifelian, Freikofel Section.)

Meischner (1964) introduced the term 'allodapic' for those limestones emplaced by turbidity currents. The beds found in the Carnic Alps fit into his definition of 'allodapic' limestones if it is somewhat broadened. Differences between Meischner's examples of the Kulmplattenkalk and Devonian Flinz from the Rheinisches Schiefergebirge can be seen in that with the limestones here described: (a) the area of deposition was not poorly supplied with oxygen but showed oxidizing conditions; (b) the authochtonous pelagic (basinal) sediments interbedded with the turbidites were carbonate oozes, and not clays; (c) the basinal deposits were not soft but largely cemented and thus inhibited erosion by the turbidity currents. Sea-floor solution produced indurated pelagic limestone clasts, which were subsequently incorporated into the turbidity-current deposits. (d) Rich benthonic faunas were present in the carbonate oozes of the Carnic Alps, which led to bioturbation of the upper parts of the limestone turbidites.

Similar graded limestone beds composed of shallow-water carbonate particles have been described by Winterer & Murphy (1960) from the Silurian of Nevada, Carozzi & Frost (1966) from the Silurian of Indiana, Tucker (1969) from the Devonian of Cornwall, Garrison & Fischer (1969), Jurgan (1969), and Flügel & Pölsler (1965) from the Jurassic of the northern Alps.

The upper parts of graded beds of the 'Lithoklastkalk' facies were exposed to sea water until emplacement of a new bed by turbidity currents. During these very long time intervals hardgrounds were formed due to lithification, solution and boring. This early diagenetic cementation in the sense of Zankl (1969) created a bottom sediment with more than one indurated layer, whose upper level consisted of more or less loose slabs. These could be picked up by the following turbidity current and incorporated into the lower part of the newly deposited bed. Water had access from almost

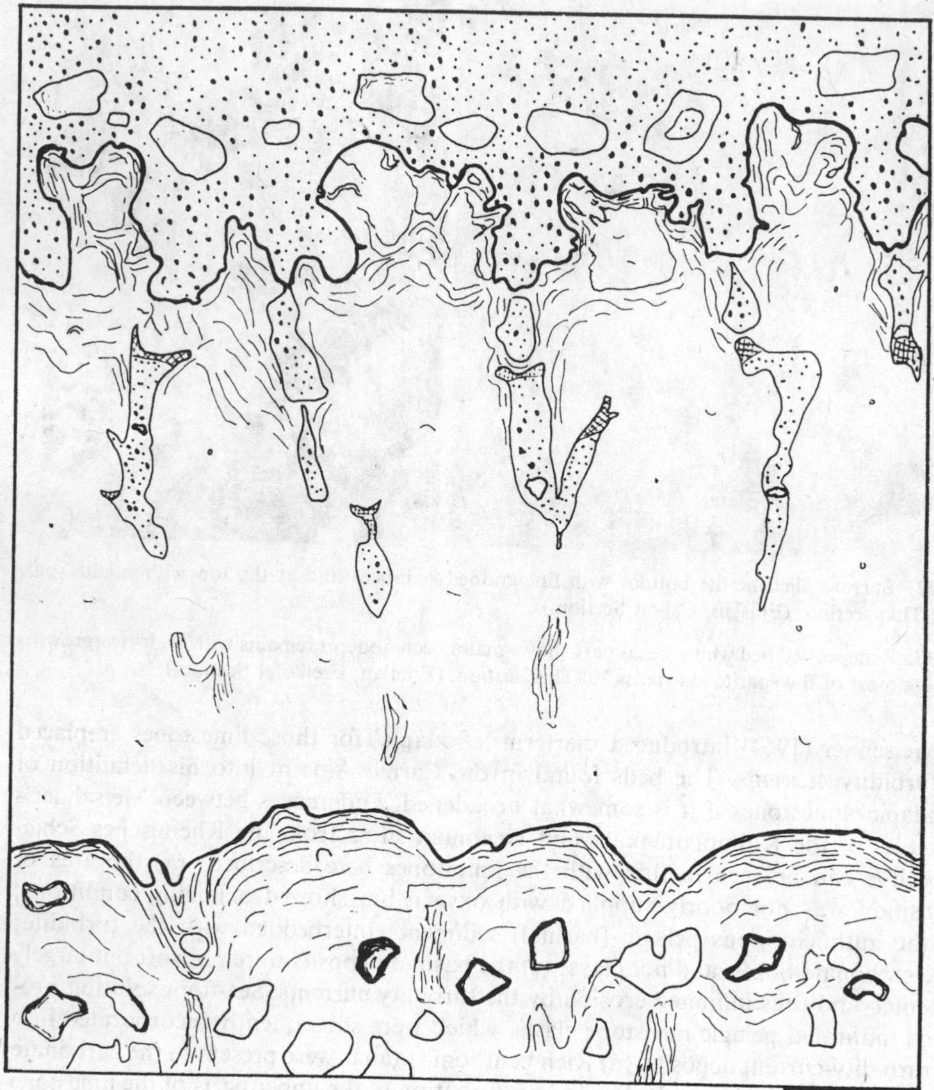

Fig. 14. Hardground surface preserved by redeposited bed. The upper surface is enriched with ferruginous material. The bed below is bioturbated and contains no solution clasts. The base is formed by another hardground surface encrusted with more ferruginous material than the upper one. Here the limestone also contains abundant solution clasts with thick encrustations of ferruginous material. Scale at the base = 1 cm.

all sides to the surface of these slabs as documented by boring and encrustation with phosphatic material. Sometimes solution and cementation continued to such an extent that the cemented upper slab layer itself was composed of earlier solution fragments (Fig. 15). Clay minerals and ferruginous materials were not enriched on the solution surfaces because (as analysis shows) only up to 0·2% of the rock consists of insolubles with most of it in the form of authigenic quartz crystals. In this respect

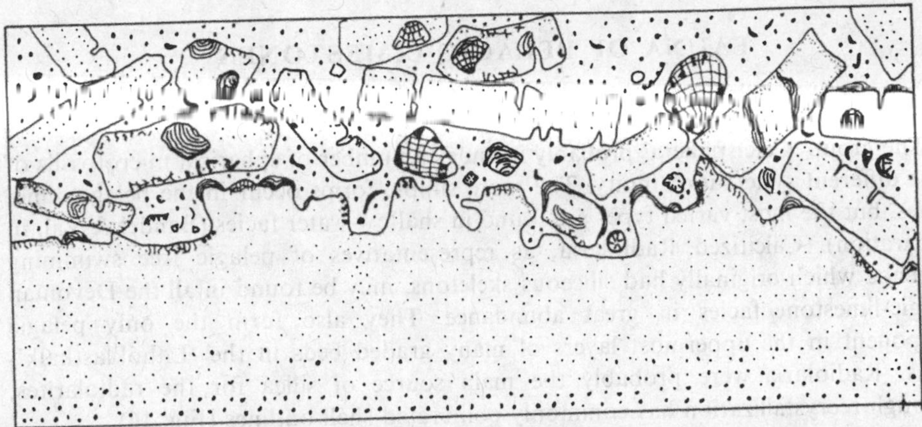

Fig. 15. Base and top of a redeposited bed in the 'Lithoklastkalk' sequence. The angular clasts are bored and encrusted by phosphatic material. Borings and encrustations are also seen on the hard-ground surface and in and on coral and stromatoporoid fragments. Many clasts are compound. Scale at the lower right side = 1 cm.

the contents of insolubles reflect the nature of the source area. Rocks from shallow-water environments contain 0·02–0·6% insolubles comprising mainly quartz crystals.

The continuation of very strong solution and the gradation into micritic pelagic sediment in the stratigraphic column in the Emsian and the Frasnian together with the interleaving radiolarites in sections similar to these in the east (Schönlaub, 1969) indicate formation of the 'Lithoklastkalk' in deep water. However, scattered colonies of stromatoporoids and tabulate corals attached to hardground surfaces may be counted as contrary evidence (H. Flügel, personal communication).

In the upper Frasnian a rapid change in the type of deposition took place, which was directly linked with the submergence of the shallow-water platform which had been the source area for the redeposited beds. From then on resedimented layers became of minor importance. However, quite a lot of thin biosparitic beds are found within the pelagic uppermost Frasnian and lowermost Famennian. The source area of these beds must have been further away at that time than the carbonate platforms active until lower Frasnian times, because the redeposited beds are thinner and show in their composition some admixture of pelagic components. The micrites between the resedimented layers compare well with the Lower Devonian 'roter Flaserkalk' in being composed mainly of differently coloured solution clasts in a matrix slightly richer in clay minerals. Therefore these sediments must be considered as deep-water deposits. Radiolarites within limestones of this composition and age are developed at the Hohe Trieb.

Limestone deposits in the Carnic Alps are continuous into the Lower Carboniferous and may only grade into radiolarites east of these sections. In small tectonically dislocated outcrops, Lower Carboniferous limestones, probably deposited on the former shallow-water platform, give way to 'Goniatiten-Flaserkalk' that shows all features so far described for pelagic limestones (solution-surfaces, rich pelagic fauna, ferruginous crusts). This indicates a subsidence after Frasnian times into considerable depth.

FAUNA OF PELAGIC LIMESTONES

Planktonic organisms

The group 'Calcisphaera' probably includes a number of spherical microfossils of quite different systematic affinity. The more simple forms occur in the pelagic limestones, but the most varied types are found in shallow-water facies (Bandel & Vai, in preparation). Calcitized Radiolaria, as representatives of pelagic free swimming Protozoa which originally had siliceous skeletons, may be found in all the Devonian pelagic-limestone facies in great abundance. They also form the only pelagic component in the uppermost layers of many graded beds in the 'Lithoklastkalk'-facies. Radiolaria were probably the main source of silica for the radiolarites, although recrystallization has commonly obliterated shell outlines (Fig. 16).

By analogy with modern forms, some of the small gastropods (few mm length) may also have had a free-swimming mode of life. The valves of most larval bivalves of today do not generally mineralize except near the end of their larval life. Therefore the small fossil bivalve shells found probably belonged to the benthos. Cephalopods are represented by orthocone shells in the Lower Devonian, orthocone and coiled shells in the middle Devonian and mainly coiled shells in the Upper Devonian. The 'Grauer Orthocerenkalk' is especially rich in orthocone cephalopods, which are also commonly found in the 'Roter Flaserkalk'. Here shell lengths of 1 m were seen. Usually goniatite shell diameters measure only up to a few centimetres and only rare specimens measure over 10 cm. Most of the orthocone nautiloids characteristic of this facies possessed smooth slender shells.

Smooth and ribbed small tentaculitids (*Styliolina*, *Nowakia* and others) were an important constituent of the pelagic deposits until their extinction in the upper Frasnian. They probably lived like the recent pteropods as free-swimming organisms in the high seas, and accumulations of their tests formed a sediment very similar to indurated modern pteropod oozes (Figs 9, 10; see also Tucker & Kendall, 1973). Their extinction must have produced a great change in marine planktonic fauna because no equivalent faunal element seems to have taken their place in the Late Devonian and Early Carboniferous. Ostracods with ornamented valves (the Entomozoidea) appeared in the Late Devonian and were probably planktonic.

Vertebrate remains in the form of fish teeth, scales, fins, spines and skeletal plates are found in all stratigraphic situations in the sections. Remains of Arthrodira are especially common in the Frasnian phosphate nodule layer. At that time remains of Arthrodira are found on a world-wide scale indicating a pelagic way of life for these fishes. Selachian teeth occur from the Famennian onwards, indicating common occurrence of pelagic sharks since that time.

A planktonic way of life probably has to be considered for the unknown animal that carried conodonts as skeletal elements, otherwise they would not be of such value to stratigraphy (for Carnic Alps, see Bandel & Schönlaub, in preparation).

Benthonic organisms

Shells of multicameral Foraminifera with calcareous shells are common but never abundant. Simple agglutinating Foraminifera are present in great numbers and can be used for correlation (Bandel, 1972). They are interpreted as representatives of the

Fig. 16. Scanning electron micrographs of organisms typical for Palaeozoic deep-water limestones. In the upper line from left to right: Radiolaria (× 120), detail of surface (× 300), agglutinating Foraminifera (*Lagenammina*) from the Lower and Middle Devonian (× 102), *Hyperammina* as found in the whole stratigraphial section (× 120). Organisms shown in the middle line all represent agglutinating Foraminifera: *Thurammina* from the whole stratigraphical section (× 60), *Psammosphaera* from the whole stratigraphical section (× 60), *Hyperammina* from the Frasnian to the Lower Carboniferous, *Astrammina* from the whole Devonian. In the lower line: *Ammodiscus* (Foraminifera from the Middle Devonian and Frasnian), *Hyperammina* from Upper Devonian and Lower Carboniferous (× 60), inarticulate branchiopod (cf. *Torynelasma*) from the Lower Devonian to Eifelian (× 120), and apex of another brachiopod (× 240).

vagile and sessile benthos. Free-moving forms lived in the uppermost levels of the substrate and perhaps accompanied burrowing animals further into the sediment. Others encrusted solution clasts and skeletal fragments lying on the sea floor.

Colonies of stromatoporoids and tabulate corals found suitable bottom substrates for settlement even in deeper waters when hardground surfaces were available. Here they settled in single colonies living scattered randomly on the sea floor at great distances from one another. Single polyps of rugose corals (*Syringoxon, Enterolasma*) occur. Some species of these small rugose corals show a very wide stratigraphic range indicating a very uniform living habitat for a long time. They have been recorded in similar Devonian pelagic limestones from the Bosporus to North Spain (Kullmann, personal communication).

Small bivalves and gastropods settled on the bottom of the sea during the whole depositional history of these limestones. As in Recent deep-water oozes they may represent a small-sized fauna with many species.

Inarticulate brachiopods with conical shapes (cf. *Torynelasma*) were common up to the Eifelian (Fig. 16), and *Lingula*-shaped forms are frequent in the whole section, particularly so in the Frasnian.

The trilobites are mostly smooth forms belonging to the Proetida family and the free-swimming spinose varieties are missing. The trilobites present had large eyes and apparently roamed across the sediment. Ostracods are one of the major biogenic benthonic components of these rocks. A rich silicified fauna was found in the lowermost to middle Devonian limestones of the Wolayer-lake sections (Bandel & Becker, in preparation). Very little species change was noted for that time which again indicates very stable environmental conditions for these many millions of years.

Crinoids must have settled on the bottom substrate in a random fashion similar to the corals, but with large distances between single animals. Sometimes large parts of the skeleton of one crinoid are found still connected, but usually single fragments are scattered about, probably as a result of bioturbation.

About 50% of the larger skeletal fragments in the pelagic limestones are bored (Figs 6–8). The range in size and shape of the borings suggests that a variety of organisms were responsible, either making tubes only a fraction of a millimetre across, or up to 1 cm wide. Communicating tunnels of equal width and bottle-shaped holes with only one exit reflect the extremes in shape. It is also likely that many different animal groups have been involved in the bioturbation which is prevalent throughout the pelagic limestones.

RECONSTRUCTION OF THE DEPOSITIONAL ENVIRONMENT

Far from the shallow-water platform purely pelagic deposits accumulated at depths between 200 and 4000 m. Periods of carbonate deposition were followed by times when sedimentation was virtually at a stand-still. Rich benthonic life destroyed all stratification and infaunal organisms produced open burrow systems that aided cementation during non-depositional intervals. After induration, bottom water rich in carbon dioxide and undersaturated with calcium carbonate led to sediment dissolution, aided by boring organisms.

Relicts of dissolved beds were coated by ferruginous crusts and incorporated into the succeeding sediments during renewed deposition; or, if non-deposition continued, were rendered into a clay residue. Hardgrounds were colonized by randomly distributed encrusting Foraminifera, corals and crinoids. Foraminifera, thin-shelled gastropods and bivalves, ostracods and trilobites lived on the sea floor. Single rugose corals and brachiopods formed the sessile benthos of soft substrates. The shells of organisms living in the upper layers of the sea upon their death formed a major part of the sediment, but were partly ground to small unrecognizable particles by the activity of the benthos.

Close to the shallow-water carbonate platform, periodic turbidity currents derived from these banks covered the sea-floor under a layer of sediments. These coarse-grained beds contained much more organic material than the extremely slowly deposited pelagic muds. Benthonic animals only occupied the top few centimetres of turbidites so that organic material was preserved, which altered previously formed ferruginous oxides into iron sulphides.

In upper Eifelian times, sedimentation of fine-grained pelagic oozes ceased, owing to the effects of current activity and dissolution of slowly deposited carbonate material. Only thick beds of shallow-water carbonate brought into this environment by turbidity currents withstood dissolution. Thickness changes from the Cellon (140 m) to Freikofel (94 m) to Woderner-Törl (34 m) to Wolayer Lake (less than 1 m) reflect distances from the source area. Cold up-welling waters undersaturated with calcium carbonate caused strong dissolution of the upper parts of graded turbidite beds. These beds were cemented in layers up to 6 cm thick and then broken up later along cracks with the aid of boring organisms. Soft layers between indurated ones were more easily dissolved leaving great solid slabs on the bottom which were colonized on all sides by boring organisms, colonial stromatoporoids and tabulate corals. These slabs were sometimes encrusted with phosphatic material, possibly supplied by currents. Endobenthonic animals, with the exception of boring organisms, did not find a very suitable substrate for life.

With the disappearance of reefs on the shallow-water platform in the late Frasnian, slow subsidence of the platform occurred and from this time turbidity currents became less important. During most of the Famennian, shallow-water material still contributed to deposition to some extent but supply of this material ceased during the latest Famennian. Furthermore, in the Wolayer Lake area from uppermost Frasnian time, very slowly deposited deep-water carbonate oozes, similar to those described from the Lower Devonian, had been laid down. By latest Devonian and earliest Carboniferous times this type of sedimentation had spread across all depositional environments in the Carnic Alps.

ACKNOWLEDGMENTS

I should like to thank Mrs Manuela Shamin-Vanagtmael and Mr Uwe Marr for looking through the manuscript. Special care was taken by Dr Maurice Tucker and Dr Hugh Jenkyns in refining style and English language.

REFERENCES

Bandel, K. (1969) Feinstratigraphische und biofazielle Untersuchungen unterdevonischer Kalke am Fuß der Seewarte (Wolayer See, zentrale Karnische Alpen). *Jb. geol. Bundesanst., Wien*, **112**, 197–234.

BANDEL, K. (1972) Palökologie und Paläogeographie im Devon und Unter-Karbon der zentralen Karnischen Alpen. *Palaeontographica Abt. A,* **141,** 1–117.

BARTLETT, G.A. & GREGGS, R.G. (1969) Carbonate sediments: oriented lithified samples from the North Atlantic. *Science,* **116,** 741–742.

BATHURST, R.G.C. (1967) Depth indicators in sedimentary carbonates. *Mar. Geol.* **5,** 447–471.

BERNER, R.A. (1969) Goethite stability and the origin of red beds. *Geochim. cosmochim. Acta,* **33,** 267–273.

BRAMLETTE, M.N., FAUGHN, J.L. & HURLEY, R.J. (1959) Anomalous sediment deposition on the flank of Eniwetok Atoll. *Bull. geol. Soc. Am.* **70,** 1549–1552.

BRINKMANN, R. (1935) Über Rotfärbung in marinen Sedimenten. *Geol. Rdsch.* **26,** 124–127.

BROMLEY, R.G. (1967) Marine phosphorites as depth indicators. *Mar. Geol.* **5,** 503–509.

BUGGISCH, W. (1972) Zur Geologie und Geochemie der Kellwasser Kalke und ihrer begleitenden Sedimente (Unteres-Oberdevon). *Abh. hess L.-Amt. Bodenforsch.* **62,** 1–68.

CAROZZI, A.V. & FROST, S.H. (1966) Turbidites in dolomitized flank beds of Niagaran (Silurian) Reef, Lapel, Indiana. *J. sedim. Petrol.* **36,** 563–573.

DAVIES, D.K. (1968) Carbonate turbidites, Gulf of Mexico. *J. sedim. Petrol.* **38,** 1100–1109.

FABRICIUS, F. (1961) Faziesentwicklung an der Trias/Jura-Wende in den mittleren Nördlichen Kalkalpen. *Z. dt. geol. Ges.* **113,** 311–319.

FISCHER, A.G. & GARRISON, R.E. (1967) Carbonate lithification on the sea floor. *J. Geol.* **75,** 488–496.

FLÜGEL, H. & PÖLSLER, P. (1965) Lithogenetische Analyse der Barmstein-Kalkbank B2 nordwestlich von St. Koloman bei Hallein (Tithonium, Salzburg). *Neues Jb. Geol. Paläont. Mh.* **1965,** 513–527.

GARRISON, R.E. & FISCHER, A.G. (1969) Deep-water limestones and radiolarites of the Alpine Jurassic. In: *Depositional Environments in Carbonate Rocks, a Symposium* (Ed. by G. M. Friedman). *Spec. Publs Soc. Econ. Paleont. Miner., Tulsa,* **14,** 20–56.

GEVIRTZ, J.L. & FRIEDMAN, G.M. (1966) Deep-sea carbonate sediments of the Red Sea and their implications on marine lithification. *J. sedim. Petrol.* **36,** 143–151.

HEEZEN, B.C. (1959) Deep-sea erosion and unconformities. *J. Geol.* **67,** 713–714.

HINZE, C. & MEISCHNER, D. (1968) Gibt es rezente Rot-Sedimente in der Adria? *Mar. Geol.* **6,** 53–71.

HOLLMANN, R. (1962) Über Subsolution und die "Knollenkalke" des Calcare Ammonitico Rosso Superiore am Monte Baldo (Malm; Norditalien). *Neues Jb. Geol. Paläont. Mh.* **1962,** 163–179

JAANUSSON, V. (1961) Discontinuity surfaces in limestones. *Bull. geol. Inst. Uppsala,* **14,** 221–241.

JENKYNS, H.C. (1970) Fossil manganese nodules from the West Sicilian Jurassic. *Eclog. geol. Helv.* **63,** 741–774.

JURGAN, H. (1969) Sedimentologie des Lias der Berchtesgadener Kalkalpen. *Geol. Rdsch.* **58,** 464–501.

MEISCHNER, K.D. (1964) Allodapische Kalke, Turbidite in Riff-nahen Sedimentations-Becken. In: *Developments in Sedimentology,* Vol. 3 (Ed. by A. H. Bouma and A. Brouwer), 156–191. Elsevier, Amsterdam.

MILLIMAN, J.D. (1967) Carbonate sedimentation on Hogsty Reef, a Bahamian atoll *J. sedim. Petrol.* **37,** 658–676.

MILLIMAN, J.D., ROSS, D.A. & KU, T.L. (1969) Precipitation and lithification of deep-sea carbonates in the Red Sea. *J. sedim. Petrol.* **39,** 724–736.

PÖLSLER, P. (1969) Stratigraphie und Tektonik im Nordabfall des Findenigkofels (Silur bis Karbon; Karnische Alpen, Österreich). *Jb. Geol. Bundesanst.* **112,** 355–398.

SCHÖNLAUB, H.P. (1969) Das Paläozoikum zwischen Bischofalm und Hohem Trieb. (Zentrale Karnische Alpen). *Jb. Geol. Bundersanst.* **112,** 265–320.

SKALA, W. (1969) Ein Beitrag zur Geologie und Stratigraphie der Gipfelregion des Poludnig (Karnische Alpen, Österreich). *Jb. Geol. Bundesanst.* **112,** 235–264.

SWINCHATT, J.P. (1969) Algal boring: a possible depth indicator in carbonate rocks and sediments. *Bull. geol. Soc. Am.* **80,** 1391–1396.

TUCKER, M.E. (1969) Crinoidal turbidites from the Devonian of Cornwall and their palaeogeographic significance. *Sedimentology,* **13,** 281–290.

TUCKER, M.E. (1973a) Ferromanganese nodules from the Devonian of the Montagne Noire (S. France) and West Germany. *Geol. Rdsch.* **62,** 137–153.

TUCKER, M.E. (1973b) Sedimentology and diagenesis of Devonian pelagic limestones (Cephalopodenkalk) and associated sediments of the Rhenohercynian Geosyncline, West Germany. *Neues Jb. Geol. Paläont. Abb.* **142,** 320–350.

TUCKER, M.E. (1974) Sedimentology of Palaeozoic pelagic limestones: the Devonian Griotte (Southern France) and Cephalopodenkalk (Germany). In: *Pelagic Sediments: on Land and under the Sea* (Ed. by K. J. Hsü and H. C. Jenkyns). *Spec. Publs int. Ass. Sediment.* **1,** 71–92.

TUCKER, M.E. & KENDALL, A.C. (1973) The diagenesis and low-grade metamorphism of Devonian styliolinid-rich pelagic carbonates from West Germany: possible analogues of Recent pteropod oozes. *J. sedim. Petrol.* **43,** 672–687.

WINLAND, H.D. (1968) The role of high Mg calcite in the preservation of micrite envelopes and textural features of aragonite sediments. *J. sedim. Petrol.* **38,** 1320–1325.

WINTERER, E.L. & MURPHY, M.A. (1960) Silurian reef complex and associated facies, Central Nevada. *J. geol.* **68,** 117–139.

ZANKL, H. (1969) Structural and textural evidence of early lithification in fine-grained carbonate rocks. *Sedimentology,* **12,** 241–256.

Spec. Publs int. Ass. Sediment. (1974) **1,** 117–148

The pelagic ooze-chalk limestone transition and its implications for marine stratigraphy*

SEYMOUR O. SCHLANGER *and* ROBERT G. DOUGLAS

Department of Earth Sciences, University of California, Riverside, California, and Department of Geology, Case Western Reserve University, Cleveland, Ohio, U.S.A.

ABSTRACT

Recovery of long sequences of cores, at Deep Sea Drilling Project sites, from Recent to Upper Jurassic pelagic ooze-chalk-limestone sections has shown that in general lithification increases with age and depth of burial. However, the relationship between degree of lithification and depth of burial in any core is not a direct one. A diagenetic model is presented that accounts for the major reduction in porosity and foraminiferal content with depth and age and the development of cement and overgrowth on those microfossils which are not dissolved. The primary diagenetic mechanism functions through the solution of less stable, very small, calcite crystals such as make up small coccolith elements and the walls of Foraminifera, and reprecipitation of calcite upon large crystals such as make up discoasters and large coccoliths. The concomitant decrease in surface energy probably provides the driving force for this process. The variation in the degree of cementation of ooze-chalk-limestone sequences when plotted as a function of depth is ascribed to initial variations in the diagenetic potential of the sediments as they are buried. The diagenetic potential of a sediment is defined as the length of the diagenetic pathway the sediment has left to traverse before it becomes a crystalline aggregate. In marine acoustistratigraphy, the concept of the diagenetic potential relates seismic reflectors to original intrastratal differences in microfossil content. Seismic reflectors in this context therefore record palaeo-oceanographic events such as changes in the calcite compensation depth, surface water temperature, plankton productivity and glacio-eustatic sea level.

INTRODUCTION

The success of the Deep Sea Drilling Project (DSDP) in recovering cores from thick sequences of strata of Recent to Jurassic age comprising oozes, chalks and limestones made up almost entirely of the remains of calcareous plankton has accelerated the study of the diagenesis of oceanic pelagic carbonates. Many chapters in the Initial Reports of the Deep Sea Drilling Project and dozens of articles in journals

*Contribution No. IGPP–UCR–73–43, Institute of Geophysics and Planetary Physics, University of California, Riverside.

have appeared covering all aspects of the diagenesis of pelagic carbonates. A brief review of the results of Legs I through IX was given by Davies & Supko (1973). Much data and many observations on the variation with depth of burial and age, of (1) physical properties—density, porosity, and sonic velocity—; (2) fossil preservation and (3) sediment textures have been published (e.g. Gealy, 1971; Moberly & Heath, 1971; Roth & Thierstein, 1972; Wise & Kelts, 1972; Douglas, 1973a, b; Bukry, 1973; Wise, 1973). One of the results of the Deep Sea Drilling Project has been the finding that in relatively pure pelagic carbonate sections there is a general progression from ooze to chalk to limestone with increasing depth of burial and age of the sediment. The exact relationship between lithification, age and depth of burial is, however, far from clear.

In this paper we develop a diagenetic model for the ooze-chalk-limestone transition using data on the physical properties, fossil preservation and textures and the geological setting of DSDP Site 167 (Schlanger *et al.*, 1973). At this Site, on the crest of the Magellan Rise in the central North Pacific, the drilling bit penetrated to what was then a record sub-bottom depth of 1185 m, 1172 of which were in a purely pelagic sequence of Quaternary to Berriasian-Tithonian (Cretaceous) strata dominated by carbonates (Figs 1, 5). From the diagenetic model we have developed the concept of a 'Diagenetic Potential' (Table 2); we have then applied this concept to the problem of the development of acoustic reflectors in the thick sections of pelagic carbonates deposited in the equatorial Pacific Basin since the period of widespread chert development in Middle to Late Eocene time.

TERMINOLOGY

As used in this paper

Lithification refers to the total effect of all diagenetic processes that serve to convert a mass of loose grains into a coherent rock; these include but are not limited to gravitational compaction, grain interpenetration and cementation.

Cementation refers to the single process in which calcite, precipitated from an intergranular solution, connects and binds grains into a rigid framework.

As stated in the lithological nomenclature conventions of the Deep Sea Drilling Project (see Winterer, Ewing *et al.*, 1973, pp. 9–10) 'Oozes have little strength and are readily deformed under the finger or the broad blade of a spatula. Chalks are partly indurated oozes; they are friable limestones that are readily deformed under the fingernail or the edge of a spatula blade. Chalks more indurated than that are simply termed limestones.' This nomenclature for oozes, chalks and limestones is followed in the present paper.

A DIAGENETIC MODEL FOR THE OOZE-CHALK-LIMESTONE TRANSITION

Any diagenetic model must be based on observations and set into a reasonable geological framework. In developing the model (pictured in Fig. 6 and discussed below) we considered the following attributes of pelagic oozes—using the carbonate sequence on Magellan Rise as an example.

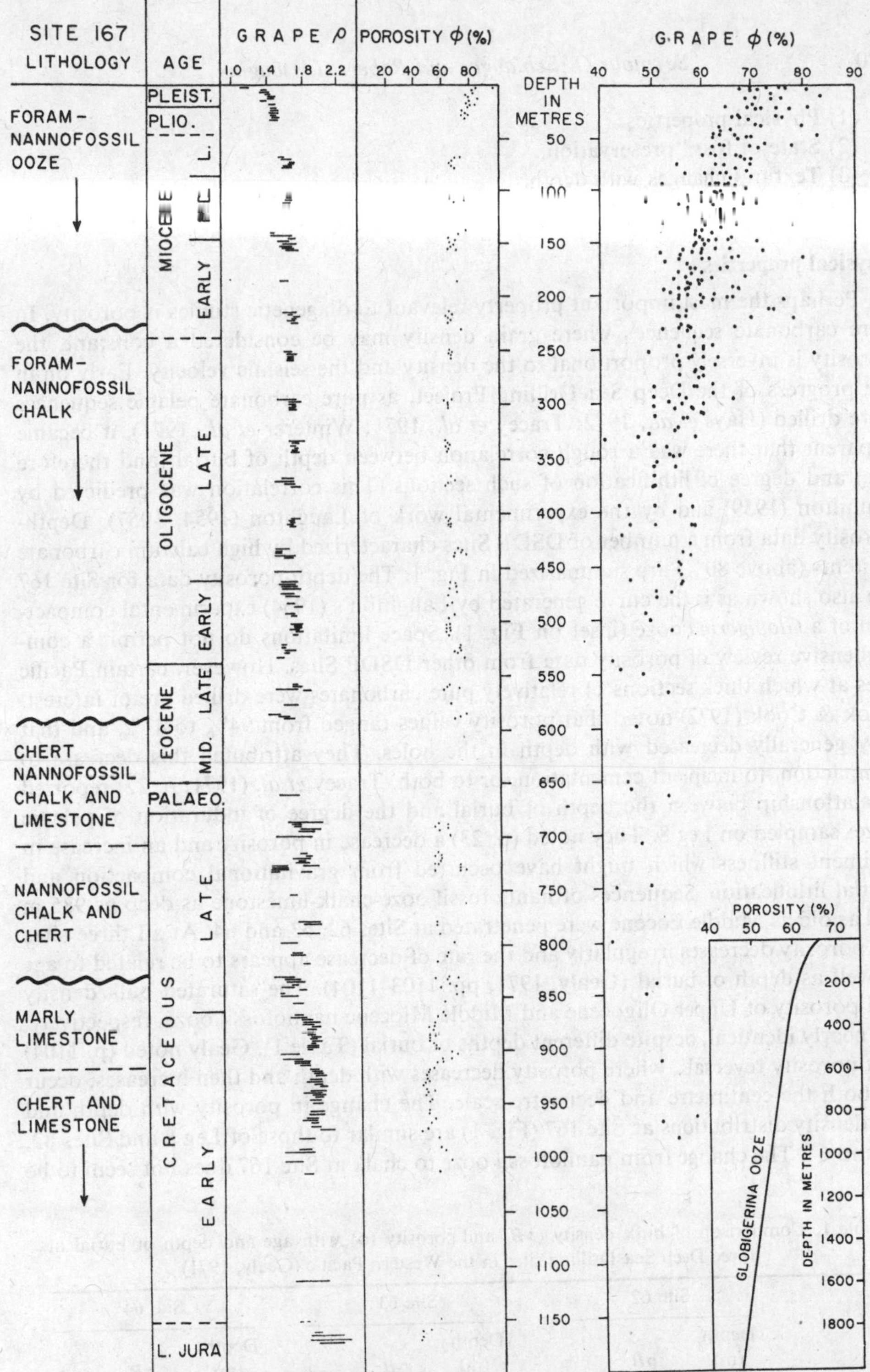

Fig. 1. Porosity-depth relations in pelagic carbonates. Plots to the left of the depth scale are GRAPE density and porosity data for DSDP Site 167 (Winterer, Ewing *et al.*, 1973). The points plotted to the right of the scale are based on averaged porosity data from DSDP Sites, 62, 63, 64, 71, 77, 78, 158, 161, 214, 216, 217. These sections are all characterized by high $CaCO_3$ (80% or higher) content. Inset is from Hamilton (1959) based on experimental data on the compaction of a *Globigerina* ooze containing 54% $CaCO_3$. Note the rapid decrease of porosity with depth in the upper 100–200 m.

(1) Physical properties.
(2) State of fossil preservation.
(3) Textural changes with depth.
(4) The geological setting and sedimentological history.

Physical properties

Perhaps the most important property relevant to diagenetic studies is porosity. In pure carbonate sequences where grain density may be considered a constant, the porosity is inversely proportional to the density and the seismic velocity. Early on in the progress of the Deep Sea Drilling Project, as pure carbonate pelagic sequences were drilled (Hays *et al.*, 1972; Tracey *et al.*, 1971; Winterer *et al.*, 1971), it became apparent that there was a rough correlation between depth of burial (and therefore age) and degree of lithification of such sections. This correlation was predicted by Hamilton (1959) and by the experimental work of Laughton (1954, 1957). Depth-porosity data from a number of DSDP Sites characterized by high calcium carbonate contents (above 80%) are summarized in Fig. 1. The depth-porosity data for Site 167 are also shown as is the curve generated by Laughton's (1954) experimental compaction of a *Globigerina* ooze (inset on Fig. 1). Space limitations do not permit a comprehensive review of porosity data from other DSDP Sites. However, certain Pacific sites at which thick sections of relatively pure carbonates were drilled are of interest. Cook & Cook (1972) noted that porosity values ranged from 94% to 41% and that they generally decreased with depth in the holes. They attributed this decrease to compaction, to incipient cementation, or to both. Tracey *et al.* (1971, p. 22) reported a relationship between the depth of burial and the degree of induration of pelagic oozes sampled on Leg 8. They noted (p. 23) a decrease in porosity and an increase in sediment stiffness which might have occurred from gravitational compaction and partial lithification. Sequences of nannofossil ooze-chalk-limestone as deep as 985 m and as old as Middle Eocene were penetrated at Sites 62, 63 and 64. At all three sites the porosity decreases irregularly and the rate of decrease appears to be related to age as well as depth of burial (Gealy, 1971, pp. 1103–1104). The saturated bulk density and porosity of Upper Oligocene and Middle Miocene nannofossil ooze, respectively, are nearly identical, despite different depths of burial (Table 1). Gealy noted (p. 1104) that porosity reversals, where porosity decreases with depth and then increases, occur on both the centimetre and decimetre scale. The change in porosity with depth and the density distributions at Site 167 (Fig. 1) are similar to those of Leg 8 and Sites 62, 63 and 64. The change from nannofossil ooze to chalk at Site 167 does not seem to be

Table 1. Comparison of bulk density (ρ*B*) and porosity (φ) with age and depth of burial at three Deep Sea Drilling sites in the Western Pacific (Gealy, 1971)

	Site 62			Site 63			Site 64		
	Depth (m)	ρ*B*	φ	Depth (m)	ρ*B*	φ	Depth (m)	ρ*B*	ρ
Recent	0	1·50	72	0	1·50	72	0	1·50	72
Middle Miocene	340	1·75	57	140	1·75	57	300	1·71	59
Upper Oligocene	520	1·90	49	350	1·91	49	560	1·85	51

caused by a change in porosity and occurs within a thick interval of almost constant densities of 1·60–1·70 (Winterer, Ewing *et al.*, 1973).

From the above the following can be seen.

(a) There is indeed a trend, over long stratigraphic intervals, towards decreasing porosity and increasing lithification in carbonate sections; the ooze to chalk to limestone transition appears, over long stratigraphic intervals, to be an irreversible one.

(b) Over shorter intervals, however, the correlation between depth and porosity is inexact.

(c) Furthermore, ooze-chalk-ooze-chalk sequences can occur in sections through which there is no discernible change in porosity. This last conclusion has important implications for acoustistratigraphy as discussed below. In terms of a diagenetic model there appear to be two stages in the reduction of porosity within a long section.

(i) An early dewatering stage in which, as may be seen in Fig. 1, half of the total porosity reduction takes place in the upper 200 m. This part of the section is termed in this paper the Shallow-burial Realm (Realm V of Table 2) where porosity is reduced from approximately 80% to 60% in most sections. Considering a sedimentation rate of 20 m/million years to encompass most sections drilled, this stage is completed in at most 10 million years. The dominant mechanism in this realm is gravitational compaction; cementation is a subordinate process.

(ii) A slower dewatering stage (lasting on the order of tens of millions of years). This is the Deep-burial Realm (Realm VI of Table 2) where porosity is reduced from approximately 65% to approximately 40% at a depth of 1000 m. The dominant process in this realm is cementation; gravitational compaction is the subordinate process.

Another important related physical property to be considered in any diagenetic model is the compressional velocity defined as:

$$V_c = \left(\frac{\lambda + 2\mu}{\rho}\right)^{\frac{1}{2}}$$

where

V_c=compressional velocity; ρ=density; λ and μ are the Lamé stress constants:

$$\lambda = \frac{\sigma}{1-2\sigma} \cdot \frac{E}{1+\sigma}\, ; \ \mu = \frac{1}{2} \cdot \frac{E}{1+\sigma}$$

σ and E being Poisson's ratio and Young's modulus respectively.

The values of the elastic constants, E and σ, are influenced in part by the degree of cementation of the rock whereas ρ may simply be a function of the degree of compaction of the sedimentary column. Thus, if a foraminiferal-nannofossil ooze were merely compacted without cementation, the velocity-depth function would be smooth (Fig. 7). However, there are marked deviations of V_c from such a curve because the processes of cementation occurring over an extended time interval should have the effect of increasing velocity (Nafe & Drake, 1963, pp. 811–812). The triangles on Fig. 7 are the interval velocities for the major lithological units at Site 167 (Winterer, Ewing *et al.*, 1973): from 0 to 220 m, 1·82 km/s; 220–600 m, 2·11 km/s: 600–820 m, 2·38 km/s; and 820–1185 m, 3·26 km/s. The divergence of these measured values from the values predicted by the smooth compaction curve indicates the degree of cementation the various intervals of the sedimentary column have undergone.

Fossil preservation

Foraminifera

The two types of Foraminifera, benthonic and planktonic, present in these pelagic carbonates contribute in somewhat different ways to the diagenetic process. On the average, benthonic Foraminifera are more preservable than are planktonic Foraminifera. Benthonic species usually have thicker walls and fewer pores than planktonic Foraminifera which tend to make the former more resistant to dissolution and, in some cases, to mechanical breakage. The preferential loss of planktonic species relative to benthonic species is well illustrated in the down-hole changes in Foraminifera at Site 167 (Fig. 2). The changes which occur in benthonic Foraminifera with increased depth of burial, dissolution and lithification (ooze to chalk to limestone) are many and include the following.

A deterioration in surface lustre and transparency. This seems to be the initial sign of dissolution. Hyaline calcareous benthonic Foraminifera from the Quaternary have smooth surfaces and a shiny lustre and are translucent. However, by Late Miocene these lustres have been lost or are considerably duller in appearance. Miliolids, which had a porcelaneous lustre, have a chalky appearance and most shells are broken or contain holes. Many hyaline shells have acquired a cloudy appearance.

A preferential removal of thin-walled, more porous rotaline species and especially miliolids. No miliolids were found below Core 8 which corresponds to the beginning of the transition from ooze to chalk.

A rapid down-hole increase in the percentage of broken tests. The increased fragmentation of planktonic species is most noticeable in the upper four cores, thereafter the percentage is fairly constant or decreases as the fragments become too small to identify or are dissolved. The percentage of broken benthonic species gradually increases from about 20% in the ooze to over 50% in the chalk. Below about Core 27 (Upper Eocene) the percentage of broken tests decreases as the total number of benthonic Foraminifera decreases. In the firm to hard chalks and limestones of Early Tertiary and Cretaceous age foraminiferal loss increases noticeably.

Calcite overgrowth and chamber infilling. Beginning in the Oligocene chalks (around Core 13) calcite overgrowth was noted on the inner chamber wall and pores of some species. Chamber infilling of calcite occurs in the lower part of the chalk sequence but was more common in the limestone and chalks of the Cretaceous. The percentage of shells with calcite overgrowth or infillings increased rapidly in the Cretaceous cores below Core 50 (Upper Campanian) and all shells were affected below Core 55.

The main effect of diagenesis on planktonic Foraminifera is a continued down-hole decrease in abundance due to dissolution. The minimum loss in foraminiferal sediment assemblages due to dissolution can be estimated from the equation $L=100$ $(1-R_0/R)$, where L is the loss necessary to increase the insoluble residue R_0 to R %, and by assuming benthonic Foraminifera are an insoluble component of the sediment assemblage and their fraction is altered only by dissolution, not by changing productivity (Berger, 1971). The estimate supposes there is no dissolution of benthonic Foraminifera, an assumption that can be shown to be incorrect although most benthonic species appear to be more resistant to dissolution than planktonic species. Arrhenius (1952) and Berger (1973b) assumed that benthonic species dissolve three times more slowly than planktonic species. Intuitively this assumption is more attractive than assuming R_0 is insoluble but in fact it has little justification as dissolution

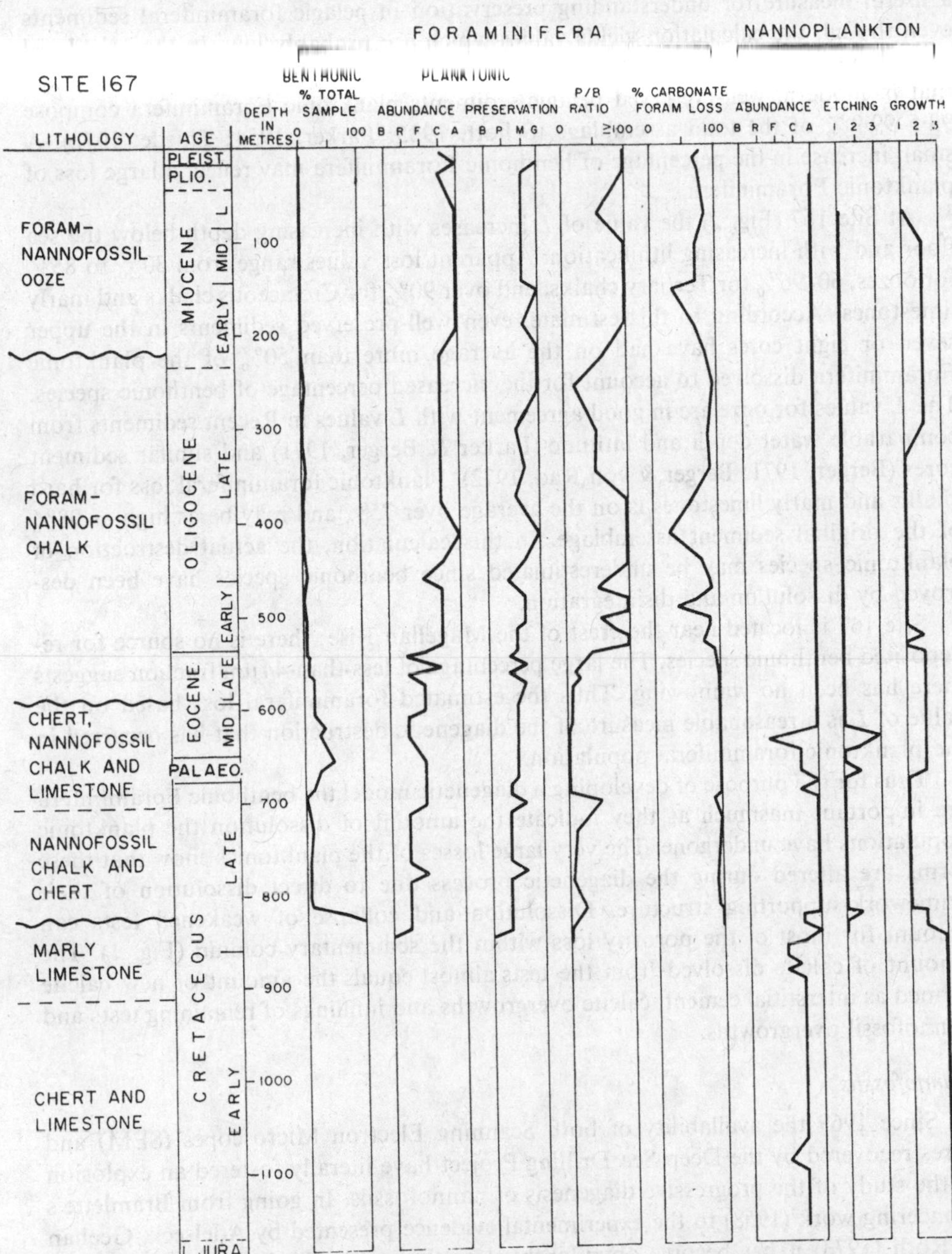

Fig. 2. Foraminifera-nannofossil preservation data for Site 167 (from Winterer, Ewing *et al.*, 1973). The shape of the planktonic/benthonic ratio (*P/B* ratio) curve is of interest in that the two sharp decreases in this ratio at horizons of Late Miocene and Late Oligocene age correspond to the position of reflectors b and d. These decreases show the extent of planktonic test dissolution at these horizons.

rates for benthonic Foraminifera have not been determined. The estimate L provides a useful measure for understanding preservation in pelagic foraminiferal sediments even though the calculation yields values which are probably low. In the calculated foraminiferal carbonate loss (L) values shown in Fig. 2, R_0 is 0·2%, based on the fact that in modern, well-preserved pelagic sediments planktonic Foraminifera compose 99·5–99·9% of the total assemblage (Schott, 1935; Parker, 1954; Thiede, 1972). A small increase in the percentage of benthonic Foraminifera may reflect a large loss of planktonic Foraminifera.

At Site 167 (Fig. 2) the value of L increases with increasing depth below the sea floor and with increasing lithification. Apparent loss values range from 30% to 85% for oozes, 60–96% for Tertiary chalks, and over 90% for Cretaceous chalks and marly limestones. According to this estimate, even well-preserved sediments in the upper seven or eight cores have had on the average more than 50% of the planktonic Foraminifera dissolved to account for the increased percentage of benthonic species. The L values for ooze are in good agreement with L values in Recent sediments from comparable water depth and latitude (Parker & Berger, 1971) and similar sediment cores (Berger, 1971; Berger & von Rad, 1972). Planktonic foraminiferal loss for hard chalks and marly limestones is on the average over 75% and may be as high as 98% of the original sediment assemblage. In this calculation, the actual destruction of planktonic species may be underestimated since benthonic species have been destroyed by dissolution and disintegration.

Site 167 is located near the crest of the Magellan Rise; there is no source for redeposited benthonic species. The large percentage of less-than-44 μm fraction suggests there has been no winnowing. Thus the estimated foraminiferal loss based on the value of L is a reasonable measure of the diagenetic destruction that has occurred in the planktonic foraminiferal population.

Thus for the purpose of developing a diagenetic model the benthonic Foraminifera are important inasmuch as they indicate the amount of dissolution the planktonic populations have undergone. The very large losses of the planktonics show that these forms are altered during the diagenetic process due to direct dissolution of their framework-supporting structure. Dissolution and collapse of weakened tests can account for most of the porosity loss within the sedimentary column (Fig. 1). The amount of calcite dissolved from the tests almost equals the amount of new calcite formed as interstitial cement, calcite overgrowths and infillings of remaining tests and nannofossil overgrowths.

Nannofossils

Since 1969 the availability of both Scanning Electron Microscopes (SEM) and cores recovered by the Deep Sea Drilling Project have literally fostered an explosion in the study of the progressive diagenesis of nannofossils. In going from Bramlette's pioneering work (1958) to the experimental evidence presented by Adelseck, Geehan & Roth (1973) it has become abundantly clear that nannofossils are an important factor in the diagenesis of pelagic carbonates. According to Roth & Thierstein (1972): 'a slight degree of secondary calcite overgrowth is found in most carbonate oozes that have been buried under about 100 metres of sediment'. Thus the initial stages of subsurface nannofossil diagenesis are within the Shallow-burial realm. As shown by Fischer, Honjo & Garrison (1967) the final stages of pelagic-carbonate diagenesis involve the remaining nannofossils. The trends in nannofossil diagenesis that have become clear through the studies of, among others, Fischer *et al.* (1967) and Adelseck *et al.* (1973) are as follows.

(i) Small coccoliths tend to dissaggregate along sutures due to dissolution, thus producing in the sediment large numbers of micron-sized crystals that are very susceptible to solution.

(ii) Discoasters tend to grow by the precipitation on them of highly euhedral calcite overgrowths; the overgrown discoaster may contain a volume of calcite several times greater than the original.

(iii) Larger coccoliths show overgrowths and persist in highly recrystallized limestones.

(iv) The very abundant micron-sized elements of coccoliths can supply some of the calcite for the overgrowths on discoasters and larger coccoliths, the interstitial cement and the calcite infillings of Foraminifera.

Textural considerations

The fact that these pelagic sediments are virtually 100% biogenic calcite means that the textural changes observed in their diagenetic transformation from ooze to chalk to limestone are the direct result of the response of the foraminiferal tests and nannofossils to burial. Numerous scanning electron micrographs in many papers have illustrated these changes (e.g. Wise, 1973; Bukry, 1973; Adelseck *et al.*, 1973; Wise & Kelts, 1972). For the purposes of this paper four scanning electron micrographs from the Magellan Rise Deep Sea Drilling site summarize the pertinent textural changes for which a diagenetic model must account (Fig. 3a–d).

(a) A Middle Miocene specimen (DSDP core 167–6, depth 103–112 m) showing intact but slightly etched tests of planktonic Foraminifera in a matrix of generally well-preserved nannofossils, mainly coccoliths. The still-whole chambers enclose large volumes of intrabiotic pore space. The abundant tests form a supporting framework. The scattered anhedral to subhedral grains of calcite are the result of disintegration of some small coccoliths and possibly Foraminifera tests. The porosity in this interval ranged from 62·8% to 67% (Winterer, Ewing *et al.*, 1973).

(b) An Upper Oligocene chalk (DSDP core 167–11–4, depth 297–306 m) shows the whole spectrum of fossil-preservation phenomena; corroded foraminiferal tests (lower right), discoasters with massive euhedral overgrowths (upper right) and severely etched coccoliths. The porosity in this interval ranged from 61% to 64%.

(c) A Valanginian-Hauterivian limestone (DSDP core 167–81CC, depth 1040 m). This shows a mass of calcite crystals, that appears to be filling a void or replacing a foraminiferal test, in a matrix of coccolith fragments and fine-grained cement. Porosity was not determined on this sample.

(d) A Berriasian-Tithonian limestone (DSDP core 167–94–2, depth 1165–1168 m) shows very severely etched coccoliths in a matrix of euhedral to subhedral cement; an etched nannoconid is seen in the lower left corner. Porosity in this interval ranged from 33% to 48%.

Geological considerations

The Magellan Rise is a relatively smooth-topped structure that reaches to within 3200 m of the sea surface and stands above a plain at a depth of 5000–6000 m (Fig. 5). The seismic profile shows that the internal stratigraphy is quite uniform. Basement

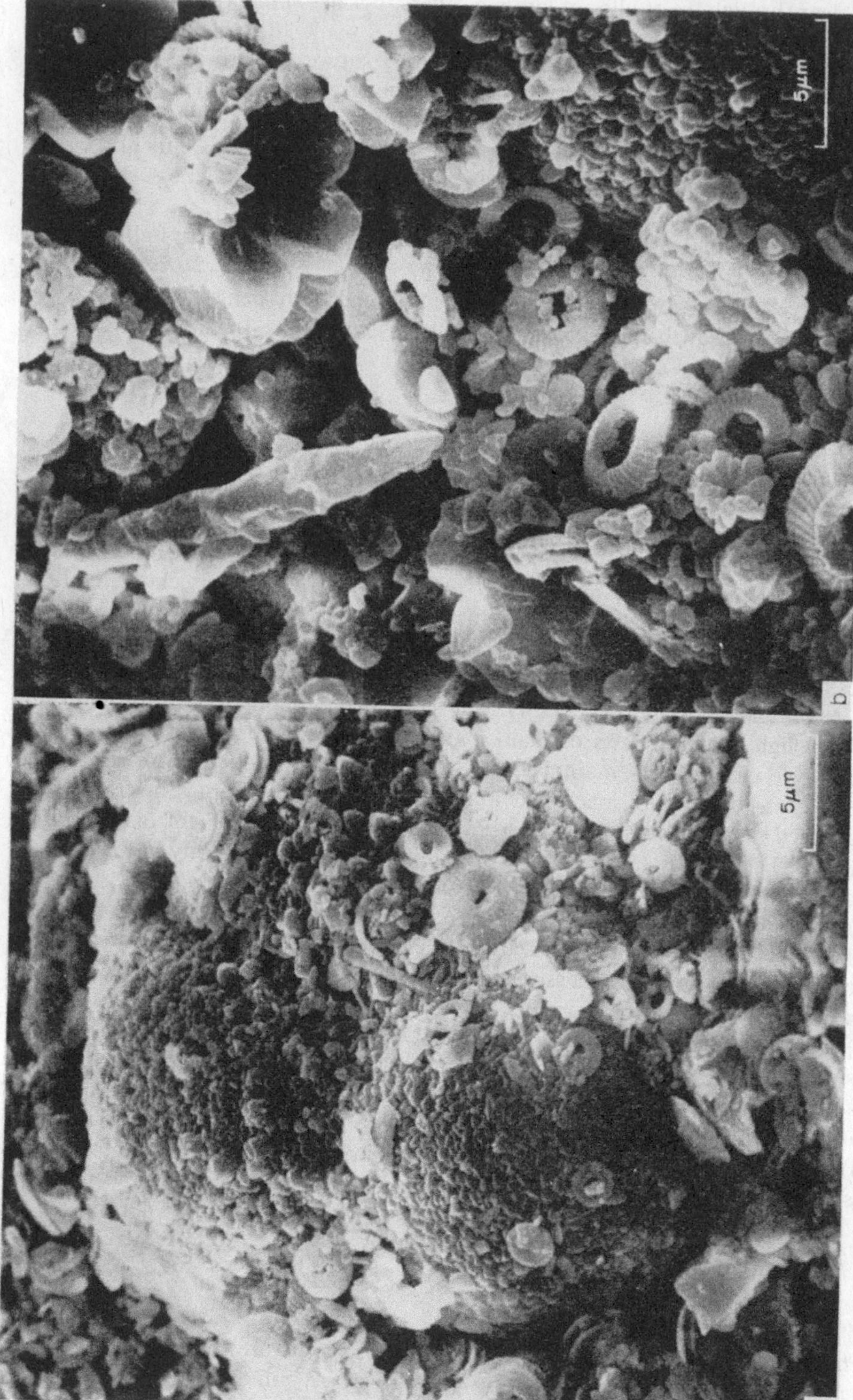

Fig. 3. Scanning electron micrographs of representative sediments from Site 167. See text for descriptions.

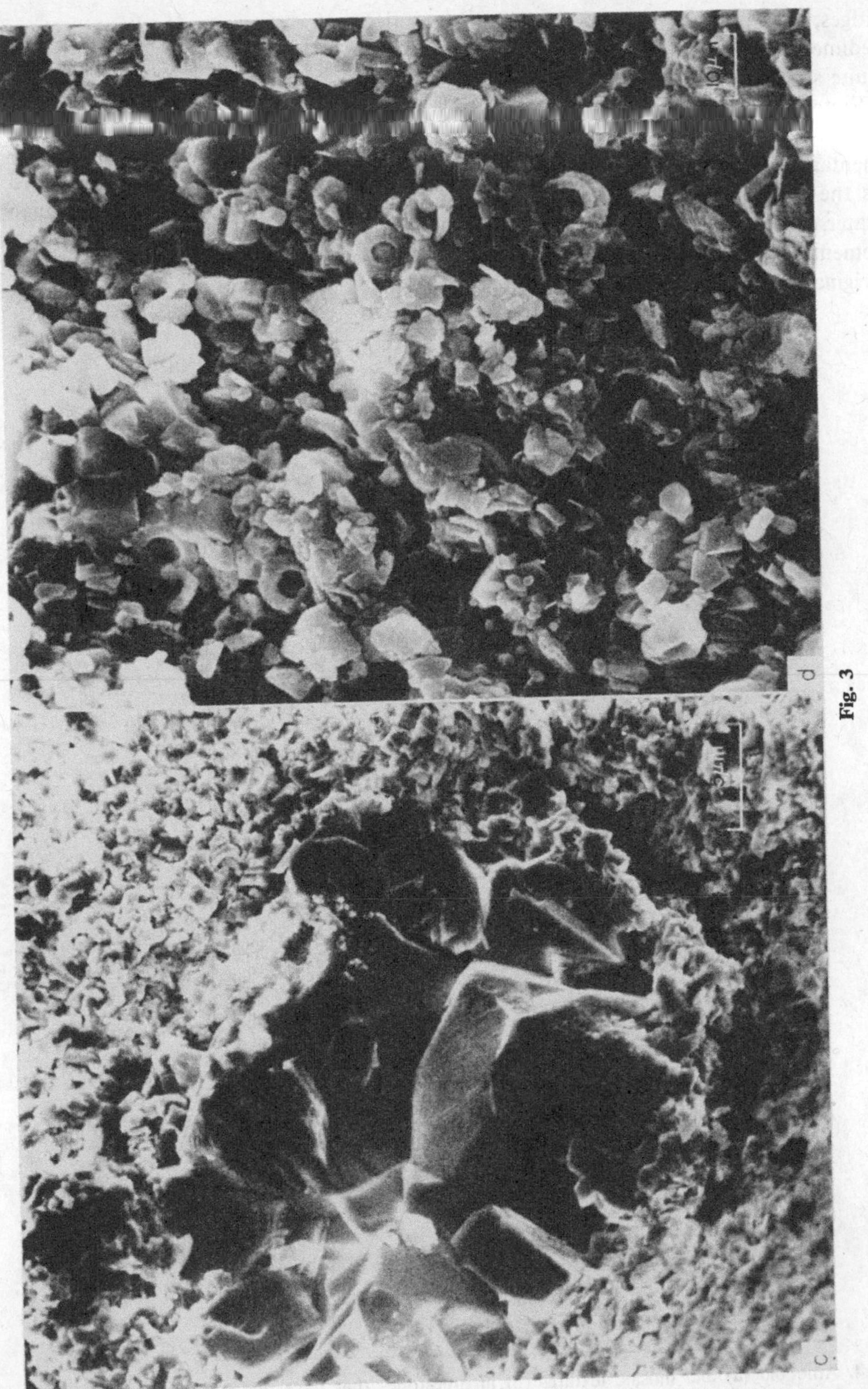

Fig. 3

ridges, that trend parallel to the long axis of the Rise, are local features where the sediments show draping effects perhaps accentuated by compaction. The Rise has, quite simply, been the recipient of pelagic sediments over the past 135 million years (Winterer, Ewing *et al.*, 1973). There is no evidence in the cores that the surface of the Rise was ever below the calcite compensation depth. The geometry of the sedimentary cap suggests that water flow within it has been primarily upward and outward as the accumulating sediments slowly compacted. It is difficult to see how much water could have flowed into the cap. Thus a diagenetic model must provide for a cement source within the sediments, i.e. a conservative system exists in which the original biogenic calcite is the sole diagenetic source material.

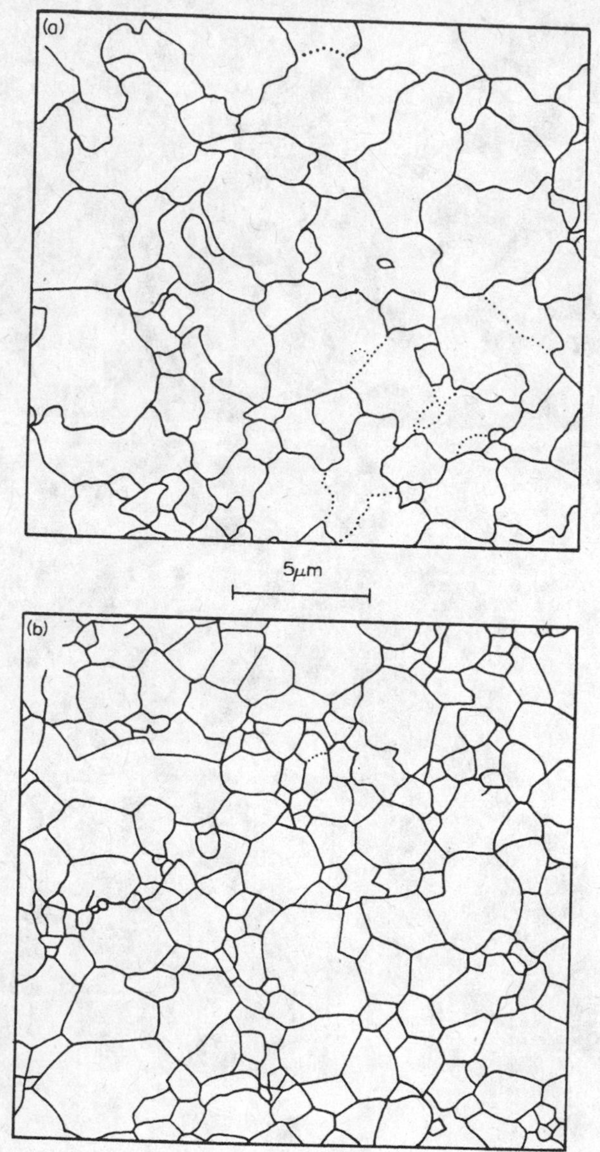

Fig. 4. Ameboid (a) and mosaic textures (b) in limestones that have undergone some degree of metamorphism (from Fischer *et al.*, 1967).

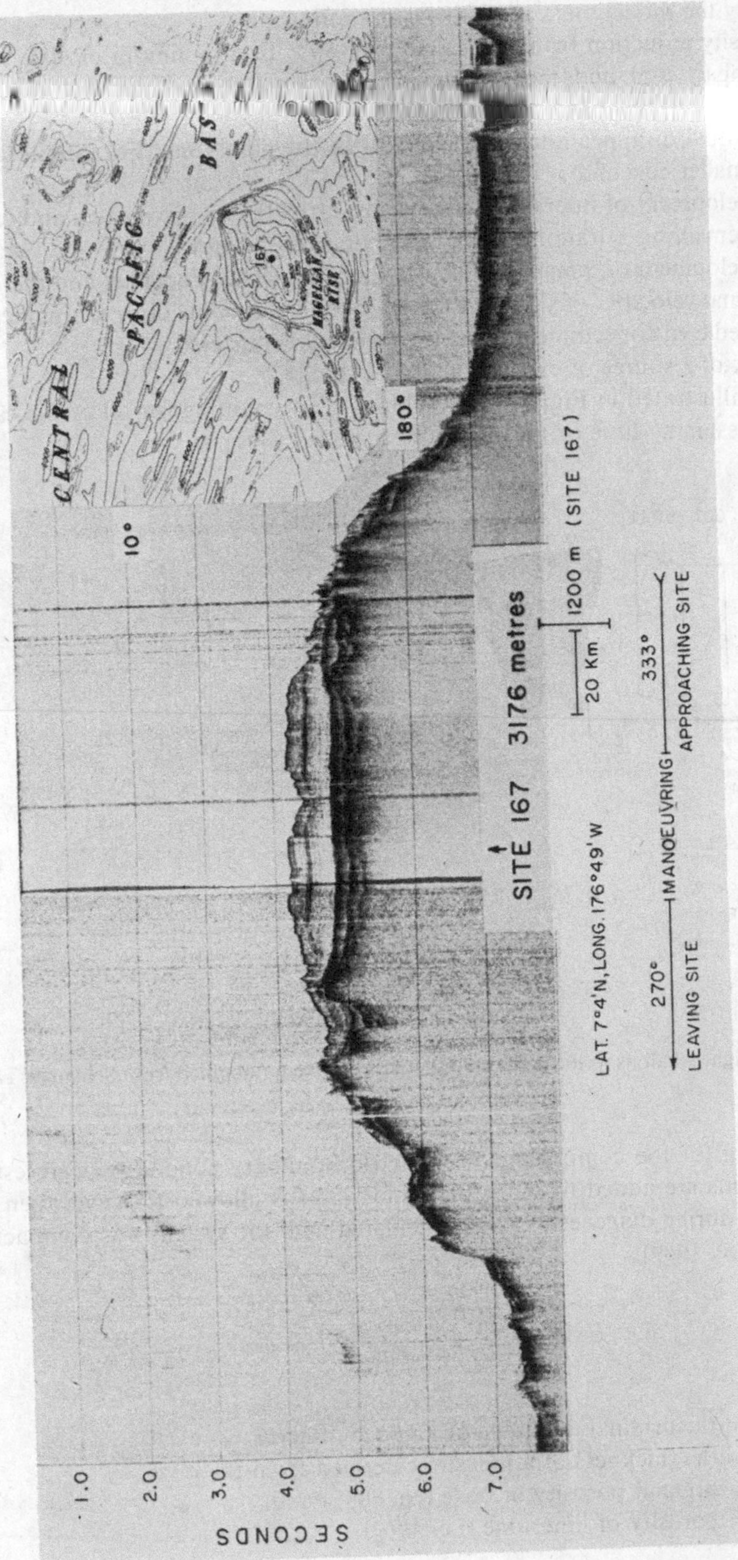

Fig. 5. Seismic profile of the Magellan Rise. The apparent 'graben' near the centre of the Rise is in part an artefact of the course changes during manoeuvring.

In summary the model must account for the following.

(a) A porosity reduction from a maximum of 80% to a minimum of 40% as the sediments compact and undergo diagenesis, approximately half of this reduction taking place in the upper 200 m.

(b) The gradual disappearance of almost all of the planktonic Foraminifera and many of the smaller coccoliths.

(c) The development of interstitial cement, overgrowths on discoasters and calcite fillings in the remaining Foraminifera (largely benthonic types).

(d) The development of cement to a degree that the elastic properties and therefore the compressional velocities of the more deeply buried sediments diverge significantly from values predicted for sediments that were merely compacted.

(e) The lack of a source of extra-formational calcite.

The model illustrated in Fig. 6 was developed by accounting for all of the calcite in the system assuming little or no loss to the expelled waters.

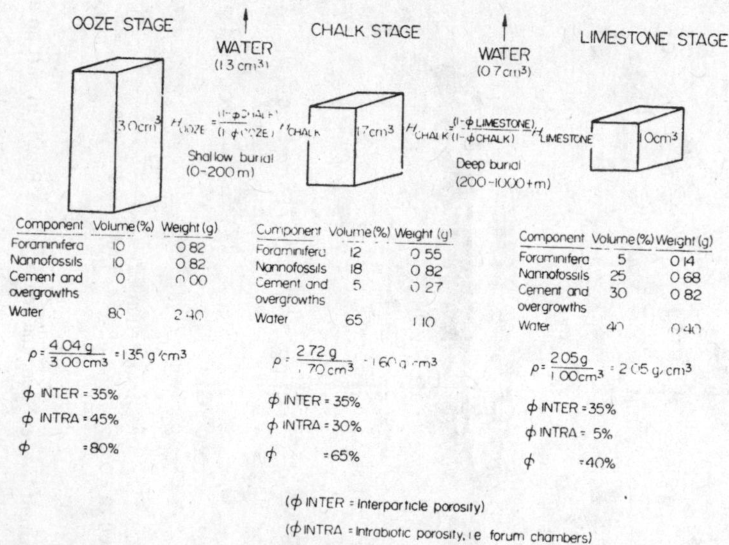

Fig. 6. Volume-weight relations in the ooze-chalk-limestone system (modified from Schlanger *et al.*, 1973).

In accounting for the components, volumetric boundary conditions were established. If no solids are added to the system and water is allowed to leave, then the volume changes during diagenesis can be calculated as if the system was compacting as follows (Moore, 1969):

$$\text{Let } H_{\text{ooze}} = \frac{(1 - \phi_{\text{limestone}})}{(1 - \phi_{\text{ooze}})} H_{\text{limestone}};$$

where

H_{ooze} = the original thickness of an ooze interval,
$H_{\text{limestone}}$ = the thickness of a limestone derived from the ooze,
ϕ_{ooze} = original porosity of ooze (i.e. 80%) and
$\phi_{\text{limestone}}$ = porosity of limestone (i.e. 40%).

Substituting the appropriate values and setting $H_{\text{limestone}}$ equal to 1 cm:

$$H_{\text{ooze}} = \frac{(1 - 0{\cdot}40)}{(1 - 0{\cdot}80)} \,.\, 1 \text{ cm}$$

$H_{\text{ooze}} = 3$ cm.

Thus, approximately 3 cm^3 of foraminiferal-nannofossil ooze with a porosity of 80% will reduce to 1 cm^3 of limestone with a porosity of 40% in going from ooze to limestone.

By using a petrographic and textural approach based on scanning electron micrographs of pelagic sediments and tests of individual Foraminifera, a model of ooze diagenesis was developed.

Scanning electron micrographs of foraminiferal tests show that the three types of primary intrabiotic porosity (chambers, foramen, and interwall) make up 80% of the volume occupied by an individual foraminiferal test. Further, a loosely packed aggregate of tests would have primary interparticle porosity. Considering a pure foraminiferal ooze as an aggregate of spheres and considering that the interparticle porosity of spherical aggregates of uniform size ranges from 26% (rhombohedral packing) to 48% (cubic packing) an original interparticle porosity of 45% is reasonable. Thus, 80% (intrabiotic porosity) of 55% (sphere-enclosed volume) equals 44% and this, plus the interparticle porosity, equals approximately 90%—that of a pure foraminiferal ooze. If an equal amount, by volume, of nannofossil calcite is added to the interparticle void space, the resulting ooze is volumetrically: 10% foraminiferal calcite, 10% nannofossil calcite, and 80% intrabiotic and interparticle pore space; the sediment would have a honeycomb structure. The ooze stage in Fig. 6 was constructed on the above considerations.

The major changes in going from ooze to limestone involve the dissolution, and eventual almost complete destruction of the framework-supporting tests of the Foraminifera and the etching and dissolution of smaller coccoliths. This decrease in the volume of biogenic calcite by dissolution is balanced by a build-up of calcite overgrowths on remaining fossils, particularly discoasters, and formation of calcite cement. The foraminiferal loss amounts to almost 50% going from ooze to chalk and nearly 100% from ooze to limestone (Fig. 6). As shown above, approximately 44% of the original ooze was intrabiotic porosity. Wholesale dissolution of the tests with reprecipitation of the test calcite as cement and nannofossil overgrowths in going from ooze to limestone would result in a porosity reduction approaching the 40% demanded by this model (Fig. 6).

THE DIAGENETIC POTENTIAL

The shallow- and deep-burial realms, discussed above, can be placed into a scheme covering all stages of diagenesis of pelagic carbonate sediments (Table 2). Such a scheme is convenient to the discussion of the concept of a diagenetic potential.

The life history of a pelagic limestone can be thought of as beginning in the upper 200 m of the ocean (Realm I) where planktonic organisms produce calcite as a highly dispersed suspension which has a very high potential for future diagenetic change. As this dispersed calcite aggregates by sedimentation to the sea floor and lithifies by compaction and cementation its diagenetic potential decreases; the sediment matures.

Table 2. Diagenetic realms

Depth	Realm	Residence Time	Petrography	Porosity (ϕ) % Velocity (V_c) km/s	Diagenetic potential O ⟷ ∞
0–200 m (surface water)	I Initial production	Weeks	Highly dispersed calcite-sea water system; 10–10^2 forams/m^3, 10^4–10^6 nannoplankton/m^3 (Lisitzin, 1971; Berger, 1971)		
200 m to sea floor	II Settling	Days to weeks for forams; months to years for coccoliths depending on pelletization (Smayda, 1971)	Pelletized coccoliths, ratio of broken to whole nannoplankton increases downward, ratio of living to empty foram tests decreasing during settling (Lisitzin, 1971; Berger, 1971)		
3000–5000 m (see Diagenetic Potential)	III Deposition	Inversely proportional to sedimentation rate and dissolution rate	'Honeycombed' structure (Tschebotarioff, 1952). Large foram tests supported by chains of coccolith discs. This surface is actually part of Realm IV	$\phi \simeq 80\%+$ $V_c \simeq 1{\cdot}45$–$1{\cdot}50$ (Nafe & Drake, 1963)	← slope at 3000 m ←Slope at 5000 m
0–1 m sub-bottom)	IV Bioturbation	50,000 years (at 20 m/10^6 years sedimentation rate)	Remoulded 'honeycomb', slight compaction, burrowing, destruction by ingestion and solution	$\phi \simeq 75$–80% $V_c \simeq 1{\cdot}45$–$1{\cdot}6$	
1 200 m (sub-bottom)	V Shallow-burial	10×10^6 years (at 20 m/10^6 years sedimentation rate)	Ooze affected by gravitational composition, establishment of firm grain contacts; dissolution of fossils and initiation of overgrowths	$\phi \simeq 75$–60% $V_c \simeq 1{\cdot}6$–$1{\cdot}8$	
200–1000 m + (sub-bottom)	VI Deep-burial	Up to $\simeq 120 \times 10^6$ years (by then either subducted or uplifted)	Chalk with strong development of interstitial cement and overgrowths; transition down to limestone with dissolution of forams, pervasion by cement and overgrowths—grain interpenetration, welding and 'ameboid mosaics' (Fischer *et al.*, 1967)	$\phi \simeq 60\%$ down to 35–40% $V_c \simeq 1{\cdot}8$ increasing to 3·3 km/s	
1–10 km sub-surface	VII Metamorphic	10^6–10^7 years	Recrystallization trending to 'pavement mosaic' (Fischer *et al.*, 1967) of completely interlocking crystals	$\phi \simeq 40\%$ down to $< 5\%$ $V_c \simeq 3+$ up to 6 km/s	

The life history of a pelagic limestone ends as the rock approaches a final state in which further diagenetic change is not probable, e.g. an aggregate of pure, unstrained crystals at a minimum free energy level. As Byrne points out (1965, p. 105):

'In three dimensions, all space can be filled so as to satisfy surface tension requirements if all grains are cubo-octahedrons of minimum surface-to-volume ratio. A cubo-octahedron is made up of eight faces which are regular hexagons and six faces which are squares . . . . If all grains had this shape, all grain boundaries would be in equilibrium and no further growth would occur.'

In discussing lithified micrites, Bathurst (1971, p. 511) states that the widespread upper crystal diameter for the groundmass of lithified micrites at 3–4 μm, 'points to the existence of a universal threshold state at which fabric evolution stops and beyond which it can, but need not, continue. A possible reason for this, which would bear further investigation, is that a stage is reached in the combined neomorphism and cementation, when the porosity and permeability are so reduced that the transport of Ca^{2-} and CO_3^{2+} from one crystal face to another becomes slow even on the geological time scale. This stage would represent virtual stability'.

He also states (Bathurst, 1971, p. 502) that:

'. . . lithification of micrite and the growth of neomorphic spar yield between them, a range of calcite fabrics which, once evolved, appear to resist strongly any further diagenetic change. These are arbitrarily classified as micrite (0·5–4 μ), microspar (5–50 μ) and pseudospar (50–100 μ) . . . .'

Fischer *et al.* (1967, Fig. 3a, b, Fig. 4a, b, p. 20) show the apparently penultimate and final life stages of nannofossil limestones from the Franciscan Formation of California as 'ameboid mosaics' and 'pavement mosaics' respectively, see Fig. 4, and Table 2. Byrne's space-filling cubo-octahedrons, Bathurst's micrite and microspar, which are the right size range from limestone made up of overgrown large coccoliths and discoasters, and the ameboid and pavement mosaics of Fischer *et al.* (1967) all represent states of near-zero diagenetic potential. Bramlette (1958, p. 126) recognized this trend to the holocrystalline aggregate stage in his studies of outcrop material from France and North Africa. He stated that the absence of coccoliths in older fine-grained limestone would be inevitable with the recrystallization that has produced the aplite texture.

Diagenetic potential may thus be defined as the length of the diagenetic pathway left for the original dispersed foraminiferal-nannoplankton assemblage to traverse before it reaches the very low free-energy level of a crystalline mosaic.

The concept of the diagenetic potential can be used to explain deviations from a strict depth-of-burial/lithification dependence demanded by a simple gravitational compaction model of diagenesis. The diagenetic potential remaining to a sediment after it passes through the critical boundary between Realms IV and V (Table 2) will determine how far cementation will proceed per unit time. Thus, if layers of very different diagenetic potential are buried sequentially, the amount of cementation in a layer per unit time will be proportional to the diagenetic potential of that layer so that chalks can form above oozes and limestones above chalks in the sedimentary column.

The diagenetic potential (DP) can be expressed as follows:

DP $= f$ (water depth, sed. rate, Temp (surface), Productivity (surface), foraminifer +coccolith: discoaster ratio; $\text{Size}_{(max)}$: $\text{Size}_{(min)}$ ratio, predation rate . . . .)

The depth of water and the sedimentation rate are very important factors as these affect the degree to which the calcite is dissolved while at the sediment/water interface; calcite dissolved at this interface is the more soluble portion of the sediment, and is not available for further diagenesis in buried strata.

The diagenetic potential of a sediment is enhanced in zones of high productivity, high lysocline and low calcite compensation depths which favour the mobilization and redeposition of skeletal calcite (Wise, 1972). The surface water temperature, and salinity and biological factors not understood, affect the $MgCO_3$ content of the pelagic sediment in a complex manner because the temperature-salinity structure of the upper water layers apparently influences the contributions made to the general population by various species and genera of Foraminifera, each of which have characteristic $MgCO_3$ contents. As Parker & Berger (1971) and Savin & Douglas (1973) point out, the $MgCO_3$ content of Foraminifera affects the solubility of these tests as they sink and as they reside at the sea floor. Since discoasters appear to be the most favoured receptors of calcite overgrowths, and foraminiferal tests and coccoliths the most called-upon donors for calcite in the cement, it seems possible that the original Foraminifera+coccolith/discoaster ratio is an important factor in the diagenetic potential because a high content of $MgCO_3$-rich tests in a buried sediment would give it a high diagenetic potential. A sorting measure should also be included since an abundance of very small grains, mixed with larger ones such as make up discoasters, appear to promote calcite transfer as discussed by Adelseck *et al.* (1973). These investigators subjected an Upper Pliocene nannofossil ooze from Site 167 to temperatures and pressures of up to 300°C and 3 kb for 1 month in order to simulate the diagenetic changes such a sediment might undergo with long and deep burial. They found that the smaller and more delicate coccoliths were more easily destroyed than the larger forms with strongly overlapping elements (crystallites). Most larger forms showed formation of secondary overgrowth though some underwent slight etching. Discoasters displayed well-developed overgrowths with crystal faces.

They further pointed out that the relatively rapid disaggregation of smaller coccoliths produced large quantities of less than 1 μm size crystals that evidently supplied the calcium carbonate for the overgrowths on the discoasters and the remaining larger coccoliths.

This growth of the larger crystals at the expense of the smaller ones, in the buried foraminiferal-nannofossil ooze, is evidently taking place through a dissolution-diffusion-precipitation mechanism. The abundant, less than 1 μm crystals, present as disaggregated coccoliths and in the tests of Foraminifera (see Fig. 3b, lower right) represent the phase that provides the calcite and the up-to-10 μm-size crystals present as discoasters and large coccoliths represent the phase that receives the calcite. The driving force for the dissolution-precipitation process is dependent on the large difference in surface area between a population of very small grains and a population of larger grains. In a dispersed system a term containing the surface energy must be taken into account in expressing the Gibbs free energy G of the system:

$$dG = -Sdt + PdV + \alpha dw + \xi \mu dn$$

where α represents the surface energy per cm^2 and dw is an element of surface area. As the small particles of calcite dissolve and the same volume of calcite is reprecipitated on larger crystals the free energy change will be proportional to the decrease in surface area of the small particles minus the increase in the surface area of the large crystals.

Further, edges and sharp corners contribute to the surface free-energy component of the total free energy of the calcite; the dissolution of very small grains lessens this contribution. In essence the appearance of the σdw term causes the dispersed system to be more soluble than the bulk system. Evidently, based on petrographic and experimental data, the apparent irreversible trend in foraminiferal-nannofossil sediments towards the ameboid and pavement mosaics of Fischer *et al.* (1967) follows a trend of decreasing free energy in the calcite system.

The predation rate of plankton-feeders may be important as Smayda (1971) showed that the accumulation of nannoplankton in faecal pellets at near-surface levels hastened the sinking of this material, lessening the chances of its being dissolved as it descended into deep waters. Faecal pellets charged with nannofossil debris sink at the rate of approximately 100 m/day whereas naked fragments of coccoliths would take 1 or more years to sink 5000 m. At this slow slow sinking rate micron-sized crystals would totally dissolve (Peterson, 1966) and these ready donors of overgrowth calcite would not be available in the buried sediment.

We postulate that variations of the diagenetic potential with depth of burial in the sedimentary column are related to original variations in basin depth, the calcite compensation depth (CCD), and the calcareous-plankton productivity of the upper water layers. This relationship is important to the discussion of diagenetic potential and acoustistratigraphy developed below.

THE DIAGENETIC POTENTIAL AND ACOUSTISTRATIGRAPHY

If the velocity-depth function in foraminiferal-nannofossil sub-sea sections was smooth and continuous, such sections would be acoustically transparent. Such would be the case if the lithification of carbonate-rich sediments were solely the result of gravitational compaction. This transparency would be due to the lack of sharp acoustic impedance gradients and discontinuities within the sedimentary column. However, refraction and Deep Sea Drilling data (see Fig. 7) show that deviations in velocity values from those predicted by compaction models are the rule rather than the exception. These deviations have generally been ascribed to cementation effects which drastically influence the values of the elastic constants. Cementation, which markedly increases the rigidity of the sediment, increases the compressional velocity. Since acoustic impedance = density × compressional velocity (Hamilton, 1959, 1971), a marked change between two layers, in compressional velocity, without marked changes in density, will cause a reflection of acoustic energy.

Pure pelagic carbonate sections are characterized by an abundance of closely spaced reflectors (acoustic artefacts aside) caused by discontinuities, marked gradients and reversals in acoustic impedance values. Data presented by Gealy (1971) for DSDP Site 64 show that both positive and negative impedance differences over distances of only a few metres, within apparently homogeneous sediments, are very common and these appear to result from very subtle changes in degree of lithification. Thus, instead of seeing a non-reversible gradual transition from ooze to chalk to limestone with increasing depth of burial, there are short range fluctuations, of ooze - chalk - ooze - chalk - ooze - chalk - limestone - chalk - limestone sequences, within the long-range trend. These alternations show that the ooze-chalk-limestone transition is not strictly time- and depth-of-burial-dependent. The fact that

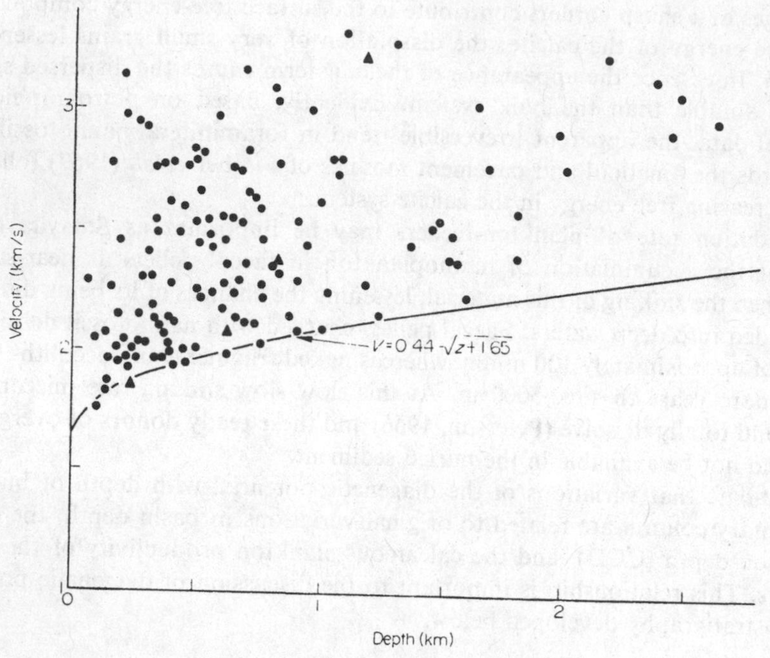

Fig. 7. Compressional velocity versus depth of burial plot. (Modified from Nafe & Drake, 1963, by addition of DSDP Site 167 interval velocity data (▲), see text for details.) The solid line is based on data from Laughton (1954) derived through compaction of *Globigerina* ooze. Other data from seismic refraction work.

uncemented oozes are found below hard chalks indicates that time and depth of burial taken alone cannot account for degree of cementation. According to arguments presented in a preceding section the degree of cementation exhibited by a carbonate pelagic sediment is in large part correlative with the diagenetic potential the sediment possessed when it finally passed below the bioturbation realm of diagenesis.

The diagenetic potential of the sediment, as discussed previously, is in large part pre-determined by oceanographic conditions prior to burial and even prior to the arrival of the sediment at the sea floor. We believe the abundant reflectors characteristic of carbonate sequences are related to the degree of cementation and that the degree of cementation is controlled by the diagenetic potential. Thus, since diagenetic potential is determined by palaeo-oceanographic conditions, it follows that an acoustistratigraphic event should correlate with a palaeo-oceanographic event.

This relationship of acoustistratigraphy to time-stratigraphy was noted by geologists during drilling operations at DSDP Site 64 on the Ontong Java Plateau (Winterer *et al.*, 1971) where many of the reflectors can be traced on profiles for tens or even hundreds of kilometres; further, the evidence of close parallelism of reflectors on the profiles suggested a time-stratigraphic control on induration. If the reflectors in these ooze-chalk-limestone sequences, so abundant in the post-Eocene-chert stratigraphy of the Pacific Basin, preserve palaeo-oceanographic events they should be correlatable and the strength of the correlation should be proportional to the magnitude and length of the event. To test this argument we compared reflection profiles and data on travel times to prominent reflectors from the Ontong Java Plateau (near DSDP Site

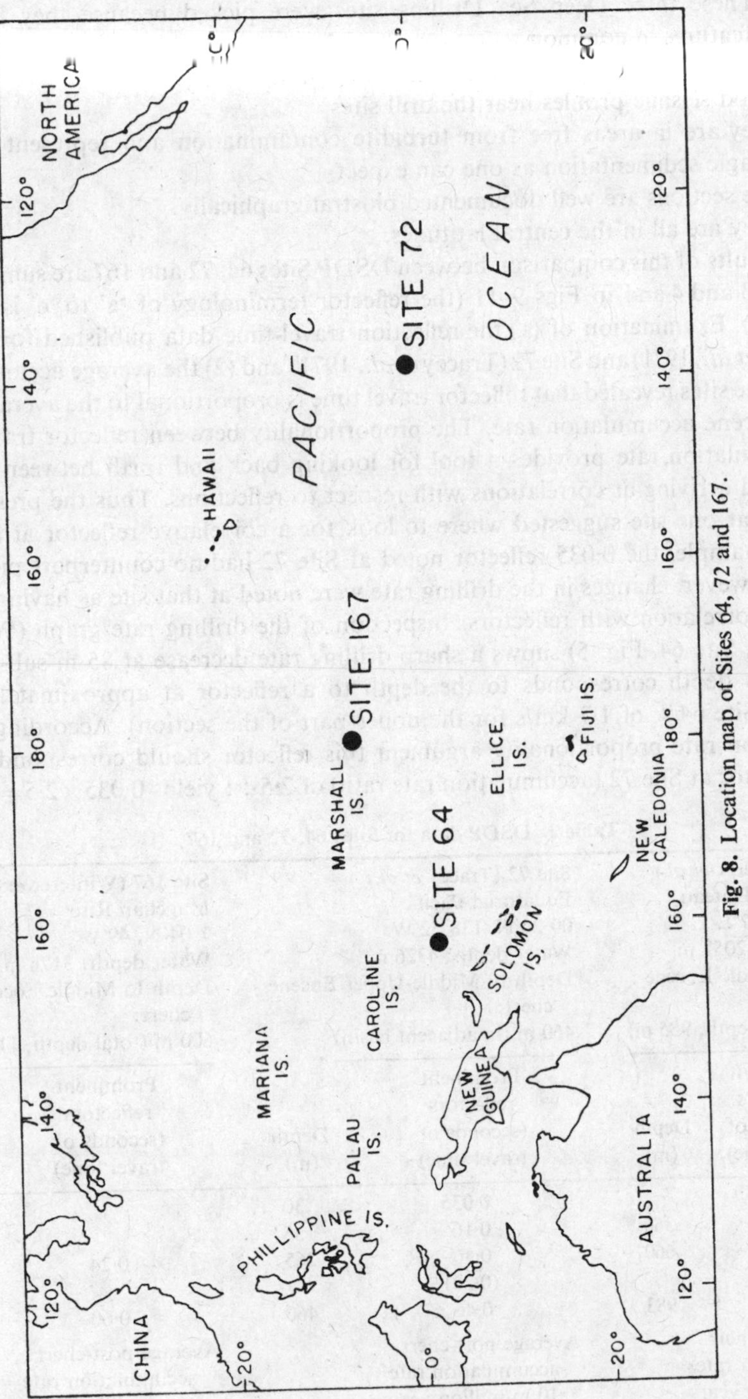

Fig. 8. Location map of Sites 64, 72 and 167.

64), the Equatorial Pacific (DSDP Site 72), and the Magellan Rise (DSDP Site 167), (Fig. 8). These three Deep Sea Drilling sites were picked because they have the following features in common.

(a) Pure carbonate sections above Eocene chert.
(b) Good seismic profiles near the drill sites.
(c) They are in areas free from turbidite contamination and represent as pure pelagic sedimentation as one can expect.
(d) The sections are well documented biostratigraphically.
(e) They are all in the central latitudes.

The results of this comparison between DSDP Sites 64, 72 and 167 are summarized on Tables 3 and 4 and in Figs 9–11 (the reflector terminology of 'a' to 'e' is for this paper only). Examination of (1) the reflection travel-time data published for Site 64 (Winterer *et al.*, 1971) and Site 72 (Tracey *et al.*, 1971) and (2) the average accumulation rates for the sites revealed that reflector travel time is proportional to the average post-middle Eocene accumulation rate. The proportionality between reflector travel time and accumulation rate provides a tool for looking back and forth between seismic profiles and arriving at correlations with respect to reflections. Thus the presence of a reflector at one site suggested where to look for a correlative reflector at another site. For example, the 0·035 reflector noted at Site 72 had no counterpart picked at Site 64. However, changes in the drilling rate were noted at that site as having a high degree of correlation with reflectors. Inspection of the drilling rate graph (Winterer *et al.*, 1971, Site 64, Fig. 5) shows a sharp drilling rate decrease at 85 m sub-bottom depth. This depth corresponds to the depth to a reflector at approximately 0·1 s (using the Site 64 $\bar{V}$ of 1·7 km/s for the upper part of the section). According to the accumulation rate proportionality argument this reflector should correspond to the 0·035 reflector at Site 72 (accumulation rate ratio of 2·5 : 1 yields $0{\cdot}035 \times 2{\cdot}5 = 0{\cdot}09$ s).

Table 3. DSDP data for Sites 64, 72 and 167

Site 64 (Winterer *et al.*) Ontong Java Plateau 1°45′N, 158°37′E Water depth: 2052 m Depth to Middle Eocene chert: 985 m (total depth, 985 m)		Site 72 (Tracey *et al.*) Equatorial Pacific 00°26′N, 138°52′W Water depth: 4326 m Depth to Middle-Upper Eocene chert: 460 m (in adjacent basin)		Site 167 (Winterer *et al.*) Magellan Rise 7°04′N, 49′W Water depth: 3176 m Depth to Middle Eocene chert: 600 m (total depth, 1185 m)	
Prominent reflectors (seconds of travel time)	Depth (m)	Prominent reflectors (seconds of travel time)	Depth (m)	Prominent reflectors (seconds of travel time)	Depth (m)
		0·035	30		
0·43	366	0·16	135		
0·71	660	0·30	265	0·24	220
		(0·28)			
0·97	983	0·46	460	0·60	600
Average post-chert accumulation rate= 25 m/million years		Average post-chert accumulation rate= 10 m/million years		Average post-chert accumulation rate= 15 m/million years Average accumulation rate from 0–220 m=9 m/million years	

Table 4. Acoustistratigraphic correlations—Sites 64, 72 and 167

Site 64					Site 72					Site 167					Reflecting horizons¶
Reflector strength	Travel time	Depth	Zone	Age (million years)	Reflector strength	Travel time	Depth	Zone	Age (million years)	Reflector strength	Travel time	Depth	Zone	Age (million years)	Designation
Weak	0·10	85 (d.b.)*	N. 20 interpol. (N. 19–22)	3	Moderate	0·035	30	N. 20–N. 21	3			Not seen			a
Moderate	0·20	170 (d.b.)*	N. 18 interpol. (N. 17–19)	5–6			Not seen			Weak	0·07 0·08	60–70 §	N. 17–N. 18	5–6	b
Strong	0·43	366	N. 12 interpol. (N. 10–14)	14	Strong	0·16	135	N. 14 interpol. (N. 12–16)	13			Not seen			c
Strong	0·71	660	NP 25 L. bipes	24–26	Strong	0·30 (0·28) †	265	P. 22	23–26	Strong	0·24	220	P.22–N. 4	21–26	d
Strong	0·97	983	P. 14 or older	43 or less	Strong	0·46	460 ‡	P. 13 or younger	44 or less	Strong	0·60	600	P. 14	43–44	e

* d.b. indicates drilling time break used to locate reflector.
† 0·28 corrected figure used in this paper. (Fig. 12, Site 72 report, Tracey *et al.*, 1971.)
‡ Projected basement depth in basin adjacent to Site 72.
§ Probably at hiatus between cores 3 and 4 in N. 17–18 Zone.
¶ Letter assigned for this paper.

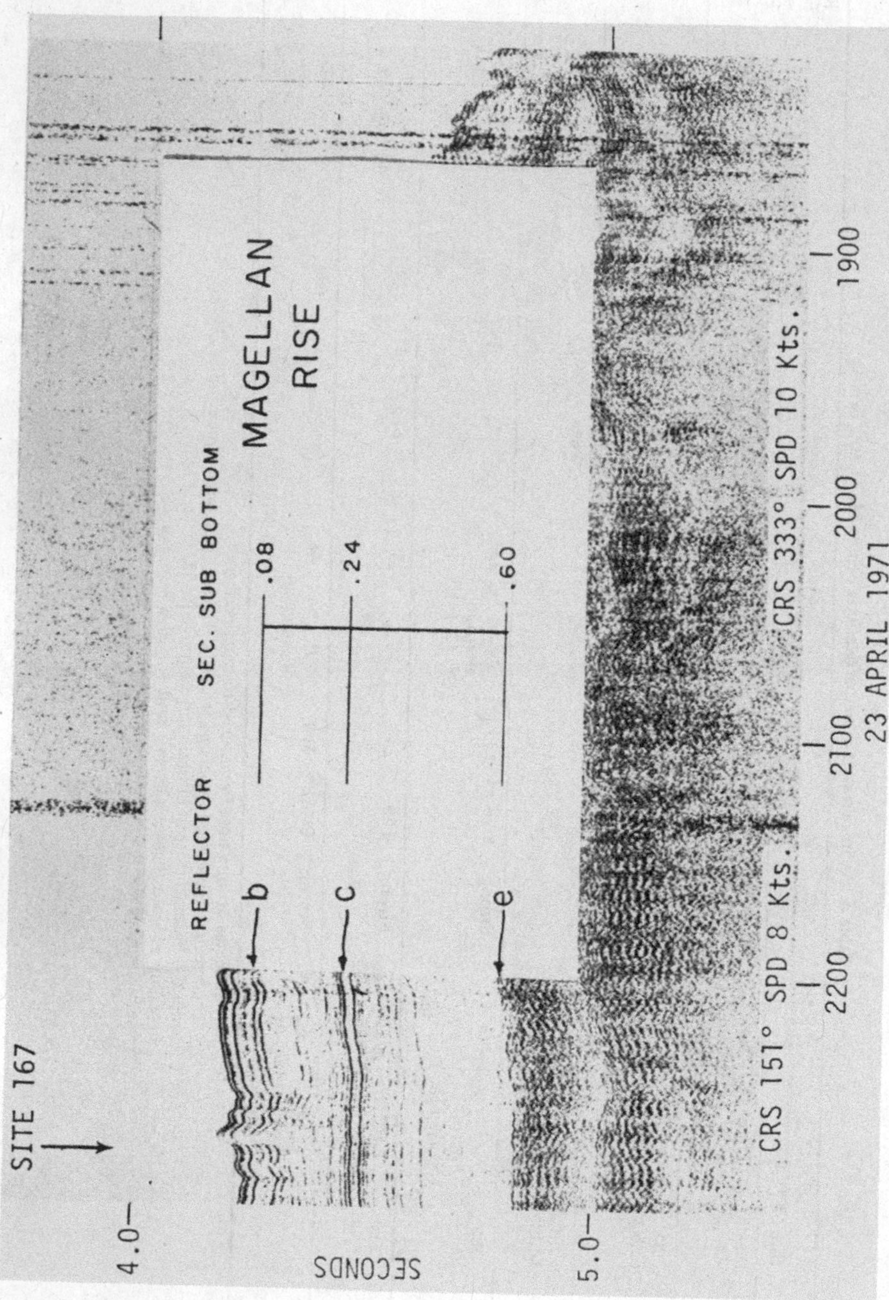

Fig. 9. Seismic profile of Site 167 and vicinity showing reflectors discussed in text.

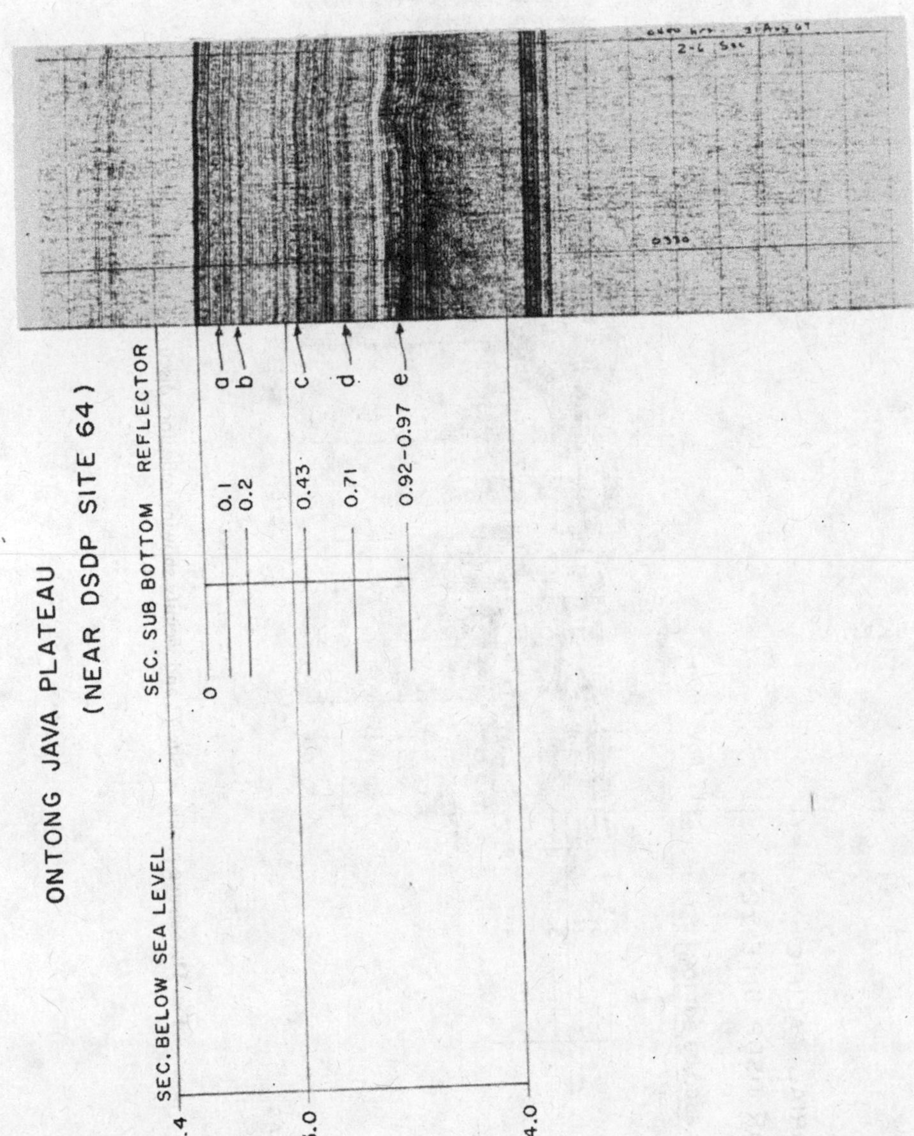

Fig. 10. Seismic profile near Site 64 showing reflectors discussed in text.

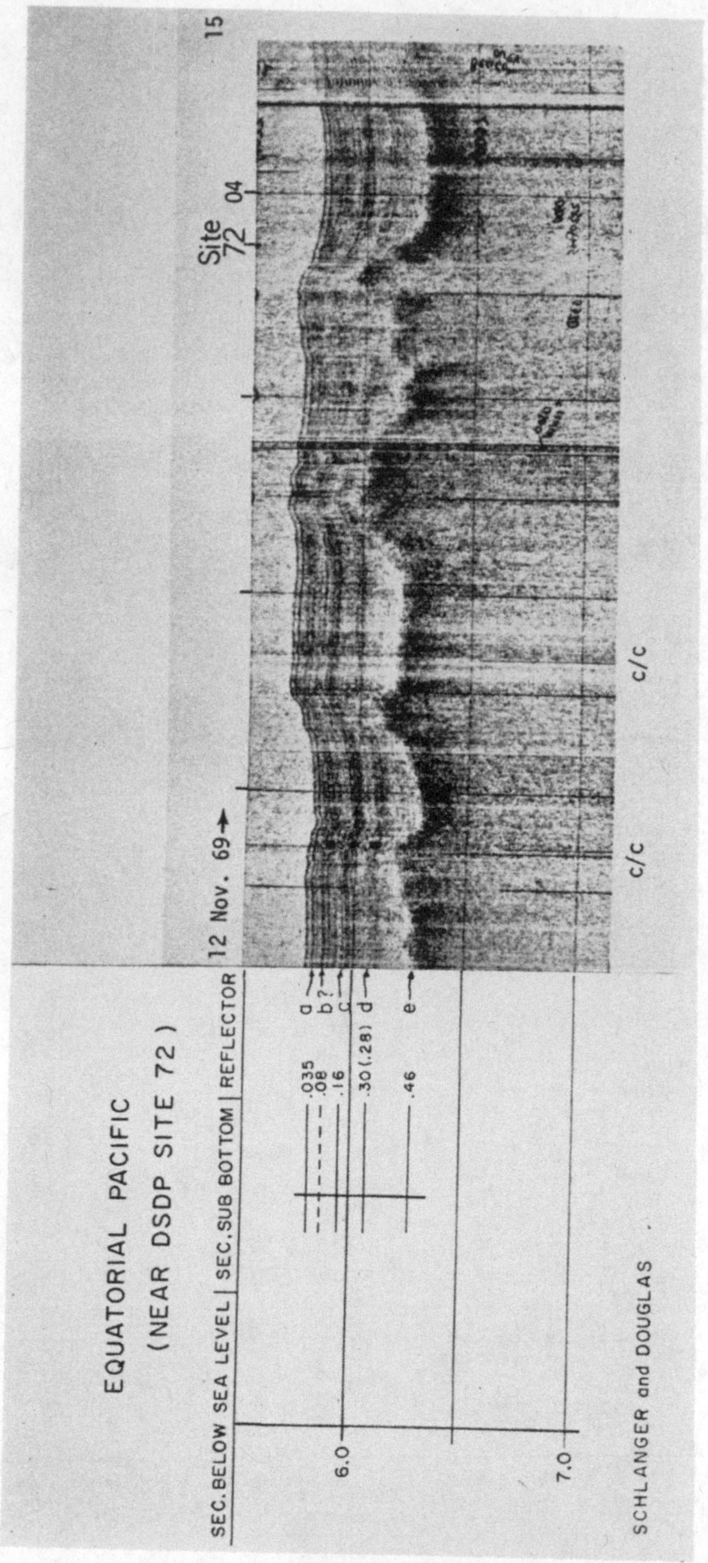

Fig. 11. Seismic profile of Site 72 and vicinity showing reflectors discussed in text.

In this manner the correlations shown on Table 3 were generated. The biostratigraphic zone assignments and ages are taken from the data published in the Deep Sea Drilling Project reports for these sites.

Certain events appear significant with respect to the 'a', 'b', 'c' and 'd' reflectors which date at approximately 3, 5–6, 12–14 and 21–26 million years B.P. respectively. Berger (1973a, Fig. 12); see Fig. 12 of this paper) points to two major fluctuations in the calcite compensation depth (CCD), that stand out above the 'noise' of shorter range events, at approximately 13 and 27 million year B.P. These two dates correlate with major temperature changes in the Pacific (Devereux, 1967; Douglas & Savin, 1971, 1973; Savin, Douglas & Stehli, in press) A rapid temperature decline in the Eocene reached a minimum in the Late Oligocene, about 27 million years B.P. and corresponds to the last and probably major phase of Palaeogene glaciation in the southern hemisphere (Savin *et al.*, in press; Hayes *et al.*, 1973) At approximately 12–14 million years B.P., the first significant temperature decline of the Neogene occurred. This event appears related to initiation of major glaciation in the northern hemisphere, glaciation having started earlier in the southern hemisphere (Hayes *et al.*, 1973). These dates also correlate very well with the periods of emergence of atolls in the Pacific (Fig. 12) described by Schlanger (1963), Emery, Tracey & Ladd (1954) and Ladd, Tracey & Gross (1970). The coincidence of these periods of atoll emergence with the 'deepenings' of the CCD, suggests that these events both correlate perhaps with glacio-eustatic changes in the depth of the Pacific Basin. Thus the 'c' and 'd' reflectors correlate well with glacio-eustatic regressions, the 'a' reflector at 3 million years B.P. correlates with a widespread cool period discussed by Ciaranfi & Cita (1973, pp.

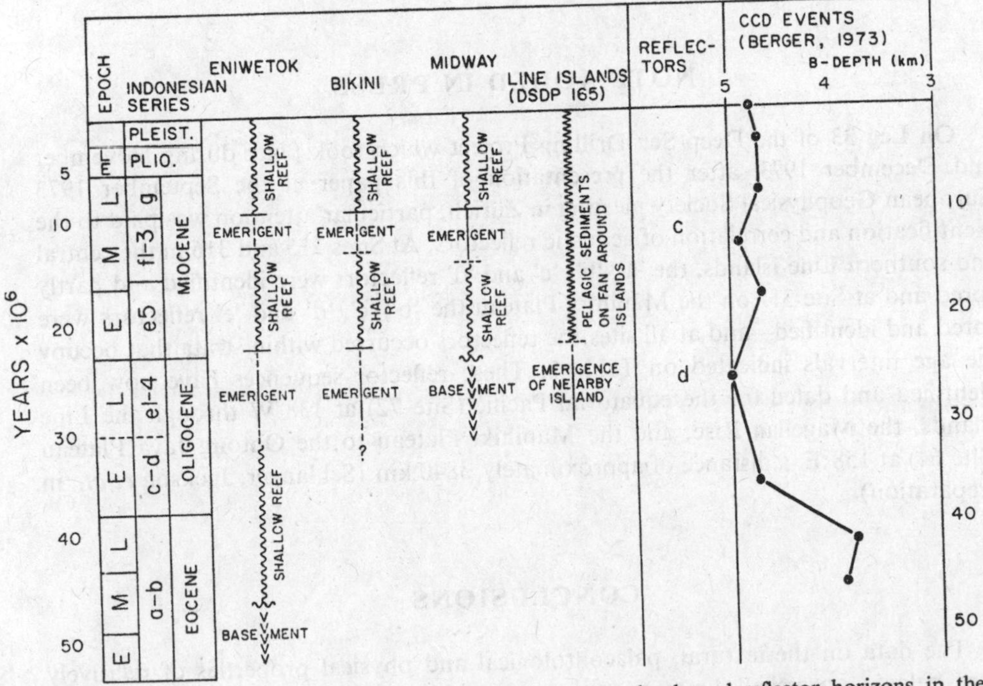

Fig. 12. Geological events, variations in calcite compensation depth and reflector horizons in the Pacific.

1396–98) and Cita & Ryan (1973, pp. 1408–09). The 'b' reflector correlates with the Messinian regression (5·5–7·0 million years B.P.) (see discussion in Ryan, Hsü *et al.*, 1973) and a glacial climax in Antarctica (Hayes *et al.*, 1973). Evidently the sediments deposited during the regression had a high diagenetic potential. Arrhenius (1952) postulated that during glacial intervals there would be a higher carbonate productivity due to increased upwelling. Thus a greater rate of carbonate deposition might account for the higher diagenetic potential and, therefore, the presence of the reflectors.

Another kind of event that appears to correlate with the reflectors is tectonic in nature—although these may be directly related to changes in basin depths as shown by Rona (1973). According to his analysis of sedimentation rates on continental shelves, marine transgression-regression cycles and sea-floor spreading rates, fast spreading correlates with transgression and slow spreading rates correlate with regression. Two major changes in either Pacific plate direction or velocity appear to have taken place, at approximately 25 million years B.P. (Dott, 1969) and at approximately 10 million years B.P. (data summarized by Hays *et al.*, 1972).

There are large numbers of minor local reflectors, e.g. the multiplicity of these on the Ontong Java Plateau. These are spaced only 10 m or so apart; at a sedimentation rate of approximately 25 m/million years (at Site 64): these would represent events that took place about 400,000 years apart. These may then represent fluctuations in diagenetic potential due to short-range changes in surface-water conditions. Short-term temperature fluctuations in surface temperatures, with a periodicity of approximately 80,000 years, have been identified in carbonate sequences in the Pacific and Atlantic (Savin *et al.*, in press). Changes of this order of time magnitude suggest correlations with events such as shifts in upper mixed-layer productivity discussed by Arrhenius (1963, Fig. 39).

NOTE ADDED IN PRESS

On Leg 33 of the Deep Sea Drilling Project which took place during November and December 1973, after the presentation of this paper at the September 1973 European Geophysical Society meeting in Zürich, particular attention was paid to the identification and correlation of acoustic reflectors. At Sites 315 and 316 in the central and southern Line Islands, the 'a', 'b', 'c' and 'd' reflectors were identified and partly cored and at Site 317 on the Manihiki Plateau the 'b', 'c', 'd' and 'e' reflectors were cored and identified—and at all sites the reflectors occurred within strata that occupy the age intervals indicated on Table 3. These reflector sequences have now been identified and dated for the equatorial Pacific (Site 72) at 138°W through the Line Islands, the Magellan Rise, and the Manihiki Plateau to the Ontong Java Plateau (Site 64) at 158°E, a distance of approximately 3840 km (Schlanger, Jackson *et al.*, in preparation).

CONCLUSIONS

The data on the textural, palaeontological and physical properties of relatively pure pelagic ooze-chalk-limestone sequences can be interpreted as indicating the following.

(1) That the general trend towards increasing lithification with length and depth of burial is interrupted by local reversals in the degree of lithification. Thus soft, plastic, oozes occur below stiff but friable chalks and the latter occur below harder, more dense limestones.

(2) That these reversals are due to variations in the amount of calcite cement in the rock and that this amount is somewhat independent of depth of burial.

(3) That pelagic carbonate oozes, because they contain an abundance of both: (a) calcite crystals, less than 1 μm in size, produced by the dissaggregation of small coccoliths and foraminiferal tests and (b) larger crystals such as discoaster segments and large coccolith elements, up to 10 μm in size, become cemented through a dissolution-diffusion-reprecipitation mechanism. The direction of diagenesis is irreversible because the dissolution of the very fine-grained calcite and the reprecipitation of the same volume of calcite on the larger crystals lowers the free energy of the system.

(4) That the degree to which a pelagic carbonate sediment becomes cemented depends on the diagenetic potential the sediment possessed on burial. The diagenetic potential is the measure of how much more diagenesis the sediment can be expected to undergo in the normal course of geological history.

(5) The diagenetic potential is a function of the following.

(a) The depth of water in which the sediment was deposited and the relation of this depth to the CCD level.

(b) The fertility of the upper mixed layers in terms of calcareous plankton production since this affects the sedimentation rate. The sedimentation rate is in part a function of the temperature and salinity structure of the upper mixed layers.

(c) The ratio of Foraminifera + small coccoliths: discoasters + large coccoliths in the sediment; this determines the size distribution characteristics of the sediment.

(d) Predation by plankton feeders as this affects the state of aggregation of the nannoplankton and therefore the sinking rate of this calcite.

(6) Acoustic reflectors are pre-determined in their characteristics and stratigraphic distribution by the diagenetic potential the reflecting horizon had upon burial. Reflectors then are related to palaeo-oceanographic events such as glacio-eustatic sea-level changes, alterations in calcite compensation depth and tectonic events that affected plate motion.

ACKNOWLEDGMENTS

This study was supported in part by the Institute of Geophysics and Planetary Physics, University of California, Riverside and the Marathon Oil Company. The authors wish to thank Professors L. H. Cohen and J. B. Combs and Samuel Savin for their suggestions and criticism. Peter Kolesar and Michael Arthur helped compile and analyse data.

REFERENCES

Adelseck, C.G., Geehan, G.W. & Roth, P.R. (1973) Experimental evidence for the selective dissolution and overgrowth of calcareous nannofossils during diagenesis. *Bull geol. Soc. Am.* **84**, 2755–2762.

ARRHENIUS, G. (1952) Sediment cores from the East Pacific. *Rep. Swed. deep Sea Exped.* (1947–1948), Parts 1–4, **5,** 1–288.

ARRHENIUS, G. (1963) Pelagic sediments. In: *The Sea*, Vol. 3 (Ed. by M. N. Hill), pp. 655–727. Interscience Publishers, New York.

BATHURST, R.G.C. (1971) *Carbonate Sediments and Their Diagenesis,* pp. 620. Elsevier Publishing Co., Amsterdam, London, New York.

BERGER, W.H. (1971) Sedimentation of planktonic Foraminifera. *Mar. Geol.* **11,** 325–358.

BERGER, W.H. (1973a) Cenozoic sedimentation in the Eastern Tropical Pacific. *Bull. geol. Soc. Am.* **84,** 1941–1954.

BERGER, W.H. (1973b) Deep-sea carbonates. Pleistocene dissolution cycles. *J. Foram. Res.* **3,** 187–195.

BERGER, W.H. & VON RAD, U. (1972) Cretaceous and Cenozoic sediments from the Atlantic Ocean. In: *Initial Reports of The Deep Sea Drilling Project,* Vol. XIV (D. E. Hayes, A. C. Pimm *et al.*), pp. 788–854. U.S. Government Printing Office, Washington.

BRAMLETTE, M.N. (1958) Significance of coccolithophorids in calcium-carbonate deposition. *Bull. geol. Soc. Am.* **69,** 121–126.

BUKRY, D. (1973) Coccolith stratigraphy, eastern Equatorial Pacific, Leg 16, D.S.D.P. In: *Initial Reports of the Deep Sea Drilling Project,* Vol. XVI (Tj. H. van Andel, G. R. Heath *et al.*), pp. 653–712. U.S. Government Printing Office, Washington.

BYRNE, J.G. (1965) *Recovery, Recrystallization and Grain Growth,* pp. 175. Macmillan, New York.

CIARANFI, N. & CITA, M.B. (1973) Paleontological evidence of changes in the Pliocene climates. In: *Initial Reports of the Deep Sea Drilling Project,* Vol. XIII (W. B. F. Ryan, K. J. Hsü *et al.*), pp. 1367–1399. U.S. Government Printing Office, Washington.

CITA, M.B. & RYAN, W.B.F. (1973) Time scale and general synthesis. In: *Initial Reports of the Deep Sea Drilling Project,* Vol. XIII (W. B. F. Ryan, K. J. Hsü *et al.*), pp. 1405–1415. U.S. Government Printing Office, Washington.

COOK, F.M. & COOK, H.E. (1972) Physical properties synthesis. In: *Initial Reports of the Deep Sea Drilling Project,* Vol. IX (J. D. Hays *et al.*), pp. 645–646. U.S. Government Printing Office, Washington.

DAVIES, T.A. & SUPKO, P.R. (1973) Oceanic sediments and their diagenesis: some examples from deep-sea drilling. *J. sedim. Petrol.* **43,** 381–390.

DEVEREUX, I. (1967) Oxygen isotope palaeotemperatures on New Zealand Tertiary fossils. *N. Z. Jl Sci.* **74,** 49–57.

DOTT, R. H. JR (1969) Circum-Pacific Late Cenozoic structural rejuvenation: implications for sea floor spreading. *Science,* **166,** 874–876.

DOUGLAS, R.G. (1973a) Benthonic foraminiferal biostratigraphy in the Central North Pacific. *Initial Reports of the Deep Sea Drilling Project*, Vol. XVII (E. L. Winterer, J. I. Ewing *et al.*), pp. 607–671. U.S. Government Printing Office, Washington.

DOUGLAS, R.G. (1973b) Planktonic foraminiferal biostratigraphy. In: *Initial Reports of the Deep Sea Drilling Project,* Vol. XVII (E. L. Winterer, J. I. Ewing *et al.*), pp. 673–694. U.S. Government Printing Office, Washington.

DOUGLAS, R.G. & SAVIN, S.M. (1971) Isotopic analysis of planktonic Foraminifera from the Cenozoic of the northwest Pacific. In: *Initial Reports of the Deep Sea Drilling Project,* Vol. VI (A. G. Fischer *et al.*), pp. 1123–1127. U.S. Government Printing Office, Washington.

DOUGLAS, R.G. & SAVIN, S.M. (1973) Oxygen and carbon isotope analyses of Cretaceous and Tertiary Foraminifera from the central north Pacific. In: *Initial Reports of the Deep Sea Drilling Project,* Vol. XVII (E. L. Winterer, J. I. Ewing *et al.*), pp. 591–605. U.S. Government Printing Office, Washington.

EMERY, K.O., TRACEY, J.I., JR & LADD, H.S. (1954) Geology of Bikini and nearby Atolls. *Prof. Pap. U.S. geol. Surv.* **260-A,** 1–255.
ment Printing Office, Washington.

FISCHER, A.G., HONJO, S. & GARRISON, R.W. (1967) *Electron Micrographs of Limestones and their Nannofossils,* pp. 137. Princeton University Press.

GEALY, E.L. (1971) Saturated bulk density, grain density and porosity of sediment cores from the Western Equatorial Pacific. In: *Initial Reports of the Deep Sea Drilling Project,* Vol. VII (E. L. Winterer *et al.*), pp. 1084–1104. U.S. Government Printing Office, Washington.

HAMILTON, E.L. (1959) Thickness and consolidation of deep-sea sediments. *Bull. geol. Soc. Am.* **70,** 1399–1424.

HAMILTON, E.L. (1971) Elastic properties of marine sediments. *J. geophys. Res.* **76,** 579–604.

HAYES, D.E., FRAKES, L.A., BARRETT, P., BURNS, D.A., CHEN, P.-H., FORD, A.B., KANEPS, A.G., KEMP, E.M., MCCOLLUM, D.W., PIPER, D.J.W., WALL, R.E. & WEBB, P.N. (1973) Leg 28 deep-sea drilling in the Southern Ocean. *Geotimes,* **18** (6), 19–24.

HAYS, J.D. *et al.* (1972) *Initial Reports of the Deep Sea Drilling Project,* Vol. IX, pp. 1205. U.S. Government Printing Office, Washington.

LADD, H.S., TRACEY, J.I., JR & GROSS, G. (1970) Deep drilling on Midway Atoll. *Prof. Pap. U.S. geol. Surv.* **680-A,** A1–A21.

LAUGHTON, A.S. (1954) Laboratory measurements of seismic velocities in ocean sediments. *Proc. R. Soc.* **222,** 336–341.

LAUGHTON, A.S. (1957) Sound propagation in compacted oceanic sediments. *Geophysics,* **22,** 233–260.

LISITZIN, A.P. (1971) Distribution of carbonate microfossils in suspension and in bottom sediments. In: *The Micropalaeontology of the Oceans* (Ed. by B. M. Funnell and W. R. Riedel), pp. 197–218. Cambridge University Press, London.

MOBERLY, R., JR & HEATH, R. (1971) Carbonate sedimentary rocks from the western Pacific: Leg 7, Deep Sea Drilling Project. In: *Initial Reports of the Deep Sea Drilling Project,* Vol. VII (E. L. Winterer *et al.*), pp. 977–986. U.S. Government Printing Office, Washington.

MOORE, D.G. (1969) Reflection profiling studies of the California continental borderland. *Spec. Pap. geol. Soc. Am.* **107,** 142 pp.

NAFE, J.E. & DRAKE, C.L. (1963) Physical properties of marine sediments. In: *The Sea* (Ed. by M. N. Hill), pp. 794–813. Interscience Publishers, New York.

PARKER, F.L. (1954) Distribution of the Foraminifera in the northeastern Gulf of Mexico. *Bull. Mus. comp. Zool.* **III,** 454.

PARKER, F.L. & BERGER, W.H. (1971) Faunal and solution patterns of planktonic Foraminifera in surface sediments of the South Pacific. *Deep Sea Res.* **18,** 73–107.

PETERSON, M.N.A. (1966) Calcite: rates of dissolution in a vertical profile in the Central Pacific. *Science,* **154,** 1542–1544.

RONA, P.A. (1973) Relations between rates of sediment accumulation on continental shelves, sea-floor spreading and eustasy inferred from the Central North Atlantic. *Bull. geol. Soc. Am.* **84,** 2851–2872.

ROTH, P.H. & THIERSTEIN, H. (1972) Calcareous nannoplankton. In: *Initial Reports of the Deep Sea Drilling Project,* Vol. XIV (D. E. Hayes, A. C. Pimm *et al.*), pp. 421–486. U.S. Government Printing Office, Washington.

RYAN, W.B.F. & HSÜ, K.J. *et al.* (1973) *Initial Reports of the Deep Sea Drilling Project,* Vol. XIII, pp. 1447. U.S. Government Printing Office, Washington.

SAVIN, S.A. & DOUGLAS, R.G. (1973) Stable isotope and magnesium geochemistry of Recent planktonic Foraminifera from the South Pacific. *Bull. geol. Soc. Am.* **84,** 2327–2342.

SAVIN, S.M., Douglas, R.G. & STEHLI, F.G. (in press) Tertiary marine paleotemperatures. *Bull. geol. Soc. Am.*

SCHLANGER, S.O. (1963) Subsurface geology of Eniwetok Atoll. *Prof. Pap. U.S. geol. Surv.* **260-BB,** 901–1066.

SCHLANGER, S.O., DOUGLAS, R.G., LANCELOT, Y., MOORE, T.C. & ROTH, P. (1973) Fossil preservation and diagenesis of pelagic carbonates from the Magellan Rise, Central North Pacific Ocean. In: *Initial Reports of the Deep Sea Drilling Project,* Vol. XVII (E. L. Winterer, J. I. Ewing *et al.*), pp. 467–527. U.S. Government Printing Office, Washington.

SCHLANGER, S.O., JACKSON, E.D. *et al.* (in preparation) *Initial Reports of the Deep Sea Drilling Project,* Vol. XXXIII. U.S. Government Printing Office, Washington.

SCHOTT, W. (1935) Die Foraminiferen in dem äquatorialen Teil des Atlantischen Ozeans. *Dt. Atl. Exped. Meteor* 1925–1927, **3,** 43–143.

SMAYDA, T.J. (1971) Normal and accelerated sinking of phytoplankton in the sea. *Mar. Geol.* **11,** 105–122.

THIEDE, J. (1972) Planktonische Foraminiferen in Sedimenten vom ibero-marokkanischen Kontinentalrand. *'Meteor' Forsch. Ergebn.* **7,** 15–102.

TRACEY, J.I., JR *et al.* (1971) *Initial Reports of the Deep Sea Drilling Project,* Vol. VIII, pp. 1037. U.S. Government Printing Office, Washington.

TSCHEBOTARIOFF, G.P. (1952) *Soil Mechanics, Foundations and Earth Structures*, pp. 645. McGraw-Hill, New York.

WINTERER, E.L. *et al.* (1971) *Initial Reports of the Deep Sea Drilling Project*, Vol. VII, pp. 1756. U.S. Government Printing Office, Washington.

WINTERER, E.L., EWING, J.I. *et al.* (1973) *Initial Reports of the Deep Sea Drilling Project*, Vol. XVII, pp. 930. U.S. Government Printing Office, Washington.

WISE, S.W., JR (1972) Calcite overgrowths on calcareous nannofossils: a taxonomic irritant and a key to the formation of chalk. *Abst. geol. Soc. Am.* **4,** 115–116.

WISE, S.W., JR (1973) Calcareous nannofossils from cores recovered during Leg 18, Deep Sea Drilling Project: biostratigraphy and observations of diagenesis. In: *Initial Reports of the Deep Sea Drilling Project*, Vol. XVIII (L. D. Kulm, R. von Heune *et al.*), pp. 565–615. U.S. Government Printing Office, Washington.

WISE, S.W., JR & KELTS, K.R. (1972) Inferred diagenetic history of a weakly silicified deep sea chalk. *Trans. Glf-Cst Ass. geol. Socs*, **22,** 177–203.

Spec. Publs int. Ass Sediment. (1974) **1**, 149–176

Some aspects of cementation in chalk*

JOACHIM NEUGEBAUER

Geologisch-Paläontologisches Institut der Universität Tübingen, Sigwartstrasse 10, *D-74 Tübingen, West Germany*

ABSTRACT

Three aspects of chalk diagenesis are discussed: the production of cement by pressure solution, the magnesium content in the pore fluid of deep-sea cores, and the distribution of syntaxial cement.

The volume of cement produced by pressure solution of low-magnesium calcite is calculated as a factor of overload. The production of cement by this source is insignificant up to 300 m and volumes of 0·5–5% cement are not generated before an overload corresponding to 1000 m is attained. Below this depth there is an enhanced probability for consolidation of chalk. Only under favourable conditions can chalk withstand a cover of 2000–4000 m sediment without total hardening.

The magnesium content of the pore fluid, which is crucial for the diagenesis of chalk, decreases only slowly with depth. The deficient magnesium is mainly consumed by substitution in carbonate minerals. Pressure solution is held to be an important process for reduction of magnesium at greater depth. A concentration of 0·01 M magnesium seems to be critical for the hardening of chalk.

Chalk cement is normally a syntaxial cement. It avoids coccolith crystals, which are handicapped because of their size and geometrical shape. The solubility of coccolith overgrowths can differ from those of other fossils by several percent due to the small size of the crystal plates. Layers rich in fossils other than coccoliths (inoceramids, *Braarudosphaera*-beds) are favoured in the hardening process.

Dissolution of Foraminifera, observed in shelf-sea chalks, is consistent with an earlier hypothesis that the Late Cretaceous chalk sea floor was undersaturated with respect to aragonite. Unlike Foraminifera, coccoliths are somewhat protected against dissolution by the nature of their crystal faces.

INTRODUCTION

Chalk is an extraordinary deposit. It is the only carbonate sediment which, under certain circumstances, can endure an overload of 1500 m, or more, without hardening to limestone.

This is usually attributed to a primary deficiency of metastable carbonate minerals, of aragonite and high-magnesium calcite. As Neugebauer (1973) demonstrated, there

* Publication No. 7 of the Research Project 'Fossil-Diagenese' supported by the Special Research Programme (Sonderforschungsbereich) 53—Palökologie, at the University of Tübingen.

is, however, one further factor: the pore fluid must be oversaturated with respect to low-magnesium calcite; if not, pressure solution indurates the chalk. A certain magnesium content of the pore fluid suffices to maintain an adequate oversaturation.

The present paper continues to discuss the diagenesis of marine low-magnesium calcite sediments. Using the model of chalk diagenesis advanced by the author in his 1973 paper, one can calculate approximately both the amount of cement which results from pressure solution under increasing overload, and the maximal overload which chalk can withstand in contact with magnesium-bearing fluids before turning into a compact limestone.

Pore solutions usually become poor in magnesium with increasing depth and the point of issue is whether chalk sediments (nannofossils + Foraminifera) can actually bear the maximal overload without hardening at lesser depths due to magnesium reduction. Pore fluid analyses from deep-sea bore holes give information about the actual alteration of the magnesium content and the dependent operative processes.

For our purposes the important question is to know what magnesium concentration in the pore fluid suffices to buffer pressure solution. Up to now there are but few and contradictory results, which we shall compare with those of pore fluid analyses from deep-sea cores.

Finally, we shall observe where in (soft) chalk the cement was deposited. The (biogenous) calcite crystals of the constituent fossils influence precipitation of the small volume of cement present in chalk. Operative factors are discussed, especially differences in free energy of the biogenous crystals.

CALCULATION OF CEMENT QUANTITIES

Our intention is to predict, by calculation, what quantity of cement originates with increasing depth as a result of pressure solution, and what overload chalk can withstand in contact with magnesium-bearing fluids before being solidified.

It is a drawback that the calculations must be simplified in a number of respects, which means that the results are subject to a large margin of error.

The method of calculation is as follows.

(1) We estimate the stress which obtains at points of contact in the chalk. This we can do in two independent ways: (a) by means of oversaturation—with the aid of the function for pressure solution; (b) by means of overload, density and texture.

(2) From the stress values we obtain the size of the areas of contact, as well as the approximate volume, which must be dissolved in order to produce the contact areas.

Oversaturation

As is well known, pressure increases the solubility of calcite. The following processes normally take place under overload: (a) solution of calcite at points of contact; (b) supersaturation of the pore fluid; (c) deposition of cement as a result of supersaturation. For low-magnesium calcite, stage (c), the deposition of cement, is kinetically hampered in those cases where the pore fluid is characterized by a certain magnesium content: numerous papers have acquainted us with the fact that magnesium specifically impedes the growth and consequently the precipitation of low-magnesium calcite. According to Lippmann (1960, 1973, p. 114) magnesium ions occupying atomic sites on the crystal surface of low-magnesium calcite block growth due to their high hydration energy.

The magnesium content results in an increase in CO_3^{2-} and Ca^{2+} until the precipitation of aragonite or high-magnesium calcite commences, which means that the solution is oversaturated for low-magnesium calcite. The pore fluids of chalk are characterized by a magnesium content sufficient for kinetic inhibition: in the case of deep-sea chalks this has been shown by the results of the Deep Sea Drilling Project (see page 158) and for chalks of shelf areas their limited induration provides indirect evidence (Neugebauer, 1973).

The extent of oversaturation in the pore fluid of chalk is therefore known: the concentration of Ca^{2+} and CO_3^{2-} corresponds to the solubility of aragonite or high-magnesium calcite. It cannot be lower, since during subsidence a sufficient volume of $CaCO_3$ is dissolved as a result of pressure solution, and it cannot be higher, as otherwise one of the phases mentioned would be deposited.

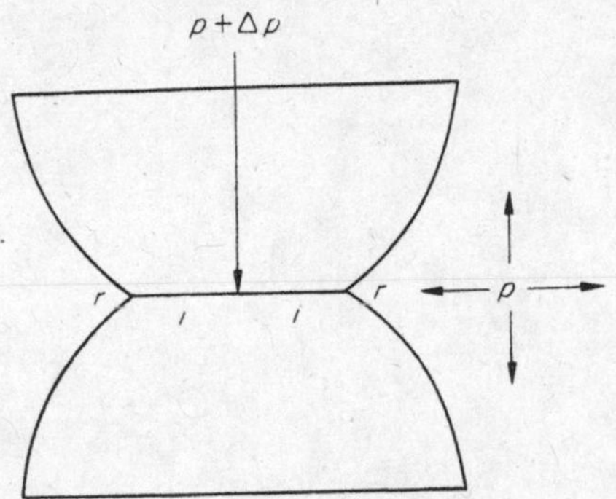

Fig. 1. Relation between stress Δp, hydrostatic pressure p, and total pressure at contact of particles $p+\Delta p$. r,i : rim and inner surface of the contact.

The effect of pressure solution

The effect of pressure on the solubility of calcite can be evaluated by the two functions (1) and (2) (cf. Neugebauer, 1973). Function (1), derived by Correns & Steinborn (1939), is valid for the rim (r) of contact and function (2), given by Owen & Brinkley (1941), for the inner surface (i) of contact (Fig. 1).

$$\ln \frac{K_{p+\Delta p}}{K_p} = \Delta p \frac{V}{RT} \qquad (1)$$

$K_{p+\Delta p}$ = solubility product at the pressure $p+\Delta p$;
K_p = solubility product at the pressure p;
Δp = stress;
V = molar volume of calcite;
R = gas-law constant;
T = absolute temperature.

$$\ln \frac{K_{p+\Delta p}}{K_p} = -\Delta \bar{V} \frac{(\Delta p - \Delta p^2 \Delta \bar{K}/2\Delta \bar{V}}{RT} \qquad (2)$$

$\Delta \bar{V}$ = partial molal volume change;
$\Delta \bar{K}$ = partial molal compressibility change;
other symbols as in equation (1).

The stress Δp is the crucial factor for pressure solution and is the difference between pressure on contact surfaces $(p+\Delta p)$ and the hydrostatic pressure p in the pore fluid (Fig. 1). Variations in the temperature have little effect and in Fig. 2a T is selected as $T_o = 298°K$. For other temperatures T the abscissa (stress Δp) in Fig. 2a must be amended by the factor T/T_o.

The reason for the pressure solution effect is the fact that the solution of calcite causes a volume reduction V and $\Delta \bar{V}$ respectively. The solution process reaches

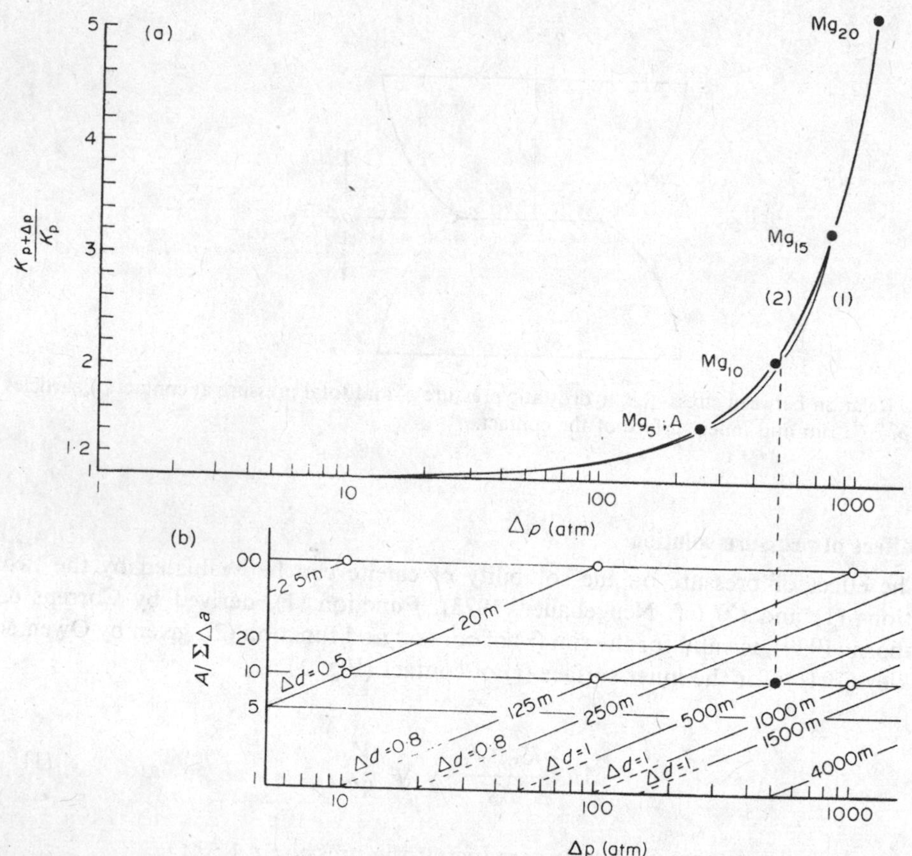

Fig. 2. Pressure solution of low-magnesium calcite (from Neugebauer, 1973). (a) Relation between the stress Δp and the ratio 'solubility product under stress $(K_{p+\Delta p})$ / solubility product without stress (K_p)'. Curve (1) after equation (1) and curve (2) after equation (2). A, Mg_{5-20}: solubility product of aragonite and high-magnesium calcite with 5–20 mol % magnesium compared to that of low-magnesium calcite. (b) Dependence of the stress Δp on superposition (2·5–4000 m), difference of density)Δd), and the ratio 'total area/contact area $(A/\Sigma\Delta a)$'. Compare equation (3).

equilibrium when the concentrations of the solution rise so far that the solubility product attains the value $K_{p+\Delta p}$ of the equations (1) and (2).

The reduction in volume for the two equations (systems) is of a different type. In system (1), where the stressed calcite is in direct contact with the pore fluid, the dissolved calcite disappears from the zone of stress and the volume reduction of equation (1) represents the (molar) volume of the calcite. Apart from an extremely small variation of the molar volume with variation of pressure, system (1) is not influenced by other factors.

In the system (2)—in the inner sphere of contact—the dissolved calcite passes into a 'solution film' (Weyl, 1959) (or into a surficial layer with marked transport characteristics), which is under the same high pressure $p+\Delta p$ as the solid calcite. In this case the volume reduction arises from the fact that the dissolved $CaCO_3$ ($+H_2O$) has a lesser volume than calcite ($+H_2O$). Equation (2) thus depends on the partial molal volume change. By chance it is of approximately the same order of magnitude as the molar volume of calcite. Both equations lead to almost the same result.*

Equation (2) has been corrected for compressibility, which is significant above about 500 atm Δp.

Both functions are shown in Fig. 2a. The abscissa of the diagram represents stress Δp (in atm); the ordinate indicates by what factor the solubility product under stress exceeds the solubility product with no stress obtaining; i.e. factor 2 thus means that the solubility product doubles itself.

The course of curve (2) continues to be influenced by the following variables, which were not taken into account in Fig. 2a: by the salt content of the pore fluid, which affects the partial molal volume and by the influence of stress on the dissociation constants of carbonic acid; the latter shifts curve (2) somewhat to the left (according to data from Culberson & Pytkowicz, 1968, p. 409). If one takes into account these variables (T included), then the same solubility product values (Fig. 2a) might emerge for stress values altered by some 10–30%.†

The above extensive discussion of sources of possible errors demonstrates that, strictly speaking, curve (2) should be understood not as a single curve, but rather as a group of curves. Calculation data are to be found in Neugebauer (1973).

Calculation of the stress Δp

With the aid of the diagram given in Fig. 2a we can calculate the stress obtaining at points of contact of the low-magnesium calcite particles. The stress results from the degree of oversaturation in chalk. In Fig. 2a the oversaturation is indicated by the points A and Mg_{5-20}, which correspond to the solubility product of aragonite and high-magnesium calcite (see p. 151).

The contact surfaces can only dissolve when the solution is undersaturated with

* The application of equation (2) to the solution film presupposes that the dissolved ions are hydrated as in a solution. Were equation (2) to be invalid, pressure solution alone would have to proceed in accordance with equation (1) at the edge of areas of contact and would advance from there inwards ('Bathurst's model', Neugebauer, 1973). Since functions (1) and (2) lead to the same results, our calculations of the pressure solution effect must, in any event, be of the correct order of magnitude.

† It is uncertain to what extent pressure affects the dissociation of $MgCO_3°$ and $CaCO_3°$. Pytkowicz, Disteche & Disteche (1967, p. 432) interpret discrepancies of some experimental data as due to this source of error (compare also Bathurst, 1971, p. 271).

respect to low-magnesium calcite. But since the solution is oversaturated, the stress must increase the solubility product of low-magnesium calcite by such an amount, that the oversaturation is compensated and exceeded. The 'compensation' stress necessary for this to happen can be read off the abscissa in Fig. 2a. The points A and Mg_{5-20} are attained by a stress of 250–1000 atm. Between these two limits, with a mean value of 500 atm, lies the compensation stress, which obtains at all points of particle contact in the chalk (during subsidence).

A higher stress value cannot maintain itself over long periods. Should, for example, the stress grow due to increased overload, the points of contact will dissolve in accordance with the increased solubility product until the stress has subsided to its compensation value.

Apart from the theoretical approach, the order of magnitude of the stress Δp can be estimated from equation (3):

$$\Delta p = \frac{t}{10}\,\Delta d\,\frac{A}{\Sigma\Delta a}\ \text{(atm)} \tag{3}$$

t = thickness of overload in metres;
Δd = difference between mean density of the sediment and density of pore fluid;
$\frac{A}{\Sigma\Delta a}$ = ratio of total area/contact area.

Besides the overload t and the difference of density between sediment and pore fluid Δd, the ratio $A/\Sigma\Delta a$ is also of great importance: A is the total area of a horizontal section and $\Sigma\Delta a$ the sum value of the contact areas of all particles bearing the overload in the section in question. Unfortunately in chalk $\Sigma\Delta a$ cannot be measured. From scanning electron micrographs we can deduce that $A/\Sigma\Delta a$ must be very large. The areas of contact may only rarely exceed 10–20% of the total horizontal section ($A/\Sigma\Delta a$=10–5), although most chalk occurrences were once overlain by many hundreds of metres. By this means we obtain a stress value in the order of several hundred atmospheres (Neugebauer, 1973).

Von Engelhardt (1960) provides us with a minimum value of stress Δp. 30–40% porosity was measured in chalk under an overload of 1500 m. The theoretical minimum value of $A/\Sigma\Delta a$ can be derived if we assume that the area of contact of each grain corresponds to its maximum horizontal cross-section (cf. columns). At 30–40% porosity (or 0·7–0·6 proportion of solid matter) $A/\Sigma\Delta a$ is 1/0·7–1/0·6=1·4–1·7 and under 1500 m overload and with Δd=1 this results in Δp=210–255 atm. The above-assumed 'columnar' mode of thought is, however, so naïve that the true value was certainly far in excess of this.

In Fig. 2b the stress Δp (abscissa) is plotted as a factor of $A/\Sigma\Delta a$ (ordinate) and various overloads t. For example: t=500 and $A/\Sigma\Delta a$=10 result in Δp=500 atm (black dot).

Theory and observation both point to a compensation stress of about 500 atm (with variation by a factor of 2), a value which serves as a basis for further estimates.

Areas of contact

The areas of contact grow under increasing overload. By how much they grow may be deduced from equation (3) (Fig. 2b).

In Fig. 2b the (mean) compensation stress of 500 atm is hatched. This vertical line denotes correlated pair-values for overload t and ratio $A/\Sigma\Delta a$ (e.g. $t=500$ m and $A/\Sigma\Delta a=10$). The reciprocal values of $A/\Sigma\Delta a$ is the sum value of the contact areas related to the total cross-section.

For the sake of clarity the correlated pair-values for overload and contact area are assembled in columns 1 and 2 in Fig. 3. Column 1 represents a section starting with 70 m overload and continuing to 4000 m overload. In this section the area of contact increases from 1% at 70 m to 100% at 4000 m. This implies that under an overload of 4000 m the chalk is totally hardened due to pressure solution; the correlation presupposes that pore fluids containing sufficient magnesium are present. We see furthermore that chalk under an overload of 250–1000 m—values which commonly occur in nature—should have a contact area of 4–20%.

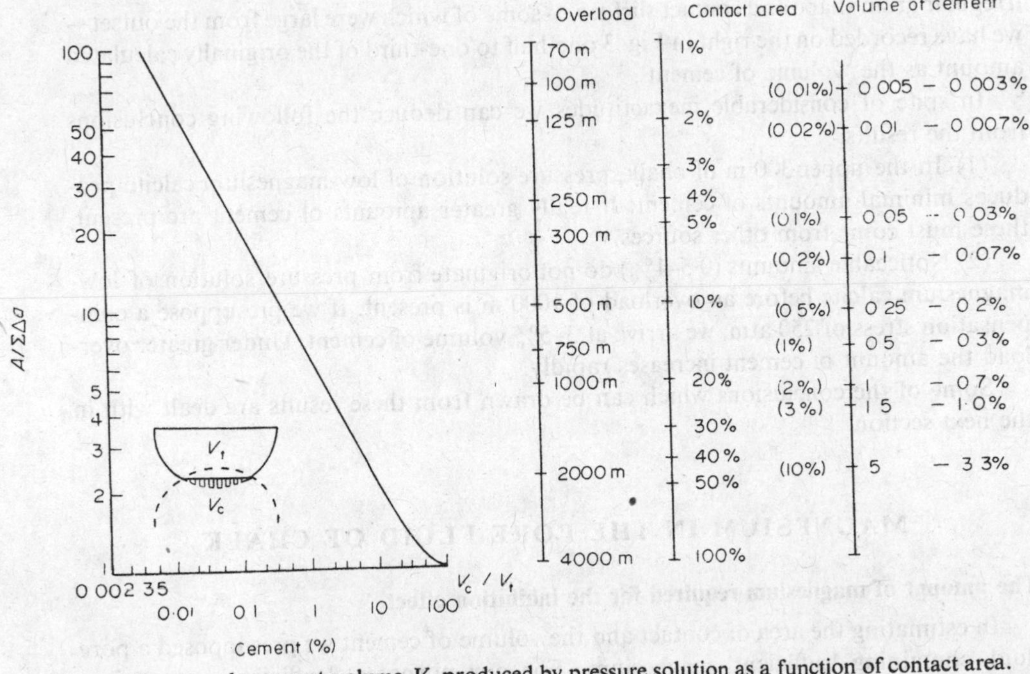

Fig. 3. Diagram: the cement volume V_c produced by pressure solution as a function of contact area. V_t=total volume of the solid matter. $A/\Sigma\Delta a$=ratio total area/contact area. Columns: relation between overload, contact area, and volume of cement at a compensation stress of 500 atm.

Let us also look at the results of a variation in values for stress Δp from 250–1000 atm (compare Fig. 2). At a compensation stress of 250 atm the values of overload (first column in Fig. 3) are approximately halved; at a compensation stress of 1000 atm they are nearly doubled (assuming $\Delta d=1{\cdot}25$ and $1{\cdot}5$). Total hardening then occurs at 2000 m and 7000 m respectively.

Cement volume

A much simplified model provides us with an approximation for the amount of cement resulting from pressure solution at various depths in the section.

We assume first of all that the sediment consists of spheres, which being densely packed touch each other in the manner shown in Fig. 3 (porosity is 48%). In Fig. 3 V_t is the volume of the hemisphere (the total volume of solid matter) and V_c represents the volume of the calotte (the volume of cement; hatched area). The abscissa V_c/V_t thus denotes the amount of cement given as a percentage of the total volume of solid matter. The cement volume increases nearly linearly with the area of contact (or with the decreasing ratio $A/\Sigma\Delta a$). The results are given in column 3 with the numerical values in brackets.

The use of this model exaggerates the amount of cement: each hemisphere has, instead of one surface of contact, on average about three upper or lower surfaces.* As the calculation shows, the volume of cement alters virtually in inverse proportion to the number of contact points; in the present case by one-third. On the other hand, the contact surfaces are no longer horizontal, but rather inclined, which increases the amount of cement somewhat. As a rough approximation, and bearing in mind the irregular grain shape and contact surfaces—some of which were large from the outset—we have recorded on the right in Fig. 3 one-half to one-third of the originally calculated amount as the 'volume of cement'.

In spite of considerable inexactitudes we can deduce the following conclusions from the results.

(1) In the upper 300 m of chalk, pressure solution of low-magnesium calcite produces minimal amounts of cement. If really greater amounts of cement are present, these must come from other sources.

(2) Noticeable amounts (0·5–1%) do not originate from pressure solution of low-magnesium calcite before an overload of 1000 m is present. If we presuppose a compensation stress of 250 atm, we arrive at 3–5% volume of cement. Under greater overload the amount of cement increases rapidly.

Some of the conclusions which can be drawn from these results are dealt with in the next section.

MAGNESIUM IN THE PORE FLUID OF CHALK

The amount of magnesium required for the inhibition effect

In estimating the area of contact and the volume of cement we presupposed a pore fluid containing 'sufficient' magnesium. For our purposes, 'sufficient' magnesium refers to a solution where the precipitation of low-magnesium calcite is inhibited to such an extent that aragonite or high-magnesium calcite appear. What is the lowest permissible magnesium content at which this effect can still occur?

Very few observations and experiments are available in this field.

(1) Müller, Irion & Förstener (1972) have investigated the conditions of formation of low-magnesium calcite, high-magnesium calcite, and aragonite in lakes. According to these authors low-magnesium calcite is formed when the Mg/Ca ratio of the solution is less than 0·8, whereas high-magnesium calcite is found where the Mg/Ca ratio exceeds 2·4. The minimum content permissible for the 'inhibition effect' lies within these limits; the above authors assume a value of about 2.

(2) In Lippmann's experiments (1960, 1973) low-magnesium calcite and aragonite

* According to von Engelhardt (1960, p. 5) the average number of contact points at 45% porosity is about 7, or 3·5 per hemisphere.

Table 1. Precipitation of $CaCO_3$ in the presence of varying amounts of Mg^{2+} (Lippmann, 1960)

Mg^{2+}/Ca^{2+} molar ratio	Calcite*	Aragonite	Mg^{2+} M
Less than 1·09	+	–	0·003
1·45	+/–	+	0·004
2·91	(–)/+	+	0·008
4·36 and above	–	+	0·012

* With maximally 5 mol % magnesium.

were precipitated from homogeneous solutions (T=20°C), in which the minerals did not appear until weeks or months had passed. Slow precipitation from homogeneous solution prevents the occurrence of aragonite solely as a consequence of momentary or local oversaturation. The results of two test series are given in Table 1.

Aragonite always appeared suddenly and in substantial amounts, indicating a considerable oversaturation with respect to aragonite (Lippmann, 1973, p. 110). When both minerals appear jointly, it is probable that the calcite was formed before the aragonite.

A third test series of similar duration, characterized by lower Ca^{2+} content and higher CO_3^{2-} rate of production, yielded the results presented in Table 2.

Table 2. Precipitation of $CaCO_3$ in the presence of varying amounts of Mg^{2+} (Lippmann, 1973)

Mg^{2+}/Ca^{2+} molar ratio	Calcite*	Aragonite	Mg^{2+} M
Less than 3·4	+	–	0·0024
5·7	+	+	0·004
11·5 and above	–	+	0·008

* With maximally 5 mol % magnesium.

Judging from the Mg/Ca ratio, it would appear that only the first two test series are compatible with our knowledge of the precipitation of aragonite and low-magnesium calcite in sea water and lakes; the third test series seem to be contradictory. They are only compatible with the first two when, instead of the Mg/Ca ratio, the absolute concentration of magnesium is taken as crucial. Lippmann suggests the alternative explanation that a magnesium concentration of '0·01 M appears to be the critical order of magnitude above which aragonite forms as the only phase at normal temperature'.

(3) Möller & Rajagopalan (1972) investigated the relationship between the Mg/Ca ratio of the surface layer of low-magnesium calcite and the Mg/Ca ratio of the solution. According to these authors, the Mg/Ca ratio at the surface (which is crucial for the inhibition effect) changes by (less than) 10% as the Mg/Ca ratio of the solution decreases from 5·5 (sea water) to 2. Down to the latter value we can thus expect almost the same inhibition effect as in sea water. Below 2 the magnesium of the surface layer rapidly becomes impoverished.

Although these data do not provide us with a clear answer to our question, they do at least define the limits within which the 'minimum content' of magnesium lies, this level determining that aragonite and high-magnesium calcite occur in place of low-magnesium calcite.

Two solutions appear feasible.

(1) The 'minimum content' occurs approximately at a Mg/Ca ratio of 1·0–2·0 (in sea water Mg/Ca = 5·0–5·5).

(2) The 'minimum content' occurs approximately at 0·01 M magnesium (in sea water Mg = 0·05–0·055 M).

Certainly, some inhibition should also occur below this limit. Thus Bischoff & Fyfe (1968) observed that concentrations of magnesium as low as 0·0001 M (Mg/Ca = 0·1) can delay the growth of low-magnesium calcite. Similar experiments of Taft (1967) indicated that magnesium concentrations from 0·001 to 0·002 M upwards (Mg/Ca = ?) can prevent the recrystallization of aragonite to low-magnesium calcite, at least within the period of observation, which was one year.

The diagenetic significance of these latter results is that they suggest that chalk does not harden abruptly, if the concentration of magnesium subsides below the discussed 'minimum content'. The compensation stress (see page 154) should only decrease rapidly below this level and the points of contact withstand less and less stress, without being dissolved. However, even very small magnesium concentrations should cause a noticeable reduction in the effect of stress.

Magnesium in the pore fluid of deep-sea chalks

The invaluable investigations of Manheim, Sayles, Chan, Waterman and other associated workers (1969–73) have yielded a large number of analyses of interstitial water from deep-sea boreholes. Here we can examine changes in the magnesium concentration and of the Mg/Ca ratio, and check whether the two alternative 'minimum contents' referred to above are exceeded in deep-sea chalks. Of special interest are those (few) cases, where chalk has hardened to limestone.

To begin with we shall consider all published geochemical studies of deep-sea boreholes in which primarily nannofossil oozes and chalks were encountered.

Pore-water composition and processes reducing the magnesium concentration

Figures 4 and 5 show those constituents of the interstitial water of chalk which are important for our purposes.* These are the molar concentrations of Mg^{2+}, Ca^{2+}, HCO_3^- and SO_4^{2-} as well as the Mg/Ca ratio, plotted against borehole depth. The concentrations of Na^+ and Cl^- and thus the salinity as such remain practically constant.

For the purpose of the following discussion it is appropriate to arrange the pore fluids of chalk into two groups according to calcium content:

(a) Fig. 4: the calcium content scarcely increases with depth.

(b) Fig. 5: the calcium content markedly increases with depth, comparable to the calcium increase of argillaceous sediments.

Between these two groups there is no definite division; in Fig. 4 a few boreholes, which are intermediate between the two groups, are represented hatched.

The interstitial waters of all deep-sea chalks (Figs 4, 5) are characterized by a relatively slight reduction of the magnesium content with increasing depth. This distinguishes them from the interstitial waters of argillaceous and volcanic successions,

* A small group of pore fluids influenced by evaporites has been excluded. Furthermore, for the sake of clarity, those series of pore-water samples which extended to depths of less than 100 m have been omitted, since their analyses show no further variation.

where the magnesium reduction is usually more pronounced (Manheim, Chan & Sayles, 1970b; Manheim & Sayles, 1971a). The bicarbonate content of the interstitial water of chalks varies remarkably little down to a depth of 500 m; even the sulphate level is scarcely reduced. This is yet another dissimilarity between the interstitial water of chalk and that of argillaceous sediments, the latter being characterized by marked changes in the SO_4^{2-} and HCO_3^- level. The interstitial water of chalk is thus seen to be relatively unreactive.

Correlations between some of the ionic displacements lead to certain inferences about the processes which alter the magnesium and calcium content of interstitial water in chalk, as shown by Manheim, Sayles, Chan, Waterman and other co-workers (1969–73).

The increase in calcium content with depth of the group in Fig. 5 cannot be attributed to increased solubility of $CaCO_3$, since otherwise the HCO_3^- content would have to show a similar increase. There are two possible sources for the additional calcium.

(1) Calcium is liberated from silicates (i.e. plagioclases, Sayles, Manheim & Waterman, 1971; Manheim, Chan & Sayles, 1970b).

(2) Calcium is released by substitution of magnesium in calcite (Manheim & Sayles, 1971b).

Processes concerning the magnesium concentration are of special interest in our context. As far as we know at present, the relatively slight magnesium reduction observed in chalks (Figs 4, 5) cannot be correlated with the formation of dolomite (Sayles, Manheim & Waterman, 1971). This is exemplified by borehole 10 (marked '*dol*' in Fig. 4), which is perhaps even characterized by magnesium increase in the dolomite-bearing section.

Silicate reactions can use up magnesium. A process in which magnesium replaces the iron of clay minerals (Drever, 1971) would appear to be important; this process is thought to take place during the formation of pyrite. It must manifest itself in the pore solution in such a way that the magnesium reduction is accompanied by an equivalent (double) sulphate reduction and bicarbonate increase.

This process can be seen in the SO_4^{2-} curve from borehole 94 (Fig. 4) which displays an unusual shape. Logically, this is accompanied by increased bicarbonate content and a slight calcium decrease. However, the decrease in magnesium concentration corresponding to the decrease of sulphate is not sufficient to explain fully the actual magnesium decrease, and is only adequate to explain the 'hump' in the magnesium curve.

In other boreholes the sulphate decrease is so minute that it is difficult to demonstrate the operation of this process. In boreholes 116 (Fig. 4) and 135 (Fig. 5) it possibly plays a supplementary role and is perhaps one of the reasons for the width of the parallel band of magnesium curves. However, we cannot in general attribute a role of decisive importance to this process in the diagenesis of chalks.

A third possible explanation for magnesium reduction is the substitution of calcium by magnesium in calcite (see above). Should calcite dissolve and reprecipitate in the form of a carbonate mineral with a higher magnesium content, we would expect an increase in calcium concentration in the pore water corresponding to the magnesium consumed.

Analyses support the hypothesis that this process takes place. The boreholes of Fig. 5—but also those of Fig. 4—show reductions in magnesium which, in the

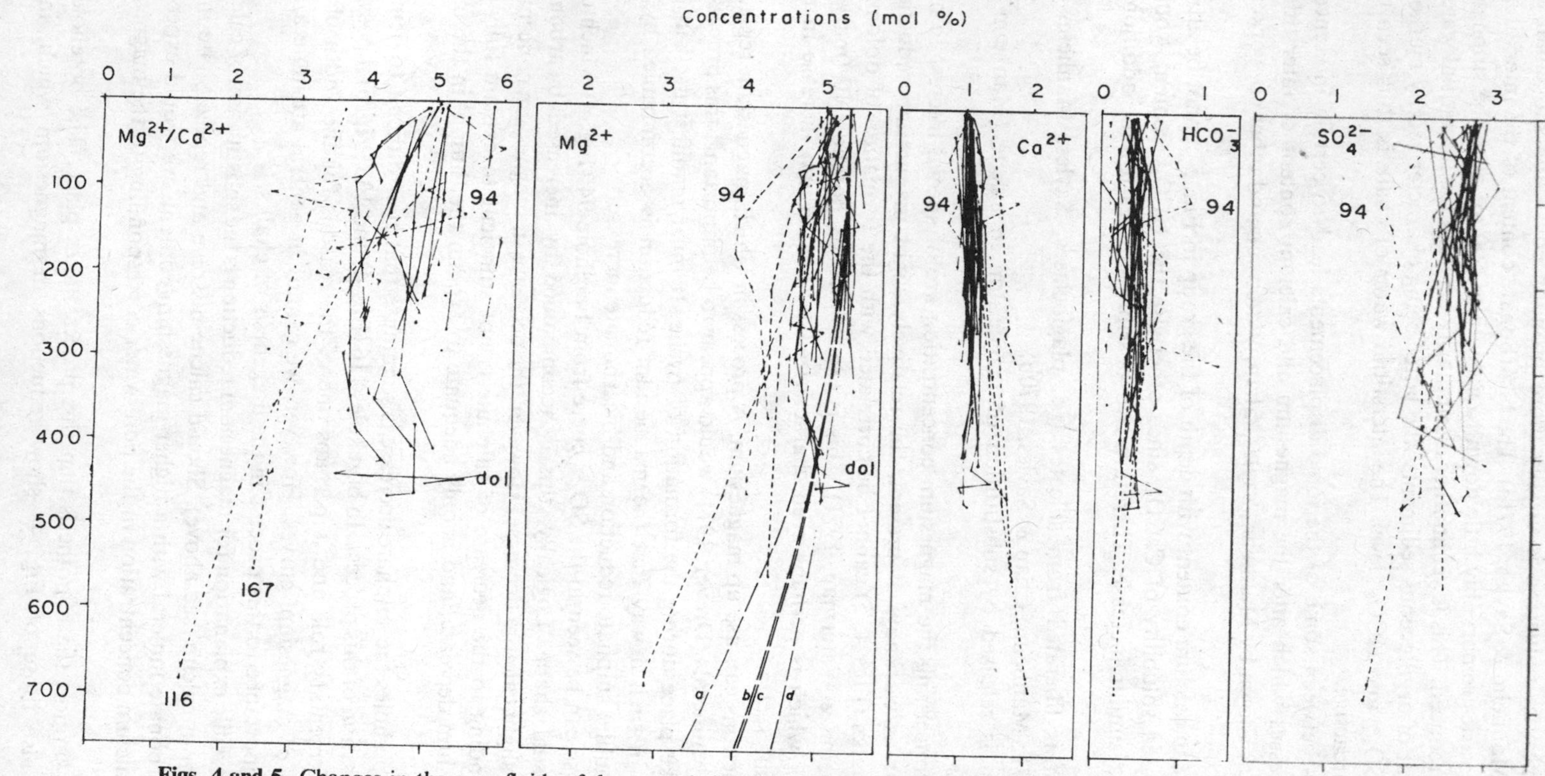

Figs. 4 and 5. Changes in the pore fluids of deep-sea chalk. Indicated ions are involved in processes consuming magnesium.

Fig. 4. Main group of pore fluids, showing small decrease of magnesium and small increase of calcium with depth.

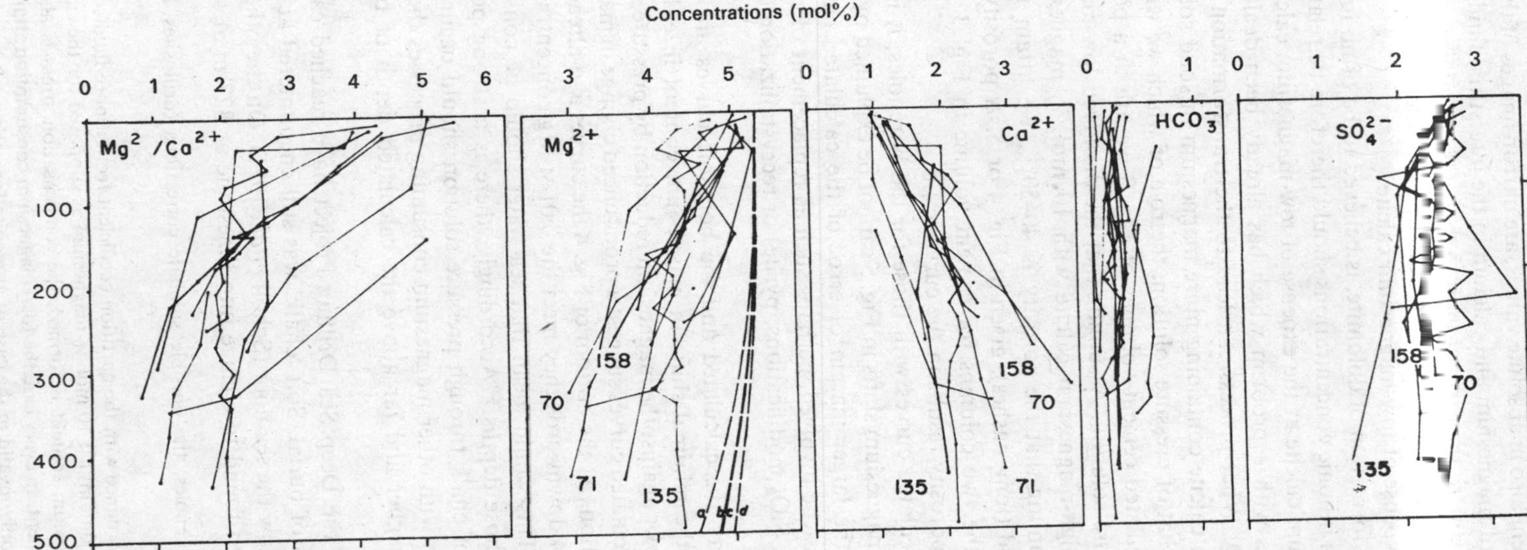

Fig. 5. More reactive pore fluids with respect to magnesium and calcium. Curves a, b, c, d: calculated decrease of magnesium caused by pressure solution. dol = dolomite occurrences; numbers refer to Deep Sea Drilling boreholes. Data in Figs 4 and 5 from Chan & Manheim (1970), Manheim *et al.* (1970a, b), Manheim & Sayles (1971a, b), Manheim, Sayles & Friedman (1969), Manheim, Sayles & Waterman (1972, 1973), Sayles & Manheim (1971), Sayles, Manheim & Chan (1970), Sayles, Manheim & Waterman (1971), Sayles, Waterman & Manheim (1972, 1973a, b) and Waterman, Sayles & Manheim (1972, 1973).

gradients and to a great extent also in absolute values, are mirror-images of the calcium increases.* The behaviour of magnesium and calcium in the interstitial fluids can thus be well explained if we postulate the formation of carbonates containing more magnesium in solid solution; dolomite at the expense of the available calcites, or high-magnesium calcite at the expense of low-magnesium calcite.

The first possibility, the formation of dolomite, is believed to be of no importance (see page 159). The changes in ionic concentrations should therefore be related to the formation of high-magnesium calcite at the expense of low-magnesium calcite.

This result confronts us with a problem which has already been dealt with by Manheim & Sayles (1971b). What process can lead to the transformation of stable low-magnesium calcite to a calcite containing more magnesium in solid solution? A possible process would be that of pressure solution, the role of which we will discuss in detail. By using the calculated cement volumes of Fig. 3 we are in a position to estimate the effect of pressure solution. Assuming that the dissolved low-magnesium calcite is precipitated as high-magnesium calcite with 10 mol % magnesium (see footnote † below) and assuming that the porosity is 40–50%, we attain the magnesium reduction curves a–d (long dashes), given in Fig. 4 or 5. A porosity of 40% combined separately with the two columns for cement volume in Fig. 3 gives the curves a + b, whereas 50% porosity results in the curves c + d.

If one compares the calculated curves with those for the boreholes, it is evident that the major calcium and magnesium shifts in Fig. 5 cannot be explained by pressure solution, even considering the larger margin of error of the calculated curves (see p. 155). Other processes must be involved, perhaps such as replacement of calcite by other growing minerals (e.g. SiO_2 modifications, pyrite) or recrystallization (see page 165).

In Fig. 4 the gradients of the calculated and the borehole curves are parallel; since this is mainly due to the minimal shifts in magnesium content in either case, this is no argument either for or against the magnesium reduction by pressure solution.

For greater depths the calculated curves suggest a pronounced change in magnesium content due to pressure solution. At the bottom of Fig. 4 the curves a–d already begin to flatten. If we extend Fig. 4 downwards they meet the 0·01 M Mg concentration line between 1100 and 1650 m. This would mean that sea water should be considerably depleted in magnesium at those depths.† Accordingly, there is a strong possibility that extensive cementation of chalk through pressure solution should occur even at these depths. In combination with other magnesium-consuming processes the critical level may be even slightly higher and far above the calculated depth of burial of 4000 m.

So far, few boreholes of the Deep Sea Drilling Project have reached carbonate sediments under such depths of burial. Soft 'chalk' was still encountered at site 192 at a depth of 930–1050 m below the sea floor (Scholl *et al.*, 1971). On the other hand, a sequence of chalks has been found to change into limestone at 827 m at site 167;

* Comparing the individual boreholes, the possible sulphate corrections (drill sites 116, 135) should be borne in mind.

† If calcite is precipitated in accordance with the partition coefficient from a pore fluid containing magnesium, it does not normally contain just 10 mol % magnesium as supposed for the curve a–d. But, if calcite with a lower magnesium content is formed, the compensation stress is also lower, resulting in a greater volume of cement. In every case the total magnesium consumption attains nearly the same level. This conclusion is only invalid in the case of aragonite formation.

this limestone remained very porous as deep as 1172 m, where it overlies basalt (Schlanger *et al.*, 1973). In boreholes 288 and 289 'limestone with interbeds of chert' and 'Radiolaria-bearing limestone, siliceous limestone, nannochalk' underlie chalks at a depth of 840–988 m and 969–1262 m respectively. Unfortunately, no analyses of the interstitial water are as yet available from such depths.

The effect of strong decrease of magnesium concentration

As a next step, we shall examine published pore-fluid data with the following questions in mind. (a) To what concentrations does the magnesium content actually decrease? (b) Is there a change from chalk to limestone when the magnesium content falls below 0·01 M or below Mg/Ca = 1–2?

All known pore fluids from sedimentary sequences dominated by chalk (Figs 4, 5) contain about 0·03 M magnesium or more. This is three times the concentration suggested by Lippmann (1973) for the inhibition effect.

Intercalations of chalk in argillaceous and volcanic sequences sometimes contain less than 0·03 M magnesium in the pore fluid.

The magnesium content was found to drop below 0·01 M at three sites.

Site 1. Borehole 2 at 140 m: 'calcite caprock' of a salt dome. In this case a sharp drop in salinity was also observed (Manheim, Sayles & Friedman, 1969; Ewing *et al.*, 1969). Measured magnesium content = 0·008 M (Mg/Ca = 0·18).

Site 2. Borehole 53 at 174 m and 193–200 m: 'altered' (partly recrystallized) chalk ooze and limestones (Pimm, Garrison & Boyce, 1971). Andesite and basalt have been found associated with the carbonate rocks below 195·4 m.* Magnesium content = 0·007 M (Mg/Ca = 0·09).

Site 3. Borehole 155 at 490 m: 'dolomitic (chalk), massive and *well* indurated' (van Andel *et al.*, 1973, p. 23). Magnesium content = 0·008 (3) M (Mg/Ca = 0·27).

These cases indicate a transition of chalk into limestone at a concentration below 0·01 M Mg.

The Mg/Ca ratio drops at a number of sites into the 1–2 range (site 116 and most of the boreholes in Fig. 5), but excessive induration of chalk was not observed. Lithification or excessive induration of chalk was not reported even from the few cores in which the Mg/Ca ratio drops below 1.

(1) Site 155 at 515 m; chalk ooze. Mg/Ca = 0·34 (Mg = 0·012 M). Thus the magnesium content increases again below the 'dolomitic well indurated chalk' mentioned above. At the same time the sediment 'is again softer' (van Andel *et al.*, 1973).

(2) Site 137 at 265–382 m; 'nanno marl to chalk ooze'; Mg/Ca = 0·63–0·35 (Mg = 0·035–0·028 M).

The very limited available data thus suggests that the magnesium content must decrease to a very low level for the change to limestone to occur with less than a few thousand metres of overlying sediments. The order of magnitude of 0·01 M magnesium, given by Lippmann (1973), seems to be critical for the premature lithification of chalk. Connate waters often contain such small amounts of magnesium (von Engelhardt, 1973).

* On the basis of the oxygen isotope composition of altered carbonates Anderson (1973) states that the partial recrystallization observed in this borehole is not the result of thermal metamorphism, but 'probably a consequence of chemical changes in ambient pore waters resulting from the submarine weathering of volcanic material'.

So far we have disregarded chalk pore fluids influenced by evaporites. These are characterized not only by increased sodium and chloride concentrations, but also by a sometimes drastic increase of calcium and magnesium. The Mg/Ca ratio frequency conforms to Figs 4 and 5. In principle, additional magnesium from the evaporites may keep the magnesium content at a sufficiently high level to allow the chalk to reach great depths in a soft state. This mechanism may be active, for instance, in the deep subsurface of the North Sea, under the influence of Permian salts.

We cannot close this discussion without mentioning that a number of other limestones have been found in the Deep Sea Drilling Project: (1) limestones derived from redeposited shallow-water sediments; (2) siliceous limestones; (3) limestones in direct contact with basalt.

They are not considered here because their pore fluids have not been analysed. However, these limestones suggest other ways in which deep-sea carbonate sediments can be lithified. (1) In the case of redeposited shallow-water sediments we can assume considerable amounts of aragonite and high-magnesium calcite in the sediment, which behave totally differently during diagenesis from low-magnesium calcite (e.g. in the case of pressure solution; cf. Neugebauer, 1973). (2) A larger proportion of siliceous skeletons presumably contributes in various ways to lithification (cf. for example page 162). (3) Near to basalt the magnesium content of the pore solution may drop markedly as indicated in borehole 53 (see page 163 and footnote).

Apart from these limestones there is a fourth group of thin lithified or excessively hardened seams intercalated in soft chalk, which occur in the deep-sea as well as in the shelf environment. The consolidation of these beds, characterized by a high content of low-magnesium calcite fossils (other than coccoliths), is discussed in the next section.

DEPOSITION OF CEMENT IN CHALK

In this third section we will discuss the problem of where the deposition of cement in chalk takes place. The amount of carbonate cement present in chalk is small as long as a sufficient magnesium concentration in the pore fluid prevents the rapid transition to limestone. It must be borne in mind that this small amount of cement must not exclusively have been produced by the pressure solution of low-magnesium calcite. In the first few hundred metres of burial depth, where the cement production by this process is very small (Fig. 3), it is imaginable and sometimes probable (see page 162 and Fig. 5) that more cement originates from other sources, for example, the highly efficient pressure solution of high-magnesium calcite particles, or perhaps the replacements of low-magnesium calcite by non-carbonate minerals (see page 162). In contrast to the first part (Fig. 3) we include here all sources and discuss the total (small) amount of cement present.

Cement in typical chalk originally containing no or very small amounts of high-magnesium calcite and aragonite exhibits two features.

(1) Normally most cement appears to be syntaxial that is, it grows in optical continuity with the available biogenic calcites.*

* Isolated idiomorphic calcite crystals are sometimes interspersed between the coccolith plates. Their origin is open to question. They are possibly entirely of biogenic origin or they may possess a biogenic core (cf. Wise & Kelts, 1972, p. 183).

(2) Different groups of fossils (coccoliths, discoasters, Foraminifera, etc.) attract different amounts of cement.

These peculiarities can be observed in deep-sea chalk (see below), in experiments with samples of deep-sea chalk subjected to elevated temperatures and pressures (Adelseck, Geehan & Roth, 1973) and will be demonstrated by the following figures of fossils from shelf-sea chalks.

The first group we will look at are the coccoliths. Cement deposition is rather unimportant on coccolith crystals (Fig. 6) and it is difficult to find crystals with demonstrable overgrowth formation. The excrescences, marked by arrows (Fig. 6b), should be interpreted as one type of diagenetic overgrowth on coccoliths. They appear on certain crystal faces, whereas adjacent faces are smooth. Similar formations are found on larger coccolith crystals of the size of one micron or more. From deep-sea chalks, and from experiments, some overgrowth on coccolith crystals was reported by Berger & von Rad (1972), Wise & Kelts (1972) and Schlanger *et al.* (1973) and by Adelseck *et al.* (1973).

The same authors noted also the common occurrence of overgrowth on the bigger crystals of discoasters.

Foraminifera regularly show a certain amount of cement, particularly on the inner surface of their chambers (Fig. 7; Pimm *et al.*, 1971; Schlanger *et al.*, 1973). These inner surfaces are smooth in the living animal and are now studded with centripetal crystals of diagenetic origin. It should be noted that the crystals in Fig. 7 chiefly exhibit the cleavage rhombohedron as a growing face and that, by and large, the 'teeth' of overgrowth on foraminiferal walls are larger than on coccolith plates (Figs 6 and 7). Cement is preferentially deposited on Foraminifera rather than on coccoliths.

Disturbing in this context is that an additional process plays a role in the diagenesis of Foraminifera. As shown in Fig. 7, the foraminiferal walls no longer have their primary structure. Originally most of the Foraminifera occurring in chalk (all of the prevalent planktonic and some of the benthonic forms) were built up of 'prisms', which are probably in turn composed of minute crystals (compare Towe & Cifelli, 1967; thickness of the crystal units scarcely 0·2 μm). The originally compact walls of the foraminifers are altered to a porous state by dissolution and 'recrystallization', possibly because of the high solubility of small crystals (Fig. 10). The dissolution should predominantly have taken place at the bottom of the chalk sea, comparable to the solution processes in the present deep-sea environment (Berger, 1967, 1972; Thiede, 1971 and others).

This process screens the starting form and size of the crystals effected by overgrowth. It appears that the top surfaces of the prisms served as the basis for the overgrowth.

Here and there larger crystals grew vertically on the wall of Foraminifera, probably via enlargement of some of the syntaxial 'teeth'.* It should be noted that, with increasing size of the cement crystals, the number of crystal forms increases also.

On prisms of inoceramids and ossicles of crinoids relatively large epitaxial crystals are found, resulting in a large amount of cement. Figure 8 shows cement crystals in the pore space between two prisms of a disintegrated *Inoceramus* shell, whilst in Fig. 9 cement can be seen to have grown into the cavity between two articulating

* The composition of the large crystal in the centre of Fig. 7b was verified as $CaCO_3$ with an EDAX microprobe.

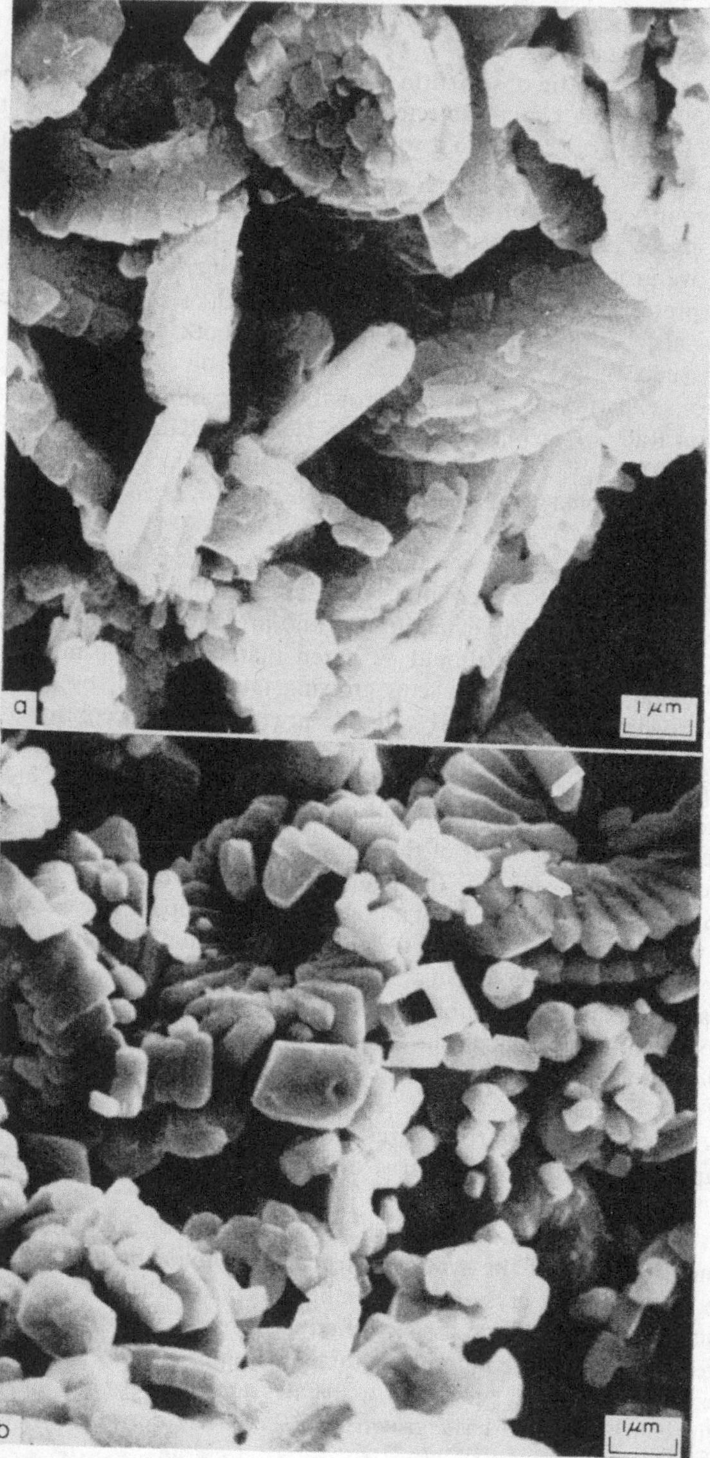

Fig. 6. Crystal size and crystal form influencing diagenesis: coccoliths. The arrows mark overgrowth. (a) Chalk of Kansas, *Uintacrinus-zone*. Scanning electron micrograph. (b) Chalk from Calais, France. Scanning electron micrograph.

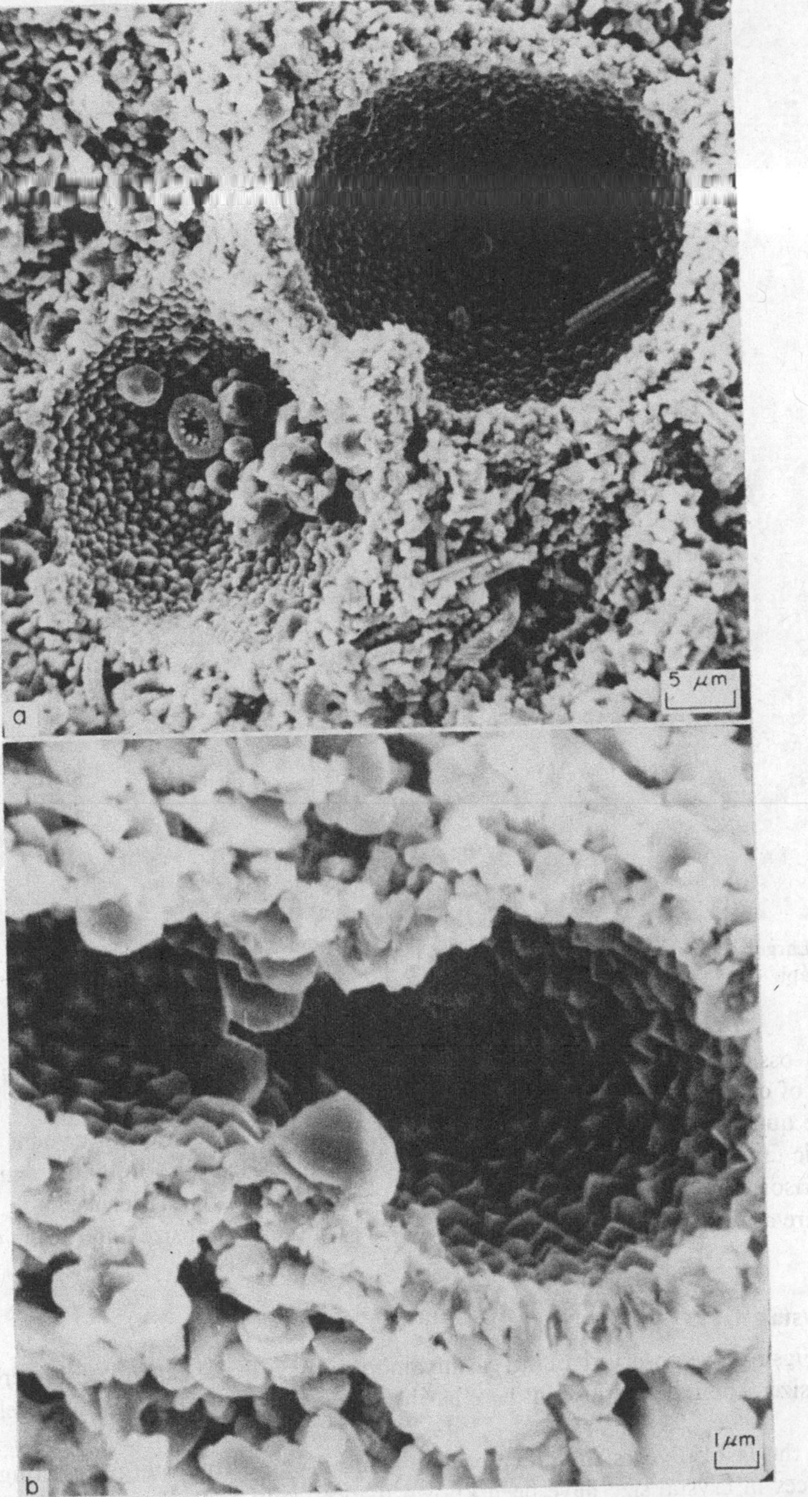

Fig. 7. Dissolution of the walls of Foraminifera and syntaxial overgrowth on the surface of their chambers. The small cement crystals chiefly exhibit the cleavage rhombohedron as a growing face. Chalk from Calais, France; shelf environment. Scanning electron micrographs. Numbers of the stereoscan electron micrographs of Figs 6–9: SEM Geol. Paläont. Inst. Tübingen Nos. 1447/31964: Fig. 6a; 1447/32963: Fig. 6b; 1447/33002: Fig. 7a; 1447/37741: Fig. 7b; 1447/33728: Fig. 8; 1447/33144: Fig. 9. The samples for the stereoscan studies (Figs 6–9) were simply broken and covered with Au/Pd or carbon.

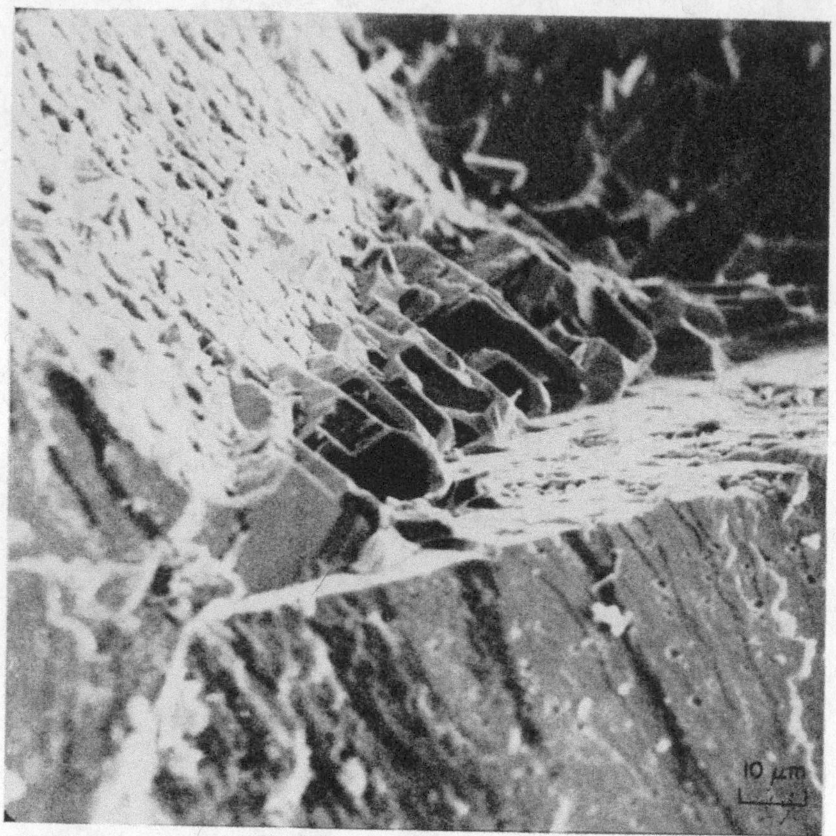

Fig. 8. Larger cement crystals on big prisms of *Inoceramus*. Chalk of Kansas. Scanning electron micrograph.

crinoid ossicles. The whole photomicrograph shows a part of a complicated single crystal of cement covering a biogenic single crystal of a crinoid ossicle.

The question now arises: what controls the selective deposition of cement on the biogenic crystals? Why is cementation on coccolith crystals discriminated against in comparison to Foraminifera, discoasters, inoceramids and echinoderms?

There are two factors of special importance: (1) the crystal size and (2) the crystal shape.

The crystal size

In Figs 6–9 we observed an increasing amount of syntaxial cement with increasing crystal size. The important point here is whether or not this is truly a causal relationship.

For the last 50 years the differences in solubility and free energy resulting from differences in crystal size have been known theoretically and experimentally. Differences in crystal size are only important for very small crystals, whilst they can be neglected if the crystals are larger (cf. Spangenberg, 1935). The size of calcite crystals in nannofossils and microfossils of chalk are just in between these two categories and we must therefore establish whether or not the differences in the free energy (solubility

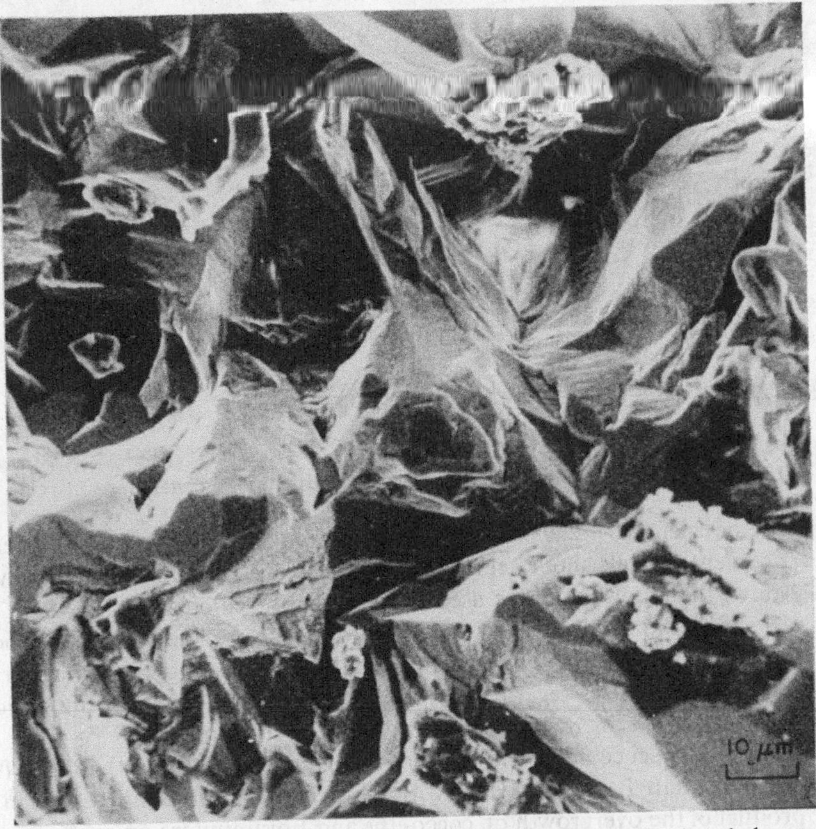

Fig. 9. Cementation on a crinoid ossicle from chalk of Rügen, Germany; the whole complicated overgrowth is optically a single crystal.

product) of the different-sized calcite crystals are great enough to effect the deposition of cement.

To begin with we will consider the theoretical principles for the dependence of the solubility product on crystal size (cf. Fig. 10). The theoretical principles for Fig. 10 were already given by Gibbs (in Spangenberg, 1935):

$$\Delta G = RT\ \ln \frac{K_d}{K_\infty} = \frac{F}{d}\ . \tag{4}$$

ΔG = difference in free energy; R = gas-law constant; T = absolute temperature; K_d and K_∞ = solubility products of crystals of the dimension d and ∞ respectively; F = factor and d = linear dimension of the crystal. According to Chave & Schmalz (1966) the factor F depends on the surface energy

$$F = s\sigma V \tag{5}$$

s = shape factor; σ = surface energy of the exposed faces and V = molar volume.

For small calcite crystallites of unknown form Chave & Schmalz determined (via ΔG) F values between 1·0 and 1·6 ·10^{-3} cal cm, the average being 1·3 ·10^{-3} cal cm. A similar value 1·22 ·10^{-3} cal cm results from the free specific surface energy of the

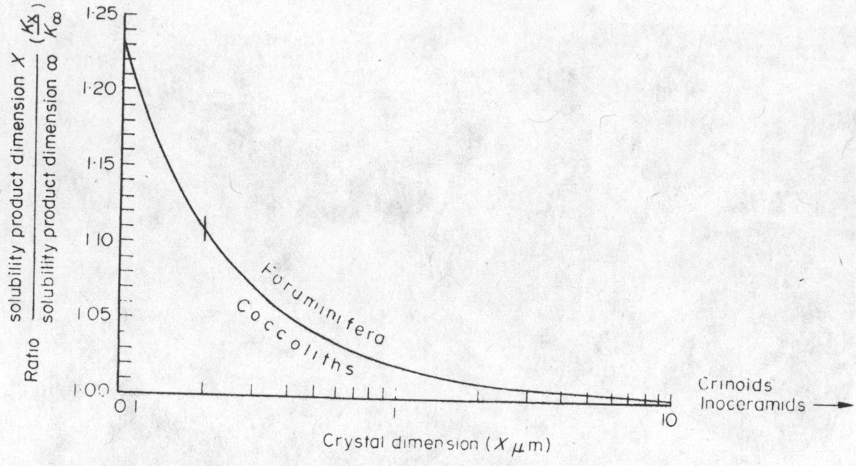

Fig. 10. Dependence of solubility product of calcite on crystal size.

cleavage plane of calcite ($\sigma = 230$ ergs/cm²), as determined by Gilman (1960). Because the smallest cement crystals should chiefly exhibit cleavage rhombohedra as growing faces (see below) the latter value was used as basis for the calculation ($T = 298°$ K).

Figure 10 shows the resulting curve for the dependence of the solubility product on the crystal dimension d of calcite (bounded by the cleavage rhombohedron). Furthermore, the crystal sizes regarded as roughly comparable to linear dimension d of formula (4) are shown for the fossil groups under discussion.

The crystal size of coccoliths and Foraminifera (after dissolution) ranges from about 0·2 to more than 1 μm (or even 2–3 μm). Accordingly, the differences in the solubility product of the overgrowth on coccoliths and Foraminifera on one hand, and on crinoids, inoceramids and on discoasters on the other hand, are about 1–10%.* The solubility product of the overgrowth on the larger coccolith crystals approaches that of the overgrowth on crinoids and inoceramids far more closely than that of the overgrowth on the small coccolith crystals.

Once again we may ask whether these differences in the solubility product caused by differences of crystal sizes are sufficiently large to affect the amount of cement deposited. The differences are small compared to differences in solubility product caused by stress (Fig. 2), but they should still have an effect: it is known from growth experiments with NaCl, KCl and other substances that supersaturations of a small fraction of a percent determine growth and growth velocities (Honigmann, 1958; Schüz, 1969 and others).

Though the substances employed in the experiments are easily soluble compared with calcite, the differences in the solubility product are most probably great enough to affect the deposition of cement. Compared to coccolith and foraminiferal crystals the larger units of crinoids and inoceramids should therefore preferentially accrete cement.

Above a crystal dimension of a few microns, grain-size differences have negligible effect on the solubility product (Fig. 10). This could be the reason why many fine-grained carbonate rocks do not recrystallize beyond a grain size of a few microns (micrite, microsparite, Folk, 1962).

* The real difference in the solubility product should be somewhat less than given in Fig. 10 since, to simplify matters, a constant (rhombohedral) form as well as a pure solution were considered.

The small crystals of coccoliths (0·2 μm) in chalk did not dissolve and form larger crystals a few microns across because the pore fluid was supersaturated with respect to low-magnesium calcite.

The crystal shape

The effect of crystal size does not explain the differences of overgrowth between equal-sized crystals of coccoliths and Foraminifera or of inoceramids and crinoids. There is another determining factor: the shape of biogenic crystals. Many biogenic crystals have a shape which is only partially, or not at all, bounded by crystal faces. For this reason these biogenic crystals are favoured in growth until crystal faces of the so-called 'equilibrium form' (Gleichgewichtsform, Honigmann, 1958) are reached. Subsequent growth is slow.

The influence of the biogenic shape on cement deposition is schematically depicted in Fig. 11. The cleavage rhombohedron $\{10\bar{1}1\}$ is used as the equilibrium form because, theoretically, and in a series of experiments with and without impurities, it has the lowest (or a very low) specific surface energy (Honigmann, 1958). Figure 7 shows that we find this form predominantly in the form of the small 'teeth' on the foraminiferal walls.

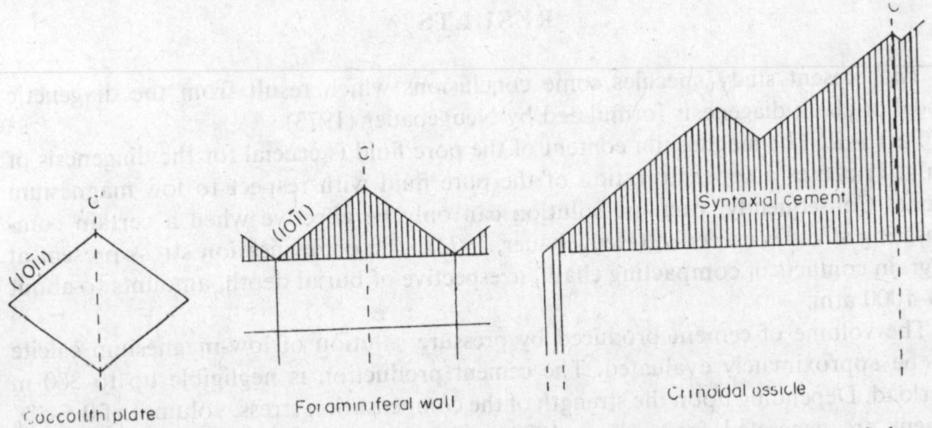

Fig. 11. Influence of the shape of biogenic crystals on the deposition of syntaxial cement (schematically).

The influence of surface energy decreases with increasing crystal size (Spangenberg, 1935); compare equations (4) and (5). This results in a larger number of crystal forms occurring on bigger crystals. In this respect the crinoids in Fig. 11 are not quite correctly represented, because the cleavage rhombohedron, though present on crinoids (Fig. 9), is no longer the dominant crystal form.

Figure 11 illustrates that the coccoliths are handicapped by the shape factor. The coccolith plates appear to be largely bounded by slowly growing crystal faces; according to Black (1963) the cleavage rhombohedron is frequently used as the habit of coccolith crystals. The shape of the coccolith plates therefore impedes the deposition of cement. This is consistent with Fig. 6, where cement-like excrescences appear only on certain faces of big crystals.

On the other hand, due to their stable form, the coccolith plates are more difficult to dissolve in comparison with other biogenic crystals of the same size. Together with the ultrastructure of the prisms (see p. 165) this could contribute to the fact that on the deep-sea bottom Foraminifera are dissolved easier than coccoliths (cf. Berger, 1972). The selective dissolution of Foraminifera and coccolith species (Bukry *et al.*, 1971; Douglas, 1971; McIntyre & McIntyre, 1971) should be examined from the aspect of the participant crystals (cf. Adelseck *et al.*, 1973).

Summarizing, we can state that the quantity of the deposited cement (and dissolved biogenic calcite?) is influenced by the size and shape of the biogenic crystals.

What are the geological implications of this statement? Cementation in chalk avoids coccoliths and concentrates on all other carbonate fossils, but favours big crystals. Layers within chalk that are rich in carbonate fossils other than coccoliths can therefore draw larger amounts of cement. This should cause partial or total hardening of such layers.

This is our interpretation of the hardening of *Inoceramus* layers in the chalk of Kansas (Fig. 8 and Neugebauer, 1973). More or less consolidated, Foraminifera-rich layers in England (Black, 1953) and the *Braarudosphaera* horizon in the South Atlantic (Wise & Hsü, 1971; Wise & Kelts, 1972) should be examined from this point of view.

RESULTS

The present study specifies some conclusions which result from the diagenetic model of chalk diagenesis formulated by Neugebauer (1973).

Part I–II. The magnesium content of the pore fluid is crucial for the diagenesis of chalk. It causes a supersaturation of the pore fluid with respect to low-magnesium calcite. Consequently, pressure solution can only be effective when a certain compensating stress is exceeded (Neugebauer, 1973). The compensation stress present at all grain contacts of compacting chalk, irrespective of burial depth, amounts to about 250–1000 atm.

The volume of cement produced by pressure solution of low-magnesium calcite can be approximately evaluated. The cement production is negligible up to 300 m overload. Depending upon the strength of the compensation stress, volumes of 0·5–5% cement are generated from about 1000 m overload onwards. The production of cement increases considerably with further burial.

Chalk can remain highly porous down to great depths. Under favourable conditions complete lithification requires an overload of 2000–4000 m or more.

Pressure solution lithifies chalk at shallower depths when the magnesium content of the pore fluid is exhausted. As evidenced by pore fluid analyses of the deep-sea cores, processes consuming magnesium are not very important in the first 500–700 m of chalk deposits. At greater depths pressure solution is held to become an important process for reduction of magnesium through the formation of magnesium-bearing cement at the cost of biogenic low-magnesium calcite. For this reason the magnesium content of the pore fluids should often be exhausted at a depth of about 1000–1600 m. A few changes from chalk to limestones have been observed at these sub-bottom depths. Where additional magnesium is supplied by evaporites, one should expect soft chalk below 1600 m burial depth.

The growth of low-magnesium calcite is influenced by very low concentrations of magnesium (0·0001–0·001 M). So far, the lowest magnesium concentration that leads to precipitation of aragonite or high-magnesium calcite instead of low-magnesium calcite is unknown. However, various lines of evidence indicate that an absolute concentration of about 0·01 M Mg or a Mg/Ca ratio from 1–2 might represent the concentrations below which low-magnesium calcite is formed. Pore fluids of chalk normally contain more than 0·01 M magnesium. In exceptional cases, where the concentration of magnesium lies below 0·01 M, stronger induration or limestones are observed. In some chalks the Mg/Ca ratio reaches values of 1–2. No special induration is observed in these cases.

Part III. Chalk is characterized by its dearth of cement. The small amounts of cement which are produced by pressure solution and by other processes are usually precipitated as syntaxial cement.

The biogenic calcite crystals of the constituent fossil groups attract different amounts of cement. Shape and size of the skeletal crystals are of crucial importance. Cement precipitation on coccoliths is handicapped by these two factors. Due to their small crystal size the supersaturation level is lowered by several percent for the overgrowth on coccoliths. Coccoliths are also discriminated against with respect to cementation by the shape factor. Consequently, cementation in chalk avoids coccoliths and concentrates on all other carbonate fossils, preferentially fossils composed of large crystals. Layers within the chalk which are rich in fossils other than coccoliths can thus draw larger amounts of cement and will be selectively lithified.

Foraminifera of chalk from the shelf environment show the same dissolution characteristics as Foraminifera from the deep-sea bottom. The dissolution of small-sized low-magnesium calcite particles indicates a very low concentration of $CaCO_3$ on the bottom of the Chalk Sea, a concentration which was below the solubility product of aragonite. This is consistent with the hypothesis of Hudson (1967).

ACKNOWLEDGMENTS

For fruitful discussions and help I am indebted to Dr W. Bay, Dr Chr. Hemleben, Dr F. Lippmann, Professor A. Seilacher, Dr N. Shrivastava and Professor K. M. Towe. I thank Professor A. Seilacher for samples of chalk and inoceramite from Kansas. Dipl.-Geol. F. Fürsich and Dr U. von Rad kindly helped to obtain copies of the papers on interstitial water analyses of the Deep Sea Drilling Project. Miss R. Freund instructed me during stereoscan studies. The impulse for my interest for chalk diagenesis was given by my friend Dr G. Ruhrmann. The translation of this manuscript was accomplished especially with the help of Dipl.-Geol. S. Chrulew, and further of Professor A. Seilacher, Dr N. Shrivastava and Professor R. D. K. Thomas. Dr A. Matter and Professor K. J. Hsü kindly reviewed the manuscript and made many useful suggestions. I would like to thank all who helped in this work.

REFERENCES

Adelseck, C.G. Jr, Geehan, G.W. & Roth, P.H. (1973) Experimental evidence for the selective dissolution and overgrowth of calcareous nannofossils during diagenesis. *Bull. geol. Soc. Am.* **84**, 2755–2762.

ANDERSON, T.F. (1973) Oxygen and carbon isotope compositions of altered carbonates from the western Pacific, core 53.0, Deep Sea Drilling Project. *Mar. Geol.* **15,** 169–180.

BATHURST, R.G.C. (1971) *Carbonate Sediments and their Diagenesis,* pp. 620. Elsevier, Amsterdam.

BERGER, W.H. (1967) Foraminiferal ooze: solution at depths. *Science,* **156,** 383–385.

BERGER, W.H. (1972) Deep-sea carbonates: dissolution facies and age-depth constancy. *Nature, Lond.* **236,** 392–395.

BERGER, W.H. & VON RAD, U. (1972) Cretaceous and Cenozoic sediments from the Atlantic Ocean. In: *Initial Reports of the Deep Sea Drilling Project,* Vol. XIV (D. E. Hayes, A. C. Pimm *et al.*), pp. 787–954. U.S. Government Printing Office, Washington.

BISCHOFF, J.L. & FYFE, W.S. (1968) Catalysis, inhibition, and the calcite–aragonite problem. I. The calcite–aragonite transformation. *Am. J. Sci.* **266,** 65–79.

BLACK, M. (1953) The constitution of the Chalk. *Proc. geol. Soc. Lond.* no. **1499,** 81–86.

BLACK, M. (1963) The fine structure of the mineral parts of Coccolithophoridae. *Proc. Linn. Soc. Lond.* **174,** 41–46.

BUKRY, D., DOUGLAS, R.G., KLING, S.A. & KRASHENINNIKOV, V. (1971) Planktonic microfossil biostratigraphy of the northwestern Pacific Ocean. In: *Initial Reports of the Deep Sea Drilling Project,* Vol. VI (A. G. Fischer *et al.*), pp. 1253–1300. U.S. Government Printing Office, Washington.

CHAN, K.M. & MANHEIM, F.T. (1970) Interstitial water studies on small core samples, Deep Sea Drilling Project, leg. 2. In: *Initial Reports of the Deep Sea Drilling Project,* Vol. II (M. N. A. Peterson *et al.*), pp. 367–371. U.S. Government Printing Office, Washington.

CHAVE, K.E. & SCHMALZ, R.F. (1966) Carbonate–seawater interactions. *Geochim. cosmochim. Acta,* **30,** 1037–1048.

CORRENS, C.W. & STEINBORN, W. (1939) Experimente zur Messung und Erklärung der sogenannten Kristallisationskraft. *Z. Kristallogr.* **101,** 117–133.

CULBERSON, C. & PYTKOWICZ, R.M. (1968) Effect of pressure on carbonic acid, boric acid, and the pH in seawater. *Limnol. Oceanogr.* **13,** 403–417.

DOUGLAS, R.G. (1971) Cretaceous Foraminifera from the northwestern Pacific Ocean: leg 6, Deep Sea Drilling Project. In: *Initial Reports of the Deep Sea Drilling Project,* Vol. VI (A. G. Fischer *et al.*), pp. 1027–1046. U.S. Government Printing Office, Washington.

DREVER, J.I. (1971) Magnesium-iron replacement in clay minerals in anoxic marine sediments. *Science,* **172,** 1334–1336.

EWING, M. *et al.* (1969) *Initial Reports of the Deep Sea Drilling Project,* Vol. I, pp. 672. U.S. Government Printing Office, Washington.

FOLK, R.L. (1962) Spectral subdivision of limestone types. In: *Classification of Carbonate Rocks* (Ed. by W. E. Ham). *Mem. Am. Ass. Petrol. Geol.* **1,** 62–84.

GILMAN, J.J. (1960) Direct measurements of the surface energies of crystals. *J. appl. Phys.* **31,** 2208–2218.

HAUSSÜHL, S. (1964) Das Wachstum großer Einkristalle. *Neues. Jb. Miner. Abh.* **101,** 343-366.

HONIGMANN, B. (1958) *Gleichgewichts- und Wachstumsformen von Kristallen,* pp. 161. Steinkopf, Darmstadt.

HUDSON, J.D. (1967) Speculations on the depth relations of calcium carbonate solution in Recent and ancient seas. *Mar. Geol.* **5,** 473–480.

LIPPMANN, F. (1960) Versuche zur Aufklärung der Bildungsbedingungen von Kalzit und Aragonit. *Fortschr. Miner.* **38,** 156–161.

LIPPMANN, F. (1973) *Sedimentary Carbonate Minerals,* pp. 228. Springer, Berlin.

MANHEIM, F.T., CHAN, K.M., KERR, D. & SUNDA, W. (1970a) Interstitial water studies on small core samples, Deep Sea Drilling Project, leg 3. In: *Initial Reports of the Deep Sea Drilling Project,* Vol. III (A. E. Maxwell *et al.*), pp. 663–666. U.S. Government Printing Office, Washington.

MANHEIM, F.T., CHAN, K.M. & SAYLES, F.L. (1970b) Interstitial water studies on small core samples, Deep Sea Drilling Project, leg 5. In: *Initial Reports of the Deep Sea Drilling Project,* Vol. V (D. A. McManus *et al.*), pp. 501–511. U.S. Government Printing Office, Washington.

MANHEIM, F.T. & SAYLES, F.L. (1971a) Interstitial water studies on small core samples, Deep Sea Drilling Project, leg 6. In: *Initial Reports of the Deep Sea Drilling Project,* Vol. VI (A. G. Fischer *et al.*), pp. 811–821. U.S. Government Printing Office, Washington.

MANHEIM, F.T. & SAYLES, F.L. (1971b) Interstitial water studies on small core samples, Deep Sea Drilling Project, leg 8. In: *Initial Reports of the Deep Sea Drilling Project*, Vol. VIII (J. I. Tracey, Jr *et al.*), pp. [illegible]. U.S. Government Printing Office, Washington.

MANHEIM, F.T., SAYLES, F.L. & FRIEDMAN, I. (1969) Interstitial water studies on small core samples, Deep Sea Drilling Project, leg 1. In: *Initial Reports of the Deep Sea Drilling Project*, Vol. I (M. Ewing *et al.*), pp. 403–410. U.S. Government Printing Office, Washington.

MANHEIM, F.T., SAYLES, F.L. & WATERMAN, L.S. (1972) Interstitial water studies on small core samples, Deep Sea Drilling Project, leg 12. In: *Initial Reports of the Deep Sea Drilling Project*, Vol. XII (A. S. Laughton, W. A. Berggren *et al.*), pp. 1193–1200. U.S. Government Printing Office, Washington.

MANHEIM, F.T., SAYLES, F.L. & WATERMAN, L.S. (1973) Interstitial water studies on small core samples, Deep Sea Drilling Project, leg 10. In: *Initial Reports of the Deep Sea Drilling Project*, Vol. X (J. L. Worzel, W. Bryant *et al.*), pp. 615–623. U.S. Government Printing Office, Washington.

MCINTYRE, A. & MCINTYRE, R. (1971) Coccolith concentrations and differential solution in oceanic sediments. In: *The Micropalaeontology of Oceans* (Ed. by B. M. Funnell and W. R. Riedel), pp. 253–261. Cambridge University Press, Cambridge.

MÖLLER, P. & RAJAGOPALAN, G. (1972) Cationic distribution and structural changes of mixed Mg–Ca layers on calcite crystals. *Z. phys. Chem.* N.F. **81,** 47–56.

MÜLLER, G., IRION, G. & FÖRSTNER, U. (1972) Formation and diagenesis of inorganic Ca–Mg carbonates in the lacustrine environment. *Naturwissenschaften,* **59,** 158–164.

NEUGEBAUER, J. (1973) The diagenetic problem of chalk: the role of pressure solution and pore fluid. *Neues. Jb. Geol. Palaont. Abh.* **143,** 223–245.

OWEN, B.B. & BRINKLEY, S.R. (1941) Calculation of the effect of pressure upon ionic equilibria in pure water and in salt solutions. *Chem. Rev.* **29,** 461–474.

PIMM, A.C., GARRISON, R.E. & BOYCE, R.E. (1971) Sedimentology synthesis: lithology, chemistry and physical properties of sediments in the northwestern Pacific Ocean. In: *Initial Reports of the Deep Sea Drilling Project,* Vol. VI (A. G. Fischer *et al.*), pp. 1131–1252. U.S. Government Printing Office, Washington.

PYTKOWICZ, R.M., DISTECHE, A. & DISTECHE, S. (1967) Calcium carbonate solubility in sea water at *in situ* pressures. *Earth Plan. Sci. Letts,* **2,** 430–432.

SAYLES, F.L. & MANHEIM, F.T. (1971) Interstitial water studies on small core samples, Deep Sea Drilling Project, leg 7. In: *Initial Reports of the Deep Sea Drilling Project,* Vol. VII (E. L. Winterer *et al.*), pp. 871–881. U.S. Government Printing Office, Washington.

SAYLES, F.L., MANHEIM, F.T. & CHAN, K.M. (1970) Interstitial water studies on small core samples, leg 4. In: *Initial Reports of the Deep Sea Drilling Project,* Vol. IV (R. G. Bader *et al.*), pp. 401–414. U.S. Government Printing Office, Washington.

SAYLES, F.L., MANHEIM, F.T. & WATERMAN, L.S. (1971) Interstitial water studies on small core samples, leg 11. In: *Initial Reports of the Deep Sea Drilling Project,* Vol. XI (C. D. Hollister, J. I. Ewing *et al.*), pp. 997–1008. U.S. Government Printing Office, Washington.

SAYLES, F.L., WATERMAN, L.S. & MANHEIM, F.T. (1972) Interstitial water studies on small core samples, leg 9. In: *Initial Reports of the Deep Sea Drilling Project,* Vol. IX (J. D. Hays *et al.*), pp. 845–855. U.S. Government Printing Office, Washington.

SAYLES, F.L., WATERMAN, L.S. & MANHEIM, F.T. (1973a) Interstitial water studies on small core samples from the Mediterranean Sea. In: *Initial Reports of the Deep Sea Drilling Project,* Vol. XIII (W. B. F. Ryan, K. J. Hsü *et al.*), pp. 801–808. U.S. Government Printing Office, Washington.

SAYLES, F.L., WATERMAN, L.S. & MANHEIM, F.T. (1937b) Interstitial water studies on small core samples, leg 19. In: *Initial Reports of the Deep Sea Drilling Project,* Vol. XIX (J. S. Creager, D. W. Scholl *et al.*), pp. 871–874. U.S. Government Printing Office, Washington.

SCHLANGER, S.O., DOUGLAS, R.G., LANCELOT, Y., MOORE, T.C. JR, & ROTH, P.H. (1973) Fossil preservation and diagenesis of pelagic carbonates from the Magellan Rise, central North Pacific Ocean. In: *Initial Reports of the Deep Sea Drilling Project,* Vol. XVII (E. L. Winterer, J. I. Ewing *et al.*), pp. 407–427. U.S. Government Printing Office, Washington.

SCHOLL, D.W., CREAGER, J.S., BOYCE, R.E., ECHOLS, R.J., FULLAM, T.J., GROW, J.A., KOIZUMI, I., LEE, J.H., LING, H-Y., SUPKO, P.R., STEWART, R.J., WORSLEY, T.R., ERICSON, A., HESS, J., BRYAN, G. & STOLL, R. (1971) Deep Sea Drilling Project, leg 19. *Geotimes,* **16,** (11) 12–15.

SCHÜZ, W. (1969) Über den Einfluß der Übersättigung auf das Wachstum von KCl-Einkristallen in reinen wässerigen Lösungen. *Z. Kristallogr.* **128,** 36–54.

SPANGENBERG, K. (1935) Wachstum und Auflösung der Kristalle. In: *Handwörterbuch der Naturwiss.* Vol. X, 362–401. Fischer, Jena.

TAFT, W.H. (1967) Physical chemistry of formation of carbonates. In: *Carbonate Rocks, Physical and Chemical Aspects* (Ed. by G. V. Chilingar, H. J. Bissel and R. W. Fairbridge), pp. 151–167. Elsevier, Amsterdam.

THIEDE, J. (1971) Planktonische Foraminiferen in Sedimenten vom ibero-marokkanischen Kontinentalrand. *Meteor Forsch.-Ergebnisse,* R.C., 7, 15–102.

TOWE, K.M. & CIFELLI, R. (1967) Wall ultrastructure in the calcareous Foraminifera: crystallographic aspects and a model for calcification. *J. Paleontol.* **41,** 742–762.

VAN ANDEL, T.H. *et al.* (1973) *Initial Reports of the Deep Sea Drilling Project,* Vol. XVI, pp. 1037. U.S. Government Printing Office, Washington.

VON ENGELHARDT, W. (1960) *Der Porenraum der Sedimente,* pp. 207. Springer, Berlin.

VON ENGELHARDT, W. (1973) *Die Bildung von Sedimenten und Sedimentgesteinen,* pp. 378. Schweizerbart, Stuttgart.

WATERMAN, L.S., SAYLES, F.L. & MANHEIM, F.T. (1972) Interstitial water studies on small core samples, leg 14. In: *Initial Reports of the Deep Sea Drilling Project,* Vol. XIV (D. E. Hayes, A. C. Pimm *et al.*), pp. 753–762. U.S. Government Printing Office, Washington.

WATERMAN, L.S., SAYLES, F.L. & MANHEIM, F.T. (1973) Interstitial water studies on small core samples, leg 16, 17 and 18. In: *Initial Reports of the Deep Sea Drilling Project,* Vol. XVIII (L. D. Kulm, R. von Huene *et al.*), pp. 1001–1012. U.S. Government Printing Office, Washington.

WEYL, P.K. (1959) Pressure solution and the force of crystallization—a phenomenological theory. *J. geophys. Res.* **64,** 2001–2025.

WISE, S.W., JR & HSÜ, K.J. (1971) Genesis and lithification of a deep sea chalk. *Eclog. geol. Helv.* **64,** 273–278.

WISE, S.W., JR & KELTS, K.R. (1972) Inferred diagenetic history of a weakly silicified deep sea chalk. *Trans. Gulf-Cst Ass. geol. Socs,* **22,** 177–203.

Spec. Publs int. Ass. Sediment. (1974) **1,** 177–210

Diagenesis of Upper Cretaceous chalks from England, Northern Ireland, and the North Sea*

PETER A. SCHOLLE

Geosciences Faculty, University of Texas at Dallas, P.O. Box 30365, *Dallas, Texas* 75230, *U.S.A.*

ABSTRACT

Diagenetic differences, especially hardness changes, between relatively closely spaced localities in Upper Cretaceous chalks of the British Isles-North Sea area have long presented a difficult interpretive problem. In the present study, outcrop samples of the Upper Chalk in England, Yorkshire, and Northern Ireland, as well as Maastrichtian and Danian chalk cores from the Ekofisk field in the North Sea have been examined by petrographic, electron microscopic, and isotopic methods.

Although all of the chalks appear to have shared similar initial composition, subsequent variations in degree or type of diagenesis have yielded a remarkable range of ultimate lithologies. The White Limestone of Northern Ireland is extremely hard (with porosites of 5–10%) and shows very strong soft-sediment compaction textures as well as late stylolitization. The micritic matrix is composed of equant, blocky calcite (often 'consuming' coccoliths) and yields oxygen isotopic values averaging – 5·6/10³. Samples from Yorkshire have porosities of about 18%, show some development of blocky micrite matrix, and have lesser amounts of soft-sediment compaction although stylolitization is still common. Oxygen isotopes from Yorkshire samples average – 4·0/10³. Chalks from southern England (Dover, Thanet, Brighton) are very soft (porosities of 40–45%) and show little compaction or recrystallization to blocky calcite. Oxygen isotopic values average – 2·9/10³. The Ekofisk chalk has strong porosity variations (2–43%, average 30%), little compaction but some stylolitization, traces of dolomitization, and rounded, corroded crystal shapes. Oxygen isotopic values average – 0·4/10³.

Progressive increase in fresh-water alteration from the North Sea to Northern Ireland (related to the rifting and uplift of the North Atlantic continental margins) could explain the isotopic data but presents many geological problems. The assumption of an increasing gradient of hydrothermal-meteoric recrystallization of chalks from the North Sea to Northern Ireland (an area of extensive Tertiary volcanism), again related to North Atlantic rifting, appears capable of resolving both the geological and geochemical data. The striking compaction of Irish chalks is explained by expulsion of original marine pore fluids and loading by basalts; isotopic values are compatible with the introduction of thermally driven meteoric fluids.

Mapping of petrographic and geochemical gradients in chalk sediments should yield valuable information about structural and geothermal patterns in the Late Cretaceous and Early Tertiary of the North Atlantic region. Where data is clearly related to rift trends it may be useful in the prediction of patterns of deposition and diagenesis in formerly contiguous areas such as the Canadian Atlantic shelf.

* Contribution number 250, The Institute for Geological Sciences, The University of Texas at Dallas.

INTRODUCTION

The diagenetic alterations of the Upper Cretaceous chalks of western Europe have long been the subject of controversy. Even in areas such as the British Isles where rather uniform depositional facies can be found in the Upper Chalk, radical differences in hardness and sediment texture can be found over relatively short distances. The White Limestone of Northern Ireland, for example, has long been known to be correlative with parts of the Upper Chalk of England, yet the Irish chalk is far harder than its English equivalent.

Numerous hypotheses have been offered to explain these relations and they are well summarized by Hancock (1963). Baking through contact with overlying basalts in Northern Ireland has been proposed frequently, but as pointed out by Black (1953) and Hancock (1963) there are numerous problems with this hypothesis. In addition to the fact that unbaked lignites occur in places between the basalts and the White Limestone (Jukes, 1868), one does not see any gradient of alteration away from the basal basalt contact. Furthermore, chalks in Yorkshire are also hardened, yet there is no associated volcanism there. Finally, calculation of the geothermal effect of lava flows, such as those in Northern Ireland, on underlying sediments indicate that significant effects should extend only a few metres away from the contact; yet again, the Irish chalks are uniformly hardened throughout.

Another hypothesis for the diagenetic differences of the Irish and English chalks was proposed by Hancock (1963). He suggests some primary sedimentary variations between the deposits of the two areas, probably the presence of finer-grained interstitial sediment in Ireland, as the cause of subsequent differential response to diagenetic influences. As Hancock himself points out, however, there is no independent supporting evidence. The original aragonite content of sediments from England and Ireland does not appear to have been significantly different as judged by the frequency of occurrence of impressions of aragonitic organisms in the two deposits.

A third explanation of differential chalk hardness was offered by Peach, Grunn & Newton (1901). They inferred that the hardening of the Irish limestone was a result of silicification. Data from Hume (1897), Hancock (1963) and from this study all indicate, however, that the White Limestone of Northern Ireland is an extremely pure carbonate with an average of less than 3% acid-insoluble material. Clearly this eliminates silicification as a major feature of the Irish chalks. Thin-section analysis confirms the lack of silica cementation.

The three major hypotheses which have been offered to explain variations in chalk hardness thus all appear inadequate. The purpose of this study is, therefore, to re-examine some of the basic data on chalks of the English-Irish-North Sea area and to provide additional petrographic and geochemical information on the problem. Our understanding of the depositional and diagenetic facies of shallow- and deep-marine sediments has progressed so rapidly and so radically in the last decade that many of the basic assumptions made by previous workers no longer appear valid. From this point of view also, a new review of the data appears to be worthwhile. This paper presents the results of the first stage of a continuing project on regional patterns of chalk diagenesis in western Europe.

REGIONAL SETTING

In the Late Cretaceous, chalks and associated facies were deposited across wide areas of western Europe. The present outcrops are shown in Fig. 1, but these represent only the erosional remnants of presumably once much wider depositional realms (e.g., Walsh, 1966). The palaeogeography and facies relations of these deposits have been covered in an extensive literature, partly summarized by Rayner (1967) and Hancock (1974), and only the briefest outline will be offered here.

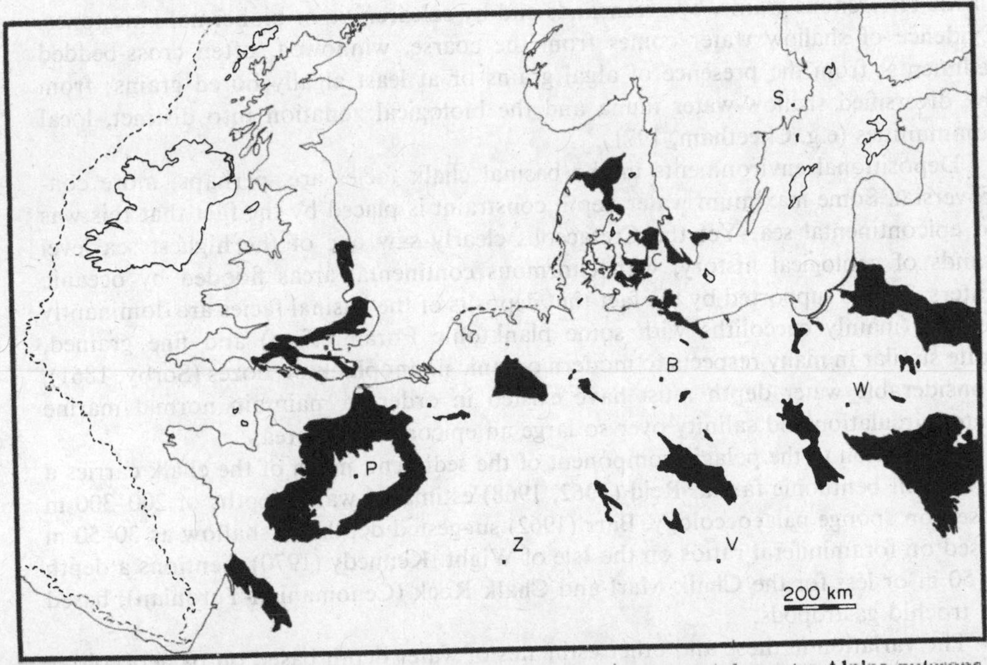

Fig. 1. Map of Central Europe and the British Isles showing present-day extra-Alpine outcrops of chalks and related facies. Subsurface, subsea, and eroded areas of chalks are not shown. Outcrops in the Aquitaine, Paris, Mons, Danish-Swedish, and other areas include calcarenitic, marginal facies as well as chalks. Letters indicate major European cities.

The Late Cretaceous represents an interval of unusually high sea-level stand, or unusually low stand of the North American and European continents. Thus, major portions (in excess of 30%) of both continents were flooded by relatively deep epi-continental seas. In Europe, Albian greensand, clay and limestone facies indicate the early phases of this major continent's flooding which was broadly continued at least until the close of the Campanian. Continued transgression from the Albian into the Turonian and Santonian apparently progressively drowned more and more of the clastic terrigenous source areas. This is evidenced by the fact that Albian sedimentation is dominated by terrigenous sands, silts, and clays; Cenomanian deposits are largely marls, and Turonian to Maastrichtian sediments are dominantly chalks over major portions of the depositional area. The central part of the Anglo-Paris Basin is charac-terized by a chalk facies through much of this time interval. Basin margins were clearly present, however, in the Aquitain Basin, the southern and eastern Paris Basin, and the Mons Basin. Another margin, with associated complex block-faulted ridge and basin structure, was developed in eastern Denmark and southern Sweden (especially in the Maastrichtian and Palaeocene). During the early Senonian and before, another margin appears to have existed in Ireland (as evidenced by Greensand facies equivalent to much of the lower part of the Upper Chalk of England). However, this margin is no longer apparent in the upper Senonian and Maastrichtian beds of Northern Ireland. Apparently, strong transgression in this area led to the establishment of basinal conditions by late Senonian (Campanian) time.

Marginal facies, wherever they are encountered, are typified by sediments rich in clastic terrigenous grains, by greensands and by calcarenitic or biohermal limestones. Evidence of shallow water comes from the coarse, winnowed, often cross-bedded sediments; from the presence of algal grains or at least algally bored grains; from the diversified shallow-water fauna and the biological zonation into distinct, local communities (e.g. Cheetham, 1971).

Depositional environments in the basinal chalk facies are, perhaps, more controversial. Some maximum water depth constraint is placed by the fact that this was an epicontinental sea. Yet the Cretaceous clearly saw one of the highest sea level stands of geological history, with enormous continental areas flooded by oceanic waters. This is supported by the fact that deposits of the basinal facies are dominantly pelagic (mainly coccoliths with some planktonic Foraminifera) and fine grained, quite similar in many respects to modern oceanic nannoplankton oozes (Sorby, 1861). Considerable water depth must have existed in order to maintain normal marine water circulation and salinity over so large an epicontinental area.

In addition to the pelagic component of the sediment, much of the chalk carries a significant benthonic fauna. Reid (1962, 1968) estimated water depths of 200–300 m based on sponge palaeoecology. Barr (1962) suggested depths as shallow as 30–50 m based on foraminiferal ratios on the Isle of Wight. Kennedy (1970) mentions a depth of 50 m or less for the Chalk Marl and Chalk Rock (Cenomanian-Turonian), based on trochid gastropods.

The variation in these and other estimates of water depth based on palaeoecology may be the result of a number of factors. By their very nature, the chalk seas were unusual. They reflect the unique (at least to the present) situation of a vast epicontinental sea coupled with the availability of a diversified calcareous planktonic fauna. Bottom sediments were very soft mud, forming difficult substrates for colonization. Clastic terrigenous input was abnormally low. All these conditions led to very specialized and unique faunas (e.g. Carter, 1972) for which palaeoecological comparisons with modern faunas are hazardous at best.

A second influence on chalk water-depth estimates has been the presence in literature of a number of sedimentological interpretations which are no longer valid and which may have influenced some palaeoecological determinations. Hardgrounds (synsedimentary lithification surfaces), once thought to be indicators of subaerial exposure, now are known to form in submarine environments (e.g. Fischer & Garrison, 1967). Also the agencies of current scouring and winnowing, formerly believed to be non-marine or shallow-marine processes, are now also known to take place in deeper waters.

A third effect operating to produce the diversity of opinions on chalk water depths is the fact that many of the descriptions of faunas come from 'atypical' facies—hardground zones such as the Chalk Rock or Melbourn Rock or from condensed zones.

The resolution of this controversy is beyond the scope of this paper. However, several observations made during this project lend support to a relatively deep-water environment of deposition. The lateral uniformity of chalk facies and the virtually complete lack of biological or ecological zonation support a uniform, deep-water environment. It is difficult to conceive of a shallow-water area the size of the Chalk basin which did not show such zonation. A second factor is the widespread lack of algal grains or algal borings in shells. While algal borings have been reported (Bromley, 1970, Kennedy, 1970) these descriptions are, again, mainly from 'atypical' hardground facies which may indeed reflect periodic lowerings of sea level. The bulk of the Chalk, however, is virtually devoid of traces of algal growth, indicating deposition below the photic zone. And with sediments as free from terrigenous detritus as the Chalk, waters must have been exceptionally clear, with a photic zone extending down at least 150–200 m. On these lines of reasoning, 200–300 m would appear to be the absolute mininum acceptable water depth for normal chalk deposition.

At the same time, the fact that algal borings have been described from hardgrounds throughout the Chalk section (which probably reflect only minor sea-level drops) indicates that water depths were probably not significantly in excess of 300 m. A maximum water depth is difficult to define because there are few sedimentological criteria that can be used to indicate depths between the photic zone and the carbonate compensation level. Earlier estimates of about 1000 m water depth by Jukes-Browne & Hill (1904) and Voigt (1929) almost certainly represent maximum figures. Estimates of water depths for the Greenhorn Chalk of the central United States, a deposit similar to European Chalk in many respects, have been made by Eicher (1969). On the basis of planktonic/benthonic foraminiferal ratios and independent palaeoslope estimates, Eicher calculated water depths in the 500–900 m range. These depths are difficult to envisage over broad continental regions. Depths of 200–500 m appear to be more in accord with all lines of evidence and such depths closely match the rise in sea level predicted by Hays & Pitman (1973) from examination of Late Cretaceous sea-floor spreading rates.

Bottom conditions in chalk seas were apparently normal marine in most respects. Although substrates were soft, the sediments (especially the Middle and Upper Chalk) indicate oxidizing, normal salinity bottom waters with abundant benthonic organisms. Outcrop examination reveals abundant, if often poorly visible, burrowing (Bromley, 1967) and common sedimentary structures such as lamination and graded bedding. Other sedimentary structures may be present, but are largely invisible due to the low-contrast potential of this pure carbonate sediment. Slump beds, scoured channels, truncation surfaces, and winnowed zones are known from some localities (e.g. Kennedy & Juignet, 1973). In thin section, virtually all samples of the Chalk show extensive grain breakage which may be partly the result of feeding activities of vagile benthos but which may also indicate transportation and deposition (or redeposition) by submarine currents. Although little work has been done in this area, many of the depositional cycles and sedimentary structures and textures in the Chalk may be the result of density currents, 'normal' currents, slumping, and other resedimentation processes.

In any case, for the purposes of this study, the Chalk provides an ideal deposit in that it shows remarkably uniform depositional facies over very large areas. Although there are local differences in the relative abundance of various skeletal organisms, by and large these variations do not appear to have produced major distinctions in average sediment grain size or mineralogical composition from area to area. Those differences which are seen today are, therefore, the result of subsequent diagenetic alterations, not primary mineralogical or textural differentiation.

SAMPLING

Field work was conducted in the summer of 1970 and samples were collected from eighteen localities in England and Ireland. Other material was obtained from three localities in Denmark and Sweden and from the Phillips Ekofisk oil field in the Norwegian part of the North Sea. Sample locality areas are shown on Fig. 2 and more detailed maps are found in Fig. 3.

In Northern Ireland, material from the Senonian to Maastrichtian White Limestone was collected from Cave Hill near Belfast, Belshaw's Quarry near Lisburn, Ulster Quarry near Moira, Magheramorne Quarry, roadcuts near Ballygalley and Glenarm, and coastal exposures at White Park Bay (Fig. 3, A). Particular attention was paid to the distance of each sample from the nearest basalt contact.

In Yorkshire, samples of hard chalk were collected from the coastal cliffs near Flamborough Head (Fig. 3, B). In this area a section of Middle and Upper Chalk extending from the *Terebratulina lata* Zone to the *Actinocamax quadratus* Zone (locally the *Inoceramus lingua* Zone, Hemingway, Wilson & Wright (1968)) was extensively sampled between Great Thornwick Bay and Bridlington.

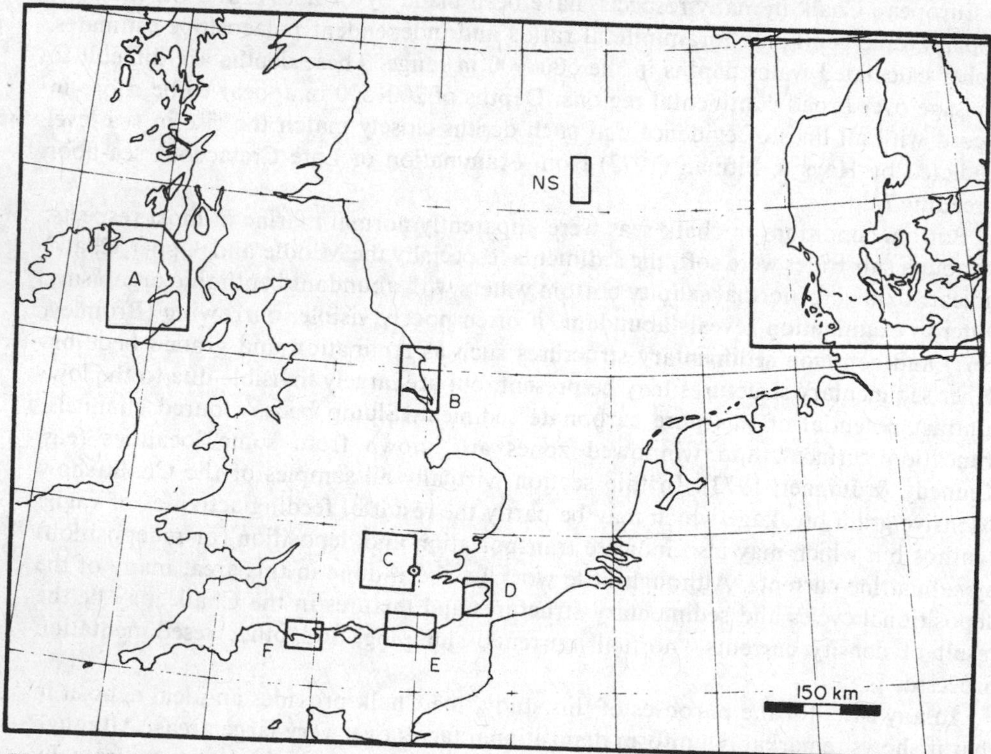

Fig. 2. Index map of chalk sample localities. Letters (A–F) indicate detailed maps of Fig. 3, and NS designates the area of subsurface samples from the Phillips Ekofisk field in the Norwegian North Sea.

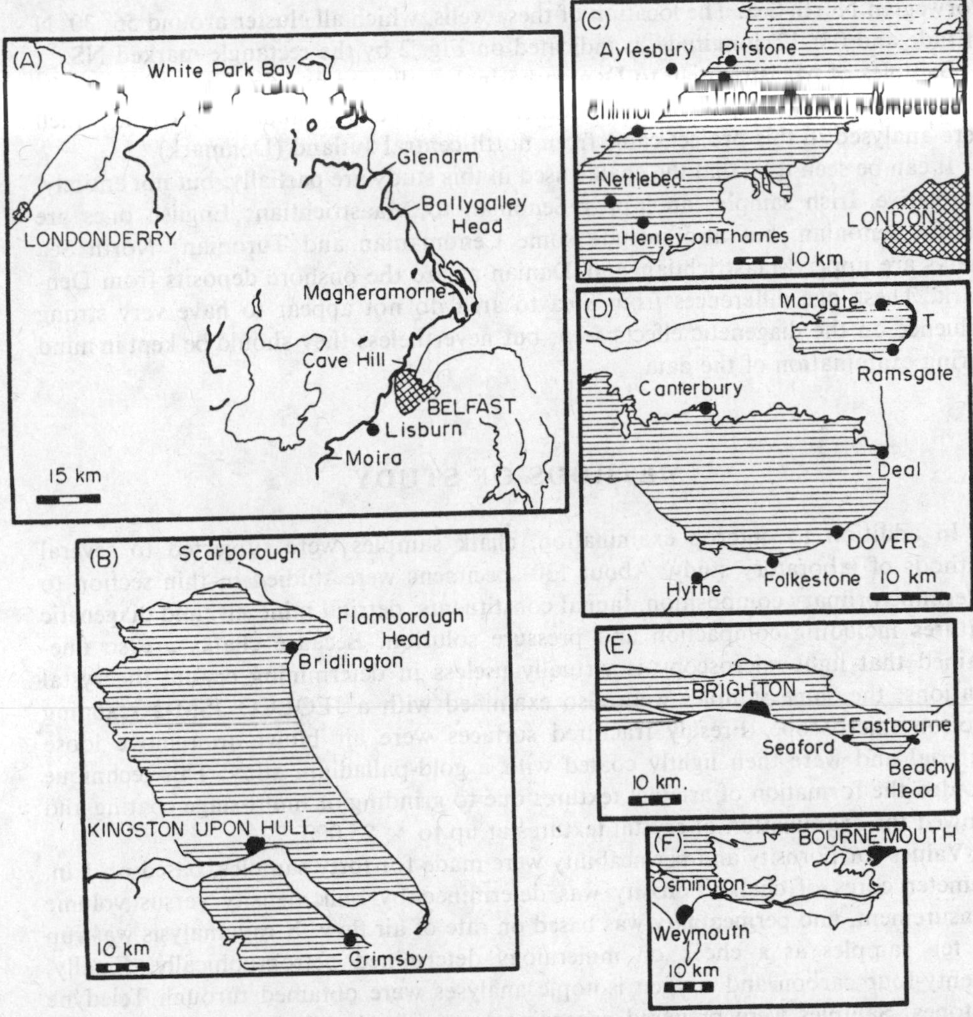

Fig. 3. Detailed maps of chalk sample localities (general index in Fig. 2). Chalk outcrops indicated by heavy line in map A and by lined pattern in maps B–F.

In England samples were obtained from four major areas. In the region between London and Aylesbury (Fig. 3, C) Cenomanian Lower Chalk was collected from quarries at Pitstone and Chinnor, and Senonian Upper Chalk was obtained from smaller outcrops near Hemel Hempstead, Nettlebed and Henley-on-Thames. Along the Kent coast (Fig. 3, D), the Lower and Middle Chalk was sampled near Folkestone; the Upper Chalk from the *Holaster planus* Zone to the *Micraster coranguinum* Zone was collected near Dover, and other Upper Chalk samples, from the *Micraster coranguinum* Zone to the *Marsupites testudinarius* Zone were collected between Broadstairs and Margate on the Thanet coast. Senonian Upper Chalk was also collected at coastal cliffs at Seaford near Brighton (Fig. 3, E). Finally, Lower Chalk was sampled in landslid masses at Osmington near Weymouth (Fig. 3, F).

Subsurface cores of upper Maastrichtian to Danian marls and chalks were obtained through Phillips Petroleum Co. from seven wells of the Ekofisk Field in the

Norwegian North Sea. The location of these wells, which all cluster around 56° 30′ N latitude and 03° 12′ longitude, is indicated on Fig. 2 by the rectangle marked NS.

Samples of Maastrichtian to Danian basinal chalk and more marginal calcarenite were obtained from Denmark and southernmost Sweden (Scania). The chalks which were analysed in this project came from north-central Jutland (Denmark).

It can be seen then that the chalks used in this study are partially, but not entirely, correlative. Irish samples are upper Senonian to Maastrichtian; English ones are mainly Senonian, but also include some Cenomanian and Turonian; North Sea chalks are upper Maastrichtian and Danian as are the onshore deposits from Denmark. These age differences from area to area do not appear to have very strong influence on the diagenetic effects seen, but nevertheless they should be kept in mind during examination of the data.

METHODS OF STUDY

In addition to outcrop examination, chalk samples were subjected to several methods of laboratory study. About 130 specimens were studied in thin section to determine primary composition, faunal constituents, detrital minerals, and diagenetic features including compaction and pressure solution. Because chalks are so fine-grained that light microscopy is virtually useless in determining crystal to crystal relations, the same samples were also examined with a JEOLCO JSM-2 scanning electron microscope. Freshly fractured surfaces were air blown to remove loose material and were then lightly coated with a gold-palladium alloy. This technique avoided the formation of artefact textures due to grinding or multi-stage coating and allowed the examination of crystal textures at up to × 25,000.

Values for porosity and permeability were made for forty samples, based on 1 in. diameter cores of chalk. Porosity was determined by bulk density versus volume measurement, and permeability was based on rate of air flow. X-ray analysis was run on ten samples as a check on mineralogy determined petrographically. Finally, seventy-four carbon and oxygen isotopic analyses were obtained through Teledyne Isotopes. Samples were prepared according to standard technique of acid decomposition (H_3PO_4 at 25°C). Both carbon and oxygen isotopic data is reported in per millilitre notation relative to the Chicago PDB-1 standard and sample correction was made according to Craig (1957). Although duplicate samples were not run, the standard reproducibility of such measurements is $0{\cdot}1/10^3$ or better.

PETROGRAPHY

Northern Ireland

The White Limestone of Northern Ireland is a virtually pure carbonate deposit. The formation is extremely hard and porosity values (to air) range between 4 and 13% with an average of 10%. Permeability averages about 0·1 millidarcys. Beds at the base of the unit which are transitional to the underlying Hibernian Greensands have significant admixture of detrital sand and silt (mainly angular quartz) as well as glauconite. Above this level, however, the White Limestone has only small amounts (generally less than 1%) of detrital silt and clay, small phosphatic fragments and chertified fossils. Discrete bands of chert nodules are found, however, throughout the unit.

In thin section, the carbonate fraction consists of sparsely fossiliferous micrites with numerous planktonic and benthonic Foraminifera and locally abundant concentrations of calcispheres (*Oligostegina*). Mollusc fragments (including *Inoceramus*), echinoderm plates, and sponge spicules are important macrofossil components of the sediment while belemnites and bryozoans are important at some horizons. Large borings, probably produced by sponges, riddle many grains.

While the primary petrographic characteristics of the White Limestone are typical for chalk-type deposits, the diagenetic characteristics are not. As pointed out by Wolfe (1968), the Irish 'chalk' has undergone extreme compaction. Although there is significant variation in the amount of compaction from sample to sample, the Irish succession as a whole shows a compaction ratio of 2 or 3 to 1. The compaction is most strikingly manifested by the crushing of microfossils (Figs 4 and 5). Where the sediment was enclosed in a rigid framework (such as a macrofossil) crushing is not seen; but where the sediment was unprotected, microfossils (especially planktonic Foraminifera and calcispheres) were extensively crushed. The deformation is bedding parallel and, because of its pervasiveness, indicates that the sediment was soft and uncemented at the time of deformation. The fact that certain zones do not show such compaction

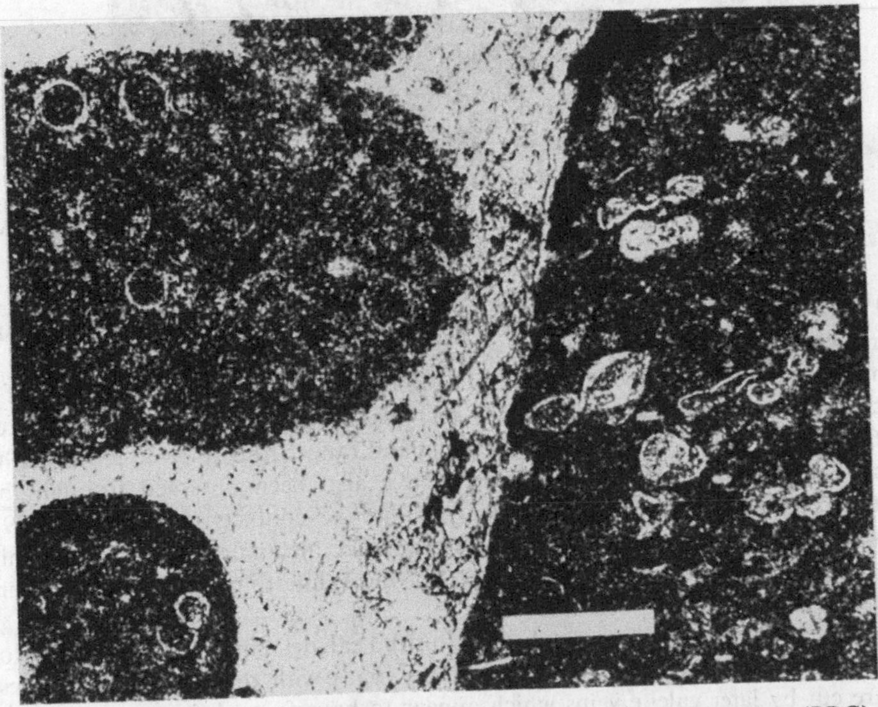

Fig. 4. Thin-section photomicrograph (plain light) of Senonian White Limestone (BDG) from Magheramorne quarry, Northern Ireland. Shows belemnite rostrum (large white area) cut by circular borings. Micritic fillings of borings within structurally rigid rostrum has spherical microfossil tests while unsupported area outside rostrum shows pervasive, bedding-plane-parallel deformation of grains. Scale bar is 0·17 mm.

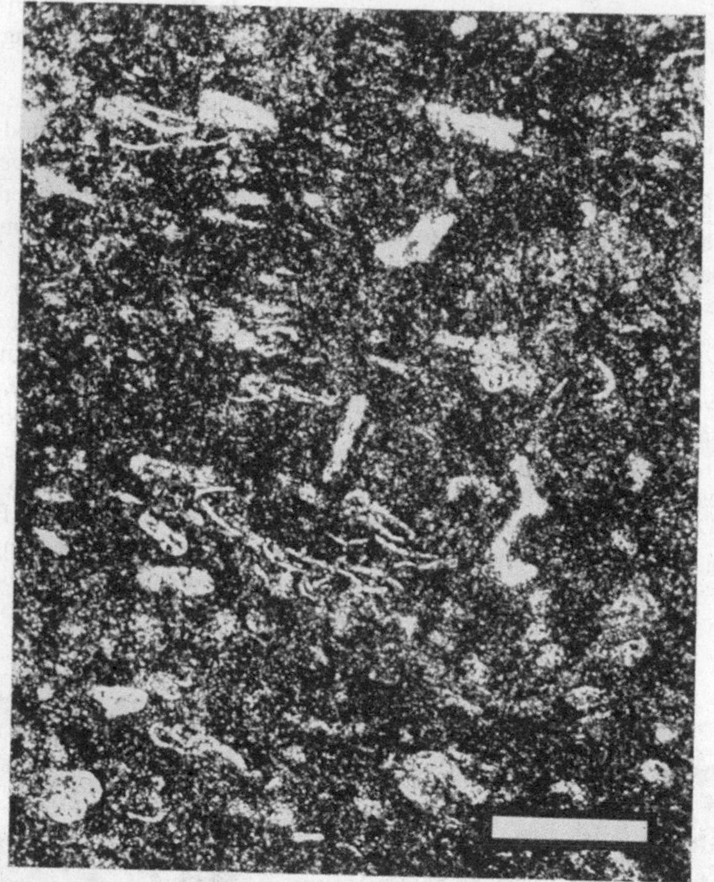

Fig. 5. Thin-section photomicrographic detail of external matrix shown in Fig. 4. Note pervasive bedding-parallel crushing of microfossil tests. Scale bar is 0·17 mm.

would indicate that those layers had been partially cemented prior to the application of compactional forces (Wolfe, 1968).

Another common diagenetic phenomenon in the Irish chalks is the formation of a contact microspar or pseudospar zone of recrystallization adjacent to basalt dykes (Figs 6 and 9). Crystals are tightly interlocking and range from 10 to 30 μm near the contact. Such contact metamorphic effects have also been observed through the formation of classic skarn assemblages at chalk-dyke contacts (e.g. Tilley, 1929). However, it must be emphasized that these contact metamorphic zones are only found in association with dykes; the contacts of the White Limestone with the overlying basalt flows show no such alterations.

A very late diagenetic effect is stylolitization and calcite-vein formation. The stylolites are bedding-parallel, low-amplitude suture zones with thin insoluble residues. They are cut by later calcite veins which appear to have formed, in many cases, by *in situ* recrystallization along possible microfractures rather than by void-filling of open fractures (e.g. Mišík, 1968).

Scanning electron microscopy has shown (Fig. 7) the remarkably equant and interlocking nature of the micritic crystals of the White Limestone. Coccoliths are still

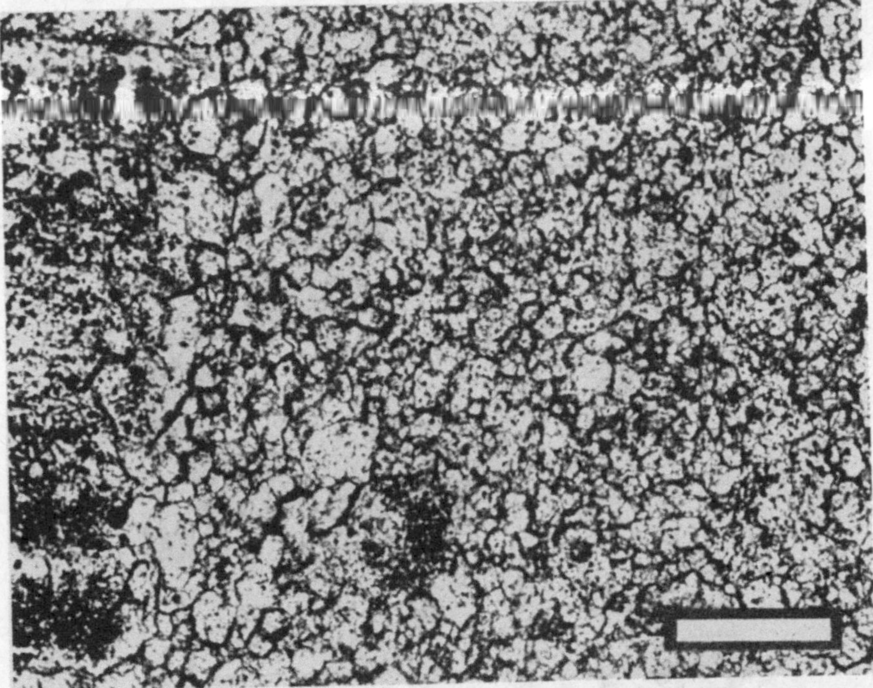

Fig. 6. Thin-section photomicrograph of Senonian White Limestone (BDC) from Belshaw's quarry near Lisburn, Northern Ireland. Shows coarse recrystallized limestone (microspar) formed adjacent to basaltic dyke. Scale bar is 0·17 mm.

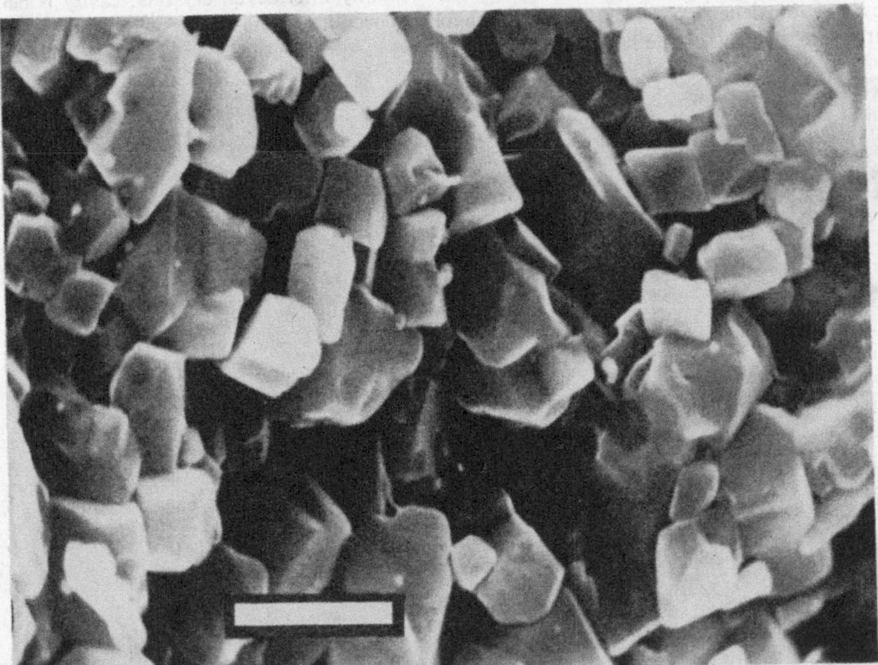

Fig. 7. Scanning electron photomicrograph of Senonian White Limestone (BDI) from Magheramorne quarry, Northern Ireland. Note extensive interlocking of crystals and euhedral crystal shapes. Scale bar is 3 μm.

Fig. 8. Scanning electron photomicrograph of Senonian White Limestone (BCY) from Belshaw's quarry near Lisburn, Northern Ireland. Matrix shows tightly interlocked crystals; cavity is being filled by coarse calcite spar which has partly engulfed coccolith. Note overgrowth on individual coccolith plate segments. Scale bar is 4 μm.

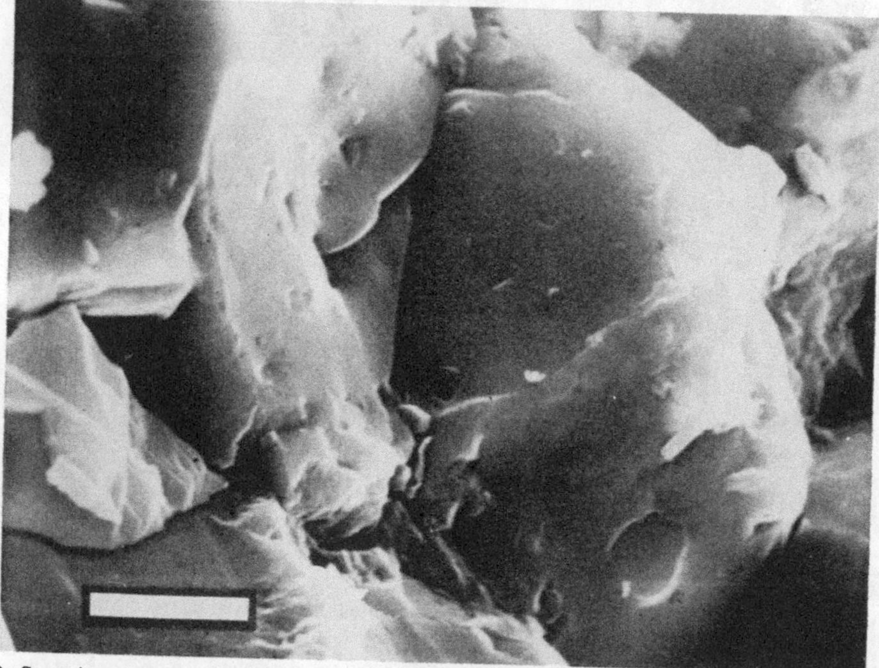

Fig. 9. Scanning electron photomicrograph of Senonian White Limestone (BDC) from Belshaw's quarry near Lisburn, Northern Ireland. Compare with Fig. 6. Coarse recrystallization adjacent to basalt dyke. Scale bar is 6·6 μm.

recognizable although in many cases they are being 'consumed' by matrix crystals (Fig. 8). Unlike the results of Hancock & Kennedy (1967), coccoliths were found to be common in the White Limestone although not so abundant as in the chalks of southern England.

The equant crystal shapes, interlocking texture and consumption of coccoliths add evidence to the idea of extensive recrystallization (in addition to compaction) of the Irish chalks. Although initial compaction may have reduced porosity considerably, subsequent growth of euhedral micrite crystals was necessary to produce the textures seen today.

Yorkshire

The Upper Chalk of Yorkshire is, again, a very pure carbonate deposit with only trace amounts of detrital clay, silt or sand and virtually no glauconite or phosphate. As with all Upper Chalk deposits, however, discrete chert nodules are distributed in bands throughout the formation. The chalks exposed near Flamborough Head are rather hard and yield porosity values of 17–20%, while the permeability averages 0·4 md.

The primary constituents of the Yorkshire chalks are in no way unusual. Planktonic Foraminifera are common; benthonic Foraminifera are considerably scarcer. Calcitic shell layers of molluscs, especially *Inoceramus*, are important constituents, as are echinoid fragments. Calcispheres (*Oligostegina*) are very abundant. Other organisms, although present in hand samples, do not appear to be important suppliers of sediment. All samples were classified as sparse biomicrites.

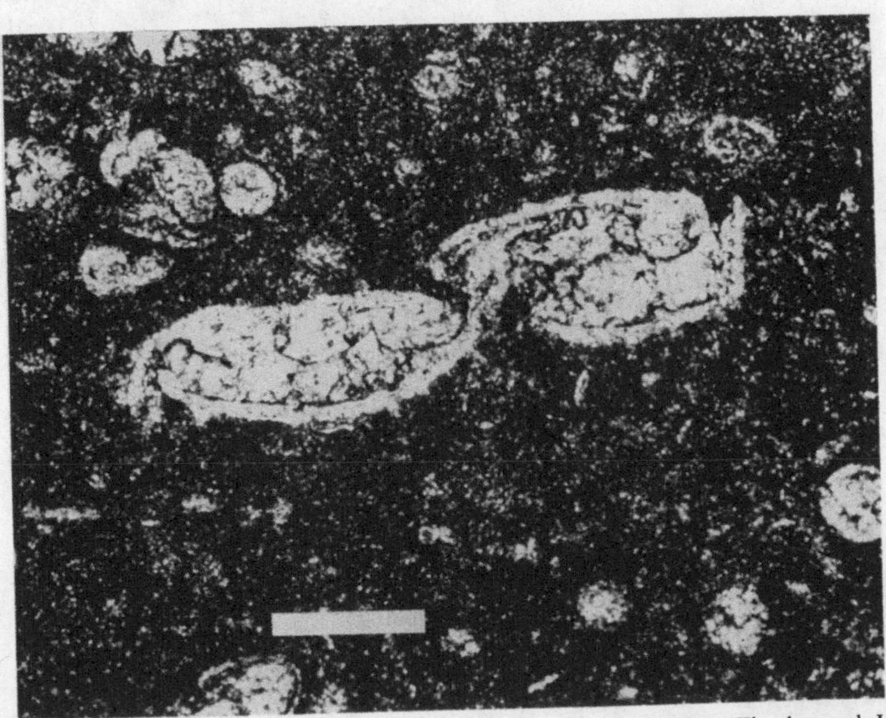

Fig. 10. Thin-section photomicrograph of Senonian Upper Chalk (BAU) from Flamborough Head, Yorkshire. Shows an occurrence of compactional crushing of microfossil test. Compaction effects are, however, not as common here as in Northern Ireland. Scale bar is 0·11 mm.

Fig. 11. Scanning electron photomicrograph of Senonian Upper Chalk (BAS) from Flamborough Head, Yorkshire. Note well-preserved coccoliths and partially interlocked crystals. Scale bar is 4 μm.

Fig. 12. Scanning electron photomicrograph of Senonian Upper Chalk (BAU) from Flamborough Head, Yorkshire. Shows euhedral, partially interlocked crystals with significant intercrystalline porosity. Scale bar is 3 μm.

The diagenetic textures of the Yorkshire chalks show strong affinities to those seen in the Irish White Limestone. Compactional crushing of microfossils is present in most samples (Fig. 10) although it is not as widely or intensively developed as in the Irish chalks. Again, however, one must deduce that the sediment was largely uncemented at the time of load-compaction.

Late diagenetic effects in Yorkshire chalks, as in Irish ones, include formation of stylolites parallel to bedding, and subsequent development of recrystallization veins perpendicular to bedding.

Scanning electron photomicrographs (Figs 11 and 12) illustrate the excellent preservation of the abundant coccoliths in the Yorkshire samples. However, they also shows that micritic matrix crystals have developed blocky, euhedral shapes and partly interlocked textures while retaining greater inter-crystalline porosity than did the Irish samples.

Southern England

A heterogeneous group of Upper, Middle, and Lower Chalk sediments were examined. The Lower Chalk is characterized by insoluble contents as high as 60% (Hancock & Kennedy, 1967). Although clay is the main contaminant, detrital quartz silt, traces of chert, phosphatic shell and bone fragments, glauconite, pyrite and marcasite also contribute significantly to the high content of impurities in the Lower Chalk. These laminated and burrowed deposits are quite soft, with porosities ranging between 27 and 42% and permeabilities averaging 2 md.

On the other hand, the Middle and Upper Chalk sediments are virtually pure carbonates, with only trace amounts of silt, clay, and phosphate. These chalks also are extremely soft, with porosities in the range of 40–46% (average 43%) and permeabilities of 6–12 md (average 8 md).

In thin section, the chalks of southern England all group as sparse or packed biomicrites (Figs 13 and 14). Lower Chalk samples have common benthonic Foraminifera, moderately abundant planktonic Foraminifera, and often very abundant calcispheres (Fig. 14). In some areas, *Inoceramus,* and to a lesser degree, echinoid fragments are important macrofossil sediment producers. In Upper Chalks, planktonic Foraminifera are more abundant than benthonic types, while calcispheres are particularly important in samples from Seaford. Molluscs, including *Inoceramus,* and echinoids are, again, the most important macrofossil sediment contributors. However, bryozoans are important in the Seaford section, perhaps indicating a slightly shallower water depth in that area.

In their faunal and textural make-up, the southern English chalks do not appear very different from their stratigraphic equivalents in Yorkshire or Northern Ireland. In diagenetic features, however, the differences are vast. Southern English chalks show virtually no compactional crushing of microfossils. Occasionally macrofossils can be found which show crushing, but the pervasive deformation of smaller skeletal constituents so common in Irish and Yorkshire samples is simply not present in southern England. Many microfossils are fragmented but, in most cases, this is clearly a pre-burial phenomenon produced by transportational or biological destruction of grains.

Later diagenetic features such as stylolites and calcite-filled fractures are also virtually absent in chalks of southern England. It appears, then, that with the exception of some loss of initial porosity through dewatering, the southern English chalks

Fig. 13. Thin-section photomicrograph of Senonian Upper Chalk (BAO) from Ramsgate area, England. Sample is unusually rich in coarse skeletal debris including planktonic and benthonic Foraminifera, echinoderms, and molluscs. Note lack of orientation and fragmentation of many grains. Scale bar is 0·42 mm.

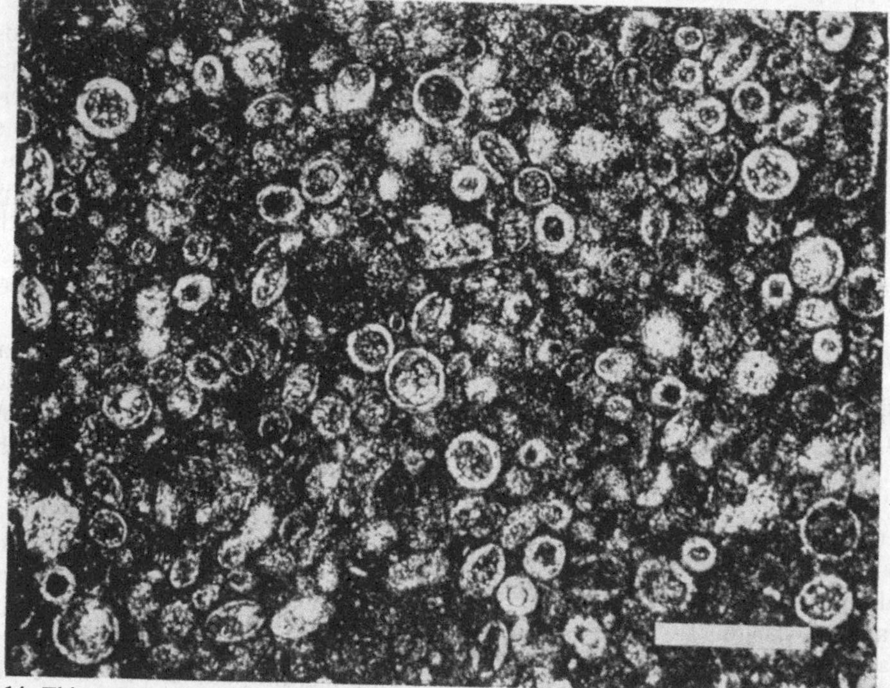

Fig. 14. Thin-section photomicrograph of Cenomanian Lower Chalk (BBO) from Osmington Mills, England. Rock contains abundant calcispheres as well as planktonic Foraminifera. Scale bar is 0·17 mm.

have received remarkably little diagenetic alteration despite a 70–90 million year history of post-depositional burial (to depths, in some cases, in excess of 500 m, as on the Isle of Wight) and uplift.

Scanning electron microscopy confirms most of these impressions. Lower Chalk (Figs 15 and 16) shows abundant clay flakes and spherulitic aggregates (Fig. 17) similar to the cristobalite described in deep-sea sediments by Wise, Buie & Weaver (1972). The pure carbonate Middle and Upper Chalk (Figs 17 and 18) has a loosely arranged fabric of subhedral to slightly rounded micrite crystals. Coccoliths are well preserved and extremely abundant. Matrix crystals and coccolith plates do show some interlocking textures and commonly have line rather than point contacts, indicating that welding of grains has occurred to some degree. But the effect is minimal when compared with Yorkshire or Northern Ireland.

Hancock & Kennedy (1967) reported that '. . . in any one field most of [the coccoliths] seem to lie more or less parallel' (p. 249). This observation was not confirmed in the present study; on the contrary, coccoliths and coccolith fragments showed a wide range of orientation, even in small areas.

North Sea

The Maastrichtian and Danian sediments from the Ekofisk field in the North Sea are mainly chalks and marls. Most come from depths of 3000–3500 m below the sea floor, with present-day temperatures at those depths averaging 128°C. Present-day pore fluids have elevated salinities; detailed analyses are not, however, available.

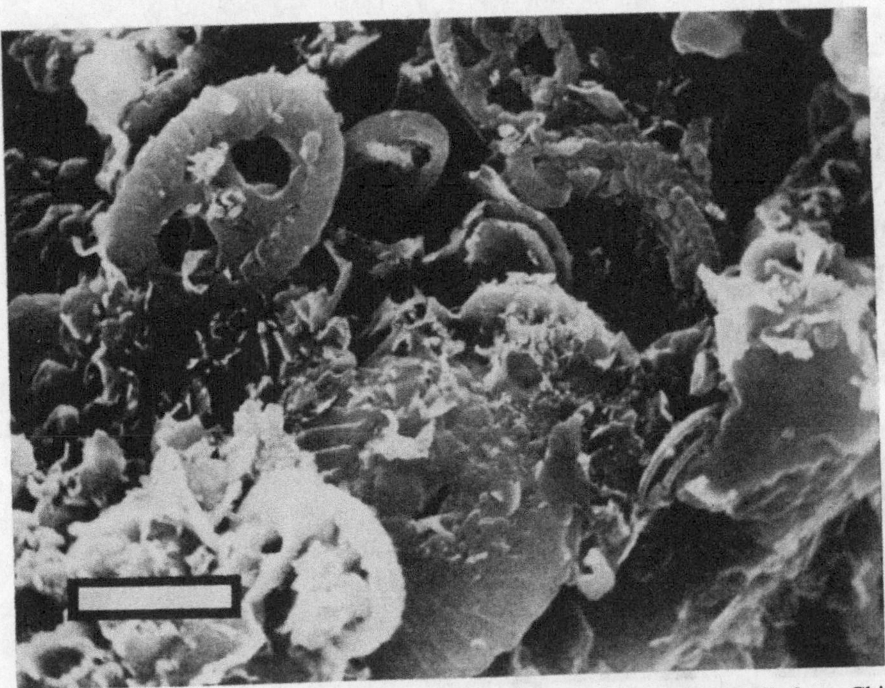

Fig. 15. Scanning electron photomicrograph of Cenomanian Lower Chalk (BGQ) from Chinnor quarry, England. Typical marly chalk shows abundant clay flakes as well as well-preserved Foraminifera. Scale bar is 4 μm.

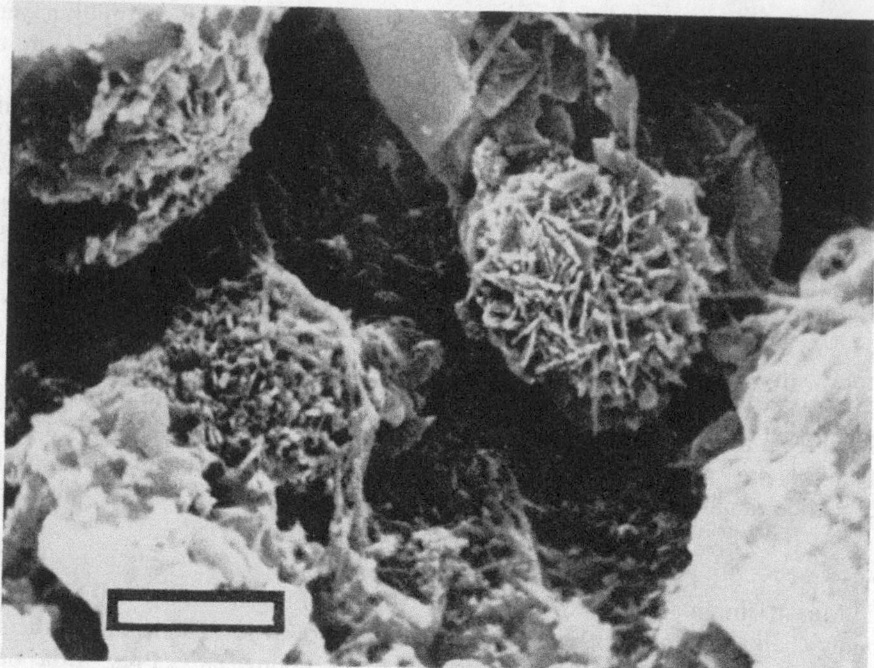

Fig. 16. Scanning electron photomicrograph of Cenomanian Lower Chalk (BGP) from Chinnor quarry, England. Abundant micro-spherules may be cristobalite 'lepispheres' like those described by Wise *et al.* (1972). Scale bar is 4 μm.

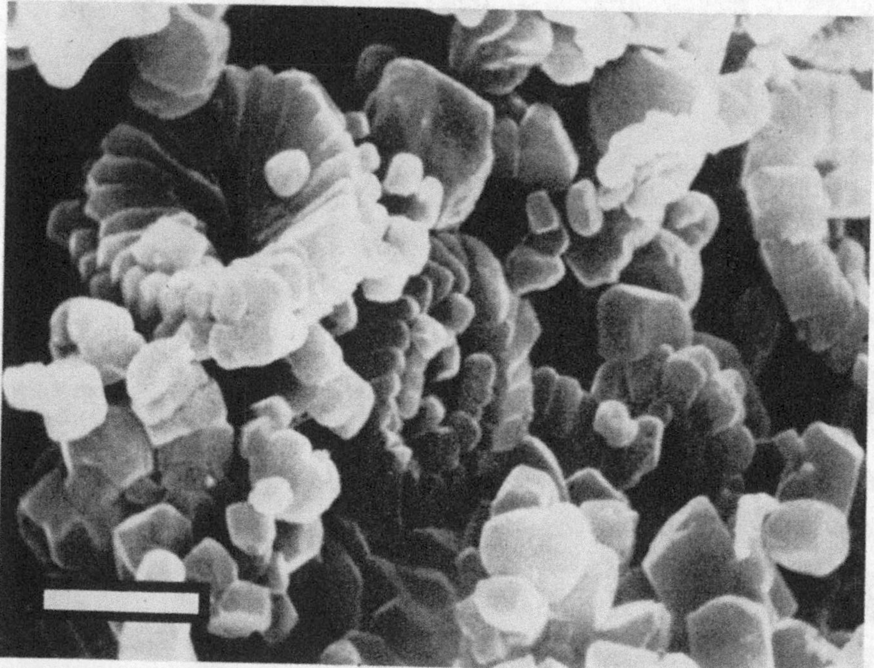

Fig. 17. Scanning electron photomicrograph of Senonian Upper Chalk (BAF) from Dover cliffs, England. Coccoliths are well preserved but show some overgrowths. Very loose crystal fabric. Scale bar is 3 μm.

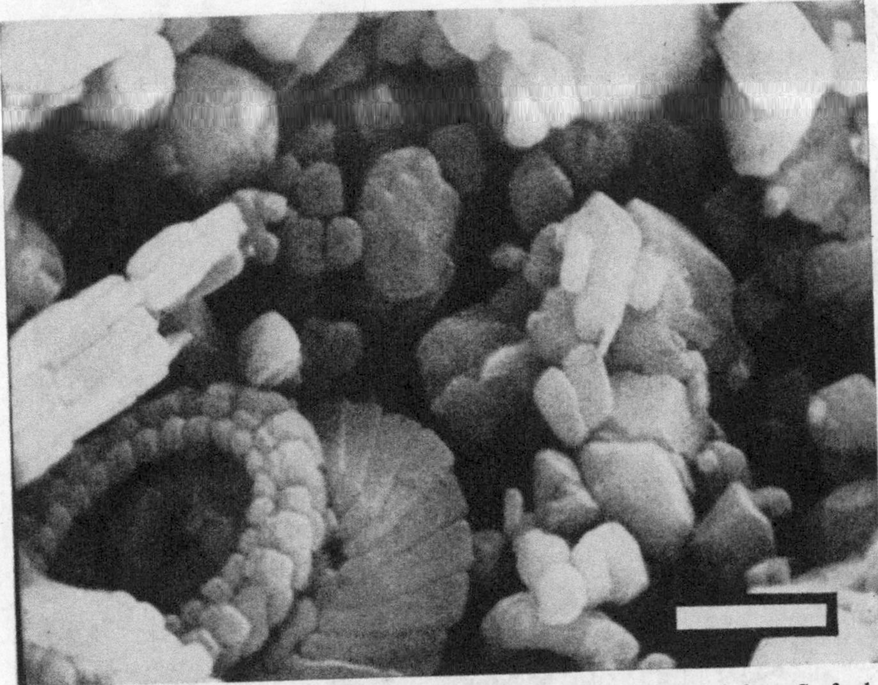

Fig. 18. Scanning electron photomicrograph of Senonian Upper Chalk (BBL) from Seaford, near Brighton, England. Note coccoliths and lack of angular, euhedral interlocking matrix crystals. Scale bar is 3 μm.

The Ekofisk samples used in this study show a wide range of hardness, with individual porosity values as low as 2% in some strata and as high as 41% in others (average 22%). Permeability of individual samples ranges from 0 md to nearly 6 md, with an average of 1·2 md. This corresponds fairly well with the published data for the Ekofisk 2/4-2X well (Owen, 1972). The logs indicate an average porosity of about 30% but with major porous zones in excess of 35% porosity alternating with tight zones with 10–20% (and in some cases even less than 5%) porosity. These alternations occur on both a coarse (tens of metres) and fine (fractions of metres) scale and no particular trend of porosity versus depth is apparent.

In thin section and insoluble residue, many of the samples could be seen to have a high percentage of non-carbonate constituents including detrital silt and clay, chert, phosphate fragments, and authigenic pyrite. However, no correlation was found between the amount of such impurities and porosity values. There is a general scarcity of carbonate skeletal grains in most samples and all the chalks are classified as fossiliferous micrites or sparse biomicrites (Fig. 19). Planktonic and benthonic Foraminifera are the major microfossil constituents, although calcispheres and Radiolaria are present in a few samples. Molluscan grains, including *Inoceramus*, and sponge spicules, plus rare echinoid fragments are the only significant macrofossil components in the sediment.

The diagenetic alteration of the North Sea samples is complex and difficult to interpret because of the discontinuous core samples and lack of outcrop control. In many samples, dolomite is present in amounts ranging from trace to over 8%. The

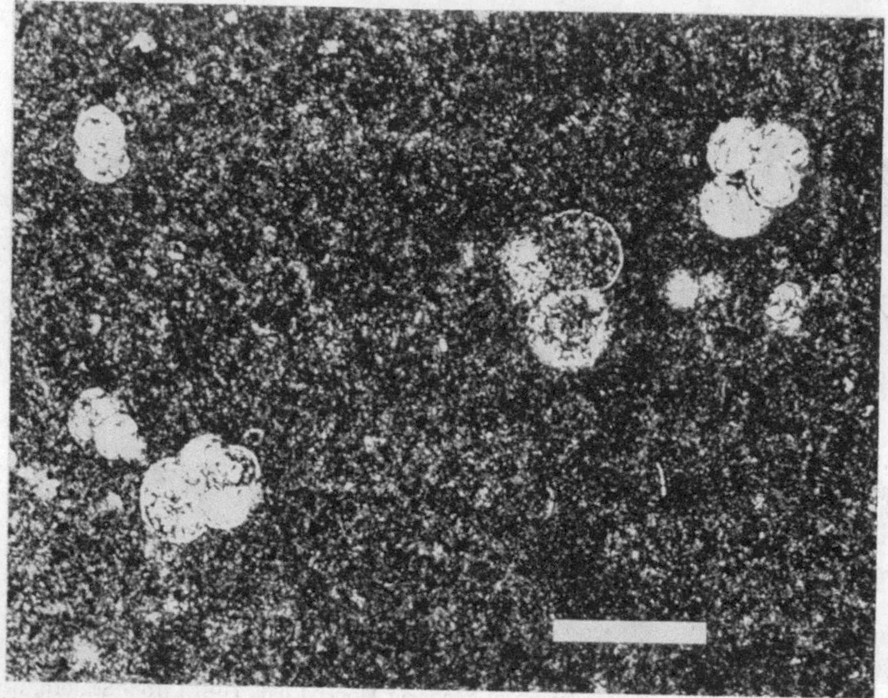

Fig. 19. Thin-section photomicrograph of Danian chalk from Phillips Ekofisk well 2/4–1x, 9722 ft depth. Typical micrite with sparse planktonic foraminiferal tests. Scale bar is 0·14 mm.

dolomite occurs as isolated single euhedral rhombs of 0·015–0·035 mm size. Some of the crystals may be detrital, but the majority are demonstrably authigenic based on cross-cutting textural relations.

Compactional crushing of skeletal fragments is present in a number of samples although it is difficult to observe in most sections because of the general scarcity of large skeletal grains. Possibly because of these problems, no correlation could be seen between sample porosity and degree of compaction.

Later diagenetic effects such as stylolitization and fracturing are prominent in many of the samples. In particular, wispy clay seams, which appear to be solution zones although they lack the undulating surface of normal stylolites, are very common.

Scanning electron microscopy (Figs 20 and 21) reveals that coccoliths are common, although in some North Sea samples they are apparently somewhat less abundant than in most southern English samples. Low porosity samples show extensive recrystallization with the formation of interlocking, equant, euhedral calcite crystals (Fig. 20) similar to those seen in samples from Yorkshire or Northern Ireland. Samples with high porosity, on the other hand, show only slight interlocking of matrix crystals and the only coarse calcite crystals are found as euhedral infillings of foraminiferal chambers. Thus, these samples are petrographically similar to ones from southern England.

A remarkable feature of many of the samples, particularly the more porous ones, is an extensive rounding of grains (Fig. 21). The rounding affects skeletal fragments, matrix grains, precipitated calcite cement, and even stylolitic surfaces and thus it appears to be a very late diagenetic effect rather than a primary, pre-burial feature.

Fig. 20. Scanning electron photomicrograph of Danian (?) chalk from Phillips Ekofisk well 2/4–5x 10,529 ft depth. Rather tightly interlocked euhedral crystal fabric with some visible coccolith debris. Still retains 24% porosity. Scale bar is 6·6 μm.

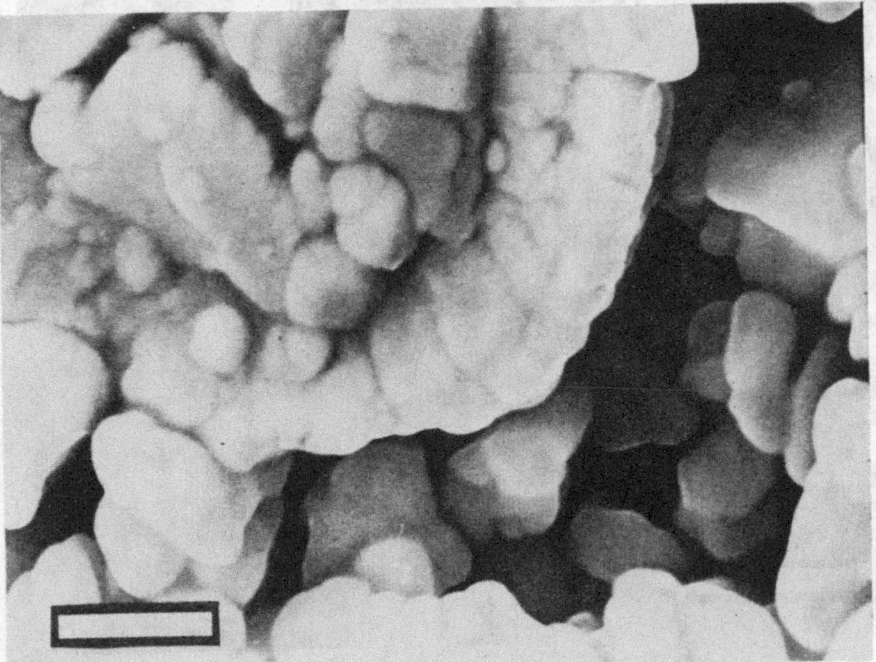

Fig. 21. Scanning electron photomicrograph of Danian chalk from Phillips Ekofisk well 2/4–2x, 10,564 ft depth. Shows rounded crystal surfaces produced either by corrosion or coating of grain surfaces late in diagenetic history. Scale bar is 2 μm.

Although there is some possibility that this phenomenon could be produced during drilling or core processing it is unlikely. The samples come from five different wells, similar effects have been seen from other areas of the North Sea (E. G. Purdy, personal communication), and no unusual drilling fluids were used in these wells. The effects could be produced by late diagenetic corrosion or by the deposition of a thick (asphaltic?) sludge on the grains. Further samples are being examined, however, to resolve this problem.

Denmark

Two Danish chalk samples (one Maastrichtian; the other Danian) were examined for comparison with the British and North Sea material. The Danish deposits are rather pure, white carbonates with a high planktonic to benthonic foraminiferal ratio, common calcispheres, and significant contributions of molluscan, bryozoan, and echinoderm fragments. The samples are extremely soft and yielded porosity values of 45% (permeability over 6 md).

These samples are diagenetically similar to the Upper Chalk of southern England. Compaction, stylolitization, and fracturing are virtually absent. Scanning electron microscopy (Fig. 22) indicates that, as in southern England, coccoliths are abundant, matrix crystals are subhedral to rounded, and inter-crystalline pore spaces are abundant. Most crystal-to-crystal boundaries appear to be line rather than point contacts. As in some of the Lower Chalk from England, spherulitic aggregates of platy material (cristobalite or clay) are locally prominent.

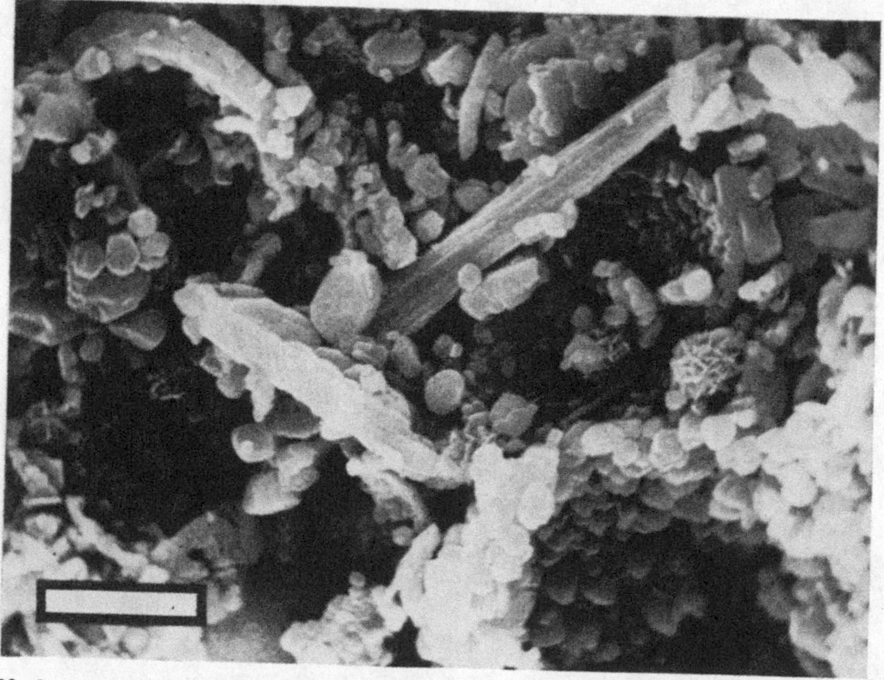

Fig. 22. Scanning electron photomicrograph of Danian chalk (BKC) from Jutland, Denmark. Has well preserved foraminiferal chambers, coccoliths, and rhabdoliths in an unrecrystallized matrix which includes some cristobalite mircospherules. Scale bar is 6·6 μm.

ISOTOPIC DATA

In an attempt to gain a better understanding of the regional differences in diagenesis which have just been described, oxygen and carbon isotopic analyses were obtained for samples from each area. Care was taken during sampling to avoid 'unusual' lithologies and it is felt that the rocks selected are reasonably representative of the main lithological types from each region.

Isotopic analyses have been made previously on material taken from European chalks (e.g. Urey *et al.*, 1951; Lowenstam & Epstein, 1954; Bowen, 1966), but most of these analyses have had palaeotemperature determination as their objective. Thus, previous workers have analysed primarily thick-walled, calcitic fossils from chalks in order to find material which has remained stable and diagenetically unaltered throughout its post-depositional history. The objective of this project, on the other hand, was to find material which was the least stable and most clearly reflected conditions of diagenetic alteration. Thus, fine-grained bulk chalk sediments, rather than fossils, were analysed.

The results of these determinations are presented in Table 1 and are graphically shown in Fig. 23. From Fig. 23 it can be seen that the samples from Ireland, Yorkshire, southern England, the North Sea, and Denmark fall into well-defined groupings. Subsurface samples from the North Sea and outcrop samples from Denmark have isotopic values which are within the range of modern normal marine carbonates (Graf, 1960). Samples from southern England show oxygen isotopic values shifted somewhat toward the negative side, but not far from normal modern marine values which commonly range between +2 and $-2/10^3$ (relative to PDB). The oxygen isotopic values (averaging $-2{\cdot}9/10^3$) for southern English chalks obtained in this study are in agreement with previously published data (Urey *et al.*, 1951; Lowenstam & Epstein, 1954) for soft, onshore European chalks. The oxygen isotopic ratios of Yorkshire chalks, by comparison, are shifted further toward the negative side δO^{18} value of $-3{\cdot}9/10^3$. Finally, samples of the Northern Irish White Limestone show markedly reduced oxygen and carbon isotopic ratios. Individual samples show δO^{18} values as low as $-8{\cdot}1$ and δC^{13} as low as $-3{\cdot}7$, values which approximate those of average freshwater limestones (Keith & Weber, 1964). The average δO^{18} value for Northern Irish samples is $-5{\cdot}61$.

DISCUSSION

The data presented in this paper indicates that there are a number of diagenetic gradients which extend from Ireland through Yorkshire to southern England. Hardness, interpreted qualitatively from outcrop examination or quantitatively through porosity and permeability measurements, decreases from Ireland to southern England with intermediate values in Yorkshire. Compactional crushing is at a maximum in Ireland, intermediate in Yorkshire, and minimal in southern England. Likewise oxygen isotopic values are most negative in Ireland and become less negative through Yorkshire to southern England. Finally, stylolitization and fracturing show a decrease in abundance from Ireland and Yorkshire toward southern England.

Table 1

No.	Area	Locality	C^{13}	O^{18}
1	Northern Ireland	Belfast	+0·77	−8·14
2	Northern Ireland	Belfast	+0·94	−5·86
3	Northern Ireland	Lisburn	+1·20	−3·30
4	Northern Ireland	Lisburn	+0·48	−5·61
5	Northern Ireland	Lisburn	−3·13	−5·53
6	Northern Ireland	Lisburn	−0·19	−6·57
7	Northern Ireland	Moira	+1·52	−5·01
8	Northern Ireland	Magheramorne	−1·22	−4·74
9	Northern Ireland	Magneramorne	+1·43	−5·22
10	Northern Ireland	Magheramorne	+1·55	−5·36
11	Northern Ireland	Magheramorne	+1·40	−4·26
12	Northern Ireland	Magheramorne	+1·49	−3·99
13	Northern Ireland	Magheramorne	+1·45	−5·82
14	Northern Ireland	Ballygalley Head	−3·72	−6·35
15	Northern Ireland	Glenarm	+1·42	−7·07
16	Northern Ireland	White Park Bay	+1·16	−6·95
17	Yorkshire	Flamborough Head	+2·10	−3·92
18	Yorkshire	Flamborough Head	+2·06	−4·44
19	Yorkshire	Flamborough Head	+2·05	−3·62
20	Yorkshire	Flamborough Head	+2·15	−4·29
21	Yorkshire	Flamborough Head	+2·32	−3·54
22	Yorkshire	Flamborough Head	+2·38	−4·16
23	Yorkshire	Flamborough Head	+2·46	−3·48
24	Southern England	Dover	+1·21	−3·52
25	Southern England	Dover	+2·14	−2·57
26	Southern England	Dover	+1·60	−2·45
27	Southern England	Dover	+1·70	−2·99
28	Southern England	Broadstairs (Thanet)	+1·22	−4·09
29	Southern England	Broadstairs (Thanet)	+1·45	−2·47
30	Southern England	Broadstairs (Thanet)	+2·20	−2·15
31	Southern England	Broadstairs (Thanet)	−3·79	−3·58
32	Southern England	Seaford	+1·85	−2·93
33	Southern England	Seaford	+2·01	−2·79
34	Southern England	Seaford	−0·09	−2·37
35	Southern England	Seaford	+1·95	−2·31
36	Southern England	Seaford	+2·06	−2·49
37	Southern England	Seaford	+1·36	−3·40
38	Southern England	Hemel Hempstead	+1·02	−3·79
39	Southern England	Henley-on-Thames	+1·72	−3·12
40	Southern England	Henley-on-Thames	+5·83	−2·56
41	Southern England	Henley-on-Thames	+2·10	−2·63
42	Southern England	Nettlebed	+2·11	−2·49
43	Southern England	Osmington	+2·04	−5·69
44	Southern England	Pitstone	+4·12	−4·02
45	Southern England	Pitstone	+1·48	−2·62
46	Southern England	Chinnor	+1·43	−3·33
47	Southern England	Chinnor	+1·47	−2·19
48	Southern England	Chinnor	+3·86	−3·86
49	Southern England	Folkestone	+1·64	−2·96
50	Southern England	Folkestone	+2·20	−2·70
51	Southern England	Folkestone	+2·47	−2·90
52	Southern England	Folkestone	+0·42	−1·36

Table 1—*continued*

No.	Area	Locality	C^{13}	O^{18}
53	North Sea	Ekofisk (2/4–1AX)	+2·33	+0·37
54	North Sea	Ekofisk (2/4–1AX)	+2·45	+0·24
55	North Sea	Ekofisk (2/4–1AX)	+2·68	−1·21
56	North Sea	Ekofisk (2/4–1AX)	+2·62	−0·66
57	North Sea	Ekofisk (2/4–2X)	+2·71	−0·25
58	North Sea	Ekofisk (2/4–2X)	+2·51	−0·47
59	North Sea	Ekofisk (2/4–2X)	+2·92	−1·19
60	North Sea	Ekofisk (2/4–2X)	+2·77	−1·21
61	North Sea	Ekofisk (2/4–4X)	+2·58	−0·06
62	North Sea	Ekofisk (2/4–4X)	+2·64	+1·61
63	North Sea	Ekofisk (2/4–4X)	+2·89	−0·96
64	North Sea	Ekofisk (2/4–5X)	+2·94	+0·71
65	North Sea	Ekofisk (2/4–5X)	+2·74	+0·49
66	North Sea	Ekofisk (2/4–5X)	+3·12	−0·45
67	North Sea	Ekofisk (2/4–7X)	+1·15	−2·76
68	North Sea	Ekofisk (2/4–7X)	+2·28	−1·67
69	North Sea	Ekofisk (2/4–7X)	+2·67	−0·88
70	North Sea	Ekofisk (2/7–1X)	+2·30	+0·22
71	North Sea	Ekofisk (2/7–1X)	+2·84	−0·28
72	North Sea	Ekofisk (2/7–2X)	+2·42	−0·79
73	Denmark	Jutland	+1·57	−1·48
74	Denmark	Jutland	+2·16	−0·60

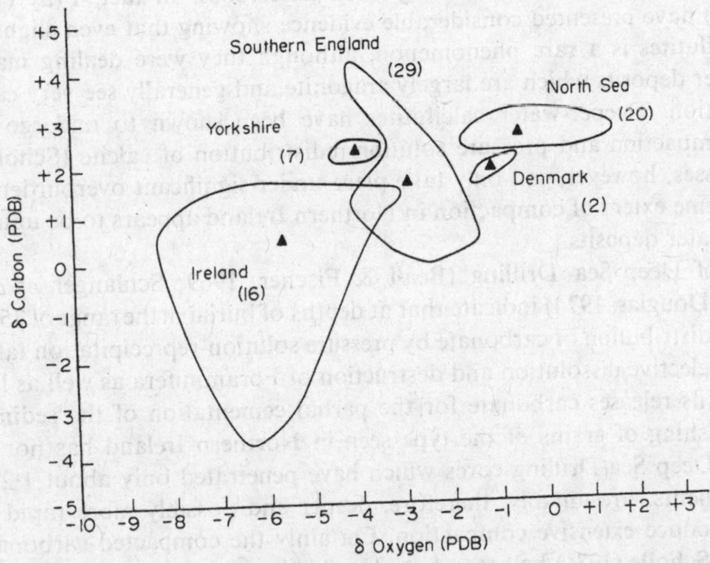

Fig. 23. Plot of sample $\delta C^{13}/C^{12}$ values (relative to PDB standard) versus sample $\delta O^{18/16}$ values (relative to PDB standard). Heavy lines enclose values for each area and numbers in parentheses indicate number of analysed samples. Triangular points indicate area averages. Southern English and North Sea envelopes each exclude one sample which fell far outside field. Data points given in Table 1.

These gradients are clearly diagenetic and do not appear to reflect primary sedimentological or mineralogical differences in the deposits. Thus, to explain these phenomena, it is diagenetic processes which must be examined. The isotopic values, as with all such data, are compatible with two possible explanations or a combination of both.

Hypothesis I

First, and geochemically simplest, is the hypothesis of fresh-water alteration. The isotopic ratios of the Irish White Limestone samples are virtually those of fresh-water carbonates. Yet the White Limestone is clearly a normal marine sediment as indicated by its extensive open marine fauna. Thus, recrystallization of the original chalk in contact with isotopically light (Craig, 1961) fresh-water pore fluids would be required to produce the observed isotopic ratios. Scholle & Kinsman (1973) postulated that there was a gradient of fresh-water input into chalks from Ireland to England related to the greater uplift of the Irish area associated with the Late Cretaceous or Early Tertiary rifting of the North Atlantic. In that hypothesis, areas such as the North Sea are considered to have received little or no fresh water input during an essentially continuously marine depositional and burial history. These sediments retained their original marine pore fluids or were exposed to marine-derived brines. Samples at the other extreme, Northern Ireland, were exposed to erosion and fresh-water input soon after deposition; these chalks then recrystallized in fresh-water pore fluids.

While this hypothesis adequately explains the geochemical observations, there are a number of geological points which argue against it. Compaction in Irish chalks preceded or accompanied recrystallization and cementation (as was argued from petrographic data). Yet bedding-parallel compaction of the magnitude seen in Northern Ireland is an unusual feature in fine-grained limestones. In fact, Pray (1960) and Zankl (1969) have presented considerable evidence showing that even slight compaction of calcilutites is a rare phenomenon, although they were dealing mainly with shallow-water deposits which are largely aragonite and generally see very early fresh-water alteration. Deeper-water calcilutites have been shown to undergo far more extensive compaction and pressure solution-redistribution of calcite (Scholle, 1971). These processes, however, can only take place under significant overburden pressure and the extreme extent of compaction in Northern Ireland appears to be unusual even for deeper-water deposits.

Results of Deep Sea Drilling (Beall & Fischer, 1969; Schlanger *et al.*, 1973; Schlanger & Douglas, 1974) indicate that at depths of burial in the range of 250 m–1 km significant redistribution of carbonate by pressure solution-reprecipitation takes place. Apparently selective dissolution and destruction of Foraminifera as well as lesser loss of nannofossils releases carbonate for the partial cementation of the sediment. But extensive crushing of grains of the type seen in Northern Ireland has not yet been observed in Deep Sea Drilling cores which have penetrated only about 1·2 km into oceanic sediments. Presumably, therefore, deeper and possibly more rapid burial is needed to produce extensive compaction. Certainly the compacted carbonate flysch discussed by Scholle (1971) had seen burial well below 1 km and sedimentation rates far higher than those seen in typical Deep Sea Drilling sections.

It is thus tentatively postulated that rapid loading and 1½–2 km of overburden were required to produce the Northern Irish compaction, although Neugebauer (1973) indicates that the loss of marine pore fluids and their replacement by meteoric or hydrothermal fluids may reduce that value significantly. This large required overburden implies that the compaction, cementation, and isotopic re-equilibration of the Irish chalks, at least, took place relatively late in the diagenetic history, after the addition of thick overburden.

Earlier workers (e.g. Hancock, 1963) postulated an early cementation of Irish chalks on the basis of the Balleycastle 'Pellet-Chalk', a Maastrichtian deposit with hard clasts of White Limestone. Because this unit with hardened clasts underlies the basalt flows of Northern Ireland, Hancock (1963, p. 157) concluded that this indicated that the basalts could not have been involved in the hardening of the Irish sediments. Hancock (1973, personal communication) no longer holds this view, and now considers the Balleycastle 'Pellet Chalk' to be a volcanic diatreme rather than a sedimentary deposit. Furthermore, Wolfe (1968) has shown that certain zones of the Irish chalks were probably cemented at or near the depositional interface. Such layers could have provided hard clasts despite the fact that the rest of the chalk was still unconsolidated.

Fresh-water alteration of the Irish chalks, if it took place, should have occurred during the time interval between the deposition of the chalks (Campanian-early Maastrichtian) and the time of outpouring of basaltic flows (latest Maastrichtian to Palaeocene according to Evans, Fitch & Miller, 1973). There is evidence for subaerial exposure during that time, including plant-bearing lignite deposits lying on the chalk, and large-scale erosional truncation and channelling of the top surface of the White Limestone (Wilson, 1972). Yet it is virtually impossible to see how major compaction could have preceded or accompanied this exposure. At any reasonable chalk sedimentation rates, only a few hundred metres of additional chalk sediments could have been deposited above the strata presently found. This would clearly be insufficient to produce the compaction observed.

Furthermore, it is difficult to understand why this stable calcitic sediment would have undergone recrystallization in fresh water under near-surface conditions. Enormous volumes of fresh water would have to move through the formation to produce such alteration, and if compaction and cementation had preceded and/or accompanied isotopic re-equilibration, then the water would have had to move through a rock of very low permeability indeed.

Finally, chalk sediments in other areas such as southern England and Denmark were exposed to erosion and fresh water during Earliest Tertiary through Recent times, both in subsurface and near-surface environments. The chalk is a major aquifer in southern England and, despite its relatively low permeabilities, shows significant rates of fresh-water flow (Edmunds, Lovelock & Gray, 1973). The southern English chalk also shows deep surface weathering with solution caverns and residual chert deposits on the surface (e.g. Havard & Stewart, 1969; Thorez *et al.*, 1971). Yet these chalks have not re-equilibrated significantly with fresh waters making it unlikely that the Irish chalks did either.

Hypothesis II

A second postulate, which appears to avoid some of these geological problems, is that the observed isotopic gradients reflect geothermal heat flow variations during Latest Cretaceous or Early Tertiary time. This would involve re-equilibration of the carbonate-water isotopic system at higher temperatures in Northern Ireland leading to lighter isotopic values of carbon and oxygen. In this hypothesis, compaction took place in Northern Ireland as a result of rapid loading of soft sediments by thick basalt flows. Wilson (1972) indicates that although later erosion has reduced basalt thicknesses in Ireland to a few hundred metres in most areas, the original thickness, as in Mull (Scotland), was probably of the order of 1800 m. Such a thick lava section would have sufficed to produce the compaction seen in Irish chalks.

In association with the intrusion of dykes and sills and the extrusion of basaltic flows the geothermal gradients in Northern Ireland would have certainly been unusually high. Although exact temperatures are impossible to determine, recent world-wide heat flow data (e.g. Wyllie, 1971; Roy *et al.*, 1968) indicates that geothermal gradients of 40–60°C per kilometre are reasonable for active volcanic rift margins. Thus at $1\frac{1}{2}$–2 km burial depth temperatures in the range of 80–120° C might be expected. Contact with basaltic dykes could provide higher temperatures locally, although the theoretical data of Jaeger (1957) and Lovering (1935) as well as the isotopic work of Deines & Gold (1969) and Tan & Hudson (1971) indicate that such direct contact effects are restricted to a distance of only centimetres to a few metres from the intrusion. Similar effects are seen in the Northern Irish White Limestone where contact aureoles of 2 m or less are found in association with cross-cutting dykes. No such effects are seen at the contacts with overlying basalt flows. Thus, overall temperatures in excess of 100°C and perhaps higher could have been maintained in the Northern Irish section for several million years.

Despite these rather high temperatures it is difficult to explain the isotopic re-equilibration of the Northern Irish limestone as a strictly thermal phenomenon. Anderson (1969), following the original work of Urey *et al.* (1951), indicates that significant solid state diffusion will not have occurred in carbonates younger than 10^8 years unless temperatures in excess of 200°C have been encountered, although the extremely small crystal size of chalk sediments may lower that value somewhat. In fact, the work of Clayton, Muffler & White (1968) in the Salton Sea area shows that carbonates reached isotopic equilibrium at 150°C and may have reached it even at temperatures as low as 100°C. Nevertheless, strictly thermal effects in a closed system (no water flow) do not appear to be entirely adequate to explain the extensive isotopic alteration found in Northern Ireland.

Hypothesis III

Although fresh-water alteration and thermal re-equilibration each fail to individually provide completely acceptable explanations for the Irish chalk alteration, a combination of the two factors appears to allow a viable solution to the problem. In this hypothesis, the uplift and erosion of the Northern Irish chalk led to the flushing of original marine pore fluids and the inflow of isotopically light meteoric waters. The lack of significant overburden, and the low near-surface temperatures prevented significant alteration of the chalks. The extrusion of thick basalt sequences during the Latest Cretaceous and Palaeocene produced extensive compaction and associated pressure-induced redistribution (solution-reprecipitation) of carbonates in meteoric pore fluids. At the same time the dykes were being intruded into the section and strong circulation cells of thermally driven meteoric waters were established. Thus, recrystallization and isotopic alteration took place under a combination of circumstances; (a) strong pressure-induced carbonate redistribution; (b) elevated temperatures; (c) very small crystal size; (d) meteoric pore fluids; and (e) strong thermally driven circulation of those fluids.

The assumption for meteoric-water circulation in a basaltic intrusive and extrusive setting is based on two main factors. First, basaltic magmas are generally dry and therefore rarely expel fluids. Second, the extensive oxygen, carbon, and deuterium isotopic studies which have been done on intrusives and their country rocks (Craig, 1963; Tan & Hudson, 1971; Taylor, 1970; and Taylor & Forester, 1971) indicate that in most cases juvenile waters (isotopically heavier than seawater) are not expelled from the magmas. Rather, one sees the extensive circulation of meteoric waters which produce isotopic lightening of both the country rock and the margins of the intrusives.

Also, Neugebauer (1973, 1974) points out that a major restricting factor in chalk pressure solution and compaction is that with retention of initial marine pore fluids (with high Mg/Ca ratios) chalk minerals are stable even at relatively high pressures. With the removal of such pore fluids, pressure solution and compaction should take place far more readily. Thus, the thermal circulation of meteoric fluids could have affected the compaction and pressure solution as well as the isotopic re-equilibration of the sediment.

The intermediate levels of hardness, compaction, and isotopic re-equilibration of the Yorkshire chalks can be explained in this same model. Although Yorkshire shows far less thermal activity than Northern Ireland, some large dyke complexes extend into Yorkshire from igneous centres in Scotland. These presumably raised geothermal gradients in Yorkshire and to some degree may have contributed to hydrothermal water circulation. Thermal anomalies trending into Yorkshire can be seen even in the plot of Recent geothermal gradients in the North Sea area (Harper, 1971). The small amount of compaction and the slight isotopic alteration seen in Yorkshire chalks can probably be explained by the presence of hundreds of metres of chalk overburden (subsequently removed by erosion), the elevated temperatures, and the expulsion of marine pore fluids.

Southern England, in contrast, shows little diagenetic alteration and little evidence of igneous-thermal activity. In that area, chalks probably never had unusually thick overburden and were subjected to lesser pore fluids circulation. Although some pressure solution took place at grain boundaries (yielding line contacts) the absence of elevated temperatures and hydrothermal fluids retarded further recrystallization.

North Sea area

The North Sea chalks are clearly a separate case. Although all the Ekofisk samples showed isotopic values close to those of modern marine carbonates, they also showed a range of hardnesses, porosities, and textures comparable to the full suite of samples from Ireland to southern England. Unlike any of the onshore samples, the North Sea chalks were deposited in a structurally controlled basin and were overlain by a rather continuously deposited cover of over 3000 m of Tertiary sediments. These cover sediments are dominantly marine, and although some minor unconformities do occur in the section, it appears likely that the centre of the North Sea Basin never saw extensive subaerial exposure. This history of deep burial in a marine setting clearly differentiates the offshore from the onshore chalks examined in this project.

The isotopic analyses confirm that the North Sea chalks probably retained marine pore waters during most of their history and saw neither significant amounts of fresh water nor hydrothermal conditions. Whatever diagenesis they underwent took place in an effectively closed system in marine or marine-derived pore fluids. The presence of extensive stylolites, solution seams, and other evidence of pressure solution indicates that, as with Deep Sea Drilling samples, pressure solution-reprecipitation may have been the major cementation mechanism.

The alternating zones of high and low porosity in North Sea chalks are poorly understood. Perhaps they reflect amplification of differences in the amount of initial sea-floor cementation. But they may also reflect differential pressure solution-reprecipitation effects. Layers which initially were finer-grained or perhaps contained more foraminiferal material may have been selectively subjected to dissolution while other layers with coarser or more stable material may have been sites of selective precipitation.

COMPARISON WITH DEEP SEA DRILLING SAMPLES

Petrographic as well as isotopic data from numerous deep-sea carbonates have been examined as part of the Deep Sea Drilling Project (e.g. Beall & Fischer, 1969; Wise & Hsü, 1971; Wise & Kelts, 1972; Lloyd & Hsü, 1972; Schlanger *et al.*, 1973; Coplen & Schlanger, 1973; Anderson & Schneidermann, 1973; Schlanger & Douglas, 1974). The samples involved have ranged from unconsolidated nannofossil oozes, through chalks, to hard limestones.

The results of these studies are highly variable and rather inconclusive at this stage. Beall & Fischer (1969) and Schlanger *et al.* (1973) concluded that much of the cementation was a result of pressure solution-reprecipitation of material. This process apparently becomes increasingly important with increasing depth of burial. Wise & Hsü (1971); Wise & Kelts (1972) and Lloyd & Hsü (1972), all working on material from Leg 3 (South Atlantic) of the Deep Sea Drilling Project, concluded, on the basis of petrographic and isotopic data, that cementation occurred at or near the sediment-water interface without volcanic thermal or hydrothermal influences. Anderson & Schneidermann (1973, as quoted in Coplen & Schlanger, 1973) explained the recrystallization and negative oxygen isotopic shift of Upper Cretaceous chalks from Leg 15 (Caribbean) as due to higher temperatures induced in the pelagic sediments by associated volcanic activity. Coplen & Schlanger (1973) explain oxygen isotopic values as negative as $-5{\cdot}46$ in Lower Cretaceous chalky limestones from Leg 17 (Pacific Ocean) as being due to changes in isotopic composition of the ocean as well as high palaeotemperatures during that period.

Thus, a complete spectrum of isotopic values and interpretations is present for oceanic chalks. Yet all these samples differ in several important respects from the chalks examined in this project. First, they were deposited in generally deeper waters. Second, unlike the British onshore chalks, they were never exposed to fresh water. Third, in general, thermal-hydrothermal sources in oceanic sediments lay below the sediment column, whereas in Northern Ireland they were injected into and above the sediment section. Fourth, oceanic sediments were deposited at slow sedimentation rates and have seen far less depth of burial relative to the North Sea chalks. Also, even where oceanic sediments are deeply buried, Deep Sea Drilling, to date, has only sampled down to about 1·2 km depth. Finally, the age range of Deep Sea Drilling oozes, chalks and limestones is far greater (Lower Cretaceous to Recent) than that of the chalks examined from Europe (Cenomanian to Danian); this may be of significance if significant variations in water temperature and composition did occur during the last 200 million years. Deep Sea Drilling sediments, then, provide interesting comparisons, but not especially good analogues, for the European chalks.

CONCLUSIONS

A geothermal-hydrothermal hypothesis in which meteoric water is heated and circulated by basaltic dykes in a regime of elevated geothermal gradients appears to be capable of reconciling the geochemical and geological data available for recrystallized British chalks. The input of these waters and the rapid loading by a thick basalt column could produce the combined compaction and recrystallization features noted in the Irish White Limestone while at the same time accounting for the isotopic re-equilibration of the unit. Lesser degrees of hardening and re-equilibration are explained by lesser amounts of geothermal-hydrothermal activity (as in Yorkshire) or by extremely deep burial in a setting in which marine pore fluids are retained (as in the North Sea) and cementation is accomplished by local pressure solution-reprecipitation.

Thus, combined examination of isotopic ratios and diagenetic features (in thin-section and scanning electron microscope) has proven most useful in understanding the diagenetic alteration of chalks. The regional patterns of this diagenesis appear closely related to trends of geothermal gradients, igneous activity, and structural lineaments (delineating deep basins) during the Late Cretaceous and Early Tertiary. These features, in turn, are probably controlled by the early rifting of the North Atlantic. Therefore, the mapping of regional patterns of chalk diagenesis could help in the interpretation of ancient thermal and structural trends and their effects on the porosity and permeability of nearby sediments. As such, in a rifted margin setting, useful predictions could even be made about structural and porosity trends on formerly adjacent shelf areas.

ACKNOWLEDGMENTS

Major support for this project has come from Cities Service Oil Company, who provided funding for field support and isotopic analyses and kindly allowed the release of this data. Partial support has come from National Science Foundation Grant GA-36696 to the author and D. J. J. Kinsman. This has allowed further field work and will enable the project to be expanded in scope in the future.

I am greatly indebted to D. J. J. Kinsman for many hours of discussion of the diagenetic problems of the chalk, and to W. J. Kennedy and R. Bromley for their excellent introductions to the geological setting of the European chalks and for criticizing early versions of the manuscript. Conversations with C. Wood, J. Hancock, J. D. Hudson, E. G. Purdy, and R. M. Lloyd stimulated or corrected many of the ideas in this paper. Finally, I would like to express my thanks to Phillips Petroleum Company for the very early release of the subsurface Ekofisk samples.

REFERENCES

Anderson, T.F. (1969) Self-diffusion of carbon and oxygen in calcite by isotopic exchange with carbon dioxide. *J. Geophys. Res.* **74,** 3918–3932.

ANDERSON, T.F. & SCHNEIDERMANN, N. (1973) Stable isotope relationships in pelagic limestones from the central Caribbean: Leg 15, Deep Sea Drilling Project. In: *Initial Reports of the Deep Sea Drilling Project,* Vol. XV (N. T. Edgar and J. B. Saunders *et al.*), pp. 795–803. U.S. Government Printing Office, Washington.

BARR, F.T. (1962) Upper Cretaceous planktonic foraminifera from the Isle of Wight, England. *Palaeontology,* **4,** 552–580.

BEALL, A.O. & FISCHER, A.G. (1969) Sedimentology. In: *Initial Reports of the Deep Sea Drilling Project,* Vol. I (M. Ewing *et al.*), pp. 521–593. U.S. Government Printing Office, Washington.

BLACK, M. (1953) The constitution of the Chalk. *Proc. geol. Soc.* no. **1499,** 81–86.

BOWEN, R. (1966) *Paleotemperature Analysis,* pp. 265. Elsevier, Amsterdam.

BROMLEY, R.G. (1967) Some observations on burrows of thalassinidean Crustacea in Chalk hardgrounds. *Q. Jl geol. Soc. Lond.* **123,** 157–177.

BROMLEY, R.G. (1970) Borings as trace fossils and *Entobia cretacea* Portlock, as an example. In: *Trace Fossils* (Ed. by T. P. Crimes and J. C. Harper). *Geol. J. Spec. Issue,* **3,** pp. 49–90. Seel House Press, Liverpool.

CARTER, R.M. (1972) Adaptations of British Chalk Bivalvia. *J. Paleont.* **46,** 325–340.

CHEETHAM, A.H. (1971) Functional morphology and biofacies distribution of cheilostome Bryozoa in the Danian Stage (Paleocene) of southern Scandinavia. *Smithson. Contr. Paleobiol.* **6,** 87 pp.

CLAYTON, R.N., MUFFLER, L.J.P. & WHITE, D.E. (1968) Oxygen isotope study of calcite and silicates of the River Ranch No. 1 well, Salton Sea geothermal field, California. *Am. J. Sci.* **266,** 968–979.

COPLEN, T.B. & SCHLANGER, S.O. (1973) Oxygen and carbon isotope studies of carbonate sediments from Site 167, Magellan Rise, Leg 17. In: *Initial Reports of the Deep Sea Drilling Project,* Vol. XVII (E. L. Winterer and J. I. Ewing *et al.*), pp. 505–509. U.S. Government Printing Office, Washington.

CRAIG, H. (1957) Isotopic standards for carbon and oxygen and correction factors for mass-spectrometric analysis of carbon dioxide. *Geochim. cosmochim. Acta,* **12,** 133–149.

CRAIG, H. (1961) Isotopic variations in meteoric waters. *Science,* **133,** 1702–1703.

CRAIG, H. (1963) The isotopic geochemistry of water and carbon in geothermal areas. In: *Nuclear Geology on Geothermal Areas,* Spoleto Conference (Ed. by E. Tongiorgi), pp. 17–53. Consiglio naz. della ricerche, Lab. di geol. nucleare, Pisa, Italy (publ. 1968).

DEINES, P. & GOLD, D.P. (1969) The change in carbon and oxygen isotopic composition during contact metamorphism of Trenton limestone by the Mount Royal pluton. *Geochim. cosmochim. Acta,* **33,** 421–424.

EDMUNDS, W.M., LOVELOCK, P.E.R. & GRAY, D.A. (1973) Interstitial water chemistry and aquifer properties in the Upper and Middle Chalk of Berkshire, England. *J. Hydrol.* **19,** 21–31.

EICHER, D.L. (1969) Paleobathymetry of Cretaceous Greenhorn sea in eastern Colorado. *Bull. Am. Ass. petrol. Geol.* **53,** 1075–1090.

EVANS, A.L., FITCH, F.J. & MILLER, J.A. (1973) Potassium-argon age determinations on some British Tertiary igneous rocks. *Q. Jl geol. Soc. Lond.* **129,** 419–443.

FISCHER, A.G. & GARRISON, R.E. (1967) Carbonate lithification on the sea floor. *J. Geol.* **75,** 488–496.

GRAF, D.L. (1960) Geochemistry of carbonate sediments and sedimentary carbonate rocks—Pt. 4-A, Isotopic composition-chemical analysis. *Circ. Ill. St. geol. Surv.* **308,** 42 pp.

HANCOCK, J.M. (1963) The hardness of the Irish chalk. *Ir. Nat. J.* **14,** 157–164.

HANCOCK, J.M. (1974) The sequence of facies in the Upper Cretaceous of northern Europe compared with that in the Western Interior. In: *The Cretaceous System in the Western Interior of North America* (Ed. by W. G. E. Caldwell). *Spec. Pap. geol. Ass. Can.* **12.** (In press.)

HANCOCK, J.M. & KENNEDY, W.J. (1967) Photographs of hard and soft chalks taken with a scanning electron microscope. *Proc. geol. Soc.* no. **1643,** 249–252.

HARPER, M.L. (1971) Approximate geothermal gradients in the North Sea Basin. *Nature, Lond.* **230,** 235–236.

HAVARD, D. & STEWART, N. (1969) The nature of the Cretaceous/Tertiary unconformity as seen at two exposures in S.E. England. *Kingston geol. Rev.* **2,** 23–25.

HAYS, J.D. & PITMAN, W.C. III (1973) A model linking Late Cretaceous extinctions and plate tectonics. *Abstr. geol. Soc. Am.* (Dallas Ann. Meet.) **5,** 660.

HEMINGWAY, J.E., WILSON, V. & WRIGHT, C.W. (1968) Geology of the Yorkshire coast. *Fld Guide geol. Ass.* **34,** 47 pp.

HUME, W.F. (1897) The Cretaceous strata of Co. Antrim. *Q. Jl geol. Soc. Lond.* **53,** 540–606.

JAEGER, J.C. (1957) The temperature in the neighborhood of a cooling intrusive sheet. *Am. J. Sci.* **255,** 306–318.

JUKES, J.B. (1868) The Chalk of Antrim. *Geol. Mag.* **5,** 345–347.

JUKES-BROWNE, A.J. & HILL, W. (1904) The Cretaceous rocks of Britain, Vol. 3. *Mem. geol. Surv. U.K.* 566 pp.

KEITH, M.L. & WEBER, J.N. (1964) Carbon and oxygen isotopic composition of selected limestones and fossils. *Geochim. cosmochim. Acta,* **28,** 1787–1816.

KENNEDY, W.J. (1970) Trace fossils in the Chalk environment. In: *Trace Fossils* (Ed. by T. P. Crimes and J. C. Harper). *Geol. J. Spec. Issue,* **3,** pp. 263–282. Seel House Press, Liverpool.

KENNEDY, W.J. & JUIGNET, P. (1973) Carbonate banks and slump beds in the Upper Cretaceous (Upper Turonian-Santonian) of Haute Normandie, France. *Sedimentology,* **21,** 1–42.

LLOYD, R.M. & HSU, K.J. (1972) Stable-isotope investigations of sediments from the DSDP III cruise to South Atlantic. *Sedimentology,* **19,** 45–58.

LOVERING, T.S. (1935) Theory of heat conduction applied to geological problems. *Bull. geol. Soc. Am.* **46,** 69–93.

LOWENSTAM, H.A. & EPSTEIN, S. (1954) Paleotemperatures of the post-Aptian Cretaceous as determined by the oxygen isotope method. *J. Geol.* **62,** 207–248.

MIŠÍK, M. (1968) Some aspects of diagenetic recrystallization in limestones. *Rep. Int. Geol. Congr.* 23rd Session (Prague), *Proc. Sect.* **8,** 129–136.

NEUGEBAUER, J. (1973) The diagenetic problem of chalk—the role of pressure solution and pore fluid. *Neues Jb. Geol. Paläont. Abh.* **143,** 223–245.

NEUGEBAUER, J. (1974) Some aspects of cementation in chalk. In: *Pelagic Sediments: on Land and under the Sea* (Ed. by K. J. Hsü and H. C. Jenkyns). *Spec. Publs int. Ass. Sediment.* **1,** 149–176.

OWEN, J.D. (1972) A log analysis method for Ekofisk Field, Norway. In: *Proc. Soc. Prof. Well Log Anal. 13th Ann. Logging Symposium* (*May* 7–10, 1972), Paper X, 22 pp.

PEACH, B.N., GUNN, W. & NEWTON, E.T. (1901) On a remarkable volcanic vent of Tertiary age in the Island of Arran, enclosing Mesozoic fossiliferous rocks. *Q. Jl geol. Soc. Lond.* **57,** 226–243.

PRAY, L.C. (1960) Compaction in calcilutites. *Bull. geol. Soc. Am.* **71,** 1946 (abstract).

RAYNER, D.H. (1967) *The Stratigraphy of the British Isles,* pp. 453. Cambridge University Press, Cambridge.

REID, R.E.H. (1962) Sponges and the Chalk Rock. *Geol. Mag.* **99,** 273–278.

REID, R.E.H. (1968) Bathymetric distribution of Calcarea and Hexactinellida in the present and the past. *Geol. Mag.* **105,** 546–559.

ROY, R.F., DECKER, E.R., BLACKWELL, D.D. & BIRCH, F. (1968) Heat flow in the United States. *J. Geophys. Res.* **73,** 5207–5221.

SCHLANGER, S.O. & DOUGLAS, R.G. (1974) The pelagic ooze-chalk-limestone transition and its implications for marine stratigraphy. In: *Pelagic Sediments: on Land and under the Sea* (Ed. by K. J. Hsü and H. C. Jenkyns). *Spec. Publs int. Ass. Sediment.* **1,** 117–148.

SCHLANGER, S.O., DOUGLAS, R.G., LANCELOT, Y., MOORE, T.C. Jr & ROTH, P.H. (1973) Fossil preservation and diagenesis of pelagic carbonates from the Magellan Rise, central North Pacific ocean. In: *Initial Reports of the Deep Sea Drilling Project,* Vol. XVII (E. L. Winterer and J. I. Ewing *et al.*), pp. 407–427. U.S. Government Printing Office, Washington.

SCHOLLE, P.A. (1971) Diagenesis of deep-water carbonate turbidites, Upper Cretaceous Monte Antola Flysch, northern Apennines, Italy. *J. sedim. Petrol.* **41,** 233–250.

SCHOLLE, P.A. & KINSMAN, D.J.J. (1973) Diagenesis of Upper Cretaceous chalks from North Sea, England and Northern Ireland. *Bull. Am. Ass. Petrol. Geol.* **57,** 803–804 (abstract).

SORBY, H.C. (1861) On the organic origin of the so-called 'crystalloids' of the chalk. *Ann. Mag. nat. Hist.*, ser. 3, **8,** 193–200.

TAN, F.C. & HUDSON, J.D. (1971) Carbon and oxygen isotopic relationships of dolomites and co-existing calcites, Great Estuarine Series (Jurassic), Scotland. *Geochim. cosmochim. Acta,* **35,** 755–767.

TAYLOR, H.P., Jr (1970) Oxygen isotope evidence for large-scale interaction between meteoric ground waters and Tertiary diorite intrusions, western Cascade Range, Oregon. *Trans. Am. geophys. Un.* **51,** 453.

TAYLOR, H.P., Jr & FORESTER, R.W. (1971) Low-O^{13} igneous rocks from the intrusive complexes of Skye, Mull, and Ardnamurchan, western Scotland. *J. Petrology,* **12,** 465–497.

THOREZ, J., BULLOCK, P., CATT, J.A. & WEIR, A.H. (1971) The petrography and origin of deposits filling solution pipes in the Chalk near South Mimms, Hertfordshire. *Geol. Mag.* **108,** 413–423.

TILLEY, C.E. (1929) On larnite and its associated minerals from the limestone contact-zones of Scawt Hill, Co. Antrim. *Min. Mag., Lond.* **22,** 77–86.

UREY, H.C., LOWENSTAM, H.A., EPSTEIN, S. & MCKINNEY, C.R. (1951) Measurements of paleotemperatures and temperatures of the Upper Cretaceous of England, Denmark, and the southeastern United States. *Bull. geol. Soc. Am.* **62,** 399–416.

VOIGT, E. (1929) Die Lithogenese der Flach und Tiefwassersedimente des jungeren Oberkreidemeeres. *Jb. halle. Verb. Erforsch. mitteldt. Bodenschatze,* **8,** 1–136.

WALSH, P.T. (1966) Cretaceous outliers in south-west Ireland and their implications for Cretaceous palaeogeography. *Q. Jl geol. Soc. Lond.* **122,** 63–84.

WILSON, H.E. (1972) *Regional Geology of Northern Ireland*, pp. 113. Geological Survey Northern Ireland, Belfast.

WISE, S.W., Jr, BUIE, B.F. & WEAVER, F.M. (1972) Chemically precipitated sedimentary cristobalite and the origin of chert. *Eclog. geol. Helv.* **65,** 157–163.

WISE, S.W., Jr & HSÜ, K.J. (1971) Genesis and lithification of a deep sea chalk. *Eclog. geol. Helv.* **64,** 273–278.

WISE, S.W., Jr & KELTS, K.R. (1972) Inferred diagenetic history of a weakly silicified deep sea chalk. *Trans. Gulf-Cst Ass. geol. Socs,* **22,** 177–203.

WOLFE, M.J. (1968) Lithification of a carbonate mud: Senonian chalk in Northern Ireland. *Sedim. Geol.* **2,** 263–290.

WYLLIE, P.J. (1971) *The Dynamic Earth,* pp. 416. John Wiley & Sons, New York.

ZANKL, H. (1969) Structural and textural evidence of early lithification in fine-grained carbonate rocks. *Sedimentology,* **12,** 241–256.

Spec. Publs int. Ass. Sediment. (1974) **1**, 211–233

Maastrichtian chalk of north-west Europe — a pelagic shelf sediment

E. HÅKANSSON, R. BROMLEY *and* K. PERCH-NIELSEN

Geological Institute, University of Copenhagen, Denmark

ABSTRACT

Towards the end of the Cretaceous the shelf sea covering much of northern Europe made a final transgression in Campanian times followed by a phased regression throughout the Maastrichtian. In late Maastrichtian times the chalk facies became restricted to Denmark and parts of the North Sea, surrounded by shallower or near-shore facies of skeletal limestones.

From a series of some hundreds of bulk samples taken from both the chalk and related facies of this basin, seven samples representing two well-defined time planes have been chosen. An analysis of these reveals that the chalk is an almost totally biogenic sediment with at least 75% planktonic components. It can therefore be classified as a pelagic sediment which is shown to have been deposited in moderately shallow water no deeper than approximately 250 m, but generally below the euphotic zone.

Diagenesis includes bioturbation, compaction, loss of aragonite and genesis of flint. However, cementation is generally absent except at hardground horizons.

INTRODUCTION

In recent years much attention has been focused on pelagic and deep-water sediments, largely inspired by the results of the Deep Sea Drilling Project.

Supposed deep-sea sediments play a very limited role in the preserved and exposed geological record. It has, however, been shown that certain rocks, especially in orogenic belts, are of a deep-water origin. One of the main characters of a deep-sea sediment is its dominantly pelagic composition, namely a very low content of terrigenous material and a dominance of either pelagic calcareous or siliceous fossils such as coccoliths, planktonic Foraminifera, diatoms etc., or of pelagic mud. This has in many instances led to the misleading use of the terms pelagic and deep-sea sediments as synonyms. A clear distinction has to be drawn between *pelagic sediments* = sediments composed of pelagic constituents, and *deep-sea sediments* = sediments deposited outside the continental shelves, on the oceanic bottom at depths of several kilometres.

The importance of this quite simple distinction is revealed by the well-known facies of the Upper Cretaceous chalks of north-west Europe. These chalks were deposited on the bottom of comparatively shallow epicontinental seas during a period of about 35 million years (Cenomanian-Maastrichtian). The primary components of the chalks are of an almost totally biogenic origin, with an absolute dominance of pelagic constituents.

As an example, we have chosen the Maastrichtian chalk of north-western Europe. Two datum planes have been selected for quantitative analyses of the sedimentary composition: the lower/upper Maastrichtian boundary, and the top of the Maastrichtian (Fig. 1).

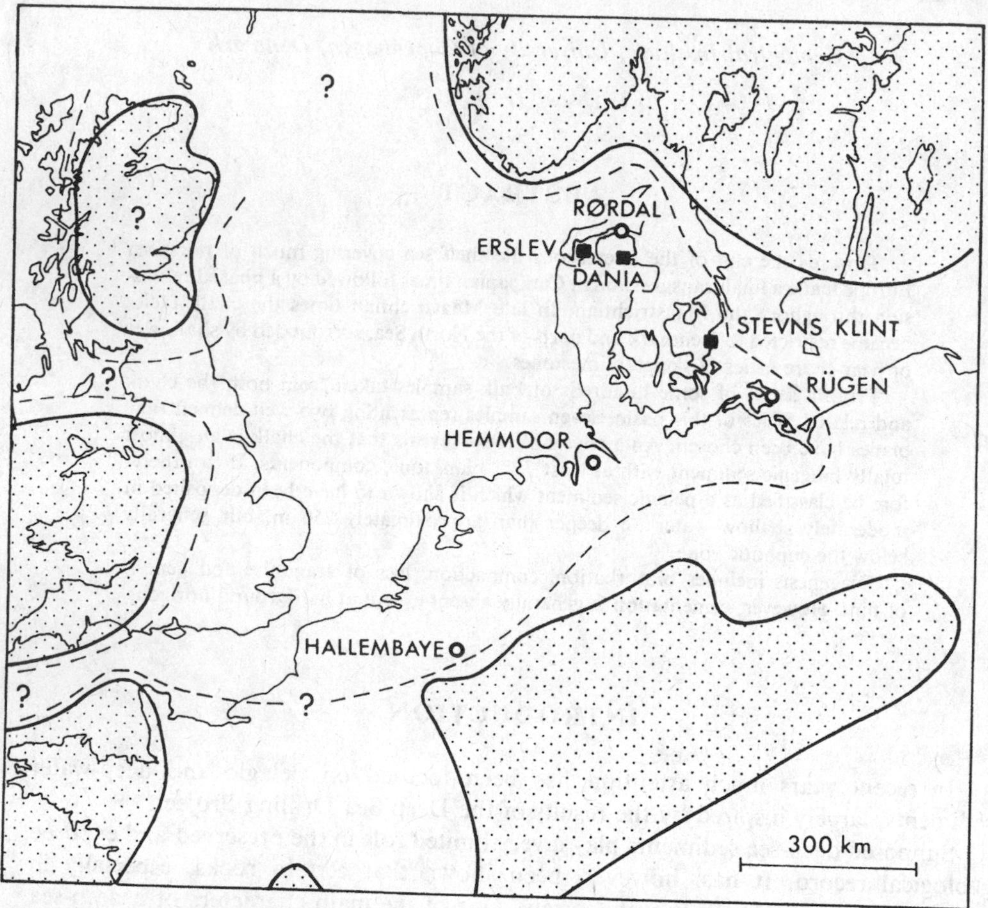

Fig. 1. Palaeogeography of the Maastrichtian sea in north-west Europe. Dashed lines indicate the approximate extension of chalk facies in the middle Maastrichtian; dotted areas are stable landmasses.

Samples from the upper Maastrichtian (■). *Stevns 2*, 2 m below the Maastrichtian/Danian boundary at Højerup, Stevns Klint; from within a bryozoan bioherm. *Dania 68*, 2 m below the Maastrichtian/Danian boundary in the quarry 'Dania' at Mariager Fjord. *Erslev 9*, 2 m below the Maastrichtian/Danian boundary in the quarry 'Erslev', Mors.

Samples from the boundary between the lower and upper Maastrichtian (○). *Hallembaye 20*, 1·5 m above the top Campanian hardground in the lowermost upper Maastrichtian in the 'Hallembaye' quarry. *Rørdal* 8, 2 m above the boundary in the quarry 'Rørdal', Aalborg. *Hemmoor 100*, 0·5 m above the boundary in the quarry 'Hemmoor'. *Rügen 1*, lowermost upper Maastrichtian, Jasmund.

STRATIGRAPHY

Only in recent years has a more detailed stratigraphical frame been worked out for the Maastrichtian chalk. Earlier workers found the facies and faunas very monotonous and a detailed zonation was not made until the studies of Troelsen (1937) and Brotzen (1945). Their zonations were purely local and a correlation within the whole basin was first made by Jeletzky (1951) and Schmid (1956) by means of belemnites. Belemnites and other cephalopods are, however, rare throughout most of the Maastrichtian chalk, especially in the upper Maastrichtian, and it was only possible to date a few localities.

On a local scale, a very detailed lithostratigraphical correlation can be made using flint layers, hardgrounds, prominent burrow horizons and different associations of trace fossils (Fig. 2). However, this lithostratigraphy is only useful for correlation within a few kilometres, partly due to outcrop failure.

Up to the present, the most detailed and reliable biozonation has been achieved by means of brachiopods. Quantitative diagrams can be used on a very local scale, as can lithostratigraphy, but the rapid vertical changes in the brachiopod assemblages make it possible to establish a detailed zonation covering the Danish-German area (Steinich, 1965; Surlyk, 1970, 1972). Pelagic belemnites and ammonites can then be used in the correlation of the brachiopod zonation on an inter-basinal level in north Europe, while the Foraminifera and coccoliths make correlation possible over greater distances. The brachiopod zonation was originally thought to represent merely local ecological successions but continuous re-checking with the pelagic groups has shown it to be quite close to the time-stratigraphic ideal (Surlyk & Birkelund, in press).

Thus, the Maastrichtian of north Europe is divided into four standard zones named after belemnites (Fig. 3) while local stratigraphical control is based on brachiopods and, to a smaller degree, on ostracods, Foraminifera, bryozoans, echinoderms and bivalves.

PALAEOGEOGRAPHY AND FACIES

The Maastrichtian chalk of north-west Europe was deposited in a seaway running from the Cretaceous Atlantic in the west to Poland in the east. Further eastwards it widened into the extensive sea which covered the Russian shelf. To the north the sea was bordered by the Fenno-Scandian Precambrian shield and to the south by the Middle European island (Fig. 1). There were connections with the Tethyan sea both to the south-east and south-west.

In near-shore areas greensands, sandstones, and especially biocalcarenites and -rudites were deposited. The nature of deposition in these marginal facies was controlled by topography and tectonic activity in the borderlands, and by latitude and climate. However, the near-shore deposits may pass laterally into biomicrites quite close to the coast (Fig. 1). These basinal biomicrites are relatively variable in composition but are normally lumped together under the term 'chalk'.

The chalk is composed dominantly of organic carbonates with coccoliths playing a major role. The content of other skeletals varies considerably. In the Maastrichtian chalks of the investigated area bryozoans form the most important group of

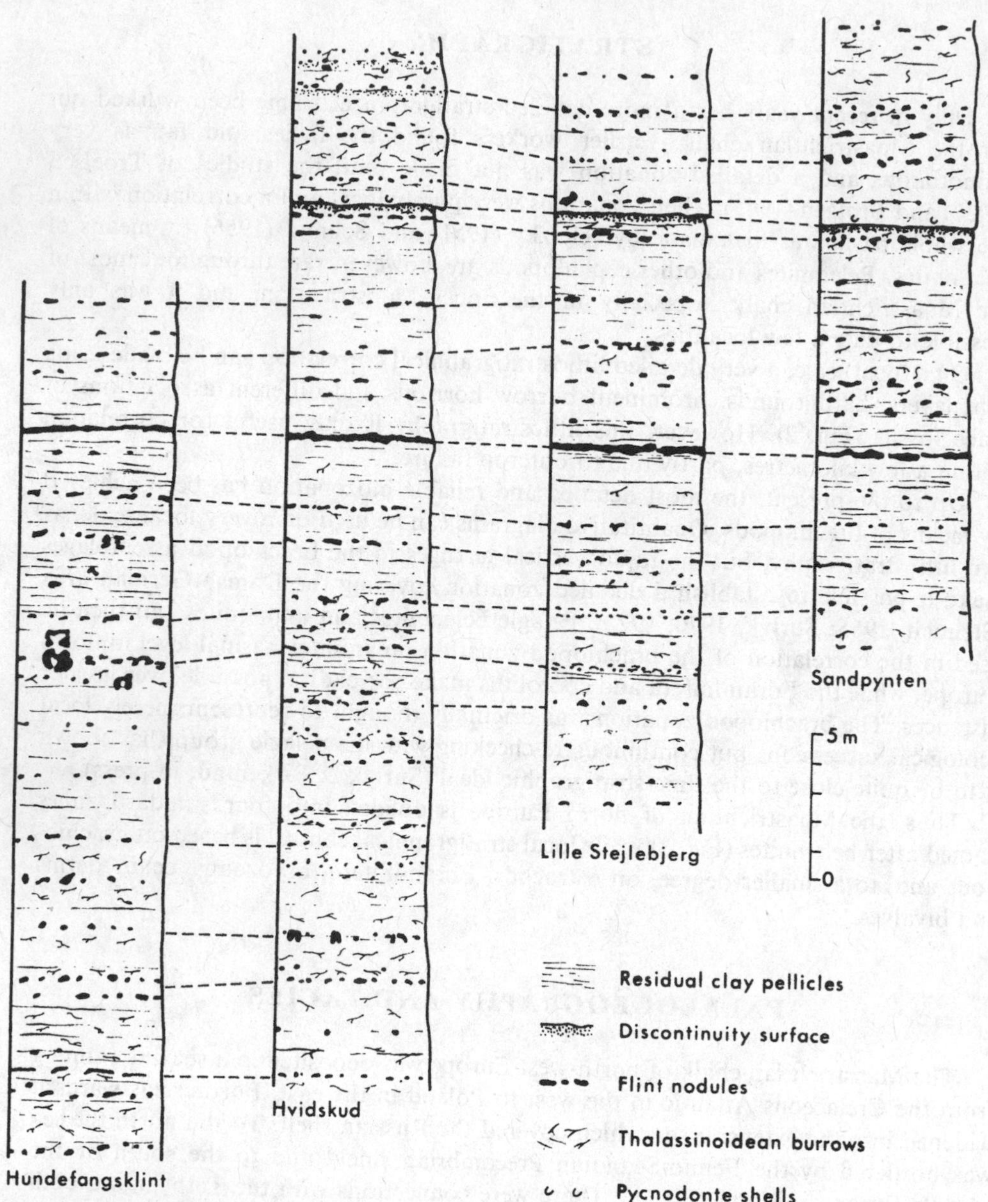

Fig. 2. Correlation of different sections from Møns Klint, Denmark, on the basis of detailed lithology (after Surlyk & Birkelund, in press).

macrofossils whereas other groups, such as inoceramid bivalves, may dominate in older Cretaceous chalks.

The thickness of the chalk of the basin varies considerably. Throughout the North Sea and most of the Danish-German area the Maastrichtian attains a thickness of about 200–400 m. In the central part of the Danish basin the thickness increases to about 700 m (Stenestad, 1972).

These 700 m of Maastrichtian chalk were deposited over a period of about 5

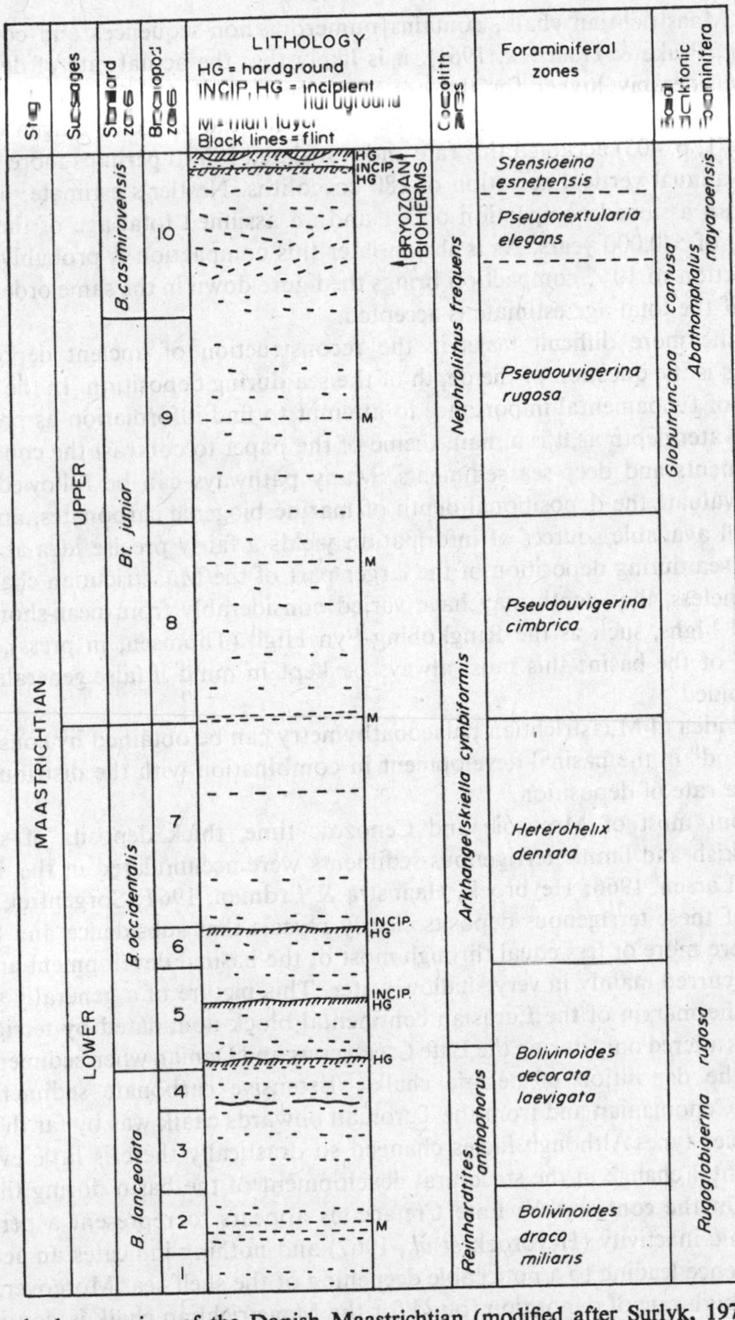

Fig. 3. Standard succession of the Danish Maastrichtian (modified after Surlyk, 1972; local foraminiferal zones based on Stenestad, 1971, 1973; coccolith zones cf. Perch-Nielsen, 1972).

million years. Accepting a compaction of 10% this gives an average depositional rate of the order of 15 cm/1000 years, which is an unexpectedly high figure for a pelagic sediment. Black (1953) found a somewhat lower depositional rate for the English chalk (Cenomanian-Campanian) of 4·5 cm/1000 years on the basis of the thickness of the preserved section. However, since the English chalk, in contrast to the major part of the Maastrichtian chalk, contains numerous non-sequences and condensed portions (e.g. Peake & Hancock, 1961), it is likely that the actual rate of deposition was also considerably higher for the pre-Maastrichtian chalk. On the other hand Nestler (1965, p. 119) estimated 50 cm/1000 years for the chalk of Rügen (Fig. 1). Bathurst (1971, p. 405) accepted this rate and translated it into perhaps more tangible terms as an annual vertical accretion of 180 coccoliths. Nestler's estimate, however, is based on an assumed compaction of 3:1 and an assumed total age of the Rügen chalk (90 m) of 540,000 years. As is shown later this compaction is probably far too great. Correction to 10% compaction brings the figure down in the same order as our figure, also if the total age estimate is accepted.

One of the more difficult tasks in the reconstruction of ancient depositional environments is the question of the depth of the sea during deposition. In the present context it is of fundamental importance to attempt to find information as precise as possible on water depth as it is a main theme of the paper to contrast the concepts of pelagic sediments and deep-sea sediments. Many pathways can be followed in the attempt to evaluate the depositional depth of marine biogenic carbonates, and compilation of all available sources of information yields a fairly precise idea about the depth of the sea during deposition of the larger part of the Maastrichtian chalk. It is clear, nevertheless, that depth may have varied considerably from near-shore areas or over local highs, such as the Ringkøbing-Fyn High (Thomsen, in press), to the central parts of the basin; this must always be kept in mind if false generalizations are to be avoided.

A general idea of Maastrichtian palaeobathymetry can be obtained by considering the major trends in the basinal development in combination with the distribution of facies and the rate of deposition.

Throughout most of Mesozoic and Cenozoic time, thick deposits of shallow marine, brackish and limnic terrigenous sediments were accumulated in the investigated basin (Larsen, 1966; Heybroek, Haanstra & Erdman, 1967; Sorgenfrei, 1969). The nature of these terrigenous deposits clearly reveals that subsidence and rate of deposition were more or less equal through most of the basinal development and that deposition occurred mainly in very shallow water. This picture of a generally subsiding basin at the margin of the Eurasian continental block dominated by terrigenous sediments was altered only during the Late Cretaceous and Danian when sedimentation changed to the deposition of pelagic chalks. Extensive carbonate sedimentation started in the Cenomanian and from the Turonian onwards chalk was by far the most important facies type. Although facies changed so drastically there is little evidence of a fundamental change in the structural development of the basin during the Late Cretaceous. On the contrary the Late Cretaceous appears to represent a period of relative tectonic inactivity (Heybroek *et al.*, 1967) and nothing indicates an acceleration of subsidence leading to a noticeable deepening of the shelf sea. Moreover, when the relatively high rate of deposition found for the Maastrichtian chalk is considered, it seems probable that deposition more or less continued to keep up with subsidence also during the Late Cretaceous and thus maintained a rather shallow water depth.

Along the north-eastern margins of the basin, Maastrichtian chalk can be followed almost to the coastline where coarser carbonates transgress a basement surface with complex topography produced by a series of horst structures. These basement structures were at least only partly submerged during deposition of the carbonates which therefore were laid down in very shallow water.

Besides the demonstration of the position of the actual coastline and its relation to facies, the best indicator of water depth below the littoral zone is the presence of light-dependent benthonic organisms. Minute borings commonly found in calcareous macrofossils have long been attributed to the activity of thallophytes. However, most of these borings seem to be the work of fungi (Bromley, 1970) while others are borings of ctenostome bryozoans (E. Voigt, personal communication, 1972). These borings are therefore of limited interest in the discussion of palaeobathymetry. The absence of borings which are distinctly attributable to algae, together with the lack of calcareous algal grains among the constituents of the chalk, may indicate that deposition generally took place below the euphotic zone. Grains of algal origin do occur in some shallow-water calcarenites in the Campanian of Scania (Voigt, 1929), and in similar Maastrichtian sediments calcareous crusts of algal origin are commonly found in Dutch Limburg.

The presence of angiosperms in the calcisiltites of the Maastricht area in Holland has been demonstrated by Voigt & Domke (1955) but no such fossil plants have yet been found in the basinal chalk facies. Many epifaunal animals, notably bryozoans and serpulids, attach themselves to plants and take on a tubular or ring-like shape, in some cases even moulding the epidermis cells of the substrate plant on their basal surface. In this way Voigt (1956) found some indications of the presence of seaweeds in parts of the Maastrichtian chalk (e.g. Hemmoor, Fig. 1). Also, in the highest parts of the Maastrichtian chalk at Stevns Klint (Fig. 1), such evidence renders the former presence of seaweeds probable, leading to the conclusion that deposition here took place within the euphotic zone. This uppermost chalk at Stevns Klint is also unusual in that it was deposited as low bioherms probably caused by a profuse growth of bryozoans.

A common approach in depth studies is the comparison with the depth distribution of the closest Recent relatives of the fossil assemblages. Basically this approach conflicts with the theory of a continuously evolving organic world but, when used critically on high diversity ecosystems of not too great an age, it can provide much useful information. The technique was used by Nestler (1965) in a study of the ecology of the fauna of the upper lower Maastrichtian chalk of Rügen. On the basis of the presumed depth ranges of the Recent representatives of a wide spectrum of the fossil groups found in this chalk, Nestler (1965, p. 111) concluded that the depth of deposition ranged from 100 to 250 m. A similar figure was reached for the pre-Maastrichtian chalk by Reid (1968) on the basis of the sponge fauna. This figure is further supported by the high diversity found among most groups of benthonic invertebrates in the Maastrichtian chalk (e.g. several hundreds of bryozoan species), a diversity which is comparable to that found in the contemporaneous shallow-water marginal biocalcarenites.

Thus, we conclude that deposition occurred in a rather shallow epicontinental sea on the broad north-west European Maastrichtian shelf, but within the euphotic zone at only a few levels in restricted areas.

COMPOSITION OF MAASTRICHTIAN CHALK

Methods

Seven representative chalk samples have been selected on the basis of on-going investigations of some hundreds of bulk samples (5–20 kg) from most parts of the Maastrichtian basin of north-west Europe (see Fig. 1). In order to facilitate comparisons, these samples originate from two well-defined time planes, namely the boundary between the lower and upper Maastrichtian (boundary between brachiopod zones 7 and 8) and the topmost Maastrichtian (top of brachiopod zone 10, see Fig. 3). From each of the bulk samples a series of part samples was taken, since different techniques had to be employed in order to determine the total composition of the chalk.

To determine the amount and composition of the coarser material, 100 g samples were disintegrated by alternately heating and freezing several times in a saturated Glauber-salt solution, following procedures described in detail by Surlyk (1972, p. 6). The samples were then washed over a 63 μm screen, dried, and divided into three fractions—>500 μm, 500–250 μm, and 250–63 μm—which were weighed. By picking and subsequent weighing, the composition of the two coarser fractions was determined for all samples. The composition of the finest fraction was calculated for one sample only, on the basis of picking a split sample. Unfortunately it seems to be impossible to completely disintegrate the chalk into its finest particles. All of the grains in the fraction >63 μm carry a film of the finest material and often cavities are filled; in some samples noticeable amounts of minute fragments of non-disintegrated chalk are even found in the two coarser fractions, while they always seem to be present in the fraction 250–63 μm. In the fraction <63 μm this tendency is very strong, making a meaningful, normal sedimentological grain-size analysis of this part of the sediment impossible. Consequently, this fraction has been studied in a semi-quantitative way on microscope slides prepared by sedimentation of disintegrated chalk directly on to the slide.

No attempts were made to find the quantitative composition of the non-carbonate material. However, the relative abundance of the clay minerals in the fraction <2 μm was analysed in 200 g samples by X-ray diffraction as described by Christensen *et al.* (1973). (These analyses were kindly made by L. Christensen.)

Composition and variation

The grain-size distribution of the Maastrichtian chalk seems to be largely identical in all samples studied (Fig. 4). The only important grain-size appears to be <63 μm, which typically amounts to somewhat around 90% of the total sample. However, to this figure should be added both the non-disintegrated chalk fragments found in the larger fractions and the unknown quantities of fine grains caught in hollows of grains >63 μm or adhering to their surfaces. In consequence, the grain-size composition given in Fig. 4 is only an approximation, the real composition having even larger dominance of the fraction <63 μm.

The non-carbonate constituents play a limited role, showing a total variation of 0·5–17% in the samples studied (Table 1). In most samples clay minerals are dominant with only negligible amounts of fine-grained quartz and pyrite and, therefore, non-carbonate material is in most samples not present in measurable amounts in the coarser fractions (Fig. 4). The extremely high figure (17%) found in Hallembaye 20, however, is due mainly to the presence of large amounts of finely distributed limonite which is considered as a late alteration of early diagenetic pyrite. Among the clay minerals montmorillonite and illite dominate, with montmorillonite typically making

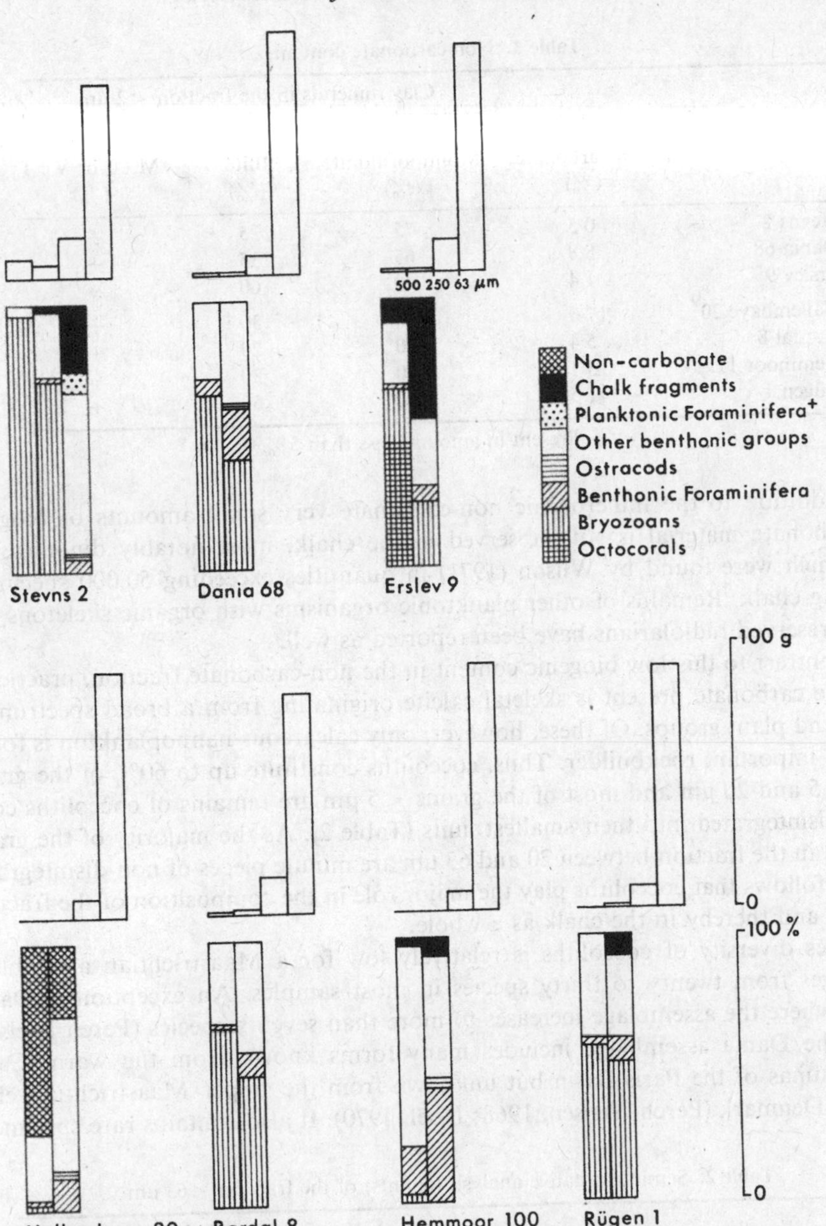

Fig. 4. Grain-size distribution and composition of seven representative chalk samples. Upper three from the topmost Maastrichtian; lower four from the boundary between the lower and upper Maastrichtian. + 'Planktonic Foraminifera' includes small amounts of calcispheres (*Palinosphaera*?, cf. Locker, 1967).

up between $^1/_2$ and $^4/_5$ of the total amount. Mixed-layer clay minerals (montmorillonite : illite and vermiculite : illite) are always present in quantities less than 5% (Table 1). The illite present is a typical low-temperature modification, and probably the clay mineral association found is neoformed during early diagenesis (Christensen *et al.*, 1973). Terrigenous material, therefore, seems to be virtually absent in the chalk samples studied.

Table 1. Non-carbonate content

	Non-carbonate (%)	Clay minerals in the fraction <2 μm: Montmorillonite (%)	Illite (%)	Mixed layer M : I or V : I (%)
Stevns 2	0·5	75	25	+
Dania 68	8·8	65	35	+
Erslev 9	1·4	40	60	+
Hallembaye 20	17·0	70	30	+
Rørdal 8	5·4	80	20	+
Hemmoor 115	3·1	70	30	+
Rügen 1	1·5	50	50	+

+ Present in amounts less than 5%.

In addition to the minerogenic non-carbonate very small amounts of biogenic non-carbonate material is still preserved in the chalk, most notably dinoflagellate cysts which were found by Wilson (1971) in quantities exceeding 50,000 specimens per 100 g chalk. Remains of other planktonic organisms with organic skeletons and rarely preserved radiolarians have been reported as well.

In contrast to this low biogenic content in the non-carbonate fraction, practically all of the carbonate present is skeletal calcite originating from a broad spectrum of animal and plant groups. Of these, however, only calcareous nannoplankton is found to be an important rockbuilder. Thus, coccoliths constitute up to 60% of the grains between 5 and 20 μm and most of the grains <5 μm are remains of coccoliths completely disintegrated into their smallest units (Table 2). As the majority of the grains observed in the fraction between 20 and 63 μm are minute pieces of non-disintegrated chalk, it follows that coccoliths play the major role in the composition of the fraction <63 μm and thereby in the chalk as a whole.

Species diversity of coccoliths is relatively low for a Maastrichtian assemblage and ranges from twenty to thirty species in most samples. An exception is Dania (Fig. 1) where the assemblage increases to more than seventy species (Perch-Nielsen, 1973). The Dania assemblage includes many forms known from the warm Campanian faunas of the Paris Basin but unknown from the upper Maastrichtian elsewhere in Denmark (Perch-Nielsen, 1968; Noël, 1970). It also contains rare specimens

Table 2. Semiquantitative analysis (counts) of the fraction <63 μm

	<5 μm (%)	5–20 μm (%)	20–63 μm (%)	Whole coccoliths 5–20 μm (%)	<63 μm of whole sample (weight %)
Stevns 2	90	10	+	53	73
Dania 68	90	10	+	55	92
Erslev 9	92	8	+	27	86
Hallembaye 20	91	9	+	60	92
Rørdal 8	89	11	+	28	83
Hemmoor 100	87	13	+	35	93
Rügen 1	88	12	+	43	91

+ Present in amounts less than 1%.

of *Tetralithus murus* which is an indicator of upper Maastrichtian in tropical seas (Worsley & Martini, 1970) and which does not occur in other Danish localities.

An explanation for this unusually large diversity may in part lie in the better state of preservation of coccoliths at this locality, due in turn to the high clay content of the sediment. However, this does not account for the local occurrence of a flora containing a decidedly warm element.

Kamptnerius magnificus is common in the coccolith assemblage of the Maastrichtian chalk. This species appears to have been missing or very rare in the open oceanic environment. On the other hand, *Braarudosphaera bigelowi*, which lives today under restricted or near-shore conditions, is absent or very rare in the Maastrichtian chalk. This suggests that the chalk was deposited in a fully marine but not fully oceanic environment.

In addition to coccoliths small amounts of planktonic Foraminifera and undeterminable fragments of various benthonic groups are found in the fraction <63 μm. Planktonic Foraminifera are also found in relatively small amounts in the fraction 63–250 μm, whereas various benthonic groups completely dominate the two coarser fractions (Fig. 4). In terms of weight, however, the benthonic constituents are of limited importance.

The diversity of planktonic Foraminifera is generally low compared to more southern Maastrichtian deposits, with very few species of *Heterohelix* and *Globigerinelloides* completely dominating. Further isolated specimens of *Hedbergella*, *Rugoglobigerina*, *Planoglobulina*, and *Globotruncana* are found in the samples here investigated. It should be noted that small amounts of calcispheres (*Palinosphaera?*) of unknown affinities but considered to be planktonic (Locker, 1967) are also included here.

The composition of the benthonic fauna varies considerably both in density and diversity throughout the Maastrichtian chalk (Fig. 4), also if larger samples than 100 g are considered. The most important benthonic groups found comprise brachiopods, various echinoderm groups, bivalves, ostracods and benthonic Foraminifera but typically with an absolute dominance of bryozoans. In general, however, the diversity is high for a monotonous, soft, fine silt bottom as in the Maastrichtian chalk sea, and consequently a generally high degree of specialization to this rather hostile environment should be expected. Apart from a few obvious examples, detailed investigations along these lines have only recently demonstrated such refined specializations in some of the major groups, but future work undoubtedly will greatly expand the number of highly specialized species.

Most groups contain some species that are completely dependent on the rare, large, hard substrates. More interesting in this connection, however, are the many species adapted to life directly on the soft bottom or relying only on the very small hard substrates which are generally available. Brachiopods and to some extent bryozoans have been investigated most intensively and a broad variety of adaptive strategies in invading the soft bottom have been demonstrated (Surlyk, 1972, 1973; Håkansson, 1974).

Among the irregular echinoids there are both shallow ploughers (e.g. *Echinocorys*) and rare deeper burrowers (*Hagenowia*).

Aragonitic bivalves, which are only preserved in cemented hardgrounds, comprise a variety of generally small, taxodont, burrowing and epibyssate forms. The latter probably lived attached to, e.g. bryozoans. This part of the fauna has not yet been worked out in any detail but, unlike other groups among the benthonic invertebrates, it shows a certain resemblance to Recent deep-sea faunas (C. Heinberg, personal communication, 1973). Throughout the chalk there are found in addition various calcitic bivalves which lived unattached, 'floating' on the soft bottom.

Although only relatively little is known about the absolute quantitative composition of the main part of the chalk (viz. the fraction <63 μm), it seems justified to conclude that all of the Maastrichtian chalk samples here studied contain more than 75% planktonic constituents, the only exception possibly being Stevns 2. However, counts of grains in different size-classes in the fraction <63 μm (Table 2) clearly indicate that the composition of this fraction is practically identical in all samples. Stevns 2, therefore, most likely represents the same type of pelagic sedimentation as found everywhere else in the chalk areas of this basin, with other environmental factors responsible for the profuse growth of bryozoans. Further, this implies that there seem to be no essential differences between the samples, although the composition of the very conspicuous benthonic elements (essentially the fraction >63 μm) varies considerably.

Therefore, a detailed comparative analysis of the variation in the benthos of the samples here dealt with is not relevant, since now we know, from our studies of the benthonic fauna in more than 200 chalk samples, that changes of the same order of magnitude may be found within short distances both laterally and vertically. However, such studies, when based on a large number of samples, have proved essential in the evaluation of the environmental changes within the frame of a rather uniform pelagic sedimentation.

DIAGENESIS

Bioturbation and slumping

At most horizons the Maastrichtian white chalk is apparently 100% bioturbated in that it shows no primary sedimentary structures and fossils are randomly orientated. The trace fossils preserved, however, are all deep structures, characterized by *Thalassinoides* and *Chondrites* with *Zoophycos* at certain horizons (Voigt & Häntzschel, 1956; Bromley, 1967). These forms are well preserved within a structureless or vaguely burrow-mottled surrounding sediment. Despite the presence of shallow-burrowing elements in the fauna (echinoids, bivalves), burrows attributable to these have not been found. This is probably due in part to the fact that the shallow burrowers are mainly highly vagile forms and their structures transient and self-destructive; in part to the initially thixotropic nature of the sediment; and in part to the rapid obliteration of the shallow burrows by an active meiobenthos (cf. Howard & Elders, 1970; Cullen, 1973). The deeper burrows were then superimposed upon this bioturbation texture at sediment depths beyond the reach of the destructive shallow burrowers.

At certain horizons the growth of bryozoan bioherms on the sea floor produced depositional slopes up to 20°. These horizons, particularly that of the topmost chalk at Stevns Klint, show slumping and structures attributable to sediment flow (Surlyk, 1972, p. 14). Both of these are rendered visible by deformation of burrows. The slumps are picked out by nodular flint layers which have formed later in the deformed burrow horizons, and the flow of sediment has streaked out burrow fills and caused a secondary lamination (Fig. 5).

Fig. 5. UV-photograph of vertical surface of a chalk sample from the topmost Maastrichtian bryozoan bioherms at Stevns Klint. Note secondary lamination and streaked out thalassinoid burrow fill caused by sediment flow. Dark lines represent clay pellicles in seams of pressure solution. Width of specimen 31 cm. (Photograph courtesy N. Svendsen.)

Compaction

The Maastrichtian chalk of England, Germany and Denmark shows only some 10% compaction, and retains a porosity of about 45–65%. In contrast, the Maastrichtian white chalk of Northern Ireland has suffered the same extreme compaction as that which Wolfe (1968) has demonstrated for the pre-Maastrichtian chalk. Wolfe recorded degrees of compaction of 15–30% and attributed it to weight of overburden.

The chalk bottom was not fluid, as is shown by the large diversity of free-living, very small, specialized benthonic animals and the presence of a rich encrusting fauna on almost all large fossils (Ernst, 1969; Surlyk, 1972). However, the adaptations of the fauna together with the vague background bioturbation texture correspond well to those associated with a thixotropic sediment (Rhoads, 1970). In contrast, the clear-cut deeper burrows superimposed on this texture demonstrate that the sediment was firm at the time of emplacement of *Thalassinoides*, *Chondrites* and *Zoophycos*. These deeper burrows were emplaced at a sediment depth of about 0·5–1·5 m. Thus the burrows record the change in the first 0·5 m from a thixotropic to a firm sediment through water loss which was no doubt largely due to the activity of the infauna itself. The deformation of the deeper burrows therefore records only the later stages of the dewatering process. *Thalassinoides*, *Chondrites* and *Zoophycos* are compressed by about 10% from their original thickness.

Zoophycos is a particularly good indicator of how little grain-to-grain movement there has been within the deeper levels of the sediment.

Immediately beneath (uncemented) omission surfaces, on the other hand, *Thalassinoides* is largely undeformed, with circular cross-section, indicating that dewatering had stabilized these surfaces and produced a fabric sufficiently rigid to withstand subsequent compaction.

Later compaction, caused by overburden, is largely represented by small, high-angle faults with throws of a few mm. These are again particularly well documented by *Zoophycos* (Fig. 6). This network of fine fractures may correspond with those recorded by Wolfe (1968) in the Irish chalk and those which were demonstrated by Steinich (1967, 1972) in the Rügen chalk. Compaction has not caused crushing of microfossils or interpenetration of grains. Macrofossils are commonly more or less crushed. However, the texture of the chalk within uncrushed echinoids is not visibly different from that outside and the flattening of many echinoids can often be attributed to incomplete filling of the test with sediment. Ammonite moulds generally register compaction of about 10%.

Thin pellicles of clay reveal the presence of seams of pressure solution. In some cases the clay pellicles continue laterally for some metres, in others they are extremely local features associated with inhomogeneities in the sediment (burrows, skeletal fragments, flint concretions). The clay pellicles, the surface of calcite fossils and the rind of flint often show a stylolitic contact. This localized pressure solution must be due to very late diagenesis.

Lithification

The exposed Maastrichtian chalk within the basin is only slightly lithified. This is possibly due to a number of factors. Black (1953) considered failure of lithification to be due to the low original aragonite content of the sediment, but the role of aragonite in submarine lithification is still not well established. The original content of aragonite

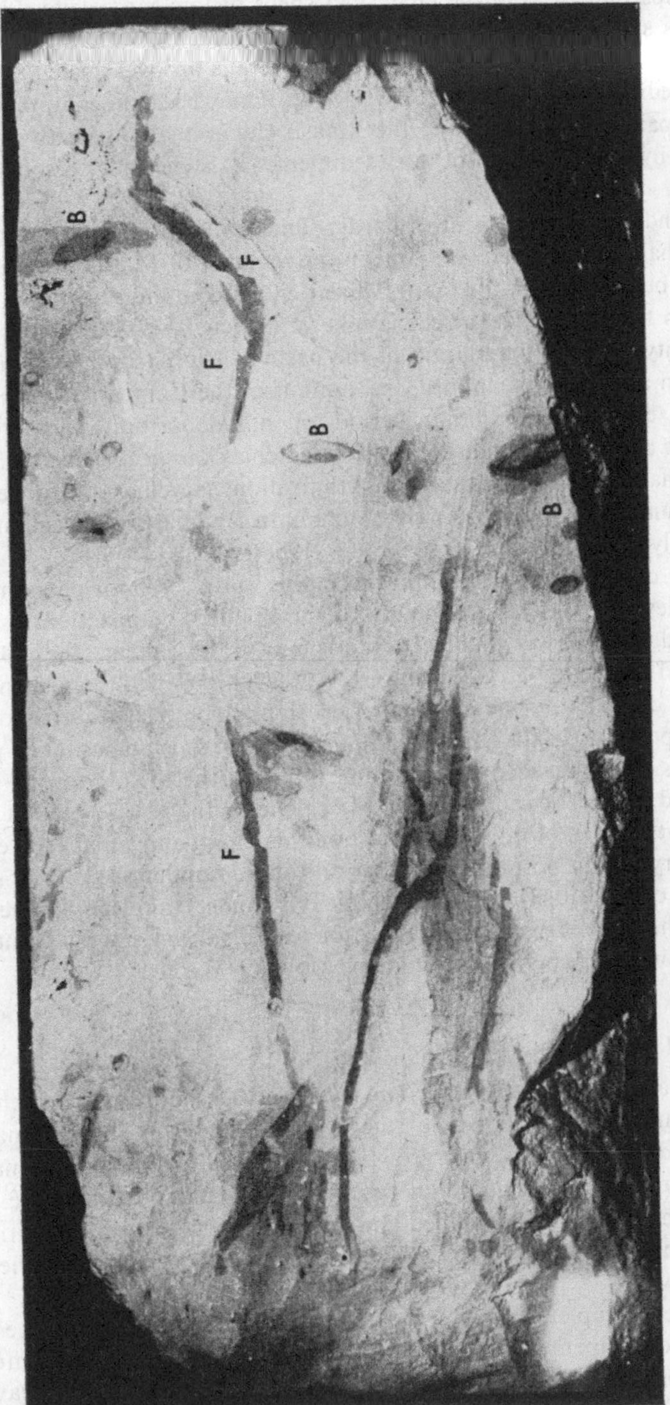

Fig. 6. UV-photograph of vertical surface of a chalk sample from the upper Maastrichtian at Stevns Klint (*c.* 2 m below the sample shown in Fig. 5). Note small-scale, high-angle compaction faults (F) cutting *Zoophycos* Spreiten. Oval grey figures to the right represent deep, more or less vertical shafts connected with the *Zoophycos* Spreiten (B), whereas the irregular light grey areas to the left are *Thalassinoides*. Width of specimen 29 cm. (Photograph courtesy N. Svendsen.)

was indeed very low (Voigt, 1929; Hancock, 1963), and practically all calcareous-shelled animal groups that lived in the chalk had shells of low-Mg calcite. Physico-chemical arguments show that in such a sediment, with a pore water with a high Mg : Ca ratio, lithification will be prevented unless there is an overburden of many hundred metres of sediment (Neugebauer, 1973, 1974). It may be significant, therefore, that in the central part of the North Sea the Maastrichtian chalk is partly lithified where it is overlain by about 2000 m of later sediments (E. Stenestad, personal communication, 1973).

On the other hand, the development of hardgrounds played an important role in the Maastrichtian chalk basin. There are none in the remnant of Maastrichtian chalk in Northern Ireland but in Norfolk the lower Maastrichtian contains several hardened omission surfaces in the Porosphaera beds (Peake & Hancock, 1961). Hardgrounds developed to maturity towards the margin of the basin at Ciply (southern Belgium) while, in the Belgian and Dutch Limburg, the summit of the Campanian chalk was lithified during the early Maastrichtian period of non-deposition and erosion. Omission surfaces in the north German Maastrichtian chalk are not lithified whereas those found in Denmark show various stages of lithification, as well as fully developed hardgrounds. The summit of the Maastrichtian chalk in Denmark was cemented as a hardground in early Danian times (Rosenkrantz, 1966).

The cementation of the Maastrichtian hardgrounds was a sea-floor phenomenon (Voigt, 1959; Bromley, 1967). The production of the lithified, compaction-resistant fabric of these hardgrounds has involved surprisingly little cement, the porosity mostly being reduced by only 5–15%. The age relationship between the cement and the parent sediment depends on the amount of erosion which has occurred during the period of non-deposition. In the Maastrichtian cases, little erosion has taken place, and cementation has thus preceded most compaction and the dissolution of much skeletal aragonite and silica. The cement is low-Mg calcite. In the best developed hardgrounds, such as that near the top of the Maastrichtian chalk at Dania, the cement occurs as euhedral crystals in both intergranular and intragranular voids. The structure of the coccoliths comprising the original sediment is almost completely obliterated and it appears that the cement represents a more or less advanced stage of syntaxial overgrowth (N. O. Jørgensen, personal communication, 1973).

Diagenesis of skeletal material

The chief modification of the primary composition of the sediment has been loss of organic matter, aragonite and opaline silica. Calcite skeletons have remained remarkably unaffected by diagenesis. Overgrowth has occurred mainly on the smallest and largest crystalline elements, namely on coccoliths and echinoderm calcite. The stereome system of the latter has been filled and sparry geodes formed in cavities in incompletely sediment-filled tests of echinoids. Other calcitic fossils are generally extremely well preserved.

Aragonite has been completely lost. Empty moulds of aragonitic shells are preserved in the cemented sediment of hardgrounds but only poorly preserved ammonite moulds are found, locally, in the soft chalk. Compaction has occluded the cavities left by the dissolution of aragonite throughout the bulk of the chalk and there is little evidence to indicate the original quantity of aragonite in the sediment.

Opaline silica has also been entirely removed from the sediment, but apparently at a later date than aragonite. Siliceous sponges are nevertheless preserved in different ways. They occur as replacements of silica in pyrite; in other cases poorly preserved cavity moulds of the lost spicules remain; or the spicules have inverted to quartz and occur with or without a chalcedonic coating. Cavities and replacement ghosts of spicules are also preserved in flint.

The calcite of belemnites, oysters and larger brachiopods is locally and incompletely replaced by chalcedonic quartz at many horizons. The date of this process is unknown. Extremely rarely, ammonites and gastropods have been found with a similar, delicate replacement of the shell by chalcedony. These ammonites, in contrast to those with normal preservation, show no trace of distortion through compaction. This suggests that the silica has directly replaced the aragonite and that the silicification was initiated, if not completed, during early diagenesis. A process similar to that leading to the emplacement of flint nodules may be envisaged.

Flint

Flints are the most conspicuous products of late diagenesis in the Maastrichtian chalk. Their field occurrence is similar to that of chert in most carbonate sequences, namely as nodular layers following bedding and as sheets along joints. As shown by ghosts of fossils and trace fossils, both types of flint have formed by a combination of replacement of calcium carbonate and precipitation within the remaining pore space of the replaced sediment. In most places the nodular flint accounts for no more than 5% of the total rock.

Nodular flint occurs rhythmically throughout most of the Maastrichtian chalk, though locally much reduced. Scattered nodules also occur, independently of the rhythmic layers. Both the layered and the scattered nodules show a close connection with trace fossils (Voigt & Häntzschel, 1956; Bromley, 1967). Moreover, many of the scattered nodules are connected with organic remains in that they partly incorporate siliceous sponges, fill echinoid tests and occur between the valves of bivalves and brachiopods. Most of the nodule layers can be shown to have originated within the fill of *Thalassinoides* systems, producing complexly shaped, anastomosing concretions. In most cases, however, silicification has continued well beyond the extent of the burrows and the layer has become more or less massive. Such layers are typically 10–20 cm thick and some 1–3 m apart, extending a variable distance laterally (Fig. 2). Exceptionally layers can be correlated over distances of tens of km.

Flint nodules have a transitional junction with the surrounding chalk, a white 'rind' of incompletely silicified chalk 1– over 10 mm thick. The bulk of the nodule is composed of dark, translucent, chalcedonic α-quartz which breaks with a conchoidal fracture and has a grain-size of 2–30 μm (Micheelsen, 1966). Areas of porcelanite usually occur within the flint, commonly ghosting bioturbation structures in the original sediment. Jensen *et al.* (1957) found Danian flints to contain up to 45% α-cristobalite but their samples of Maastrichtian flint were found to consist entirely of α-quartz. Buurman & van der Plas (1971), however, traced low tridymite in flint from upper Maastrichtian calcarenites from Holland.

Thin sheets of flint lie along joints in chalk at many horizons. Horizontal cracks produce sheets approximately parallel to bedding and oblique sheets are common along small-scale faults and tension cracks. Sheet flints vary considerably in thickness and tend to be lenticular, dying out and reappearing. The flint is characteristically uniformly dark and vitreous.

There has been considerable speculation over the problem of the genesis of flint in chalk, but progress in this area has been hindered by the lack of evidence of silicification beneath recent seas. This deficiency is now being made good by data from the Deep Sea Drilling Project (Heath & Moberly, 1971), and it is becoming feasible to piece together a reasonable hypothesis of flint genesis on the basis of this new evidence and the chemical and field data available. Problems still outweigh solutions but, as Buurman & van der Plas (1971) aptly put it: 'these questions are for a large part due to the slow diffusion of physico-chemical data obtained by our colleagues in the chemistry and physics departments into our own files'. Considerable strides forward were made when it was discovered that silica solubility is hardly dependent on pH within the range of pH to be expected in the pore water of sediments (Alexander, Heston & Iler, 1954; Krauskopf, 1959) and that the presence of an organic phase considerably reduces silica solubility (Siever, 1962). On the basis of these data, Siever (1962) offered a model for chert-nodule genesis in limestones and it is through modification of this model, on the basis of deep-sea data and field relationships, that a new model may be offered for chalk flints.

Authors are generally agreed that the silica comprising flint concretions in chalk derived largely or entirely from organic silica in the sediment (Cayeux, 1929; Buurman & van der Plas, 1971). We have no evidence for augmentation by submarine volcanic activity in the case of the present basin but it should be noted that diagenesis of clay minerals (alteration of montmorillonite to illite) releases some silica (Siever, 1962). In the chalk the desilicified remains of siliceous sponges are common, especially at certain horizons (not necessarily those rich in flint). In some cases the tests of Radiolaria are preserved in pockets of chalk enclosed in flint and as ghosts within the flint itself (Wetzel, 1971, pl. 1), suggesting that these were originally a widespread component of the sediment. Diatom frustules have not been observed.

Several authors have emphasized the preservation of soft organic parts in flint as evidence of a very early date of formation. Thus, Wetzel (1971, pl. 2) has recorded flagellates with well preserved flagellae and Steinich (1965) showed the preservation in flint of adductor muscles in large brachiopods. However, dinoflagellates are, in fact, well preserved and abundant in the white chalk (Wilson, 1971). Thus, in the case of Wetzel's flagellates, it may be argued that the organic matter was somehow fixed, perhaps as a carbon or pyrite replica, and later incorporated in the flint.

Much evidence indicates a later diagenetic origin for flint. Thus, even in areas where local deep channelling has produced considerable intraformational conglomerates, as in the Maastrichtian of Ciply, no reworked flint is ever recorded. Also, the extreme rarity of preservation of aragonitic fossils in flint suggests that in almost all cases aragonite had been lost before flints developed.

On the other hand, this late date would seem to be contradicted by the fact that flint formed preferentially at the site of organic matter. Müller (1956), for example, has pointed out that flint formed preferentially between the shells of *Pycnodonte* at the site of the adductor muscle. However, data from deep-sea cherts indicates that silicification of carbonate sediments is a double process, partly early and partly late. Silica derived from the dissolution of organic opaline skeletons is first precipitated as lepispheres of cristobalite and this is altered much later to quartz, partly as a solid-solid inversion and partly by addition of new silica from solution, to produce dense chert nodules (Heath & Moberly, 1971; Wise & Kelts, 1972). A similar history has been suggested for bedded cherts (Ernst & Calvert, 1969; Wise & Weaver, 1974).

The petrography of Maastrichtian flint supports this mode of origin (Micheelsen, 1966).

The following model for the genesis of flint attempts to account for the replacement of $CaCO_3$ by quartz during late diagenesis at sites connected with organic matter which is lost during early diagenesis.

Siever (1962) suggested that organic decay locally lowers pH enough to increase the solubility of carbonates. The resulting gradient would cause migration of HCO_3^- and Ca^{2+} away to surrounding areas where CO_2 concentration is normal. Emery & Rittenberg (1952) believed that SiO_2 solubility was lowered at sites of organic decay by adsorption on organic matter. This was supported by Siever's experimental results. The gradient so produced would cause further dissolved SiO_2 to migrate into the area, this in turn being immobilized by producing insoluble organic silica complexes (Siever, 1962, Fig. 4). Rittenberg, Emery & Orr (1955, p. 43) found that the silica content of pore waters increased regularly with sediment depth until, at about 2 m, it was ten times more concentrated than in sea water. This silica was derived chiefly from organic skeletons. These processes would build up concentrations of silica in the vicinity of decaying organisms.

The less stable organic substances would have been oxidized rapidly but the more stable would survive until buried to such a depth that reducing conditions prevailed, when their rate of decomposition would decrease. Adsorbed silica released during this breakdown would cause local saturation and be precipitated as cristobalite lepispheres, in voids or as replacement of $CaCO_3$. Reworking of the sediment by erosion at this stage would disperse the lepispheres. Left undisturbed, however, their numbers would increase until decay of the last organic matter had released the last silica complexes.

During late diagenesis the embryo flint was converted slowly into a flint nodule by two processes: the inversion of the metastable cristobalite to quartz and the accretion of additional silica (cf. Heath & Moberly, 1971). The reason for the rhythmicity of the flint nodule layers is unexplained. The layers parallel bedding so accurately that their location must have been related to, and controlled by successive sea floors and,

Table 3. Chronology of diagenesis of Maastrichtian chalk (Modified after Wolfe, 1968)

	Depositional	Early	Late
Deposition and surface reworking			
Shallow bioturbation			
Deep burrows			
Slumping and sediment flow			
Dewatering			
Lithification (hardgrounds only)			
Loss of aragonite			
Loss of opaline skeletals			
Compaction			
Nodular flint genesis			
Stylolites			
Sheet flint genesis			

Time →

therefore, due to early diagenetic processes. It is remarkable that the nodule layers are restricted to, and follow, individual beds even where these have been crumpled and deformed by slumping (Kennedy & Juignet, 1974).

Siever (1962) envisaged, for the late stage completion of the flints, that certain beds act as semipermeable membranes to upward migrating water during compaction. Beds in which burrow fills contain quantities of cristobalite lepispheres might act in this manner. Build-up of electrolytes beneath such beds would reduce silica solubility and encourage its precipitation. Thus, horizons of embryo flint, developed in originally organic-rich burrow fills, would act as nucleation sites for migrating silica and would consequently create a saturation gradient. The embryo flints would therefore be consolidated and grow beyond the boundaries of the burrow.

CONCLUDING REMARKS

The Maastrichtian chalk of north-west Europe is a pelagic sediment composed largely of coccoliths. It was deposited in parts of the extensive shallow sea which covered large areas of the Eurasian continental block in latest Cretaceous times. In restricted areas chalk deposition probably took place within the euphotic zone, but generally at somewhat greater depths ranging down to 250 m.

Rate of deposition averaged 15 cm/1000 years. Considering the composition of the sediment, this means that the primary productivity of calcareous nannoplankton must have been considerably higher in the Maastrichtian chalk sea than in modern seas. A combination of this explosion in primary productivity, with a simultaneous reduction of terrigenous sedimentation, brought about the deposition of chalk in a shelf-sea setting.

No counterpart to this Maastrichtian pelagic sedimentation is known in modern shelf seas. On the other hand, Recent deep-sea nannofossil oozes are widespread and Cretaceous and Tertiary pelagic chalks have been commonly encountered by the Deep Sea Drilling Project. These deep-sea chalks show a striking overall resemblance to the Maastrichtian shelf-sea chalk here described.

Similarities include the dominance of pelagic organisms and insignificant content of terrigenous matter; the occurrence of flint horizons; and the presence of *Zoophycos* and *Chondrites* among the trace fossil assemblages of both deep-sea and shelf-sea chalks.

Differences fall into two groups: those related to the sediment and those related to the geological setting. The former are mainly related to the benthonic fauna, notably its relative abundance in the Maastrichtian chalk and sparseness in deep-sea chalks. The composition of the benthos also differs, e.g. with benthonic Foraminifera abundant in the Maastrichtian shelf-sea chalk. *Thalassinoides* is lacking from the trace fossil assemblages in deep-sea chalks (van der Lingen, 1973; Kennedy, in press), while this form is dominant in all shelf-sea chalks. In addition, the benthonic fauna of the Maastrichtian chalk in places yields evidence of euphotic conditions.

The geological setting is the most obvious difference between the two types of chalk. Shelf-sea chalks occur within successions containing typical clastic shelf sediments, while deep-sea chalks overlie or are overlain by deep-sea sediments such as abyssal clays and radiolarites, or overlie oceanic basalts. In the case of a subsiding continental block, such as the Rockall Plateau (Laughton *et al.*, 1972), shelf sediments are replaced by deep-sea sediments as the block reaches greater depths.

The Maastrichtian chalk basin of north-west Europe was connected to the west with the opening young Atlantic. An increase and subsequent decrease in plate motion has been put forward as an explanation of the major transgression and regression of the Cretaceous (Hays & Pitman, 1973). Hydrodynamic and climatic changes resulting from these plate movements may be sufficient to explain the sudden appearance of chalk sediments in a tectonically quiet shelf setting.

ACKNOWLEDGMENTS

Our best thanks go to Dr F. Surlyk for valuable discussion during the preparation of this paper. We are also indebted to Drs G. Steinich, Greifswald, DDR, and F. Schmid, Hannover, BRD, who provided the important samples from Rügen and Hemmoor, and to L. Christensen, Århus, who made the clay mineral analysis. S. Piasecki, N. Svendsen, I. Nyegaard and H. Egelund are thanked for technical assistance. The Carlesberg Foundation supported our studies of north European chalk.

REFERENCES

Alexander, G.B., Heston, W.M. & Iler, R.K. (1954) The solubility of amorphous silica in water. *J. phys. Chem.* **58,** 453–455.

Bathurst, R.G.C. (1971) *Carbonate Sediments and their Diagenesis,* pp. 620. Elsevier, Amsterdam, London, New York.

Black, M. (1953) The constitution of the Chalk. *Proc. geol. Soc. Lond.* no. **1499,** 81–86.

Bromley, R.G. (1967) Some observations on burrows of thalassinidean Crustacea in chalk hard-grounds. *Q. Jl geol. Soc. Lond.* **123,** 157–182.

Bromley, R.G. (1970) Borings as trace fossils and *Entobia cretacea* Portlock, as an example. In: *Trace Fossils* (Ed. by T. P. Crimes and J. C. Harper). *Geol. J. Spec. Issue,* **3,** 49–90.

Brotzen, F. (1945) De geologiska resultaten från borringarna vid Höllviken. *Sver. geol. Unders. Afh.* C **396,** 204 pp.

Buurman, P. & van der Plas, L. (1971) The genesis of Belgian and Dutch flints and cherts. *Geol. Mijnb.* **50,** 9–27.

Cayeux, L. (1929) Les roches sédimentaires de France. Roches siliceuses. *Mém. Carte géol. Fr.* 774 pp.

Christensen, L., Fregerslev, S., Simonsen, A. & Thiede, J. (1973) Sedimentology and depositional environment of Lower Danian Fish Clay from Stevns Klint, Denmark. *Bull. geol. Soc. Denmark,* **22,** 193–212.

Cullen, D. (1973) Bioturbation of superficial marine sediments by interstitial meiobenthos. *Nature, Lond.* **242,** 323–324.

Emery, K.O. & Rittenberg, S.C. (1952) Early diagenesis of California basin sediments in relation to origin of oil. *Bull. Am. Ass. Petrol. Geol.* **36,** 725–806.

Ernst, G. (1969) Zur Ökologie und Biostrationomie des Schreibkreide-Biotopes und seiner benthonischen Bewohner. *Z. dt. geol. Ges.* **119,** 577–578.

Ernst, W.G. & Calvert, S.E. (1969) An experimental study of the recrystallization of porcelanite and its bearing on the origin of some bedded cherts. *Am. J. Sci., Schairer Volume,* **267A,** 114–133.

Håkansson, E. (1974) *Adaptive strategies among soft bottom cheilostomes from the Danish chalk (Maastrichtian),* 92 pp. Unpublished Ph.D. thesis, University of Copenhagen.

Hancock, J.M. (1963) The hardness of the Irish chalk. *Irish Nat. J.* **14,** 157–164.

Hays, J.D. & Pitman, W.C. (1973) Lithospheric plate motion, sea level changes and climatic and ecological consequences. *Nature, Lond.* **246,** 18–22.

HEATH, G.R. & MOBERLY, R. (1971) Cherts from the western Pacific, Leg 7, Deep Sea Drilling Project. In: *Initial Reports of the Deep Sea Drilling Project*, Vol. VII (E. L. Winterer *et al.*), pp. 991–1007. U.S. Government Printing Office, Washington.

HEYBROEK, P., HAANSTRA, U. & ERDMAN, D.A. (1967) Observations on the geology of the North Sea area. *Proc. 7th World Petroleum Congress*, **2**, 905–916.

HOWARD, J.D. & ELDERS, C.A. (1970) Burrowing patterns of haustoriid amphipods from Sapelo Island, Georgia. In: *Trace Fossils* (Ed. by T. P. Crimes and J. C. Harper). *Geol. J. Spec. Issue*, **3**, 243–262.

JELETZKY, J. (1951) Die Stratigraphie und Belemnitenfauna des Obercampan und Maastricht Westfalens, Nordwestdeutschlands und Dänemarks sowie einige allgemeine Gliederungs-Probleme der jüngeren borealen Oberkreide Eurasiens. *Beih. geol. Jb.* **1**, 142 pp.

JENSEN, A.T., WØHLK, C.J., DRENCK, K. & ANDERSEN, E.K. (1957) A classification of Danish flints etc. based on X-ray diffractometry. *Prog. Rep. Comm. Alkali Reactions in Concrete*, Dl, 37 pp. Copenhagen.

KENNEDY, W.J. (in press) Trace fossils in carbonate rocks. In: *The Study of Trace Fossils* (Ed. by R. W. Frey). Springer, New York.

KENNEDY, W. J. & JUIGNET, P. (1974) Carbonate banks and slump beds in the Upper Cretaceous (Upper Turonian-Santonian) of Haute Normandie, France. *Sedimentology*, **21**, 1–42.

KRAUSKOPF, K.B. (1959) The geochemistry of silica in sedimentary environments. In: *Silica in Sediments* (Ed. by H. A. Ireland). *Spec. Publs Soc. econ. Paleont. Miner., Tulsa*, **7**, 4–19.

LARSEN, G. (1966) Rhaetic-Jurassic-Lower Cretaceous sediments in the Danish Embayment. *Geol. Surv. Denmark, II Ser.* **91**, 127 pp.

LAUGHTON, A.S., BERGGREN, W.A. *et al.* (1972) Sites 116 and 117. In: *Initial Reports of the Deep Sea Drilling Project*, Vol. XII (A. S. Laughton, W. A. Berggren *et al.*), pp. 395–671. U.S. Government Printing Office, Washington.

LOCKER, S. (1967) Die Sphären der Oberkreide und die sogenannte Orbulinaritfazies. *Geologie*, **16**, 850–859.

MICHEELSEN, H. (1966) The structure of dark flint from Stevns, Denmark. *Meddr dansk geol. Foren.* **16**, 285–368.

MÜLLER, A.H. (1956) Die Knollenfeuersteine der Schreibkreide, eine frühdiagenetische Bildung. *Ber. geol. Ges.* **1**, 136–146.

NESTLER, H. (1965) Die Rekonstruktion des Lebensraumes der Rügener Schreibkreide-Fauna (Unter-Maastricht) mit Hilfe der Paläoökologie und Paläobiologie. *Geologie*, **14**, 147 pp.

NEUGEBAUER, J. (1973) The diagenetic problem of chalk—The role of pressure solution and pore fluid. *Neues Jb. Geol. Paläont. Abh.* **143**, 223–245.

NEUGEBAUER, J. (1974) Some aspects of cementation in chalk. In: *Pelagic Sediments: on Land and under the Sea* (Ed. by K. J. Hsü and H. C. Jenkyns). *Spec. Publs int. Ass. Sediment.* **1**, 149–176.

NOËL, D. (1970) Coccolithes Crétacés—la craie Campanienne du Bassin de Paris. *Editions Centre National Recherche Scientifique*, 129 pp. Paris.

PEAKE, N.B. & HANCOCK, J.M. (1961) The Upper Cretaceous of Norfolk. *Trans. Norfolk Norwich Nat. Soc.* **19**, 293–339.

PERCH-NIELSEN, K. (1968) Der Feinbau und die Klassifikation der Coccolithen aus dem Maastrichtian von Dänemark. *Biol. Skr.* **16**, 96 pp.

PERCH-NIELSEN, K. (1972) Remarks on Late Cretaceous to Pleistocene coccoliths from the North Atlantic. In: *Initial Reports of the Deep Sea Drilling Project*, Vol. XII (A. S. Laughton, W. A. Berggren *et al.*), pp. 1003–1069. U.S. Government Printing Office, Washington.

PERCH-NIELSEN, K. (1973) Neue Coccolithen aus dem Maastrichtian von Dänemark, Madagascar und Agypten. *Bull. geol. Soc. Denmark*, **22**, 306–333.

REID, R.E.H. (1968) Bathymetric distributions of Calcarea and Hexactinellida in the present and the past. *Geol. Mag.* **105**, 546–559.

RHOADS, D.C. (1970) Mass properties, stability, and ecology of marine muds related to burrowing activity. In: *Trace Fossils* (Ed. by T. P. Crimes and J. C. Harper). *Geol. J. Spec. Issue*, **3**, 391–406.

RITTENBERG, S.C., EMERY, K.O. & ORR, W.L. (1955) Regeneration of nutrients in sediments of marine basins. *Deep Sea Res.* **3**, 23–45.

ROSENKRANTZ, A. (1966) Die Senon/Dan-Grenze in Dänemark. *Ber. dt. Ges. geol. Wiss. A, Geol. Paläont.* **11**, 721–727.

SCHMID, F. (1956) Jetziger Stand der Oberkreide-Biostratigraphie in Nordwestdeutschland: Cephalopoden. *Paläont. Z.* **30,** Sonderheft, 7–10.

SIEVER, R. (1962) Silica solubility, 0°–200°, and the diagenesis of siliceous sediments. *J. Geol.* **70,** 127–150.

SORGENFREI, T. (1969) Geological perspectives in the North Sea area. *Bull. geol. Soc. Denmark,* **19,** 160–196.

STEINICH, G. (1965) Die artikulaten Brachiopoden der Rügener Schreibkreide (Unter-Maastricht). *Paläeontographica,* **A2** (1), 220 pp.

STEINICH, G. (1967) Sedimentstrukturen der Rügener Schreibkreide. *Geologie,* **16,** 570–583.

STEINICH, G. (1972) Endogene Tektonik in der Unter-Maastricht-Vorkommen auf Jasmund (Rügen). *Geologie,* **20,** Beiheft 71/72, 207 pp.

STENESTAD, E. (1971) Øvre Kridt i Rønde nr. 1. In: *The Deep Test Well Rønde No.* 1 *in Djursland, Denmark* (Ed. by L. B. Rasmussen). *Geol. surv. Denmark, III Ser.* **39,** 54–60.

STENESTAD, E. (1972) Træk af det danske bassins udvikling i Øvre Kridt. *Dansk geol. Foren., Årsskrift for* 1971, 63–69.

STENESTAD, E. (1973) Øvre Kridt i Nøvling nr. 1. In: *The Deep Test Well Nøvling No.* 1 *in Central Jutland, Denmark* (Ed. by L. B. Rasmussen). *Geol. surv. Denmark, III Ser.* **40,** 86–99.

SURLYK, F. (1970) Die Stratigraphie des Maastricht von Dänemark und Norddeutschland aufgrund von Brachiopoden. *Newsl. Stratigr.* **1,** 7–16.

SURLYK, F. (1972) Morphological adaptations and population structures of the Danish Chalk Brachiopods (Maastrichtian, Upper Cretaceous). *Biol. Skr.* **19,** 57 pp.

SURLYK, F. (1973) Autecology and taxonomy of two Upper Cretaceous craniacean brachiopods. *Bull. geol. Soc. Denmark,* **22,** 219–243.

SURLYK, F. & BIRKELUND, T. (in press) An integrated stratigraphical study of fossil assemblages from the Maastrichtian White Chalk of NW Europe. In: *Concepts and Methods in Biostratigraphy* (Ed. by E. G. Kauffman and J. E. Hazel). *Spec. Publ. Pal. Soc.*

THOMSEN, E. (in press) Maastrichtian and Danian facies pattern on the Ringkøbing-Fyn High. *Bull. geol. Soc. Denmark,* **23.**

TROELSEN, J. (1937) Om den stratigrafiske inddeling af skrivekridtet i Danmark. *Meddr dansk geol. Foren.* **9,** 260–263.

VAN DER LINGEN, G.J. (1973) Ichnofossils in deep-sea cores from the southwest Pacific. In: *Initial Reports of the Deep Sea Drilling Project,* Vol. XXI (R. E. Burns, J. E. Andrews *et al.*), pp. 693–700. U.S. Government Printing Office, Washington.

VOIGT, E. (1929) Die Lithogenese der Flach- und Tiefwasser-sedimente des jüngeren Oberkreidemeeres. *Jb. halle. Verb.* **8,** 165 pp.

VOIGT, E. (1956) Der nachweis des Phytals durch Epizoen als Kriterium der Tiefe vorzeitlicher Meere. *Geol. Rdsch.* **45,** 97–119.

VOIGT, E. (1959) Die ökologische Bedeutung der Hartgründe ('Hardgrounds') in der oberen Kreide. *Paläont. Z.* **33,** 129–147.

VOIGT, E. & DOMKE, W. (1955) *Thalassiocharis bosqueti* DEBEY & MIQUEL ein strukturell erhaltenes Seegras aus der holländischen Kreide. *Mitt. geol. StInst. Hamb.* **24,** 87–102.

VOIGT, E. & HANTZSCHEL, W. (1956) Die grauen Bänder in der Schreibkreide Nordwestdeutschlands und ihre Deutung als Lebensspuren. *Mitt. geol. StInst. Hamb.* **25,** 104–122.

WETZEL, O. (1971) Der gemeine Feuerstein als Fundquelle mannigfälter 'Kleinwunder'. *Grondboor Hamer,* **3,** 58–77.

WILSON, G.J. (1971) Observations on European Late Cretaceous Dinoflagellate cysts (1). In: *Proceedings of the II Planktonic Conference* (Ed. by A. Farinnaci). *Edizioni Tecnoscienza, Roma,* **2,** 1259–1275.

WISE, S.W., JR & KELTS, K.R. (1972) Inferred diagenetic history of a weakly silicified deep sea chalk. *Trans. Gulf-Cst Ass. geol. Socs,* **22,** 177–203.

WISE, S.W., JR & WEAVER, F.M. (1974) Chertification of oceanic sediments. In: *Pelagic Sediments: on Land and under the Sea* (Ed. by K. J. Hsü and H. C. Jenkyns). *Spec. Publs int. Ass. Sediment.* **1,** 301–326.

WOLFE, M.J. (1968) Lithification of a carbonate mud: Senonian chalk in Northern Ireland. *Sedim. Geol.* **2,** 263–290.

WORSLEY, T. & MARTINI, E. (1970) Late Maastrichtian nannoplankton provinces. *Nature, Lond.* **225,** 1242–1243.

Spec. Publs int. Ass. Sediment. (1974) **1,** 235–247

Magnesian-calcite nodules in the Ionian deep sea: an actualistic model for the formation of some nodular limestones

JENS MÜLLER *and* FRANK FABRICIUS

Institut für Geologie, Technische Universität, München, West Germany

ABSTRACT

Cores from the Mediterranean Ridge (Ionian Sea) locally contain magnesian-calcite nodules occurring in a micritic matrix comprising mainly magnesian calcite. The nodules vary in size, shape, and degree of consolidation. Borings and subsequent overgrowths indicate formation at or close to the sediment/water interface. Dissolution of aragonite—thought to be the main source of nodule-forming carbonate in some Tethyan nodular limestones—is insufficient to account for the large amount of carbonate cement present. Therefore, it is assumed that most of the carbonate is directly precipitated as magnesian calcite from Mediterranean sea water. The triggering mechanism is still unknown, but the conditions necessary for formation of magnesian-calcite nodules seem to be high salinity and temperature, and slow rates or absence of sedimentation. Formation of nodules apparently also involves a partial diagenetic transformation of calcitic material (coccoliths etc.) to magnesian calcite.

Although the nodules are not stable with respect to their mineralogy, it is conceivable that a similar type of formation may account for nodules which occur in Tethyan Jurassic facies.

INTRODUCTION AND PREVIOUS WORK

Lithification of carbonates on the deep-sea floor is a long established fact (Milliman, 1966; Cifelli, Bowen & Siever, 1966; Fischer & Garrison, 1967). The mechanisms leading to cementation and the factors defining the mineralogy of the cementing carbonate are, however, still under discussion. Although common in almost every ocean basin, lithification appears to be restricted locally. At present, only two areas are known where conditions are conducive for large-scale precipitation of carbonates: the Red Sea (Herman, 1965; Gevirtz & Friedman, 1966; Milliman, Ross & Ku, 1969) and the eastern Mediterranean Sea (Müller & Fabricius, 1973; Milliman & Müller, 1973). In these partly barred basins higher bottom temperatures and salinities compared to the open oceans have been considered a pre-requisite for the precipitation of carbonates (Milliman & Müller, 1973). The precipitate, in the case of the eastern

Mediterranean Sea magnesian calcite (8–12 mol-% $MgCO_3$), is concentrated in the lutite. Occasionally, this magnesian calcite-rich lutite is lithified.

Such examples of induration are the crusts described by Fischer & Garrison (1967), the lithic fragments (Milliman *et al.*, 1969; Müller & Fabricius, 1973), and the aggrégats calcaires reported by Chamley (1971) in the eastern Mediterranean Sea. More recently, the term nodule was used in this context (Milliman & Müller, 1973), a practice which will be followed in this study. The term 'nodule' here refers to concretions composed mainly of magnesian calcite that are harder than the surrounding sediment. Although the criterion of hardness becomes problematic with decreasing size, 'nodule' is used as well for silt-sized (63–2 μm) concretions, whereas clay-sized (<2 μm) magnesian calcite will be termed magnesian-calcite lutite.

The list of terms used for identical bodies illustrates the uncertainty as to their mode and site of formation. This uncertainty is further increased by the ignorance of their original size, position or orientation prior to sampling. Coring may well destroy crusts, leaving fragments, perhaps displaced into lower strata. In addition, coring provides only a limited picture of the lateral extension of lithification.

Milliman & Müller (1973) have shown that the precipitation of magnesian-calcite lutite is a widespread phenomenon in the eastern Mediterranean Sea; however, formation of nodules appears to be restricted both regionally and stratigraphically. The objective of this study is to delineate the factors which lead on one hand to the precipitation of magnesian-calcite lutite and on the other hand to the formation of nodules.

METHODS

Cores were obtained during cruises No. 17 (1969) and No. 22 (1971) of R.V. *Meteor*. Core locations are given in Table 1. Carbonate content was determined volumetrically with a LECO-analyser. Carbonate mineralogy, the proportion of carbonate minerals, and the composition of the magnesian calcite were determined by techniques described elsewhere (Milliman & Müller, 1973). Texture and composition of nodules were examined in thin sections stained with Alizarin Red S. In addition, different size fractions, including the freshly broken surfaces of nodules, were examined under a scanning electron microscope.

Table 1. Location and depth of cores

Core number	Lat. (°N)	Long. (°E)	Water depth (uncorrected)
17 M 17	36° 27·0′	20° 39·2′	2680 m
22 M 36	36° 10·6′	20° 50·7′	3336 m
17 M 14	36° 11·5′	19° 45·7′	3115 m
22 M 49	34° 18·3′	20° 02·5′	3050 m

LITHOLOGY OF CORES

Nodules were found in cores 17 M 14, 22 M 36 and 22 M 49. Core sites are situated on the Mediterranean Ridge: 17 M 14 in the central part, 22 M 36 on the eastern flank bordering the Hellenic Trough, and 22 M 49 on the western slope of the Syrte Abyssal Plain. All three cores exhibit an unusual stratigraphy compared to other cores of the same area (Fig. 1). While 'normal' cores of the Mediterranean Ridge show a similar sequence of tephra and sapropel layers (example: core 17 M 17), one or both of these features are missing in the above-mentioned cores.

According to Hieke, Sigl & Fabricius (in press), the upper core portion of core 22 M 36—comprising sediments of the Upper Quaternary—is missing, probably due to tectonic events. The boundary of nannoplankton zones NN 20/21 occurs at 170 cm, and of zones NN 19/20 at 540 cm. Nodules up to 5 mm across were found in the upper 4 cm of the core.

Core 17 M 14 contains two tephra layers at 55—60 cm and 105 cm, both of unknown age. Based on their mineralogy, J. Keller (personal communication) attributed them to the middle Italian province. Eruptions in this area are known to have occurred periodically during the last 1·2 million years. Investigation of nannoplankton showed that the base of the core is still within NN 21 (C. Müller, personal communication). Nodules are encountered within four horizons: 25–27 cm, 62–86 cm, 92–94 cm, 98–106 cm. The maximum size of nodules observed was 4 cm. The larger nodules are surrounded by abundant silt-sized nodules (Fig. 2) and a clay-sized magnesian-calcite matrix (Fig. 3).

Core 22 M 49 (NN 21 present throughout the entire core length) has neither tephra layers nor sapropels. It contains nodules (up to 5 cm in diameter) throughout the whole core length. The lack of bedding and the chaotic orientation of nodules indicate a sedimentary breccia, due to slumping of a nodule-rich sediment from a higher part of the Mediterranean Ridge. At the time of slumping, nodules must have been hardened not to have been deformed.

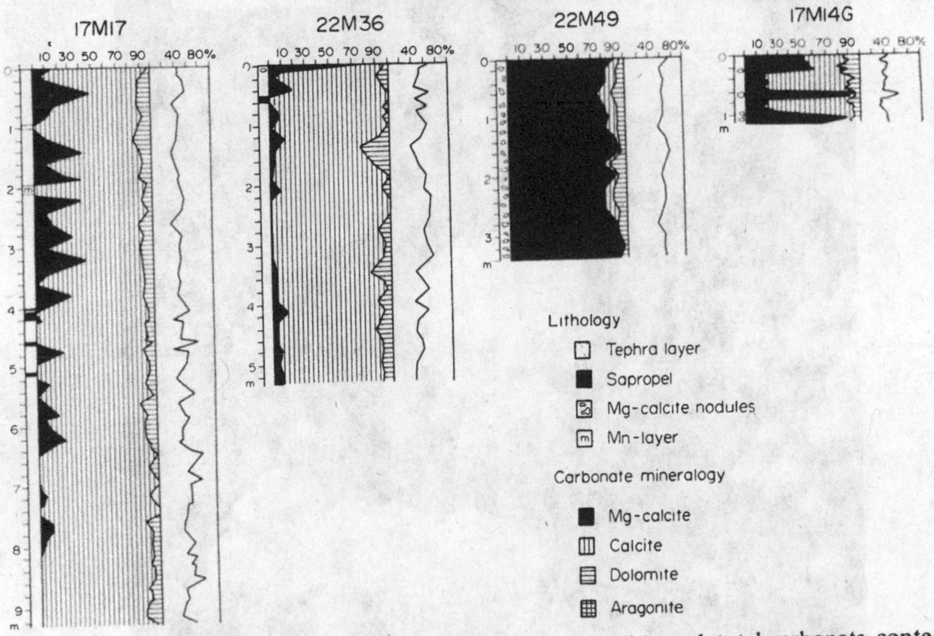

Fig. 1. Synopsis of lithology (left), carbonate mineralogy (middle), and total carbonate content (right) within cores from the Mediterranean Ridge.

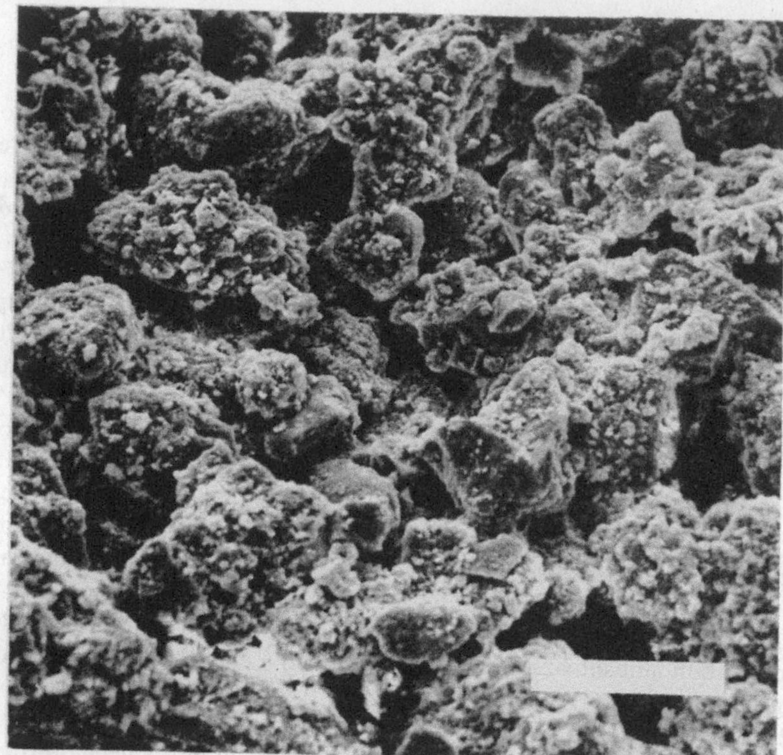

Fig. 2. Silt-sized magnesian-calcite nodules composed of lutite-sized magnesian-calcite crystals. Note dolomite crystal and shell pragment in upper central part. Bar represents 50 μm. (22 M 49; Scanning electron micrograph 26179.)

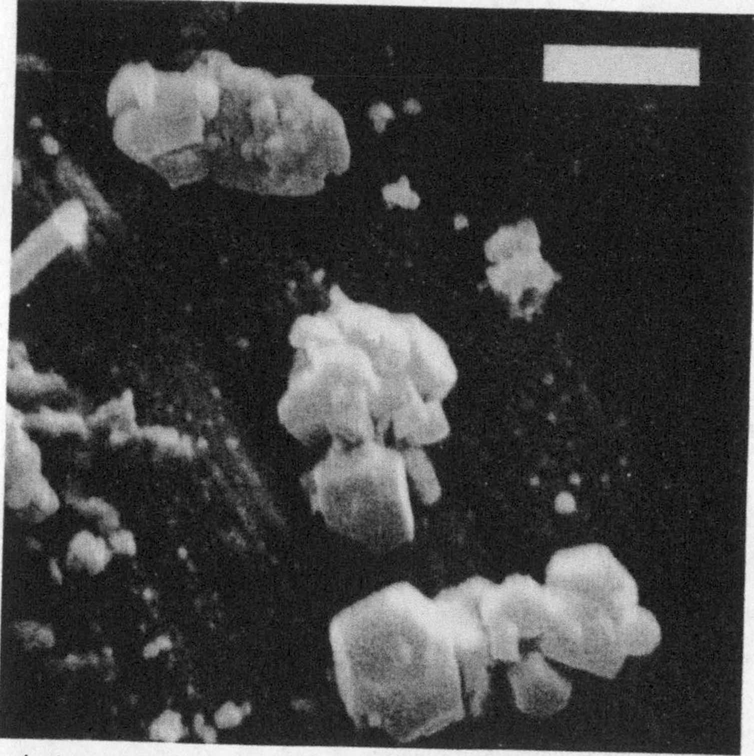

Fig. 3. Clay-sized magnesian-calcite crystals. Bar represents 1 μm. (17 M 14 : 20–22 cm; Scanning electron micrograph 4316.)

DESCRIPTION AND COMPOSITION OF NODULES

As mentioned above, the nodules vary in size from silt through to gravel. Their shape varies from angular to more or less rounded, larger nodules tending to be more flattened. Their colour ranges from light brown to grey or white. Their surface is usually uneven, sometimes pitted. Most of the larger specimens show abundant borings, some have benthonic overgrowths of unknown organisms resembling serpulid tubes (Figs 4, 5). The hardness or degree of lithification ranges from hardly perceptible to that of a ringing limestone. Differences in hardness with respect to the upper or lower side—as described by Fischer & Garrison (1967)—were not observed in our specimens. Thin sections of nodules show a micritic matrix with interspersed Foraminifera, mainly globigerinid forms. Less abundant are quartz, feldspar, tephra material, scattered dolomite rhombs, and spherical particles (Radiolaria?). In most nodules two types of burrows are observed. One type (mean diameter 0·3 mm) is filled with a carbonate-poor fine-grained material, while the other type has a coarser filling with grain diameters up to 0·5 mm. Both types can also be observed within the unconsolidated

Fig. 4. Borings, grazing traces, and 'serpulid' overgrowth on a nodule surface. Bar represents 4 mm. (22 M 49: 135 cm.)

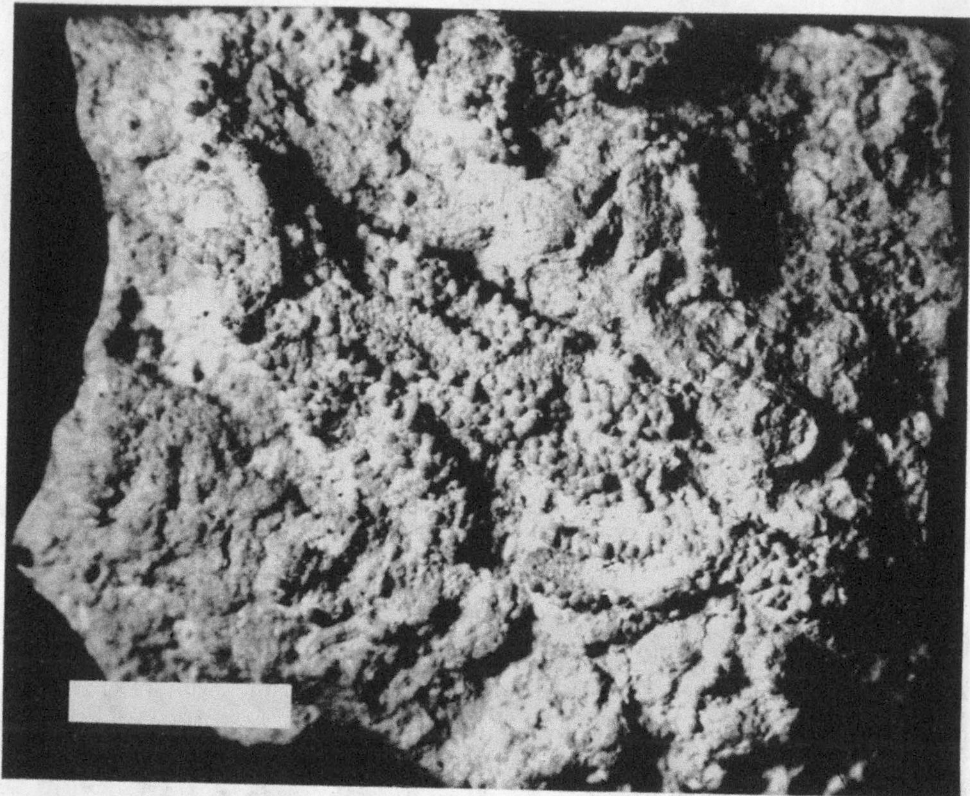

Fig. 5. Remnants of tubes composed of Foraminiferal tests on a nodule surface. Bar represents 5 mm. (22 M 49: 150 cm.)

sediment. The traces observed in nodules thus represent the preservation of textures of the unconsolidated sediment. A third type, however, indicates boring after formation of nodules. These borings penetrate from the outside and are partly filled with micrite or Foraminifera; their side walls often have a thin iron-manganese coating.

The composition of nodules is rather uniform, but is distinctly higher in total carbonate content (70–85%) than the surrounding unconsolidated sediment (30–60%). The dominant carbonate mineral is magnesian calcite (90–100%). Calcite and dolomite are subordinate, the former often beyond detection range. The composition of the magnesian calcite within the nodules differs from the lutite in having a higher magnesium content (lutite: 8–12 mol% $MgCO_3$; nodules: >12 mol% $MgCO_3$). Within nodules, there appears to be a relation between the magnesium content and hardness: nodules displaying the highest degree of lithification contain up to 16 mol% $MgCO_3$. A similar observation has been made by Chamley (1971).

PRESERVATION OF SKELETAL MATERIAL

Foraminifera, mainly globigerinid forms, show all stages of preservation from unaltered to completely micritized (Fig. 6). In this regard, little can be added to the

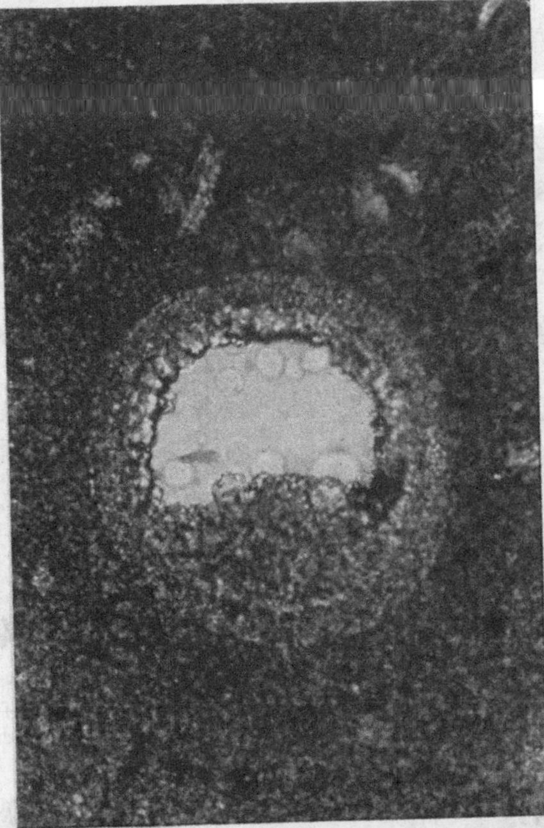

Fig. 6. Completely micritized *Orbulina* test within a nodule, interior partly filled with micrite. Note rim of drusy (magnesian?) calcite. Thin section, plane polarized light; diameter of the test 75 μm. (17 M 14.)

detailed description of Fischer & Garrison (1967). Many of the foraminiferal tests are filled with micrite, some show the presence of drusy (magnesian?) calcite with small crystals at the inner surface increasing in size towards the interior of the chamber (Fig. 7). As with calcitic foraminiferal tests, aragonitic pteropod and heteropod shells exhibit all stages of alteration, which can be often observed in single nodule specimens. The destruction of their shells seems to be caused by boring organisms. Figures 8 and 9 show the typical boring traces 1μm in diameter (fungi?) as well as a patchy obliteration of the shell wall.

Comparison of skeletal preservation within nodules and unconsolidated sediment shows that hardly any micritization of Foraminifera occurs within the unconsolidated sediment. Most tests have clearly defined walls. Aragonitic shells, however, display the same preservation stages in the unconsolidated sediment as within nodules. Unaltered aragonitic shells can be observed down to core depths of 8 m, while locally only casts were left. The dissolution of aragonite is therefore no function of burial, but limited to biological or chemical micro-environments (see also van Straaten, 1967).

Preservation of coccoliths is highly variable within the core material studied. While abundant ($\geqslant$20% of total sediment) and unaltered in lutite layers free of

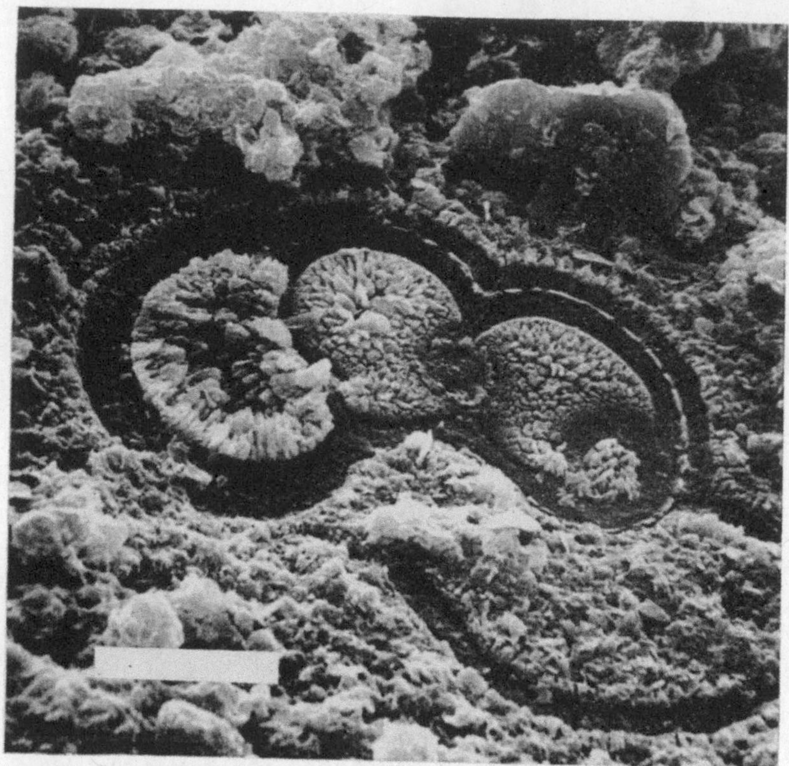

Fig. 7. Drusy (magnesian?) calcite and micrite filling (lower right chamber) of foraminiferal test within a nodule. Bar represents 20 μm. (17 M 14: 65 cm; Scanning electron micrograph 8314.)

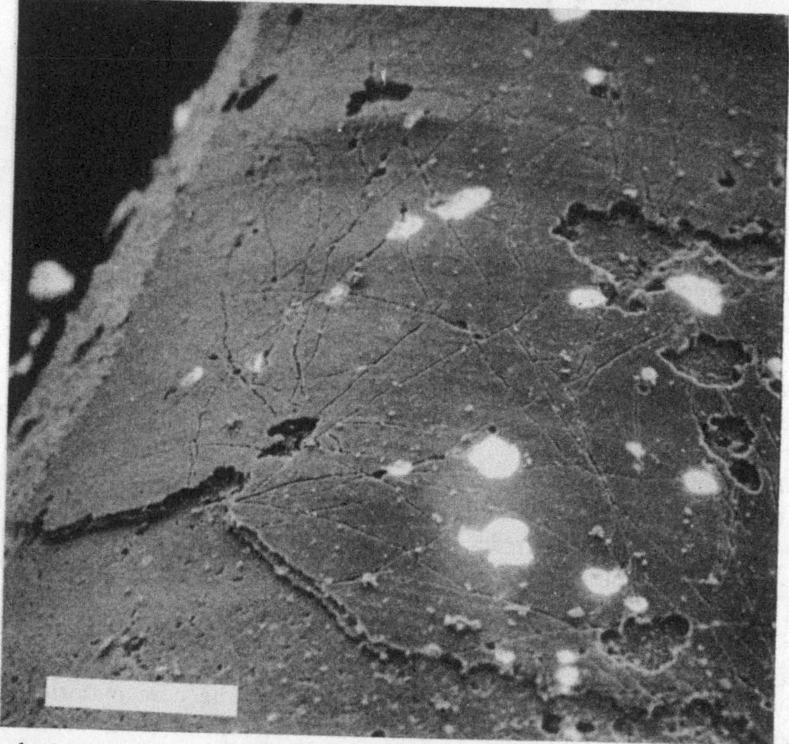

Fig. 8. Partly destroyed pteropod shell. Note boring traces and patchy obliteration of parts of the shell. Bar represents 50 μm. (22 M 49: 130–135 cm; Scanning electron micrograph 26184.)

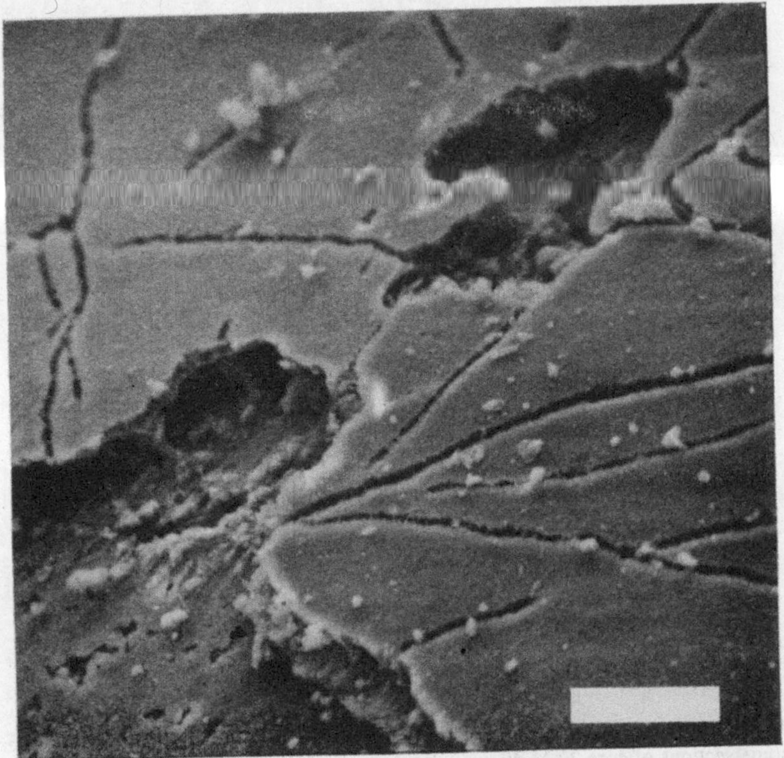

Fig. 9. Close-up of central part of Fig. 8. Bar represents 20 μm. (Scanning electron micrograph 26185.)

Fig. 10. Recrystallization and overgrowth on coccoliths. Bar represents 5 μm. (22 M 49: 190–195 cm; Scanning electron micrograph 38396.)

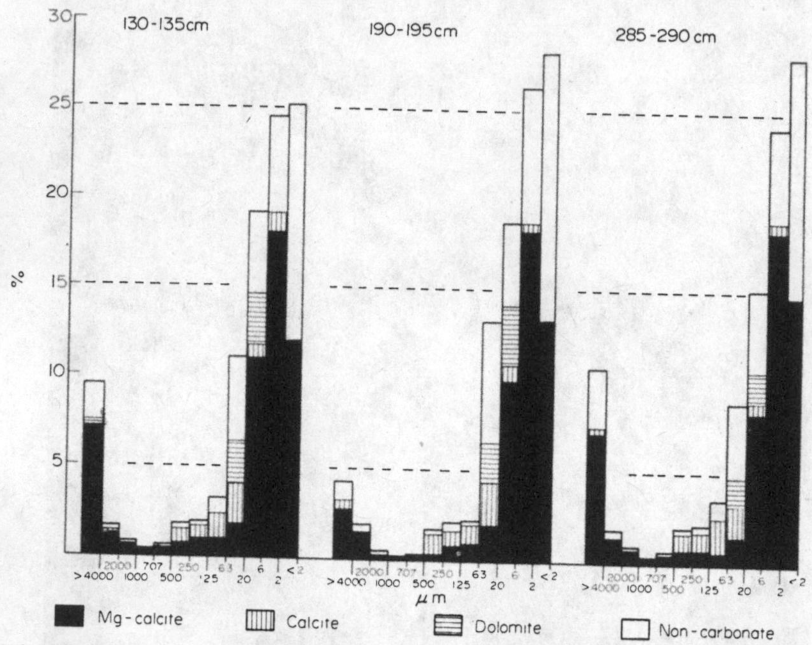

Fig. 11. Distribution of magnesian calcite, calcite, dolomite, and non-carbonate within different size fractions of some sections of core 22 M 49.

magnesian calcite, they show all stages of recrystallization and crystal overgrowth in the layers containing magnesian-calcite lutite and also within the nodules (Fig. 10).

The analysis of clay-size and fine silt (6–2 μm) fractions—where calcite should be expected due to the presence of coccoliths—showed almost pure magnesian calcite (Fig. 11). The low content of calcite cannot equate with the amount of coccoliths observed with the scanning electron microscope. This suggests at least a partial alteration of coccolithophorid calcite to magnesian calcite.

Summarizing the observations with respect to preservation of skeletal material we can state the following.

(1) Aragonitic pteropod shells are dissolved only locally, such dissolution not being related to depth of burial or formation of nodules. This, together with their scarcity within the sediment (<2% of total sediment), rules them out as a potential source of carbonate for the formation of nodules.

(2) Calcitic foraminiferal tests are extensively micritized within the nodules. Our estimates of their abundance therein (⩾10%) are not equated by the low amount or even absence of calcite. Thus, the process of micritization apparently leads to a partial replacement of calcite by magnesian calcite as apparently happens with the coccoliths. However, this alteration of coccoliths is not limited to nodules, but also occurs within the magnesian-calcite lutite.

FORMATION OF NODULES

From the occurrence of nodules within the magnesian-calcite lutite we can assume that the precipitation of magnesian-calcite lutite is the initial phase of nodule formation. Magnesian calcite within the lutite appears to be the normal phase of precipitation in the eastern Mediterranean Sea (Milliman & Müller, 1973), while the formation of nodules seems to be restricted both regionally and stratigraphically. Milliman & Müller (1973) have concluded that present-day bottom salinities (38–39°/$_{oo}$) and temperatures (14–16°C) facilitate the precipitation of magnesian calcite in the eastern Mediterranean Sea. Of critical importance is the fact that precipitation occurs at the sediment-water interface, as shown by the presence of magnesian calcite in the uppermost surface of the bottom sediments. The precipitation apparently involves sea water, being the only plausible and available source of magnesium, since dissolution of aragonite or underlying carbonate consisting of Foraminifera and coccoliths could not yield magnesium.

The role of sedimentation rate, i.e. the possibility of long-term exposure of the uppermost sediments to sea water has been considered a critical factor controlling the precipitation of cementing carbonate, forming nodules or crusts (Milliman, 1966). In our core material nodules appear to be bound to sedimentary irregularities such as stratigraphical gaps or omission horizons. Core 22 M 36 contains nodules in its surface layer. In this case, it is uncertain whether nodule formation pre-dated the displacement of younger strata or whether these concretions were formed afterwards as a result of exhumation. The nodules of core 17 M 14 are restricted to certain strata which probably indicate discontinuous sedimentation during the Quaternary at this location.

The presence of borings and benthonic overgrowths as well as the observation by Natterer that some borings contain living animals on the crusts, and the iron-manganese coatings (Fischer & Garrison, 1967) suggest that the formation of nodules occurs at the sediment surface. We cannot, however, exclude completely the possibility of nodule formation under burial, exhumation by erosion or tectonic events and subsequent attack by boring organisms and benthonic overgrowth. However, core descriptions by Chamley (1971) show the presence of millimetre-sized nodules from above a sapropel layer (deposited 8000 years ago) up to the sediment surface, indicating rather undisturbed sedimentation. It is therefore safe to assume that the formation of nodules occurs within the upper few decimetres of the sediment column, where sedimentation rates are between 3 and 5 cm/1000 year. The possibility of communication between the overlying sea water and the uppermost decimetres of the sediment has been shown by Thorstenson (1971) and is also indicated by the identical composition of the interstitial water with that of sea water within the uppermost core section (Milliman & Müller, 1973).

We therefore deduce the following tentative model for the formation of magnesian-calcite nodules: magnesian calcite is precipitated as a lutite at the sediment surface. Provided the rate of cement-precipitation is higher than the sedimentation rate, the cemented lutite gradually forms concretions of increasing size. If this process is allowed to continue, nodules can coalesce to form coherent crusts (or hardgrounds) or a nodule pavement. However, if the sedimentation rate exceeds the rate of precipitation, further cementation of the magnesian-calcite lutite is inhibited due to the upward migration of the sediment surface and concomitant separation from the precipitating brine (Mediterranean bottom water).

It should be pointed out that the abundance of magnesian-calcite lutite or the presence of nodules in a given layer does not necessarily represent a relative measure of the sedimentation rate. Abundance of magnesian calcite may also be affected by changes in the brine concentration, in temperature, or by the hydrographic regime.

GEOLOGICAL IMPLICATIONS

Nodular limestones of Jurassic age are a typical lithofacies of the Tethyan belt, especially well known from the Alpine-Mediterranean region as 'Ammonitico Rosso', 'Roter Knollenkalk' or 'Adneter Kalk'. The typical red-stained sediments (Fabricius, 1966) contain a fauna of pelagic molluscs. Often, aragonitic shells of cephalopods show partial dissolution of their upper surfaces (Hollmann, 1962) or traces of boring (Fabricius, 1968).

The formation of these fossil nodules has been much discussed because up till now no Recent counterpart was known. The interpretations range from 'algal nodules' (Schmidt, 1939) and 'Subsolution' structures (Hollmann, 1962), to diagenetic formation (Jurgan, 1969; Jenkyns, 1974). Also, the palaeobathymetry of the red nodular limestones is still under discussion (Jenkyns, 1974). There is only general agreement that the fossil nodules must have formed under conditions of very slow sedimentation.

In both the fossil and the Recent cases, the main environmental factors of nodule formation are similar or even identical: slow rate of sedimentation, oxygenated environment, partial dissolution and boring of aragonitic shells, and marine conditions. Therefore, it seems very likely that the nodules in the red Tethyan limestones of Jurassic age might have also been formed by the precipitation of magnesian calcite, which was altered to calcite in the course of the later diagenesis (see Jenkyns, 1974).

As to the present knowledge of palaeogeography, the Tethys was—at least in the Atlantic-Mediterranean region—a semi-enclosed seaway during Jurassic time. Therefore, it might be possible to think of 'Mediterranean' conditions for this part of the Tethys; that is, an area characterized by restricted circulation, higher bottom temperatures and salinities compared to large open oceans and—especially in areas of slow sedimentation—precipitation of magnesian calcite.

ACKNOWLEDGMENTS

This study was financed by the Deutsche Forschungsgemeinschaft. We thank our colleagues Drs U. Franz, W. Hieke and W. Sigl for their help during the execution of this work and Professor K. J. Hsü, Professor R. Trümpy and Dr J. D. Milliman for their critical reviews of the manuscript.

REFERENCES

CHAMLEY, H. (1971) *Recherches sur la sédimentation argileuse en Méditerranée*, pp. 401. PhD thesis, Université d'Aix-Marseille.

CIFELLI, R., BOWEN, V. & SIEVER, R. (1966) Cemented foraminiferal oozes from the Mid-Atlantic Ridge. *Nature, Lond.* **209**, 32–34.

FABRICIUS, F. (1966) Beckensedimentation und Riffbildung an der Wende Trias/Jura in den bayerisch-tiroler Kalkalpen. *Int. Sedim. Petrogr. Ser.* **9**, pp. 143. E. J. Brill, Leiden.

FABRICIUS, F. (1968) Calcareous Sea Bottoms of the Raetian and Lower Jurassic Sea from the West Part of the Northern Calcareous Alps. In: *Recent Developments in Carbonate Sedimentology in Central Europe* (Ed. by G. Müller and G. M. Friedman), pp. 240–249. Springer, Berlin, Heidelberg, New York.

Fischer, A.G. & Garrison, R.E. (1967) Carbonate lithification on the sea floor. *J. Geol.* **75,** 488–496.

Gevirtz, J.L. & Friedman, G.M. (1966) Deep-sea carbonate sediments of the Red Sea and their implications on marine lithification. *J. sedim. Petrol.* **36,** 143–152.

Herman, Y. (1965) *Études des sédiments Quaternaires de la Mer Rouge,* pp. 341–415. PhD. thesis, Université de Paris.

Hieke, W., Sigl, W. & Fabricius, F. (in press) Morphological and structural aspects of the Mediterranean Ridge SW off the Peloponnesos (Ionian Sea). *Bull. geol. Soc. Greece.*

Hollmann, R. (1962) Über Subsolution und die Knollenkalke des Calcare Ammonitico Rosso Superiore im Monte Baldo. *Neues Jb. Geol. Paläont. Mh.* 1962, 163–179.

Jenkyns, H.C. (1974) Origin of red nodular limestones (Ammonitico Rosso, Knollenkalke) in the Mediterranean Jurassic: a diagenetic model. In: *Pelagic Sediments: on Land and under the Sea* (Ed. by K. J. Hsü and H. C. Jenkyns). *Spec. Publs int. Ass. Sediment.* **1,** 249–271.

Jurgan, H. (1969) Sedimentologie des Lias der Berchtesgadener Kalkalpen. *Geol. Rdsch.* **58,** 464–501.

Milliman, J.D. (1966) Submarine lithification of carbonate sediments. *Science,* **153,** 994–997.

Milliman, J.D. & Müller, J. (1973) Precipitation and lithification of magnesian calcite in the deep-sea sediments of the eastern Mediterranean Sea. *Sedimentology,* **20,** 25–49.

Milliman, J.D., Ross, D.A. & Ku, T.L. (1969) Precipitation and lithification of deep-sea carbonates in the Red Sea. *J. sedim. Petrol.* **39,** 724–736.

Müller, J. & Fabricius, F. (1973) Carbonate mineralogy of deep-sea sediments from the Ionian Sea. *Rapp. Comm. int. Mer Medit.* **21,** 855–859.

Schmidt, H. (1939) Bionomische Probleme des deutschen Lias-Meeres. *Geologie Meere Binnengewäss.* **3,** 239–256.

Thorstenson, D. (1971) A chemical model for early diagenesis in Devil's Hole, Harrington Sound, Bermuda. In: *Carbonate Cements* (Ed. by O. P. Bricker), pp. 285–291. *Studies in Geology No.* 19. Johns Hopkins University, Baltimore, Maryland.

van Straaten, L.M.J.U. (1967) Solution of aragonite in a core from the southeastern Adriatic Sea. *Mar. Geol.* **5,** 241–248.

Spec. Publs int. Ass. Sediment. (1974) **1,** 249–271

Origin of red nodular limestones (Ammonitico Rosso, Knollenkalke) in the Mediterranean Jurassic: a diagenetic model

HUGH C. JENKYNS

*Department of Geological Sciences, University of Durham, Durham DH*1 3*LE*

ABSTRACT

Red, nodular and marly limestones (Ammonitico Rosso, Knollenkalke) are a stratigraphically condensed pelagic facies widely distributed in the Alpine-Mediterranean Jurassic. The rock comprises lime-rich nodules set in a darker red marly matrix: nodule-rich and marl-rich layers generally alternate. Previous authors have related the nodular structure to irregular solution of a deep-marine calcareous sea-bottom.

The following observations have been made with respect to Ammonitico Rosso: (1) gradational contacts occur between nodules and matrix, (2) large fossils may cross the nodule-matrix boundary; microfossils generally do not, (3) there is a greater frequency of skeletal calcite in the matrix, (4) dolomitized matrix is present in certain of these facies, (5) formerly siliceous microfossils are preserved only in the nodules, (6) calcareous nannofossils are extremely scarce or absent, (7) the nodules apparently form very early, (8) such nodular limestones were developed during the Late Jurassic on the Atlantic ocean floor. Collectively these points suggest an early diagenetic origin for Ammonitico Rosso in an environment that is not commonly found today. The model presented here has as its driving force the sub-surface dissolution of aragonite (something demonstrated by the widespread occurrence of ammonite moulds) together with some fine-grained calcite, and subsequent precipitation as lime-rich (? high-magnesian) nodules in an originally marly sediment. Slow sedimentation is also considered to have been a critical factor for the formation of the nodules.

Alternate nodule-rich and clay-rich layers are explained thus: as the sediment-water interface migrated upwards there was a drop in the concentration gradient and a concomitant decrease in the rate of withdrawal of calcium carbonate by the last-formed set of nodules; consequently, with continued supply of soluble carbonate near the sediment-water interface, the activities of Ca^{2+} and HCO_3^- rose enough to initiate a new set of nodules around skeletal nuclei and micritic intraclasts higher in the sediment column. This process continued rhythmically.

Compaction effects, particularly important in the marly interstices of the nodules, account for later diagenetic history.

INTRODUCTION

Red nodular limestones are particularly widespread in Jurassic pelagic deposits of the Alpine-Mediterranean region. This lithology has been variously christened; among the names currently in use are: Knollenkalk, Fausse Brèche, and Ammonitico Rosso. These names underline two striking features of this kind of sediment: its extreme nodularity and its richness in ammonites. This facies, because of its attractive colour and texture, has long been used as a decorative marble throughout Europe; its attraction, however, has not been solely ornamental. Some geologists (e.g. Garrison & Fischer, 1969 and references therein) have claimed these rocks as fossil analogues of deep-sea sediments. Herein lies their intrinsic fascination.

The purpose of this paper is to describe, with particular reference to western Sicily, the salient features of this lithology, and to explain its genesis. Two main problems confront us with this facies: depth of deposition and origin of the nodular structure. Both of these questions are examined in the following account, but it is on the second that I propose to concentrate.

The Jurassic palaeogeographic history of western Sicily has been summarized by several workers (e.g. Wendt, 1969a; Jenkyns, 1970; Jenkyns & Torrens, 1971). By Late Jurassic time western Sicily was the depositional site of a shallow-water pelagic 'oolite', limited to a few structural highs (Jenkyns, 1972) and deeper-water facies. These deeper-water facies include red nodular limestones.

DATA FROM WESTERN SICILY

During the Middle Jurassic, extending into the Callovian, a series of stratigraphically condensed limestones were laid down, the faunal and lithological features of which were described in Jenkyns (1971a). These condensed sequences contain ferromanganese nodules and algal stromatolites; their clay content is low and they are never nodular. When traced upward, however, with increasing clay content, the typical nodular texture makes its appearance. This calcareous Ammonitico Rosso is itself strikingly condensed (see Wendt, 1963, 1965), with assessed depositional rates in the order of a few mm/10^3 years. It is well developed on Monte Kumeta, Montagna Grande, Monte Bonifato and Monte Inici (Figs 1 and 2) although its age at these localities is variable. Fossils include ammonites, preserved as moulds, belemnites and aptychi; the microfauna is dominated by the planktonic crinoid *Saccocoma* and the spores of *Globochaete*, with Radiolaria, rare (benthonic) crinoid fragments, a little shell debris, occasional sponge spicules, and some Foraminifera.

The Monte Inici section (MI 1 of Wendt, 1963) is particularly instructive; the Middle Jurassic condensed bed contains algal laminae, and stromatolitic overgrowths on the upper side of corroded ammonites extend up into the nodular facies (Fig. 3). About 1 m above the condensed horizon the stromatolitic coatings on the ammonite shells disappear. This may give a clue to water depth at the time of development of at least some Ammonitico Rosso limestones—and could be taken as evidence of a deepening sequence that passes below the photic zone. Much depends, of course, on the correct interpretation of the ammonite-overgrowths as algal in origin. From the Garrison & Fischer (1969) standpoint—with estimated depths of several thousand

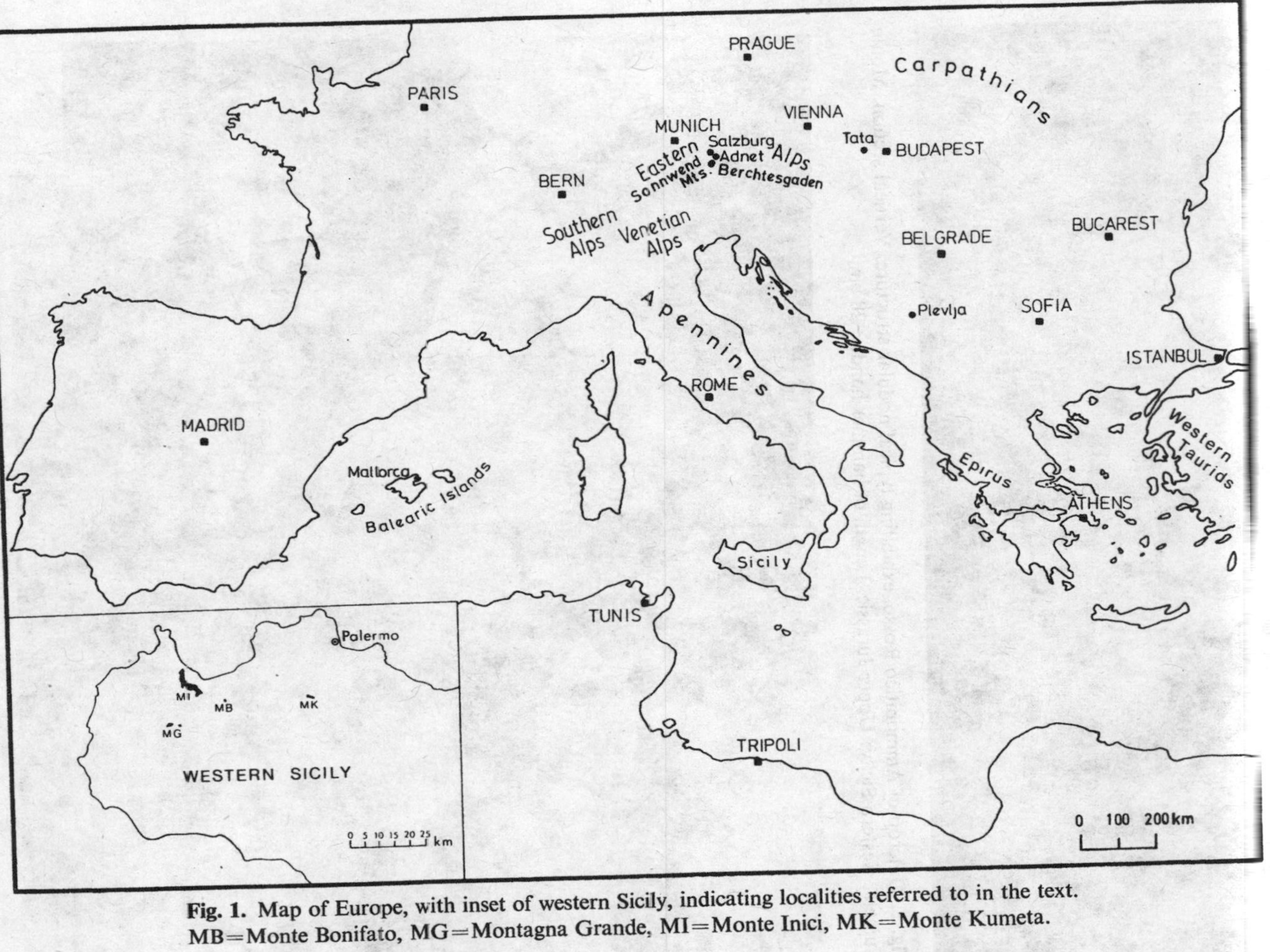

Fig. 1. Map of Europe, with inset of western Sicily, indicating localities referred to in the text. MB=Monte Bonifato, MG=Montagna Grande, MI=Monte Inici, MK=Monte Kumeta.

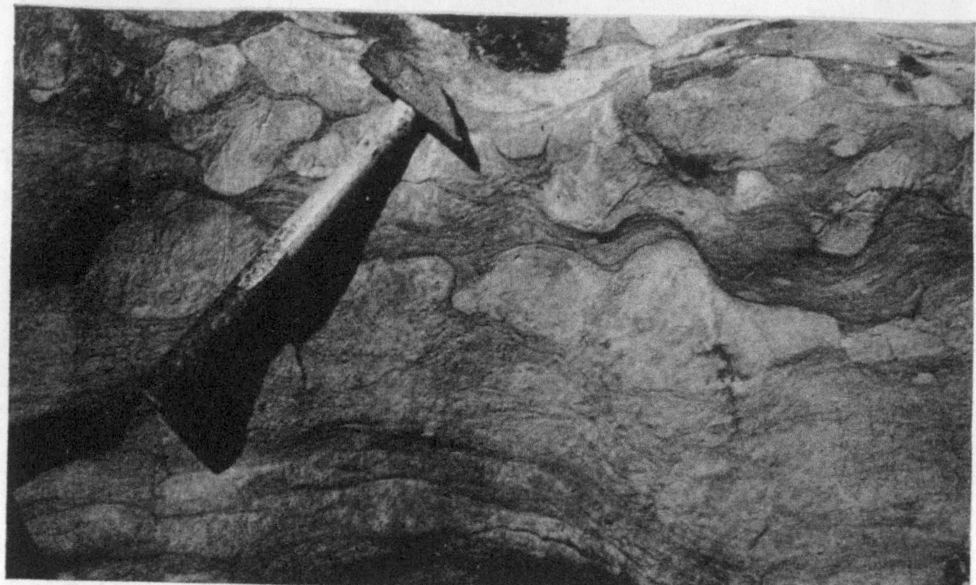

Fig. 2. Outcrop of Ammonitico Rosso, exhibiting typical nodular structure. Vertical section. Monte Kumeta, western Sicily: Upper Jurassic. Length of hammer handle: 28 cm.

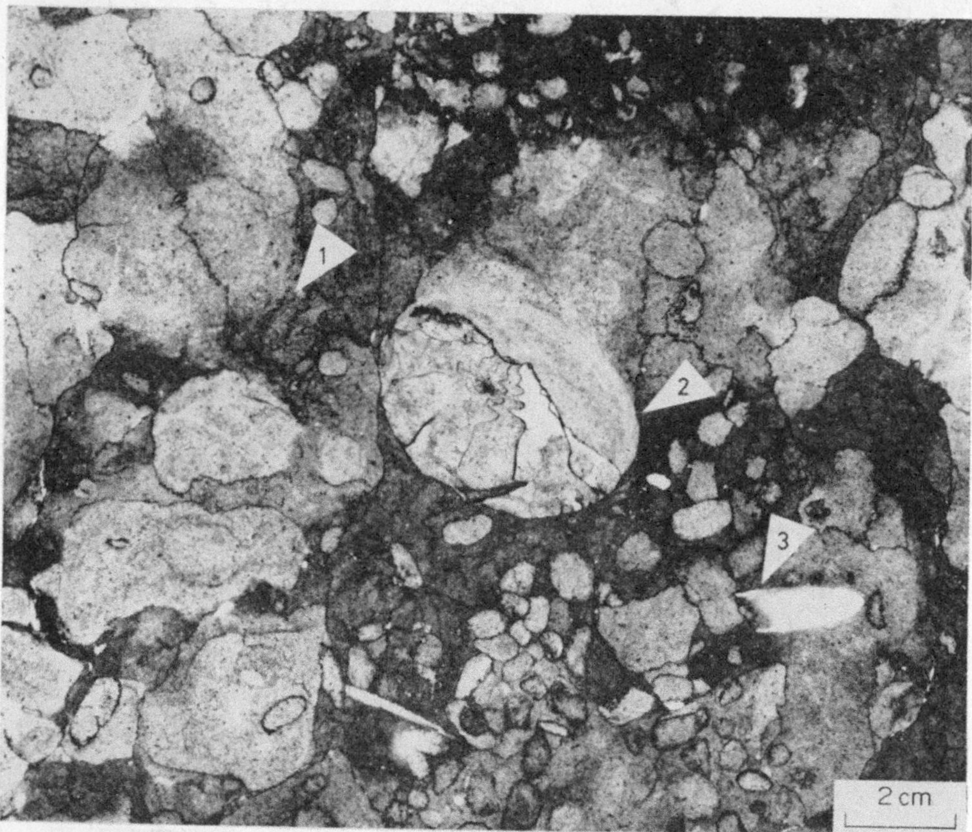

Fig. 3. Polished section, cut approximately normal to bedding, of lime-rich Ammonitico Rosso. Monte Inici, western Sicily: Upper Jurassic. Note (1) the gradation between nodules and matrix; (2) the partially dissolved ammonite, now preserved as a mould, capped by a finely laminated domed structure here interpreted as a stromatolitic head; note also the zone of disseminated calcite above the algal cupola; (3) the belemnite guard that crosses the nodule-matrix boundary. Compound nodules are widespread; some of the included bodies are coated with thin ferromanganiferous rinds and must have been present as distinct entities on the sea floor before later reburial and centrifugal precipitation.

metres for the Tethyan red nodular limestone facies—such a claim is bound to be viewed with scepticism. Nevertheless, the uniform fine-grained laminated structure of these overgrowths coupled with their close stratigraphical association with undoubted stromatolites (see illustrations in Jenkyns, 1971a) suggest that an algal origin is likely.

In general the nodules of the west Sicilian Ammonitico Rosso are of centimetre scale and lie parallel to and elongated along the bedding, interleaved with more marly layers (Fig. 4). Nodules may grade laterally into more continuous irregularly sculptured beds. Ammonite moulds always act as nodules (Fig. 3) and are invariably covered with a glistening clay patina. Some nodules are compound, containing smaller brethren (Fig. 3).

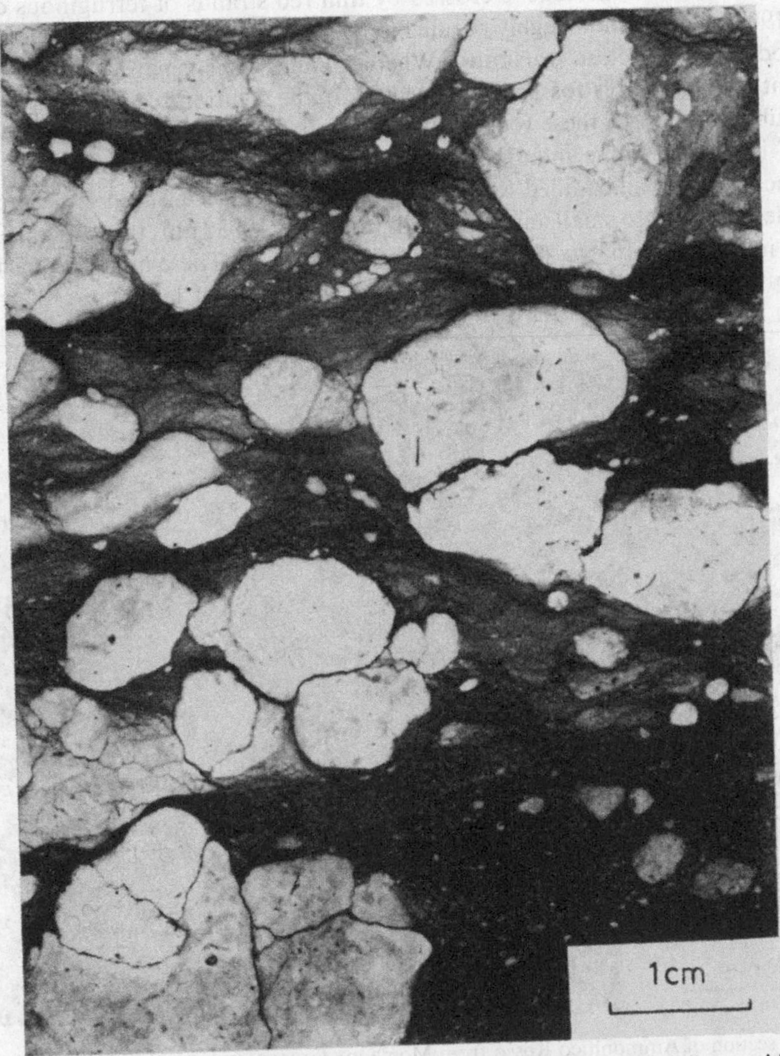

Fig. 4. Polished section, cut approximately normal to bedding, of marl-rich Ammonitico Rosso. Montagna Grande, western Sicily: Upper Jurassic. Note ferruginous strands (dark) in marly matrix, concentrated where nodules are in close vertical proximity. Small nodules, of millimetre scale, are common.

The degree of nodularity—i.e. the degree of differentiation between nodule and matrix—is roughly proportional to the clay content of the whole rock. If the total clay content is high, the nodules can often be prized out with a finger (as in Fig. 4); if the clay content is low the nodules may merge gradually into the matrix (Fig. 3; see also Lucas, 1955a; Fabricius, 1966, p. 52).

The marly interstices of these rocks are always richer in fossil material than the nodules (Fig. 5), something that has been commented upon by several workers (e.g. Lucas, 1955a; Szulczewski, 1965; Jurgan, 1969). Calcite-replaced Radiolaria, or their micritic ghosts are, however, confined to the nodules (Fig. 5). Fossils do not generally cross the nodule-matrix boundary; exceptions to this are found in the case of calcitic macrofossils when the marl content of the rock is low (Fig. 3). The marl matrix of the west Sicilian nodular limestone is crossed by thin red strands of ferruginous clay (cf. Jurgan, 1969) which trend roughly parallel to the bedding; these are particularly dense where nodules are in close proximity. Where nodules inter-penetrate, stylolites may be present; in some cases fossils are truncated by the solution seams.

Insoluble residues of these rocks usually vary between 5 and 20%; and are composed of illite, with some montmorillonite and quartz. The quartz content can be traced back to a few silica-filled Radiolaria and silica-replaced *Saccocoma*, present in the nodules, and also to small angular shards (diameter 10–30 μm) that are particularly common in the marly matrix (see also Szulczewski, 1965). These angular quartz fragments do not necessarily imply the immediate proximity of landmass; shards of a

Fig. 5. Thin section of Ammonitico Rosso from Montagna Grande, western Sicily: Upper Jurassic. Sparsely fossiliferous limestone nodule in richly fossiliferous marl matrix. Much of the skeletal debris is constituted by fragments of the pelagic crinoid *Saccocoma* and the spores of *Globachaete*. Spherical objects (arrows) within the nodules are micritic ghosts of Radiolaria. Note the geopetal fill in the largest of these.

similar size range have been recorded by Rex & Goldberg (1958) from pelagic clays—and are of aeolian origin. The clay mineral assemblage of the west Sicilian nodular limestones is also consistent with an 'oceanic' environment (Griffin, Windom & Goldberg, 1968).

Calcareous nannofossils have not been recognized in these nodular limestones.

DATA FROM OTHER JURASSIC LOCALITIES

Redeposited nodular limestones

The distinction that Aubouin (1964) makes between the two types of Ammonitico Rosso is important: the Ammonitico Rosso Calcaire (as exemplified by most west Sicilian occurrences) does not often exhibit evidence of soft-sediment deformation. The Ammonitico Rosso Marneux, a more basinal presumably deeper-water facies, commonly does. Red nodular and marly limestones showing clear evidence of redeposition have been illustrated by Bernoulli (1964, 1971) from the Southern Alps and Apennines, and by Bernoulli & Jenkyns (1970) from the Eastern Alps. The fact that the nodules may be plastically deformed and streaked out in these retextured sediments shows that the process of nodule formation had begun before consolidation of the sediment. It furthermore demonstrates that nodule formation did not proceed by migration of a solid surface; there was an intermediate soft stage.

The Adnet 'Scheck' (see Schlager, 1961, 1966, 1970; Hallam, 1967; Hudson & Jenkyns, 1969; Jurgan, 1969; Garrison & Fischer, 1969; Bernoulli & Jenkyns, 1970; Wendt, 1971) and its Grecian equivalent (Bernoulli & Renz, 1970) are two known examples of redeposition of a nodular limestone in a swell or seamount area. These are spar-cemented nodular limestones (Fig. 6): what is particularly noteworthy is the complete lack of deformation of the nodules, which must have been completely lithified when exhumed and transported. It is difficult to know how deeply currents might have excavated the sea-bottom, but it seems probable that complete consolidation of limestone nodules must have taken place at or only a little way below the sediment–water interface. Further support for this is the record by Lucas (1955b, c) of a small coral colony growing on nodules.

Dolomitized nodular limestones

Farinacci (1967), after reviewing the origin of Jurassic Tethyan nodular limestones, concluded that the Ammonitico Rosso Superiore of the Umbro-marchigian region (north-central Apennines) was intertidal: this is at odds with all previous interpretations, but it does serve to illustrate the great disparity of views on the origin of this lithology. Farinacci based her interpretation on the presence of spar-filled fissures, apparent spatial association of the Ammonitico Rosso with varved sediments containing charophytes and planorbids, and the presence of euhedral dolomite crystals in the marly matrix.

Crinoidal nodular limestones

Nodularity is also exhibited in the Pliensbachian (Lower Jurassic) crinoidal limestones exposed at Tata, Hungary (Jenkyns, 1971b), and a similar partition of fossil-poor nodules and marly matrix rich in skeletal material (crinoid ossicles, mollusc

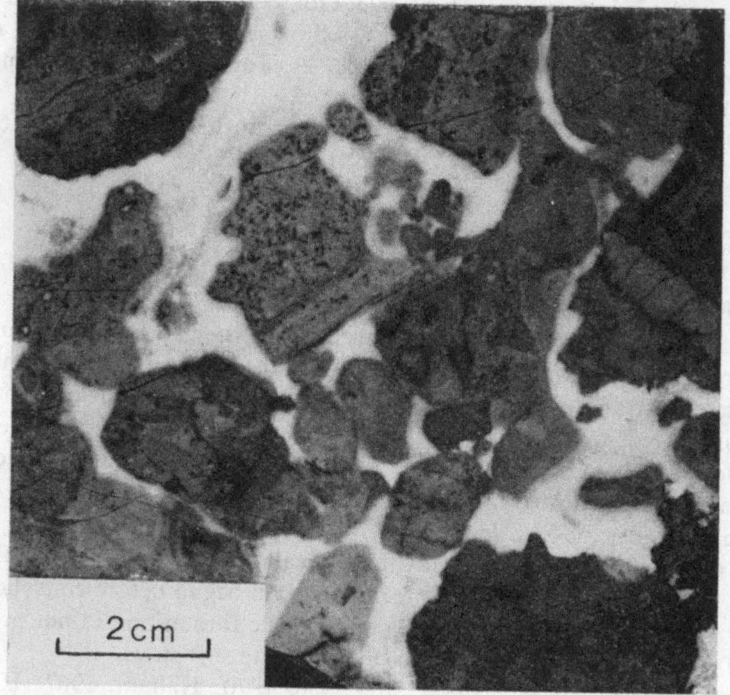

Fig. 6. Polished section of spar-cemented nodular limestone known locally as 'Scheck'. Adnet, Austria: Lower Jurassic. Note the corroded ammonite encased in one of the nodules.

shells, Foraminifera) is apparent. Tethyan crinoidal limestones, which usually overlie carbonate platforms or reefs, are generally considered shallow-water deposits (see Jenkyns, 1971b and references therein); this, coupled with the evidence of stromatolitic overgrowths in the west Sicilian facies, suggests that the formation of red nodular limestones is not necessarily a deep-water phenomenon.

Sponge spicules in Liassic nodular limestones

Ammonitico Rosso is commonly developed in the Lower and Upper Jurassic of the Tethyan zone being labelled as either 'Inferiore' or 'Superiore'. Sponge spicules are not frequent in the Ammonitico Rosso Superiore, which accounts for their scarcity in the west Sicilian facies. They are, however, tolerably common in the Liassic. Sponge spicules, together with Radiolaria, provide the exception to the rule that skeletal debris is always more abundant in the marly matrix than in the nodules: according to Mišík (1964) and Čepek (1970), dealing with Carpathian occurrences, and Jurgan (1969) dealing with those from the Berchtesgaden Alps, carbonate-replaced sponge spicules occur only in the nodules.

Nodular limestones of Toarcian age

Nodular limestones of Toarcian (Late Lower Jurassic) age are very rare in western Sicily (see Jenkyns & Torrens, 1971). Generally, however, wherever the Toarcian occurs in the Tethyan zone it is developed as a nodular and marly Ammonitico Rosso facies. This applies not only to basinal successions but also, strangely enough, to condensed sections [e.g. the Mihajlovici section near Plevlja, Yugoslavia, described by Aubouin *et al*. (1964); a section in the middle Louros valley, Epirus, Greece (Bernoulli & Renz, 1970); the Case Canapine section in the central Apennines, Italy, mentioned by Colacicchi, Passeri & Pialli (1970); sections in the Sonnwend Mountains of Austria illustrated by Wendt (1969b), a section in the western Taurids, Turkey (Gutnic & Monod, 1970)]. Tethyan condensed limestones are usually very poorly endowed with marl (e.g. Jenkyns, 1971a): this widespread nodular and marly facies might therefore represent some Toarcian 'event' characterized by influx of clay (or reduced lime mud input) over an area of considerable geographical extent.

One is led to suspect therefore that the formation of limestone nodules in a marl matrix might have been simply dependent on a sufficient quantity of clay in the original sediment. In entirely calcareous lithologies early nodule formation would presumably have been obliterated later by diagenetic addition of calcite.

Calcareous nannofossils in Tethyan red nodular limestone

As mentioned above, calcareous nannofossils have not been recognized in the west Sicilian Ammonitico Rosso facies, even though they may be tolerably common in rocks stratigraphically above and below: even the Tithonian (Late Upper Jurassic) facies on Montagna Grande which contains tintinnids and might be expected to contain abundant nannocones (see Fischer, Honjo & Garrison, 1967), has proved barren. Garrison & Fischer (1969, and personal communication) also found that calcareous nannoplankton were absent in the Austrian nodular Adnet Limestone facies, although they were common in the more calcareous lithologies. In the Glasenbach Gorge, Salzburg, Austria (Bernoulli & Jenkyns, 1970) the Ammonitico Rosso has yielded calcareous nannofossils only in redeposited horizons: in mud pebbles and fine-grained turbidite fractions. One wonders, therefore, if the process that forms nodular limestones somehow destroys or renders unrecognizable calcareous nannofossils.

RED NODULAR LIMESTONES ON THE OCEAN FLOOR

One of the discoveries made during Leg XI of the Deep Sea Drilling programme in the western central Atlantic was the presence of Upper Jurassic ammonite-bearing red nodular and marly pelagic sediments. These rest on a basement that is apparently oceanic (Lancelot, Hathaway & Hollister, 1972). Bernoulli (1972) has stressed the similarity of these central Atlantic facies with the basinal more marly Ammonitico Rosso of the Alps, Apennines and Greece. Resedimentation features are common in these central Atlantic facies. Coccoliths occur rarely. Depositional rates were in the order of 3 mm/10^3 years.

This central Atlantic occurrence, however, is unique in that no other pelagic sediment from an open ocean, whether ancient or Recent, resembles the red nodular limestones of the Mediterranean Jurassic.

SALIENT POINTS

The most important feature of Jurassic red nodular limestones can be summarized thus.

(1) Gradational contacts occur between certain nodules and matrix (Figs 3 and 4).

(2) Large fossils may cross the nodule matrix boundary (Fig. 3). Microfossils generally do not.

(3) There is a greater frequency of skeletal calcite in the matrix (Fig. 5).

(4) Dolomitized matrix is present in certain of these facies.

(5) Formerly siliceous microfossils are preserved *only* in the nodules (Fig. 5).

(6) Calcareous nannofossils are extremely scarce or absent.

(7) The nodules apparently form very early.

(8) Such nodular limestones were developed during the Late Jurassic on the Atlantic ocean floor, but do not seem to have been widely developed since.

FACTS AND THEORIES FROM PALAEOZOIC FORMATIONS

The Jurassic Ammonitico Rosso has equivalents not only in the Triassic Hallstatt facies of the Eastern Alps and elsewhere, but also in the Devonian-Carboniferous of parts of Europe and North Africa. This is the so-called 'griotte': a red, green or grey pelagic nodular limestone usually rich in goniatites and conodonts.

Numerous possible origins for Palaeozoic nodular limestones have been suggested: the comprehensive theory of Gründel & Rosler (1963), in particular, has gained a certain popularity. These authors proposed that decaying organic matter in the sediment, by releasing hydrogen sulphide and ammonia in the reduction zone, caused the solution and upward migration of carbonate which precipitated on reaching the oxidation zone—with the consequent formation of limestone nodules. High carbonate saturation values can occur in anoxic marine sediments (e.g. Presley & Kaplan, 1968; Berner, Scott & Thomlinson, 1970), although, due presumably to the inhibiting effect of organic matter, or other agencies, precipitation does not generally take place. The upper part of the sediment column would contain unoxidized organic matter; transfer of carbonate-supersaturated water into this oxidation zone would not necessarily result in precipitation. One wonders, also, how earlier-formed nodules survived when deeper burial forced them, too, inexorably into the reduction zone: although a moderately sized nodule might have a certain resistance to solution, some corrosion should take place. This is not observed. Furthermore, as Gründel & Rosler (1963) themselves point out, in areas of slow sedimentation where all but the most resistant organic matter is destroyed before burial, their suggested process cannot be valid. The Ammonitico Rosso, with its prevailing red colour—assuming this is a valid index of diagenetic oxidizing conditions—cannot therefore have arisen by this process.

Silica-replaced fossils occur only in the limestone nodules of the Upper Devonian facies studied by Gründel & Rosler (1963): this is a parallel observation to that on Jurassic nodular limestones.

THE 'SUBSOLUTION' THEORY: A CRITIQUE

Perhaps the most influential papers on the origin of Tethyan nodular limestones have been those of Hollmann (1962, 1964), who suggested that 'Subsolution' or submarine solution was the prime cause of the nodularity. Hollmann envisaged irregular limestone corrosion on the sea floor acting in such a way that isolated 'solution remnants' were left as nodules in the insoluble marly residue. He furthermore suggested that this process was cyclic and periods of limestone deposition had alternated with periods of solution, thus imposing another control on the limestone-marl ratio. Although the exact causes of these processes were never adequately explained, Hollmann's papers have gained widespread acceptance, particularly in German literature. And Garrison & Fischer (1969), after considering various origins for the nodular facies of the Austrian Adnet Limestone, finally concluded: 'the only alternative is Hollmann's interpretation'.

A necessary concomitant of the sea-floor solution that Hollmann postulated was an environment of considerable depth; clearly the presence of algal stromatolites, indicating levels probably not in excess of 200 m (see discussion in Jenkyns, 1971a), is at odds with this. Although one could make a case for aragonite solution having operated at lesser depths in the geological past than at present (e.g. Hudson, 1967) large-scale solution of carbonate sediments is difficult to envisage in waters no deeper than a few hundred metres. One could also argue that the very survival of aragonite into the burial stage—as witnessed by ammonite moulds—is incompatible with sea-floor solution of a low-magnesian calcite ooze.

The area in which Hollmann worked—the Trento Swell in the Venetian Alps—seems to have remained shallow for much of the Jurassic, and stromatolites are well developed in certain levels of the Ammonitico Rosso (Sturani, 1971): the corroded ammonite figured by Hollmann (1964, pl. 8, 8) seems, in fact, to be capped by a stromatolitic cupola! Partial sea-floor (not subsurface) solution of aragonitic ammonite shells is indisputable (Fig. 3); but Hollmann did not take into account the important differences in solution behaviour of these two calcium carbonate polymorphs. *Indisputable* evidence of sea-floor calcite solution is not common (cf. Schlager, 1974). Furthermore, the detailed observations on nodular limestones, summarized above, are completely incompatible with the concept of a solution rubble.

Hollmann's 'Subsolution' mechanism implies that the marl content of Knollenkalk is derived from solution of former limestone beds. In my view the carbonate-clay ratio is not altered by this kind of process, but is a primary sedimentary characteristic, generally governed by the submarine topography: sediments formed on topographic highs containing relatively small amounts of clay, sediments in basins much greater amounts. The Ammonitico Rosso Calcaire, as Aubouin (1964) pointed out, is characteristically related to ridges, the Ammonitico Rosso Marneux to troughs. Such a sediment fractionation would be caused by currents sweeping clay mineral particles and siliceous tests across submarine high ground into neighbouring basins. As submarine relief lessened the strong partition between clay-rich and lime-rich sediments would decrease.

AN ALTERNATIVE ORIGIN

Any alternative theory on the origin of red nodular and marly limestones must explain the eight points listed above. The most obvious mechanism is early diagenetic segregation, championed by Lucas (1955b), Hallam (1967) and Hudson & Jenkyns (1969). This is the only serious contender to 'Subsolution'. Certainly some physico-chemical process acting within the sediment might explain these critical eight points. The segregation theory, however, also suffers from the lack of a well-documented

Fig. 7. Bedding surface of Ammonitico Rosso Superiore. By Lake Garda, Northern Italy: Upper Jurassic. Length of hammer handle: 30 cm. The formation of this kind of surface was attributed by Hollmann (1962, 1964) to irregular solution of a hardened calcareous sea bottom.

mechanism and an explanation of the necessary chemical gradients. What, for instance causes the early diagenetic migration of calcium carbonate?

The great attraction of Hollmann's theory is that, on looking at a bedding plane of Ammonitico Rosso, it is difficult to avoid the impression that some kind of solution process has taken place (Fig. 7). I should like to rephrase this; replacing the word 'solution' by 'redistribution', and further suggest that this redistribution process could have taken place below, rather than at, the sediment-water interface. Could, therefore, the marly interstices simply represent zones in which calcium carbonate was not precipitated or from which it was actually withdrawn? Put another way, do the limestone nodules represent ancient loci of calcium carbonate precipitation? Surely a process of this type would account for the 'solution aspect' of red nodular limestones.

Nodule initiation: significance of aragonite

Siever, Beck & Berner (1965) have shown that the interstitial waters of Recent pelagic carbonate sediments such as *Globigerina* ooze are often slightly undersaturated with calcite; red clays are even more undersaturated. However, these saturation values may be underestimates (Bischoff, Greer & Luistro, 1970). The composition of a 'pre-segregation' Ammonitico Rosso sediment would perhaps lie between *Globigerina* ooze and red clay; one might expect therefore that its interstitial waters would have been slightly undersaturated or just saturated with calcite. Here, however, the factor of depth is important: Recent red clays and *Globigerina* oozes accumulate at depths of several thousand metres (Murray & Renard, 1891), whereas I would interpret the depositional milieu of the Ammonitico Rosso as not in excess of, say, a thousand or so metres—for reasons outlined above. This is significant as, in shallower zones, aragonite will not be completely eliminated in the water column or on the sea bottom but will survive into the burial stage (e.g. Friedman, 1965; Turekian, 1965; Harriss & Pilkey, 1966). That this indeed took place in Jurassic red nodular limestones is proved by the frequency of ammonite moulds that must have resulted from sub-surface solution (Schlager, 1974). Such moulds may also exhibit irregular truncation (Fig. 3) which may be attributable to partial pre-burial dissolution at the sea floor. Throughout the Tethyan zone fossil moulds are common in the Ammonitico Rosso—hence the name of the facies. Garrison & Fischer (1969) suggest that 'aragonite perhaps formed an appreciable part of the micritic mud in the Adnet sediments'; but they do not suggest sources (apart from ammonites) for this mineral. The nektoplanktonic bivalve *Bositra buchi,* which ranges in age from Toarcian to Oxfordian, has an aragonitic inner valve (Jefferies & Minton, 1965); this would have been another contributor for some Jurassic limestones. And there are several records of Jurassic pteropods which, by analogy with their modern counterparts, would have possessed aragonitic shells. Pteropods have been claimed by Sujkowski (1932) from Jurassic limestones in the Carpathians, although his photomicrographs chiefly illustrate thin-shelled bivalves; and by Colom (1970) from the Liassic of Mallorca, Balearic Islands. Recognition of ancient pteropods in thin section requires a combination of faith and imagination, and the significance of these two records should not be over-stressed. It seems safe enough to conclude, however, that all original Ammonitico Rosso sediments contained significant quantities of aragonite.

Pore waters of this sediment would thus have had a two-fold source of calcium carbonate. Aragonite is less stable and more soluble than low-magnesian calcite (see, for example, Kendall, 1912; Chave *et al.,* 1962; Curl, 1962); solution of significant quantities of this mineral would have rendered interstitial waters supersaturated with respect to calcite, resulting in precipitation of the less soluble polymorph. The precipitation site would probably have been governed by the location of large calcite crystals, presumably those of skeletal origin, and lime-rich intraclasts. The nodules in Ammonitico Rosso could have been initiated in this way.

Significance of sedimentary rate

Redistribution of the aragonitic component is an attractive enough mechanism, although it is certainly too simple. Slow rates of sedimentation, a prerequisite for Recent submarine lithification in pelagic environments (Milliman, 1966; Fischer & Garrison, 1967) must have also been a critical factor. As mentioned above, only a few millimetres of Ammonitico Rosso sediment accumulated every thousand years. Minimal sedimentary rates ensure that slow reactions involving a fluid phase are not terminated by rapid compaction due to increasing overburden; they also ensure that reactions dependent on the long-standing proximity of the sediment-water interface are allowed to come to fruition; they furthermore ensure that the bulk of organic material (which can interfere with the solution behaviour of calcium carbonate, see Hall & Kennedy, 1967; Berner *et al.,* 1970; Suess, 1970) is oxidized before burial. Too much should not perhaps be made of this latter point since Recent *Globigerina* oozes and red clays which have positive Eh values nevertheless commonly contain up to 2% organic matter (Correns, 1939; Revelle, 1944) and support an abundant bacterial flora particularly in surface layers (Morita & Zobell, 1955; Zobell & Morita, 1957; Kriss, 1963). It is thus possible that certain protective organic monolayers on carbonate particles might survive into the early burial stage before bacterial oxidation removed them, if indeed they were removed at all (see Honjo, 1969).

The inevitable presence of the magnesium ion in near-surface interstitial waters would have drastically slowed the precipitation of calcite (e.g. Fyfe & Bischoff, 1965; Pytkowicz, 1965; De Groot & Duyvis, 1966; Bischoff & Fyfe, 1968). However, in environments of slow sedimentation, this effect would probably have been counterbalanced. Furthermore, sorption of magnesium ions by clay minerals (e.g. Whitehouse & McCarter, 1958; Carroll & Starkey, 1960) or some kind of chemical reconstitution utilizing Mg^{2+} (Shishkina, 1964; Bischoff & Ku, 1970) would have helped counteract this inhibitory effect. If, as seems likely, the formation of these limestone nodules is related to the processes that form Recent hardgrounds in pelagic environments, this early calcite cement may have been magnesium rich (Friedman, 1965; Milliman, 1966, 1971; Fischer & Garrison, 1967; Marlowe, 1971; see also Weyl, 1967 and Bischoff, 1968 for experimental data). Interestingly enough a magnesian calcite 'lithic fragment' from the Red Sea (Milliman, Ross & Ku, 1969) strongly resembles, in general microfacies, the nodules of Ammonitico Rosso.

In carbonate-undersaturated bottom waters, very slow deposition might have completely eliminated aragonite before burial: clearly there are limiting values for sedimentary rate.

Solution–deposition of calcite

Another question which can be posed is this: could the formation of nodular limestone have resulted only from a redistribution of the original aragonitic component, or could some fine-grained calcite have been removed from the zones now represented by marly interstices? Since the solubility of calcite is a function of its grain size (see, for example, Curl, 1962; Pantin, 1965; Chave & Schmalz, 1966) very fine-grained calcite might have been selectively dissolved within the sediment. At nodule sites migration would have been minimal; however, outside the zones of calcite precipitate, solution gradients would have been engendered, resulting from interstitial waters being supersaturated with respect to limestone nodules when saturated with respect to very fine-grained calcite. One might expect coccoliths, possessing low-magnesian calcite plates a micron or so in diameter, to have been particularly prone to selective solution. This could explain the extreme paucity of these calcareous nannofossils in Jurassic nodular limestones. The presence of occasional coccoliths in the western central Atlantic facies does not disqualify the assumption, since the nodular structure is only feebly developed. However, ocean-floor nannoplankton oozes do not generally evince marked solution-deposition effects (e.g. Cita, 1970; Maxwell *et al.*, 1970; McIntyre & McIntyre, 1971; Pimm, Garrison & Boyce, 1971; Wise & Hsü, 1971; see also Honjo, 1969 on ancient micrites). It may be that once nodule formation was initiated, the solution gradient established was strong enough to eliminate coccoliths: if this is so, it emphasizes the importance of the survival of aragonite into the burial stage, which then acts as a trigger for processes involving only calcite. Recrystallization of coccoliths does seem to be linked to precipitation of micritic high-magnesian calcite as investigations on the Red Sea lithic fragments have revealed (Milliman *et al.*, 1969).

The mode of calcite addition outlined above is similar to that suggested by Pantin (1958) for the formation of a Recent carbonate concretion. Pantin was able to show that the concretion's carbonate matrix was considerably older than the shells that it contained—and must have been drawn from the enclosing sediment.

Rhythmic precipitation

Red nodular limestones exhibit a certain rhythm in their stratigraphic arrangement, nodules being interleaved with more marly beds (cf. Sujkowski, 1958). Such rhythms are more noticeable in some sequences than others (cf. Figs 2, 3 and 4). The problem of the origin of this rhythm can be expressed thus: the sediment-water interface probably migrated upward more or less continually as new material was added to it. Nodule growth, however, only took place at certain levels or at certain sites. At some point therefore there must have been a cessation of nodule growth at one horizon, and a new series, higher in the sediment column, was engendered. The mechanism of the cut-off and transfer of depositional sites is the crux of the problem.

The source of carbonate for the growth of a nodule layer must, for the most part, have come from above, since older nodules should have been at or near equilibrium with their interstitial waters. If organic material was inhibiting solution of carbonates, the supply of these ions might not have come from immediately below the sediment-water interface, but at a level deeper in the sediment pile where bacterial action had removed protective organic coatings. The level of carbonate solution may perhaps have been several centimetres, or tens of centimetres, below the sediment-water interface.

The fact that nodules formed at all, rather than a precipitate of disseminated calcite around all skeletal nuclei, shows that growing nodules were generally more competitive as depositional sites than the majority of shell and test fragments; pieces of *Saccocoma*, for example, occur scattered through the matrix of nodular limestones where no deposition has taken place (Fig. 5). Nodules were therefore capable of withdrawing calcium carbonate as calcite from weakly supersaturated solutions; whereas a higher degree of supersaturation was necessary to initiate precipitation around a skeletal particle or an intraclast. Initially, when the sites of nodule growth were close to the supply of calcium carbonate, rapid diffusion to a variety of sites must have ensured that the critical supersaturation level for precipitation on skeletal calcite was not reached. As the sediment-water interface migrated upward, the concentration gradient and hence the rate of ionic diffusion downwards would have decreased (following Fick's Law of Diffusion). Thus with reduced downward diffusion of Ca^{2+} and HCO_3^- the concentration of calcium carbonate in the immediately subsurface interstitial waters would have gradually risen towards the saturation level for aragonite, probably aided by the fact that the rate of solution of aragonite is generally greater than the rate of precipitation of calcite (Schmalz, 1967). At a certain point, therefore, the critical supersaturation level for precipitation around skeletal calcite must have been reached and a new set of nodules engendered higher in the sediment column. Such a process would have continued rhythmically. With abundant excess carbonate more or less continuous beds would have resulted; with a more restricted supply nodules would have been formed. With red nodular limestones the amount of mobile carbonate must have been limited; this again was probably a function of slow sedimentation, which ensured pre-burial dissolution of a certain amount of aragonite.

Significance of differential preservation in nodules and matrix

The absence of originally siliceous microfossils in the clay-rich matrix, coupled with the occasional presence of dolomite therein, suggests that these interstitial zones have, at one time, been passage-ways for fluids. Clays are not renowned for their permeability; and, on looking at a specimen of Ammonitico Rosso, it might seem unlikely that the marly interstices could have acted as conduits. What one sees today, however, has been strongly influenced by compaction; originally the matrix must have had a more 'spongy' nature and undoubtedly contained voids left by solution of aragonitic shells.

After early formation of limestone nodules, gradual compaction of the sediment would have forced pore waters up between these zones of calcite precipitate. Such waters may well have been slightly undersaturated with respect to amorphous silica (Siever *et al.*, 1965; Bischoff & Ku, 1970; cf. Fanning & Pilson, 1971) and would have eliminated readily soluble sponge spicules and Radiolaria, whilst leaving more resistant quartz shards unscathed (e.g. Siever, 1957). Those siliceous microfossils encased in nodules would, on dissolution, leave a supported void which could be filled later with calcite. Micro-environments within nodules—unaffected by streaming pore fluids—would also have been favourable for silica/carbonate and carbonate/silica replacements (see Buurman & van der Plas, 1971, for a discussion of the replacement problem).

Late diagenetic effects

Compaction has accounted for certain features of red nodular limestones. Solutional voids in the matrix were presumably closed. The marly interstices were forced around the cemented nodules, often causing differential movement between nodule and matrix in the more marly lithologies, so that the aspects of a flow structure were created. This squeezing of matrix relative to nodules must explain the absence of microfossils crossing the nodule-matrix boundary. Differential compaction of the unlithified marly interstices must also account for the greater frequency of skeletal calcite therein.

Jurgan (1969) was inclined to consider all the marly interstices of nodular limestones as the result of late-diagenetic pressure solution. Certainly the ferruginous clay-rich strands that are particularly abundant where nodules are in close vertical proximity seem best interpreted as small-scale solution seams. Stylolites are another manifestation of the same phenomenon. Nevertheless, selective pressure solution implies inhomogeneity in the sediment (e.g. Trurnit, 1967; Park & Schot, 1968), which requires that differentiation into lime-rich nodules and clay-rich interstices had already taken place when compaction commenced. This suggests that the role of pressure solution in developing nodular limestones is relatively minor.

TEST OF DIAGENETIC MODEL: RECAPITULATION AND EXPLANATION

At this juncture it seems relevant to see whether or not this proposed diagenetic model fully explains the crucial eight points listed above. Clearly, gradational contacts between certain nodules and matrix and the presence of large fossils crossing the nodule-matrix boundary are perfectly understandable if the nodules have been formed by sub-surface precipitation of calcium carbonate. The absence of less robust microfossils crossing this boundary can be explained by differential movement between nodules and matrix, such movement resulting from vertical load. Greater frequency of skeletal calcite in the matrix can also be explained by preferred compaction of the lime-poor marly matrix. The presence of dolomitized matrix in some examples shows that fluids could pass easily through these marly zones, but not through the nodules; such fluids, if undersaturated with respect to amorphous silica, would have dissolved sponge spicules and Radiolaria from the matrix but left them unscathed where they were encased in impermeable nodules. The absence of calcareous nannofossils can be attributed to the interstitial calcite solution that is part of the nodule-forming process: micritic high magnesian calcite cements in the Red Sea comprise minute rhombs partly derived from the total recrystallization of coccoliths (Milliman *et al.*, 1969).

Early formation of the nodules in these Jurassic limestones, coupled with their absence from Recent deep-sea sediments, suggests that the critical conditions for their formation are not realized in the oceans of today. However, in Recent pelagic environments the requisite mixture of clay minerals, aragonite and calcite are generally not found: i.e. zones of red clay accumulation and pteropod ooze accumulation are mutually exclusive (Murray & Hjort, 1912). Such a condition will generally prevail where pelagic sedimentation is confined to ocean basins. The exception of the Ammonitico Rosso-type facies in the western central Atlantic can be related to its formation in a basinal part of an early shallow Mid-Atlantic ridge (Lancelot *et al.*, 1972). The Mediterranean Ammonitico Rosso, however, is a continental-margin deposit floored by continental crust (Bernoulli & Jenkyns, 1974) and in this setting environments into which clay minerals could be fractionated were developed at depths sufficiently shallow to allow the preservation of aragonite into the burial stage.

MAGNESIAN CALCITE NODULES ON THE MEDITERRANEAN RIDGE

The foregoing account was written before I was able to read the work of Müller & Fabricius (1974). These authors describe high-magnesian calcite nodules from non-depositional environments at depths of 2–3 km on the Mediterranean Ridge (Ionian Sea). The nodules are commonly of centimetre scale, may be soft or hard, and possess a micritic matrix dotted with more or less altered globigerinid Foraminifera. Aragonitic pteropods and heteropods may also be included; coccoliths are locally made over to a magnesian-calcite precipitate. The nodules are considerably more calcareous than their matrix. They apparently formed at, or near, the sediment-water interface.

This occurrence of a Recent nodular sediment may be used as a parallel for Ammonitico Rosso (Müller & Fabricius, 1974) and may be seen as confirmation for some of the ideas expressed above.

SUMMARY OF PROPOSED ORIGIN

It is assumed that the pre-segregation Ammonitico Rosso sediment comprised a roughly homogeneous mixture of clay minerals, calcareous nannofossil ooze, and

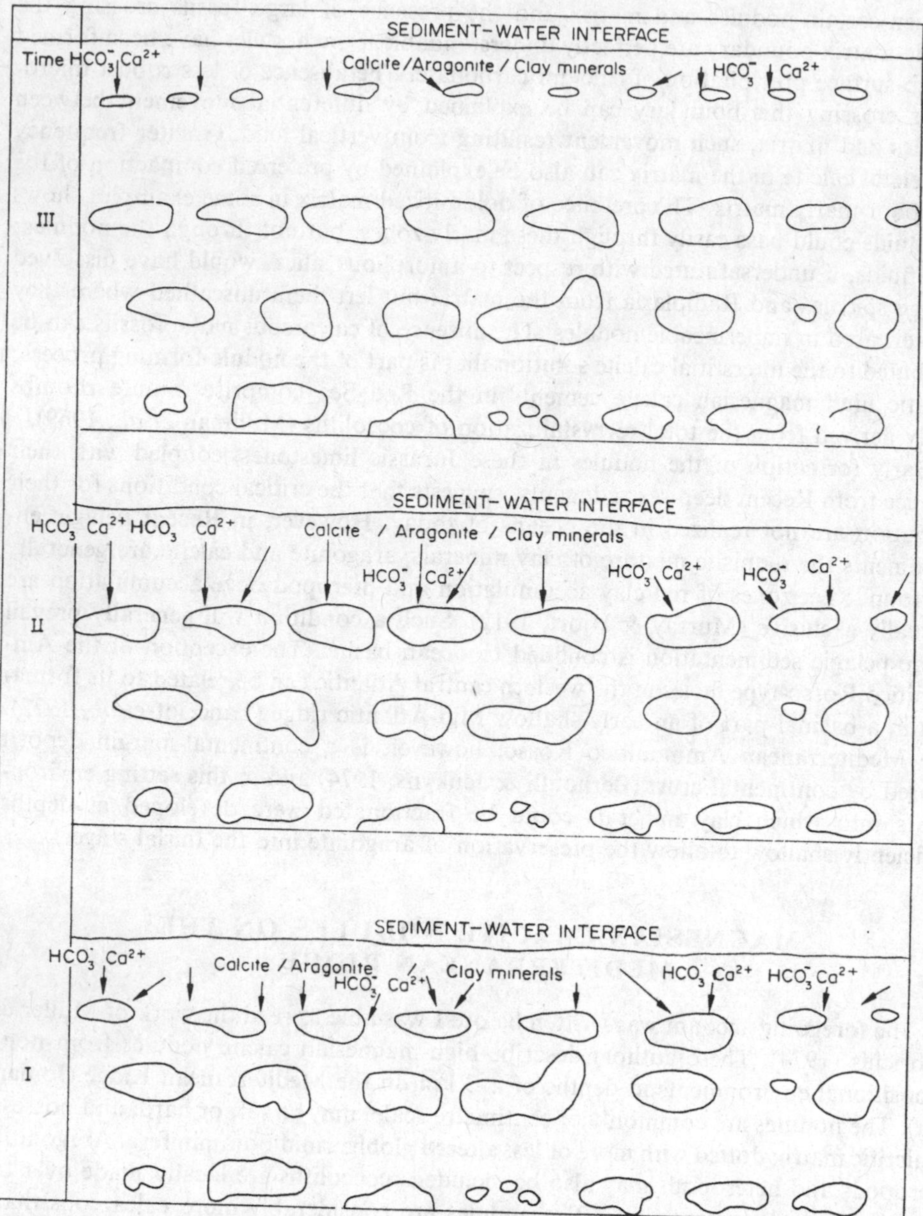

Fig. 8. Diagram to illustrate proposed rhythmic formation of nodules. Distance between nodules is of centimetre scale. At Time I Ca^{2+} and HCO_3^- are being supplied by dissolution of aragonite and very fine-grained calcite near the top of the sediment column; the latest set of nodules are also close to the sediment-water interface, and hence the diffusion paths of the ions are short, and transfer rapid. At Time II the sediment-water interface has migrated upward, lengthening the distances between solution level and depositional sites, and thus slowing diffusion rates. At Time III a new series of nodules closer to the sediment-water interface are being formed. Downward diffusion rates to the previous set of nodules became so low that Ca^{2+} and HCO_3^- were no longer removed as fast as they were supplied by the solution of aragonite and fine-grained calcite. An increase in the calcium carbonate concentration of the immediately sub-surface interstitial waters resulted, initiating a new set of nodules around skeletal calcite and micritic intraclasts.

appreciable aragonite. Accumulation rates were very slow. Owing to the solubility differential of these two forms of calcium carbonate, dissolution of aragonite caused interstitial waters to become supersaturated with respect to calcite, thus causing precipitation of the less soluble polymorph probably as a high magnesian form. Solution-transfer of very fine-grained calcite may also have taken place. The rate of nodule growth decreased as the sediment-water interface (and level of maximum carbonate supply) migrated upwards, with a concomitant drop in the concentration gradient. As the rate of withdrawal of calcium carbonate by the last formed set of nodules decreased, with continued supply of soluble carbonate near the sediment-water interface, the activities of Ca^{2+} and HCO_3^- must have risen enough to initiate a new set of nodules around skeletal nuclei and intraclasts higher in the sediment column. This process continued rhythmically (Fig. 8). Minimal sedimentary rates must also have been an important factor, ensuring virtually complete destruction of organic matter on the sea floor, and allowing slow reactions to come to equilibrium, before compaction eliminated a fluid phase. Ascending pore waters flowing between nodules dissolved readily soluble opaline skeletons. As the sedimentary pile built up compaction of the marly interstices took place, sometimes involving differential movement between nodules and matrix.

ACKNOWLEDGMENTS

I would particularly like to thank Dr D. Bernoulli for drawing my attention to the geographical extent of Toarcian Ammonitico Rosso and for supplying extra information on the distribution and preservation of calcareous nannofossils. Dr R. E. Garrison also gave me additional data on coccolith distribution in the Adnet Limestone. Drs J. D. Hudson and G. V. Middleton kindly read and criticized an early draft of the manuscript. Thanks are also due to Drs W. J. Kennedy, J. D. Milliman and B. W. Sellwood for informal discussion, and Drs E. G. Purdy and W. Schlager for helpful review. I am also grateful to Drs J. Müller and F. Fabricius for sending me a preprint of their paper on the high-magnesian calcite nodules from the Ionian deep sea.

Although this account relies a good deal on literature most of the localities mentioned I have visited myself. Travel funds for this field work were obtained from the Natural Environment Research Council of Great Britain and the Swiss National Science Foundation. These awards are acknowledged with thanks.

REFERENCES

AUBOUIN, J. (1964) Refléxions sur le faciès "ammonitico rosso". *Bull. Soc. géol. Fr.* sér. 7, **6,** 475–501.

AUBOUIN, J., CADET, J.-P., RAMPNOUX, J.-P., DUBAR, G. & MARIE, P. (1964) A propos de l'âge de la série ophiolitique dans les Dinarides yougoslaves: la coupe de Mihajlovici aux coufins de la Serbie et du Monténégro (region de Plevlja, Yougoslavie). *Bull. Soc. géol. Fr.* sér. 7, **6,** 107–112.

BERNER, R.A., SCOTT, M.R. & THOMLINSON, C. (1970) Carbonate alkalinity in the pore waters of anoxic marine sediments. *Limnol. Oceanogr.* **15,** 544–549.

BERNOULLI, D. (1964) Sur Geologie des Monte Generoso (Lombardische Alpen). *Beitr. geol. Karte Schweiz,* **118,** 134 pp.

BERNOULLI, D. (1971) Redeposited pelagic sediments in the Jurassic of the central Mediterranean area. In: *Colloque du Jurassique méditerranéen* (Ed. by E. Végh-Neubrandt). *Annls Inst. geol. publ. hung.* **54/2,** 71–90.

BERNOULLI, D. (1972) North Atlantic and Mediterranean Mesozoic facies: a comparison. In: *Initial Reports of the Deep Sea Drilling Project,* Vol. XI (C. D. Hollister and J. I. Ewing *et al.*), pp. 801–871. U.S. Government Printing Office, Washington.

BERNOULLI, D. & JENKYNS, H.C. (1970) A Jurassic basin: The Glasenbach Gorge, Salzburg, Austria. *Verh. geol. Bundesanst. Wien,* **1970,** 504–531.

BERNOULLI, D. & JENKYNS, H.C. (1974) Alpine, Mediterranean and central Atlantic Mesozoic facies in relation to the early evolution of the Tethys. In: *Modern and Ancient Geosynclinal Sedimentation* (Ed. by R. H. Dott, Jr and R. H. Shaver). *Spec. Publs Soc. econ. Paleont. Miner., Tulsa,* **19,** 129–160.

BERNOULLI, D. & RENZ, O. (1970) Jurassic carbonate facies and new ammonite faunas from western Greece. *Eclog. geol. Helv.* **63,** 573–607.

BISCHOFF, J.L. (1968) Kinetics of calcite nucleation: magnesium ion inhibition and ionic strength catalysis. *J. geophys. Res.* **73,** 3315–3321.

BISCHOFF, J.L. & FYFE, W.S. (1968) Catalysis, inhibition and the calcite-aragonite problem. 1. The aragonite-calcite transformation. *Am. J. Sci.* **266,** 65–79.

BISCHOFF, J.L., GREER, R.E. & LUISTRO, A.O. (1970) Composition of interstitial waters of marine sediments: temperature of squeezing effect. *Science,* **167,** 1245–1246.

BISCHOFF, J.L. & KU, T.-L. (1970) Pore fluids of Recent marine sediments: 1., Oxidizing sediments of 20°N, continental rise to mid-Atlantic ridge. *J. sedim. Petrol.* **40,** 960–972.

BUURMAN, P. & VAN DER PLAS, L. (1971) The genesis of Belgian and Dutch flints and cherts. *Geologie Mijnb.* **50,** 9–28.

CARROLL, D. & STARKEY, H.C. (1960) Effect of sea water on clay minerals. In: *Clays and Clay Minerals, Proc. 7th natn. Conf. on Clays,* pp. 80–101. Pergamon Press, Oxford.

ČEPEK, P. (1970) To the facies characterization of the neritic and bathyal sedimentation of the Alpine-Carpathian geosyncline. *Rozpr. csl. Akad. Ved, Rada MPV,* **80,** 78 pp.

CHAVE, D.E. & SCHMALZ, R.F. (1966) Carbonate-sea water interactions. *Geochim. cosmochim. Acta,* **30,** 1037–1048.

CHAVE, K.E., DEFFEYES, K.S., WEYL, P.K., GARRELS, R.M. & THOMPSON, M.E. (1962) Observations on the solubility of skeletal carbonates in aqueous solutions. *Science,* **137,** 33–34.

CITA, M.B. (1970) Observations sur quelques aspects paléoécologiques de sondages subocéaniques effectués dans l'Atlantique nord. *Revue Micropaléont.* **12,** 187–201.

COLACICCHI, R., PASSERI, L. & PIALLI, G. (1970) Nuovi data sul Giurese Umbro-Marchigiano ed ipotesi per un suo inquadramento regionale. *Memorie Soc. geol. ital.* **9,** 839–874.

COLOM, G. (1970) Estudio litológico y micropaleontológico del Lías de la Sierra Norte y porción central de la isla de Mallorca. *Mems R. Acad. Cienc. exact. fis nat. Madr.* **24,** 7–87.

CORRENS, C.W. (1939) Pelagic sediments of the North Atlantic Ocean. In: *Recent Marine Sediments, a Symposium* (Ed. by P. D. Trask). *Spec. Publs Soc. econ. Paleont. Miner., Tulsa,* **4,** 373–395.

CURL, R. L. (1962) The aragonite-calcite problem. *Bull. nat. speleol. Soc.* **24,** 57–73.

DE GROOT, K. & DUYVIS, E.M. (1966) Crystal form of precipitated calcium carbonate as influenced by adsorbed magnesium ions. *Nature, Lond.* **212,** 183–184.

FABRICIUS, F.H. (1966) Beckensedimentation und Riffbildung an der Wende Trias/Jura in den Bayerisch-Tiroler Kalkalpen. *Int. sedim. petrogr. Ser.* **9,** 143 pp. E. J. Brill, Leiden.

FANNING, K.A. & PILSON, M.E.Q. (1971) Interstitial silica and pH in marine sediments: some effects of sampling procedures. *Science,* **173,** 1228–1231.

FARINACCI, A. (1967) La serie giurassico-neocomiana di Monte Lacerone (Sabina). Nuove vedute sull'interpretazione paleogeografica delle aree di facies umbro-marchigiana. *Geol. rom.* **6,** 421–480.

FISCHER, A.G. & GARRISON, R.E. (1967) Carbonate lithification on the sea floor. *J. Geol.* **75,** 488–496.

FISCHER, A.G., HONJO, S. & GARRISON, R.E. (1967) Electron micrographs of limestones and their nannofossils. *Monogr. Geol. Paleont* Vol. 1 (Ed. by A. G. Fischer), 141 pp. Princeton University Press, Princeton, New Jersey.

FRIEDMAN, G.M. (1965) Occurrence and stability relationships of aragonite, high-magnesian calcite, and low-magnesian calcite under deep-sea conditions. *Bull. geol. Soc. Am.* **76,** 1191–1196.

FYFE, W.S. & BISCHOFF, J.L. (1965) The calcite-aragonite problem. In: *Dolomitization and Limestone Diagenesis, a Symposium* (Ed. by R. C. Murray and L. C. Pray). *Spec. Publs Soc. econ. Paleont. Miner., Tulsa,* **13,** 3–13.

GARRISON, R.E. & FISCHER, A.G. (1969) Deep-water limestones and radiolarites of the Alpine Jurassic. In: *Depositional Environments in Carbonate Rocks, a Symposium* (Ed. by G. M. Friedman). *Spec. Publs Soc. econ. Paleont. Miner., Tulsa,* **14,** 20–56.

GRIFFIN, J.J., WINDOM, H. & GOLDBERG, E.D. (1968) The distribution of clay minerals in the world ocean. *Deep Sea Res.* **15,** 433–459.

GRÜNDEL, J. & ROSLER, H.J. (1963) Zur Entstehung der oberdevonischen Kalkknollengesteine Thuringens. *Geologie,* **12,** 1009–1038.

GUTNIC, M. & MONOD, O. (1970) Une série mésozoique condensée dans les nappes du Taurus occidental. *C.r. somm. Séanc. Soc. géol. Fr.* **1970,** 166–167.

HALL, A. & KENNEDY, W.J. (1967) Aragonite in fossils. *Proc. R. Soc. B,* **168,** 377–412.

HALLAM, A. (1967) Sedimentology and palaeogeographic significance of certain red limestones and associated beds in the Lias of the Alpine region. *Scott. J. Geol.* **3,** 195–220.

HARRISS, R.C. & PILKEY, O.H. (1966) Interstitial waters of some deep marine carbonate sediments. *Deep Sea Res.* **13,** 967–969.

HOLLMANN, R. (1962) Über Subsolution und die 'Knollenkalke' des Calcare Ammonitico Rosso Superiore im Monte Baldo (Malm; Norditalien). *Neues Jb. Geol. Paläont. Mh.* **1962,** 163–179.

HOLLMANN, R. (1964) Subsolutions-Fragmente (Zur Biostratinomie der Ammonoidea im Malm des Monte Baldo/Norditalien). *Neues Jb. Geol. Paläont. Abh.* **119,** 22–82.

HONJO, S. (1969) Study of fine-grained carbonate matrix: sedimentation and diagenesis of 'micrite'. In: *Litho- and Bio-facies of Carbonate Sedimentary Rocks, a Symposium* (Ed. by T. Matsumo). *Spec. Pap. palaeont. Soc. Japan,* **14,** 67–82.

HUDSON, J.D. (1967) Speculations on the depth relations of calcium carbonate solution in Recent and ancient seas. *Mar. Geol.* **5,** 473–480.

HUDSON, J.D. & JENKYNS, H.C. (1969) Conglomerates in the Adnet limestones of Adnet (Austria) and the origin of the 'Scheck'. *Neues Jb. Geol. Paläont. Mh.* **1969,** 552–558.

JEFFERIES, R.P.S. & MINTON, P. (1965) The mode of life of two Jurassic species of 'Posidonia' (Bivalva). *Palaeontology,* **8,** 156–185.

JENKYNS, H.C. (1970) The Jurassic of western Sicily. In: *Geology and History of Sicily* (Ed. by W. Alvarez and K. H. A. Gohrbandt), pp. 245–254. Petroleum Exploration Society, Libya, Tripoli.

JENKYNS, H.C. (1971a) The genesis of condensed sequences in the Tethyan Jurassic. *Lethaia,* **4,** 327–352.

JENKYNS, H.C. (1971b) Speculations on the genesis of crinoidal limestones in the Tethyan Jurassic. *Geol. Rdsch.* **60,** 471–488.

JENKYNS, H.C. (1972) Pelagic 'oolites' from the Tethyan Jurassic. *J. Geol.* **80,** 21–33.

JENKYNS, H.C. & TORRENS, H.S. (1971) Palaeogeographic evolution of Jurassic seamounts in western Sicily. In: *Colloque du Jurassique méditerranéen* (Ed. by E. Végh-Neubrandt). *Annls Inst. geol. publ. hung.* **54/2,** 91–104.

JURGAN, H. (1969) Sedimentologie des Lias der Berchtesgadener Kalkalpen. *Geol. Rdsch.* **58,** 464–501.

KENDALL, J. (1912) The solubility of calcium carbonate. *Phil. Mag.* **23,** 958–976.

KRISS, A.E. (1963) *Marine Microbiology* (trans. J. M. Shewan and Z. Kabata), pp. 536. Oliver and Boyd, Edinburgh and London.

LANCELOT, Y., HATHAWAY, J.C. & HOLLISTER, C.D. (1972) Lithology of sediments from the western North Atlantic, leg XI, Deep Sea Drilling Project. In: *Initial Reports of the Deep Sea Drilling Project,* Vol. XI (C. D. Hollister and J. I. Ewing *et al.*), pp. 901–949. U.S. Government Printing Office, Washington.

LUCAS, G. (1955a) Caractères pétrographiques de calcaires noduleux, à faciès *ammonitioco rosso,* de la région méditerranéenne. *C. r. hebd. séanc. Acad. Sci., Paris,* **240,** 1909–1911.

LUCAS, G. (1955b) Caractères géochimiques et méchaniques du milieu generateur des calcaires noduleux à faciès *ammonitico rosso. C. r. hebd. séanc. Acad. Sci., Paris,* **240,** 2000–2002.

LUCAS, G. (1955c) Signification paléocéanique des calcaires noduleux à faciès *ammonitico rosso. C. r. hebd. séanc. Acad. Sci., Paris,* **240,** 2342–2344.

MARLOWE, J.I. (1971) High-magnesian calcite cement in calcarenite from Aves Swell, Caribbean Sea. In: *Carbonate Cements* (Ed. by O. P. Bricker), pp. 111–115. *Studies in Geology No.* 19. Johns Hopkins University, Baltimore, Maryland.

MAXWELL, A.E., VON HERZEN, R.P., HSÜ, K.J., ANDREWS, J.E., SAITO, T., PERCIVAL, S.F., MILOW, E.D. & BOYCE, R.E. (1970) Deep-Sea drilling in the South Atlantic. *Science,* **168,** 1047–1059.

McIntyre, A. & McIntyre, R. (1971) Coccolith concentrations and differential solution in oceanic sediments. In: *The Micropalaeontology of Oceans* (Ed. by B. M. Funnel and W. R. Riedel), pp. 253–261. Cambridge University Press, London.

Milliman, J.D. (1966) Submarine lithification of carbonate sediments. *Science*, **153,** 994–997.

Milliman, J.D. (1971) Examples of lithification in the deep sea. In: *Carbonate Cements* (Ed. by O. P. Bricker), pp. 95–102, *Studies in Geology No.* 19. Johns Hopkins University, Baltimore, Maryland.

Milliman, J.D., Ross, D.A. & Ku, T.L. (1969) Precipitation and lithification of deep-sea carbonates in the Red Sea. *J. sedim. Petrol.* **39,** 724–736.

Mišík, M. (1964) Lithofazielles Studium des Lias des Grossen Fatra und des westlichen Teils der Niederen Tatra. *Sb. Geol. Vied, Zapadne Karpaty, RAD ZK,* **1,** 9–92.

Morita, R.Y. & Zobell, C.E. (1955) Occurrence of bacteria in pelagic sediments collected during the Mid-Pacific expedition. *Deep Sea Res.* **3,** 66–73.

Müller, J. & Fabricius, F. (1974) Magnesian-calcite nodules in the Ionian deep sea—an actualistic model for the formation of some nodular limestones. In: *Pelagic Sediments: on Land and under the Sea* (Ed. K. J. Hsü and H. C. Jenkyns). *Spec. Publs int. Ass. sediment.* **1,** 235–247.

Murray, J. & Hjort, J. (1912) *The Depths of the Ocean,* pp. 821. Macmillan, London.

Murray, J. & Renard, A.F. (1891) Report on deep-sea deposits based on the specimens collected during the voyage of H.M.S. *Challenger* in the years 1872–1876. In: *'Challenger Reports'*, pp. 525. H.M.S.O, Edinburgh.

Pantin, H.M. (1958) Rate of formation of a diagenetic calcareous concretion. *J. sedim. Petrol.* **28,** 366–371.

Pantin, H.M. (1965) The effect of adsorption on the attainment of physical and chemical equilibrium in sediments. *N.Z. Jl Geol. Geophys.* **8,** 453–464.

Park, W.C. & Schot, E.H. (1968) Stylolites: their nature and origin. *J. sedim. Petrol.* **38,** 175–191.

Pimm, A.C., Garrison, R.E. & Boyce, R.E. (1971) Sedimentology synthesis: lithology, chemistry and physical properties of sediments in the northwestern Pacific Ocean. In: *Initial Reports Deep Sea Drilling Project*, Vol. VI (A. G. Fischer *et al.*), pp. 1131–1252. U.S. Government Printing Office, Washington.

Presley, B.J. & Kaplan, I.R. (1968) Changes in dissolved sulfate, calcium and carbonate from interstitial water of near shore sediments. *Geochim. cosmochim. Acta,* **32,** 1037–1048.

Pytkowicz, R.M. (1965) Rates of inorganic calcium carbonate nucleation. *J. Geol.* **73,** 196–199.

Revelle, R.R. (1944) Marine bottom samples collected in the Pacific Ocean by the *Carnegie* on its seventh cruise. *Publs. Carnegie Instn.* **556,** 1–180.

Rex, R.W. & Goldberg, E.D. (1958) Quartz contents of pelagic sediments of the Pacific Ocean. *Tellus,* **10,** 153–159.

Schlager, M. (1961) Bericht 1960 über geologische Arbeiten auf Blatt Strasswalchen (64). *Verh. geol. Bundesanst., Wien,* **1961,** A61–A67.

Schlager, M. (1966) Bericht 1965 über geologische Arbeiten auf den Blattern Berchtesgaden (93) und Hallein (94). *Verh. geol. Bundesanst., Wien,* **1966,** A50–A54.

Schlager, M. (1970) Bericht 1969 über geologische Arbeiten auf Blatt Hallein (94). *Verh. geol. Bundesanst., Wien,* **1970,** A52–A59.

Schlager, W. (1974) Preservation of cephalopod skeletons and carbonate dissolution on ancient Tethyan sea floors. In: *Pelagic Sediments:on Land and under the Sea* (Ed. by J. K. Hsü and H. C. Jenkyns). *Spec. Publs int. Ass. Sediment.* **1,** 49–70.

Schmalz, R.F. (1967) Kinetics and diagenesis of carbonate sediments. *J. sedim. Petrol.* **37,** 60–67.

Shishkina, D.V. (1964) Chemical composition of pore solutions in oceanic sediments. *Geochem. Int.* **3,** 522–528.

Siever, R. (1957) The silica budget in the sedimentary cycle. *Am. Miner.* **42,** 821–841.

Siever, R., Beck, K.C. & Berner, R.A. (1965) Composition of interstitial waters of modern sediments. *J. Geol.* **73,** 39–73.

Sturani, C. (1971) Ammonites and stratigraphy of the 'Posidonia alpina' beds of the Venetian Alps (Middle Jurassic, mainly Bajocian) *Memorie Ist. geol. miner. Univ. Padova,* **28,** 1–190.

Suess, E. (1970) Interaction of organic compounds with calcium carbonate. 1. Association phenomena and geochemical implications. *Geochim. cosmochim. Acta,* **34,** 157–168.

Sujkowski, Z.L. (1932) Radiolarites des Karpates Polonaises Orientales et leur comparaison avec les radiolarites de la Tatra. *Biul. panst. Inst. geol.* **7,** 97–168.

Sujkowski, Z.L. (1958) Diagenesis. *Bull. Am. Ass. Petrol. Geol.* **42,** 2692–2717.

SZULCZEWSKI, M. (1965) Observation sur la genèse des calcaires noduleux des Tatras. *Roczn. pol. Tow. geol.* **35,** 243–261.

TUREKIAN, K.K. (1965) Some aspects of the geochemistry of marine sediments. In: *Chemical Oceanography* (Ed. by J. P. Riley and G. Skirrow), pp. 81–126. Academic Press, London.

TRURNIT, P. (1967) Morphologie und Entstehung diagenetischer Druck-Lösungserscheinungen. *Geol. Mitt.* **7,** 173–204.

WENDT, J. (1963) Stratigraphisch-paläontologische Untersuchungen im Dogger Westsiziliens. *Boll. Soc. paleont. ital.* **2,** 57–147.

WENDT, J. (1965) Synsedimentäre Bruchtektonik im Jura Westsiziliens. *Neues Jb. Geol. Paläont. Mh.* **1965,** 266–311.

WENDT, J. (1969a) Die stratigraphisch-paläogeographische Entwicklung des Jura in Westsizilien. *Geol. Rdsch.* **58,** 735–755.

WENDT, (1969b) Stratigraphie und Paläogeographie des Roten Jurakalks im Sonnwendgebirge (Tirol, Österreich). *Neues Jb. Geol. Paläont. Abh.* **132,** 219–238.

WENDT, J. (1971) Die Typlocalität der Adneter Schichten (Lias, Osterreich). In: *Colloque du Jurassique méditerranéen* (Ed. by E. Végh-Neubrandt). *Annls Inst. geol. publ. hung.* **54/2,** 105–116.

WEYL, P.K. (1967) The solution behaviour of carbonate materials in sea water. *Stud. trop. Oceanogr. Univ. Miami,* **5,** 178–228.

WHITEHOUSE, U.G. & MCCARTER, R.S. (1958) Diagenetic modification of clay minerals in artificial sea water. *Clays and Clay Minerals, 5th Natn. Conf. Clays. Publs. nat. Acad. Sci.* **566,** 81–119. National Research Council.

WISE, S.W., JR & HSÜ, K.J. (1971) Genesis and lithification of a deep-sea chalk. *Eclog. geol. Helv.* **64,** 273–278.

ZOBELL, C.E. & MORITA, R.Y. (1957) Barophilic bacteria in some deep-sea sediments. *J. Bacteriol.* **73,** 563–568.

Spec. Publs int. Ass. Sediment. (1974) **1,** 273–299

Deposition and diagenesis of silica in marine sediments

S. E. CALVERT

Institute of Oceanographic Sciences, Wormley, Godalming, Surrey

ABSTRACT

Silica accumulates in observable quantities in marine sediments as planktonic debris (opal) in areas of high primary production and as ash and rock-alteration products in areas of extensive submarine volcanism. Diagenetic transformation of the silica leads to (a) solution and equilibration of opal and incipient precipitation of silicates in Recent sediments, and (b) the formation of chert in pre-Recent sediments. Rocks described as chert recovered during the Deep Sea Drilling Project are composed of cristobalite (=porcelanites) and quartz (=cherts *sensu stricto*). Cristobalitic porcelanites are also known in Mesozoic and Cenozoic siliceous formations on land. The cristobalite is highly disordered (=lussatite) and is formed by a solution step. The persistence of such a metastable phase in Mesozoic and Cenozoic rocks, although it does eventually recrystallize to quartz, is promoted by low ionic mobilities and/or low-energy conditions. The cristobalite precipitates from silica derived from biogenous opal and possibly also from volcanic debris. This conclusion is supported by the composition of the sediments associated with the porcelanites although the evidence is equivocal. More definitive criteria for distinguishing between these sources of silica are required.

INTRODUCTION

The marine geochemistry of silicon, with respect to its supply, circulation, deposition and diagenesis, has received a great deal of interest recently following the work of Sillén (1961), Garrels (1965) and Siever (1968), and as a result of some of the discoveries of the Deep Sea Drilling Project (DSDP) (Calvert, 1971a,b; Heath & Moberly, 1971; Heath, 1973). Information on the behaviour of silicon in the ocean is not as plentiful as that available for calcium (Turekian, 1965; Olausson, 1967), an element of similar biochemical importance in the marine production cycle. This may be attributed to the great palaeontological utility of calcareous microfossils and a comparative lack of research on siliceous microfossils, combined with the much more extensive data on the $CaCO_3$ content of pelagic sediments, due in part to the relative ease with which this component can be determined compared with that of biogenous SiO_2.

The complex feed-back process which regulates the composition of the oceans and atmosphere over geological time through the genesis and/or reconstitution of aluminosilicates, in which silicon plays a key role, has been forcefully argued by

Mackenzie & Garrels (1966a,b) and Garrels & Mackenzie (1971, 1974). The importance of biological and oceanographic factors in controlling the silicon balance in the modern ocean has, on the other hand, been stressed by Harriss (1966), Calvert (1968), Grill (1970) and Heath (1974). The relative importance of these two processes is still under debate (Edmond, 1973; Burton & Liss, 1973).

The main masses of silicon in Recent marine sediments are present as detrital aluminosilicates and the skeletons and tests of micro-organisms. This latter category represents a measurable reservoir of solid SiO_2 in the ocean whose distribution in space and time can unequivocally be related to known oceanographic factors. In addition, it is a significantly labile component of marine sediments, part of which is regenerated by way of the sediment pore water, and part transformed or recrystallized within oceanic sediments to produce cherts of a variety of compositions. Extensive and thickly bedded cherts, a class of puzzling rocks preserved on the continents, are cogent testimony to the extent of silicon accumulation under marine conditions in the past.

Non-biological extraction and deposition of silicon in the ocean, by means of reactions between dissolved silicon and degraded aluminosilicates, referred to above, and as a consequence of volcanic activity within the oceans, may also be important. The budget for dissolved silicon in the ocean has been re-examined by Burton & Liss (1973) who concluded that there is a large excess of input over removal and that the excess must be balanced by inorganic reactions of the type proposed by Sillén (1961) and also perhaps by Harder (1965). Although this process probably does control the composition of seawater over geological time periods, its immediate effect is evidently too small for the reaction products to be directly identified.

The deposition of silicon from volcanic sources is brought about almost entirely by the sedimentation of pyroclastic debris. Quantitative estimates of the amount of silicon introduced into marine sediments in this manner are not available but certain areas of the ocean are supposedly sites of extensive volcanic sediment supply, for example the central south Pacific (Menard, 1964). In addition, evidence of substantial volcanic input into some oceanic areas comes from mineralogical data on marine sediments and volcanic alteration products (Peterson & Goldberg, 1962; Peterson & Griffin, 1964; Nayudu, 1962; Bonatti, 1965). The diagenetic alteration of volcanic ash may consequently be an important secondary source of silicon in marine sediments.

This paper will review the mechanisms of extraction and deposition of silicon in the ocean, principally by biological agencies, and the diagenesis of this material over relatively recent geological time periods. I will emphasize oceanographic aspects of the problem, but will not discuss the interesting problem of the budget of dissolved silicon. It will also be clear that I draw heavily upon the results of the Deep Sea Drilling Project in discussing the diagenesis of silicon in marine sediments.

DISSOLVED SILICON IN THE OCEAN

Silicon is present in natural waters predominantly as the undissociated monomeric acid (Bruevich, 1953; Sillén, 1961). It is referred to as silicate or reactive silicate in most oceanographic reports and as silica in many geochemical publications. For the purpose of this review, it will be convenient to refer to the dissolved form as *dissolved silicon* and the solid phases (mainly skeletal) as *silica*.

The concentration of dissolved silicon in open oceanic water is everywhere less than about 170 μmol/l (~10 ppm SiO_2). The vertical distribution of dissolved silicon (Fig. 1) characteristically shows a surface minimum and increasing concentrations with depth. Such curves are produced by extraction of dissolved silicon in surface waters by phytoplankton and regeneration of silicon at deeper levels. The actual

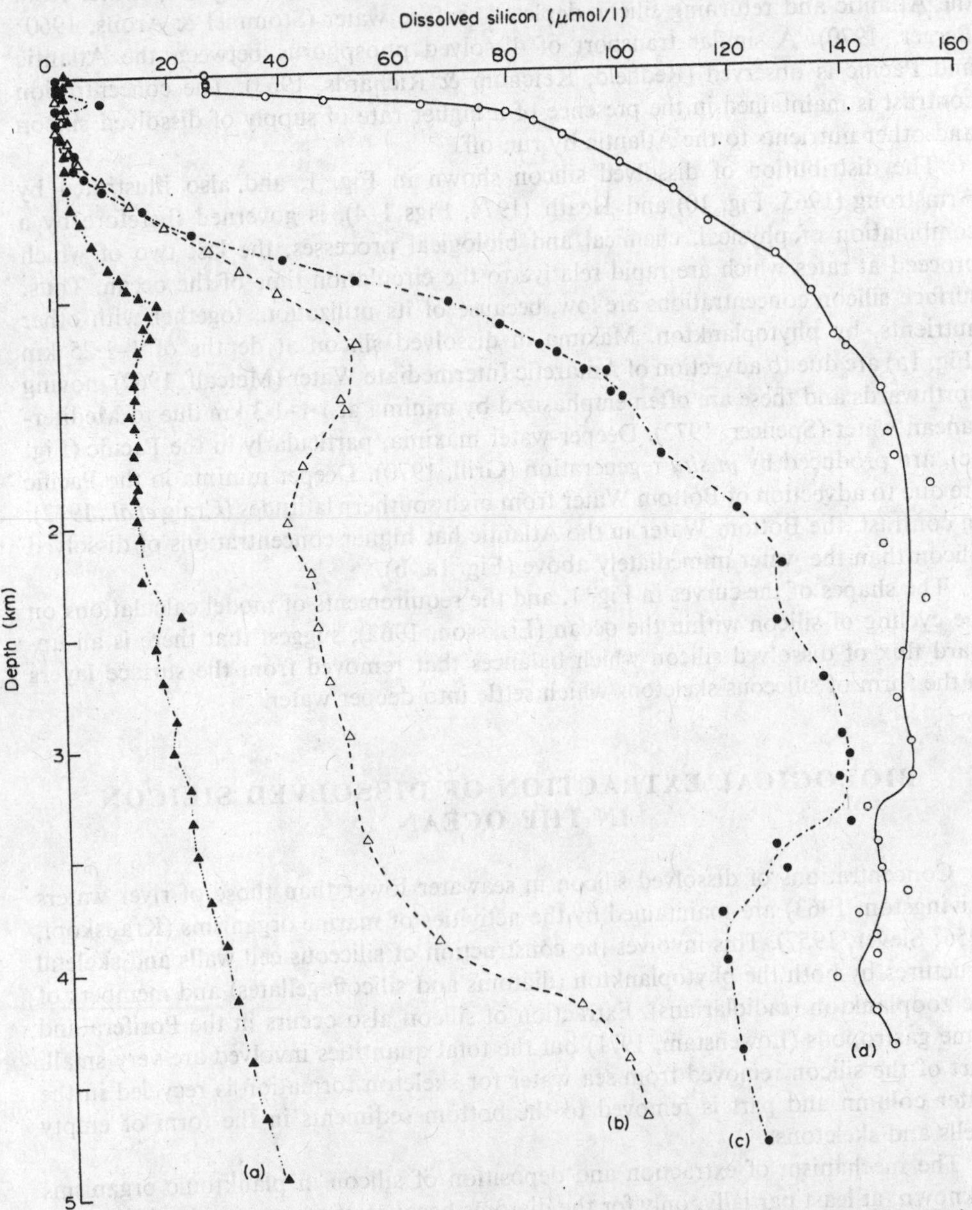

Fig. 1. Vertical distribution of dissolved silicon in the Atlantic and Pacific Oceans. (a) North Atlantic, Geosecs II, R/V Knorr Cruise 9, 35° 48′ N, 67° 59′ W. (b) South Atlantic, Circe (Scripps Institution of Oceanography) Station 133, 28° 31′ S, 7° 34′ E. (c) South Pacific, Antipode 15 (Scripps Institution of Oceanography) Station 10, 17° 58′ S, 172° 01′ W. (d) North Pacific, Zetes (Scripps Institution of Oceanography) Station 27, 52° 22′ N, 154° 56′ W.

concentrations of dissolved silicon in deep water depend upon location, as shown in Fig. 1, and may or may not show mid-depth maxima depending on regeneration or extraction of silicon in the water column (Grill, 1970) and advection of deep water having different dissolved silicon concentrations (Craig, Chung & Fiadeiro, 1972).

The marked differences in the concentrations of silicon in the Atlantic and Pacific Oceans is explained by the abyssal circulation, the Pacific gaining deep water from the Atlantic and returning silicon-depleted surface water (Stommel & Arons, 1960; Berger, 1970). A similar transport of dissolved phosphorus between the Atlantic and Pacific is observed (Redfield, Ketchum & Richards, 1963). The concentration contrast is maintained in the presence of a higher rate of supply of dissolved silicon and other nutrients to the Atlantic by run-off.

The distribution of dissolved silicon shown in Fig. 1, and also illustrated by Armstrong (1965, Fig. 10) and Heath (1974, Figs 1–4), is governed therefore by a combination of physical, chemical and biological processes, the last two of which proceed at rates which are rapid relative to the circulation time of the ocean. Thus, surface silicon concentrations are low, because of its utilization, together with other nutrients, by phytoplankton. Maxima in dissolved silicon at depths of 1–1·25 km (Fig. 1a) are due to advection of Antarctic Intermediate Water (Metcalf, 1969) moving northwards and these are often emphasized by minima at 1·1–1·3 km due to Mediterranean water (Spencer, 1972). Deeper-water maxima, particularly in the Pacific (Fig. 1c), are produced by *in situ* regeneration (Grill, 1970). Deeper minima in the Pacific are due to advection of Bottom Water from high southern latitudes (Craig *et al.*, 1972). In contrast, the Bottom Water in the Atlantic has higher concentrations of dissolved silicon than the water immediately above (Fig. 1a, b).

The shapes of the curves in Fig. 1, and the requirements of model calculations on the cycling of silicon within the ocean (Eriksson, 1962), suggest that there is an upward flux of dissolved silicon which balances that removed from the surface layers in the form of siliceous skeletons which settle into deeper water.

BIOLOGICAL EXTRACTION OF DISSOLVED SILICON IN THE OCEAN

Concentrations of dissolved silicon in seawater lower than those of river waters (Livingston, 1963) are maintained by the activities of marine organisms (Krauskopf, 1956; Siever, 1957). This involves the construction of siliceous cell walls and skeletal structures by both the phytoplankton (diatoms and silicoflagellates) and members of the zooplankton (radiolarians). Extraction of silicon also occurs in the Porifera and some gastropods (Lowenstam, 1971) but the total quantities involved are very small. Part of the silicon removed from sea water for skeleton formation is recycled in the water column and part is removed to the bottom sediments in the form of empty shells and skeletons.

The mechanism of extraction and deposition of silicon in planktonic organisms is known, at least partially, only for the diatoms because of the ease with which some species can be cultured in the laboratory. Thus it is well established from a great deal of work on the physiology and biochemistry of diatoms by Lewin (1954, 1955a,b, 1957, 1962), Coombs, Spanis & Volcani (1967a), Coombs *et al.* (1967b,c,d), Coombs & Volcani (1968), Darley & Volcani (1969), Healey, Coombs & Volcani (1967) and on the ultrafine structure of the silica shells of newly divided diatoms by Reimann (1964), Reimann, Lewin & Volcani (1965, 1966), Stoermer & Pankratz (1964) and Stoermer, Pankratz & Drum (1964, 1965) that silicon is a fundamental constituent of the diatom cell wall and is an essential requirement for cell division to take place. Lack of silicon in the culture medium leads to lower rates of synthesis of nucleic acids, proteins, carbohydrates, chlorophylls and fucoxanthin and lower rates of oxygen evolution, CO_2 fixation and phosphorus uptake. Moreover, during silicon starvation, DNA synthesis is blocked and this is expressed by an inhibition of mitosis. Silicon therefore plays a fundamental role in the metabolism of the diatom cell as well as being required for cell-wall construction. Silicon in the form of the monomeric undissociated acid is the only form that can be utilized, simple substituted silicic acids (e.g. $H_3SiO_3CH_3$) being inactive (Lewin, 1962). Furthermore, silicon cannot be replaced by any element of similar physical or chemical properties.

The degree of silicification of diatom cells varies over rather wide limits (Einsele & Grim, 1938), from less than 1% to approximately 50% of the dry weight (Vinogradov, 1953; Lewin, Lewin & Philpott, 1958). Moreover, within a given species, the amount of silicon in the cell varies according to the concentration of silicon in the growth medium and the rate of cell division. Rapidly dividing cells produce thinner walls (Lund, 1950; Jørgensen, 1955).

Provided sufficient nutrients, including phosphorus, nitrogen, trace elements, etc. are available for cell growth and division, diatoms in culture can deplete the silicon levels to vanishingly small concentrations (Jørgensen, 1953; Hughes & Lund, 1962). During this growth period, silicon uptake is constant throughout growing and dividing periods (Lewin, 1962). When a nutrient other than silicon is limiting, cell division ceases although silicon uptake may continue to produce thickened cell walls (Lewin, 1957).

Composition and structure of silica cell walls

The cell walls of diatoms are composed of hydrated amorphous SiO_2 which is here referred to as *opal*. This material is reportedly very pure (Lewin, 1962). However, a number of chemical analysis of opals from diatoms (Rogall, 1939; Desikachary, 1957) and from several species of mono- and dicotyledonous plants (Lanning Ponnaiya & Crumpton, 1958; Sterling, 1967; Jones & Milne, 1963) reveal highly variable compositions with K_2O, Na_2O, Fe_2O_3 and Al_2O_3 often prominent and SiO_2 contents of the ash frequently less than 100%. Many of these analyses are suspect because of inadequate precautions being taken during sampling and sample preparation to avoid contamination. Nevertheless, Martin & Knauer (1973) have shown that rather pure samples of the residues of siliceous plankton contain measurable amounts of Ti and Al, elements which would normally be considered indicators of aluminosilicate contamination. In view of the very stringent requirements of diatoms for silicon for metabolism and for constructing cell walls it is tempting to conclude that the shells are indeed very pure SiO_2 but this must await confirmation by an unequivocal microchemical assay.

The specific gravity of biogenous opal is 2·00–2·07 (Einsele & Grim, 1938; Hull *et al.*, 1953). High-resolution electron microscopy reveals a sponge-like wall microstructure (Helmcke, 1954), and a specific surface area of 123 $m^2\ g^{-1}$ for shells of *Navicula pelliculosa* indicates an extremely high porosity (Lewin, 1961).

X-ray diffraction spectra of diatom and radiolarian shells show a single, broad, low-intensity reflection centred at about 4 Å (Calvert, 1966; Mizutani, 1966; Kamatani, 1971) which is very similar to silica glasses and gels (Warren & Biscoe, 1938; Cartz, 1964) and precious opals (Jones, Sanders & Segnit, 1964). The structure may be considered a more or less three-dimensional random network of tetrahedral SiO_4 units, similar in fact to the structure of vitreous silica (Bell & Dean, 1972). Reports of crystalline phases in biogenous opals (Brandenberger & Frey-Wyssling, 1947; Lanning *et al.*, 1958; Sterling, 1967), including quartz, α-cristobalite and low-tridymite, are in all probability produced either by contamination or by sample preparation procedures (Jones & Milne, 1963).

ORGANIC PRODUCTION AND RECYCLING OF SILICON IN THE OCEAN

The principal areas of high primary organic production in the ocean are found in the subarctic, subantarctic and equatorial regions and in areas of coastal upwelling (Sverdrup, 1938) in the eastern boundary currents (Wooster & Reid, 1963). In these areas, nutrient concentrations at the surface are higher than they are elsewhere because of the supply of deeper, nutrient-rich water brought about by physical mixing processes. Around Antarctica, the effects of the Circumpolar Current, which in places reaches to the sea floor in abyssal depths (Gordon, 1971), and the intense vertical exchange of surface and deep water (Deacon, 1963), combine to produce very fertile surface water at all times of the year. Thus, concentrations of dissolved silicon (Bogoyavlenskiy, 1967) and phosphorus (Reid, 1962) in Antarctic waters are several times greater than those in surface waters elsewhere. Conditions are therefore adequate for high, sustained rates of production of phytoplankton wherever solar radiation is sufficiently intense. In the North Pacific, winter mixing together with a slow upwelling of deep water (Knauss, 1962) produces similar conditions to those found in the subantarctic region. Although the North Atlantic is a source of deep water (Reid & Lynn, 1971) winter mixing produces high nutrient concentrations in the surface waters. In the equatorial regions, divergence of surface waters, produced by the asymmetrical disposition of surface winds about the geographical equator, leads to extensive upwelling.

Along the western coasts of the continents, in moderately low latitudes, conditions for coastal upwelling are produced by the transport of surface water away from the coastline. Upwelling here is a seasonal phenomenon and the upwelled water originates from a few hundred metres depth (Sverdrup, 1938). The California, Peru, Berguela and Canary Currents are well-known examples of highly productive coastal upwelling regions.

Elsewhere in the ocean, wherever the stability of the upper part of the water column is sufficient to preclude vertical water exchange, production rates and plankton standing crops are very low. These areas coincide with the centres of the large oceanic gyres in the northern and southern hemispheres.

The regional variation in the extraction of dissolved silicon by diatoms, estimated from primary production rates and the SiO_2/C ratios in the suspended material in the water column (Lisitzin, 1964), is shown in Fig. 2. This closely reflects the regional variation in carbon fixation (Gessner, 1959, Plate 8) and the abundance of suspended

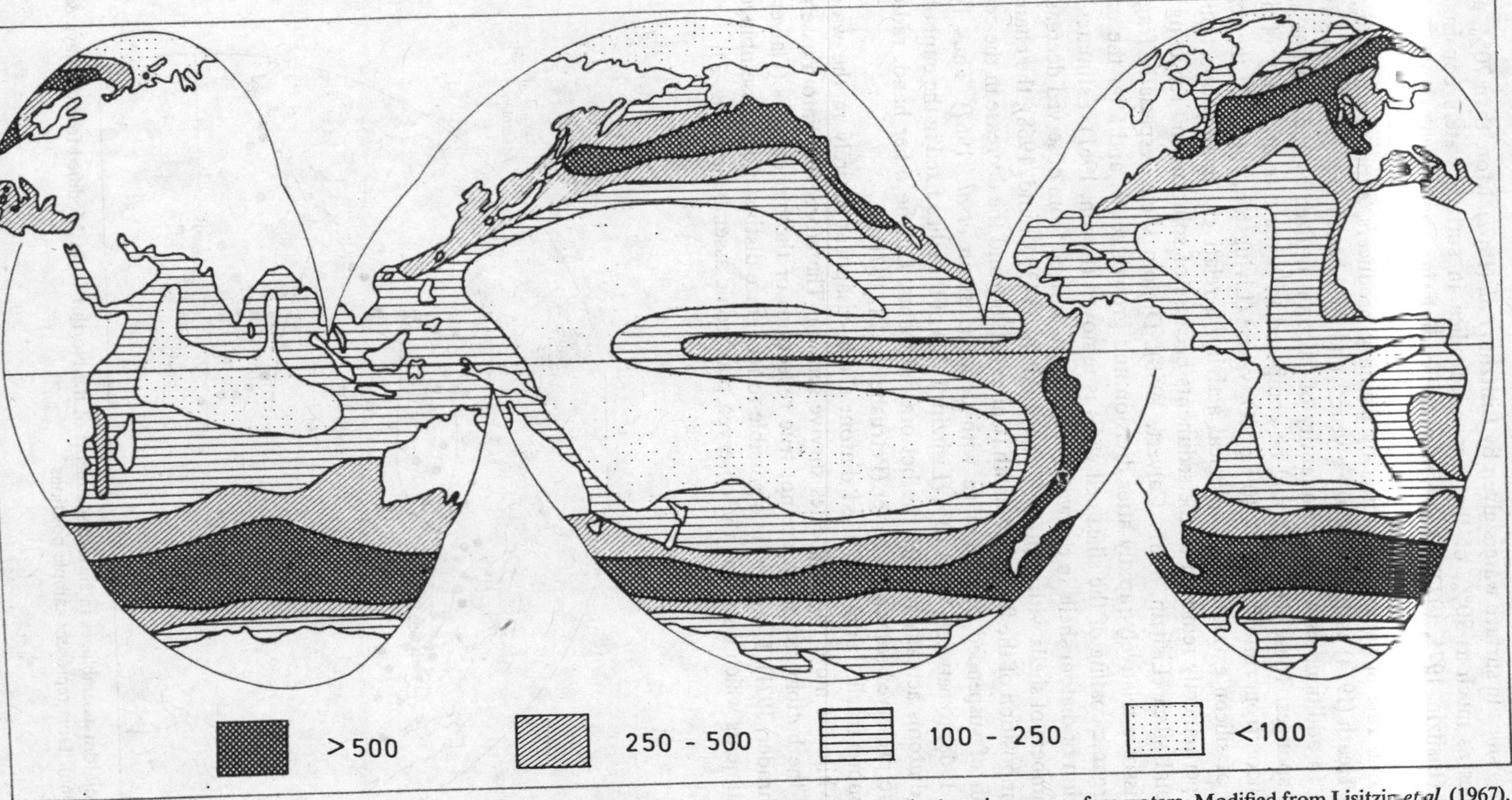

Fig. 2. Regional variation in the rate of extraction of dissolved silicon (g SiO_2 m^{-2} $year^{-1}$) by phytoplankton in near-surface waters. Modified from Lisitzin *et al.* (1967).

biogenous silica in surface waters given by Lisitzin *et al.* (1967). More than 70%, and sometimes as much as 90%, of the suspended silica in surface waters consists of diatoms (Lisitzin, 1971, 1972). In deeper water, radiolarian shells form a larger proportion of the total suspended material.

The total rate of silicon extraction by primary producers, expressed as SiO_2, is given by Heath (1974) as $1{\cdot}7$–$3{\cdot}2 \times 10^{16}$ g year^{-1} and by Lisitzin (1967) as 8–16 $\times 10^{16}$ g year^{-1}. In addition, an unknown quantity of silicon is utilized by radiolarians and siliceous sponges. These figures should be compared with the rate of supply of dissolved silicon from run-off of $4{\cdot}3 \times 10^{14}$ g SiO_2 year^{-1} (Livingston, 1963). Of the total quantity of silicon extracted by biological activity, a relatively small proportion is actually permanently removed to the sediments because of solution and recycling of the skeletal silica (Lisitzin, 1967; Calvert, 1968). Diatom shells, especially fragile forms, dissolve relatively rapidly after division and growth cease, and after the protective organic coating of the silica cell walls is removed (Lewin, 1961). Estimates of the solution of diatom shells in the water column are derived from observed decreases in the numbers of shells with depth (Gilbert & Allen, 1943; Round, 1968), the change in the composition of the flora with depth (Calvert, 1966) and the decrease in the concentration of suspended opaline silica with depth (Lisitzin *et al.*, 1967). Thus 1% (Calvert, 1968) to between 1 and 10% (Lisitzin, 1971) of the silica fixed in the euphotic zone by diatoms actually reaches the bottom sediments. On the other hand, radiolarians are preserved somewhat better (Petrushevskaya, 1971).

The more fragile, thinly silicified diatom shells disappear entirely in the water column while the more robust species survive settling. Therefore, the diatom assemblages in the bottom sediments are not true reflections of the biocoenoses (Calvert, 1966; Schrader, 1971). However, it is possible to recognize distinct floral assemblages in the sediments which do reflect the observed planktonic assemblages (Fig. 3).

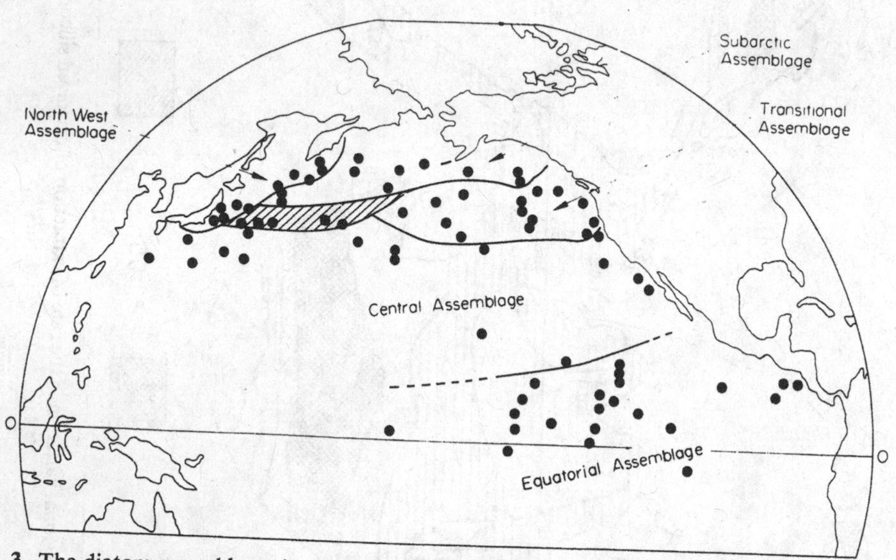

Fig. 3. The diatom assemblages in surface sediments of the North Pacific. Modified from Kanaya & Koizumi (1966). Dots represent sample positions.

Radiolarian faunas in bottom sediments are also quite different from live assemblages in the plankton, thickly silicified forms surviving settling more readily than thinner-shelled species (Petrushevskaya, 1971). Nevertheless, here too definite assemblages can be recognized in the sediments (Casey, 1971).

The fact that faunal and floral assemblages of siliceous plankton are preserved in bottom sediments in abyssal water depths implies that the settling of the shells is relatively rapid. The greatly decreased transit times are probably produced by aggregation of empty shells, most likely in faecal pellets (Riedel, 1963; Calvert, 1966; Schrader, 1971).

DISTRIBUTION OF BIOGENOUS OPAL IN MARINE SEDIMENTS

Several methods for determining opal in marine sediments have been developed using chemical (Aleksina, 1960; Hurd, 1973), X-ray diffraction (Goldberg, 1958; Calvert, 1966; Eisma & van der Gaast, 1971) and infrared absorption spectrophotometric methods (Chester & Elderfield, 1968). Unfortunately, a complete intercomparison of all methods has not been reported although it is known that X-ray and infrared methods agree (Chester & Elderfield, 1968, Table 4).

The concentration of opal in Recent sediments is shown in Fig. 4. This distribution is based on opal determinations using the chemical extraction method. The highest values are found in the deep-water areas around Antarctica in the Atlantic, Indian and Pacific Oceans. Approximately 80% of the total biogenous silica deposited in the oceans accumulates in this region and opal values in the sediments reach more than 70% by weight. The opal is overwhelmingly diatomaceous. In the northern Pacific, diatomaceous sediments occur in the Bering Sea and in the deep-water area south of the Aleutian Trench. Concentrations of opal are much lower than those around Antarctica because of dilution by terrigenous material, and seldom reach 30% by weight. In the equatorial Pacific, opal values are not usually greater than 10% by weight on a $CaCO_3$-free basis. These sediments are radiolarian. In the Atlantic Ocean, opaline sediments are not found in the northern hemisphere or along the equator, while in the Indian Ocean a few relatively small areas of equatorial radiolarian sediments are known.

In addition to the extensive, deep-water areas shown in Fig. 4, biogenous opal also accumulates in several near-shore areas at relatively high rates, for example Saanich Inlet, British Columbia (Gucluer & Gross, 1964), the Gulf of California (Calvert, 1966) and the south-west African continental shelf (Calvert & Price, 1971). In these areas, opal is present as diatom shells and concentrations reach 70% by weight of the total sediment.

The distribution of biogenous opal in marine sediments as described here is a direct reflection of the regional variation in the organic productivity of the oceans (Heath, 1974). This in turn is controlled by purely physical factors governing the stability, or lack thereof, of the near-surface waters. A detailed analysis of the oceanographic factors responsible for silicon deposition in the Gulf of California is given by Calvert (1966).

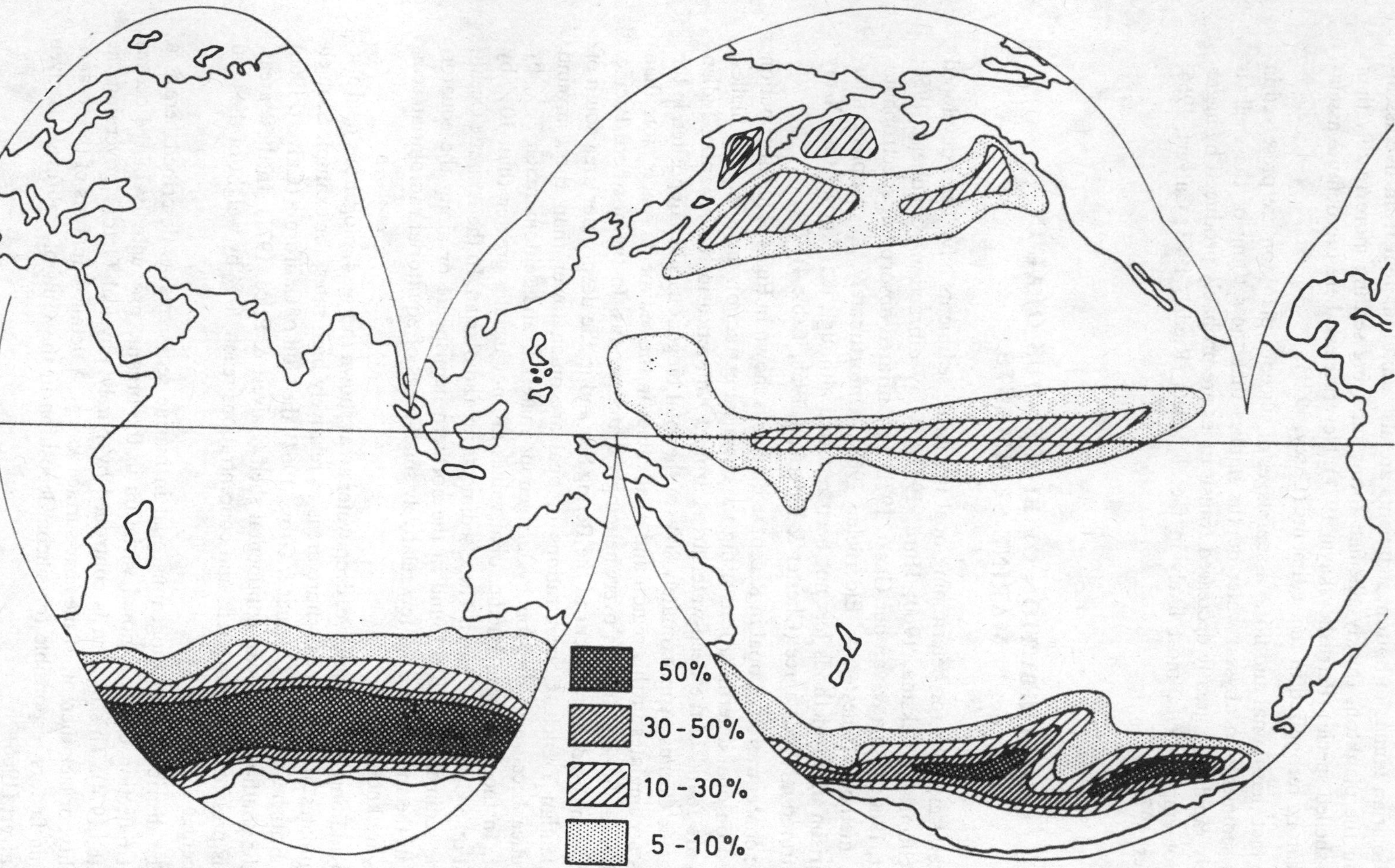

Fig. 4. Distribution and concentration of biogenous opal in surface sediments of the Pacific and Indian Oceans, on a $CaCO_3$-free basis. Modified from Lisitzin (1967).

DISSOLVED SILICON IN THE PORE WATERS OF MARINE SEDIMENTS

The concentration of dissolved silicon in the pore waters of marine sediments (Emery & Rittenberg, 1952; Bruevich, 1953; Siever, Beck & Berner, 1965; Brooks, Presley & Kaplan, 1968; Fanning & Schink, 1969; Bischoff & Ku, 1970, 1971; Fanning & Pilson, 1971; Bischoff & Sayles, 1972; Hurd, 1973) is higher than in the associated bottom water. The data are confused by problems associated with sample collection and treatment which can produce anomalously high values if the pore water is sampled at laboratory rather than *in situ* temperatures (Fanning & Pilson, 1971). However, this effect does not entirely remove the concentration contrast between bottom and pore waters. For example, differences of between 120 and 350 μmol Si l^{-1} are found (Bischoff & Ku, 1971; Bischoff & Sayles, 1972; Hurd, 1973) with a sharp downward increase in concentration in the uppermost tens of centimetres of the sediment (Fig. 5). In relatively long sediment cores, silicon concentrations are observed to decrease with depth from a maximum to a relatively constant value (Fig. 5).

The higher concentrations of dissolved silicon in the pore waters of marine sediments are produced largely by the solution of siliceous planktonic debris. Thus, in addition to a considerable amount of solution of silica in the water column as discussed previously, further solution of silica shells reaching the sea floor reduces the amount of opal permanently buried in the sediments. The concentration levels reported are much less than equilibrium solubility levels for amorphous silica (Alexander,

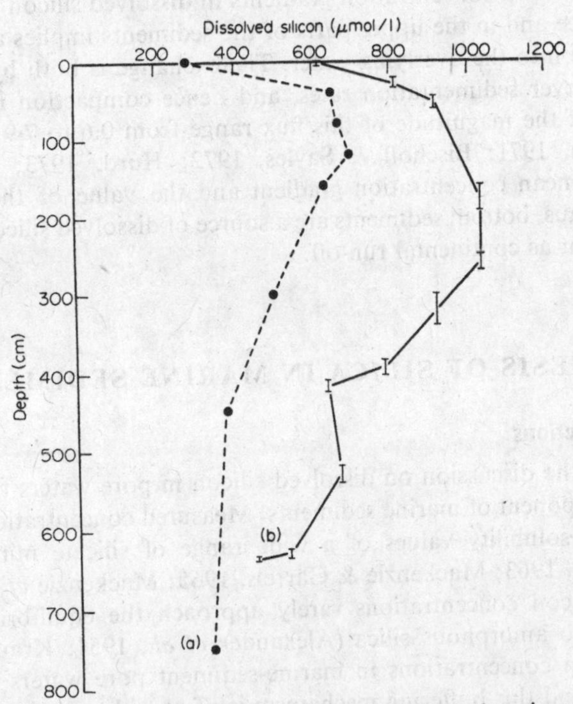

Fig. 5. Vertical distribution of dissolved silicon in the pore waters of marine sediments. (a) Eastern Pacific, Core 24 (Bischoff & Sayles, 1972). The pore water sampled at *in situ* temperature. (b) Northern Atlantic, Core 5 (Siever *et al.*, 1965). The pore water sampled at laboratory temperature.

Heston & Iler, 1954; Krauskopf, 1956) but about five to seven times the equilibrium level with respect to quartz (Siever, 1957).

Schink, Fanning & Pilson (1973) have pointed out that there is a regional variation in the concentration of dissolved silicon in the pore waters of Atlantic sediments. High concentrations are found in areas where primary production and sedimentation rates are high, while relatively low concentrations characterize areas of low production in the centres of the main open oceanic gyres. This supports a suggestion by Calvert (1968) that pore water silicon concentrations are controlled by the balance between the supply of siliceous and non-siliceous material.

Siliceous pyroclastic debris is also a possible source of dissolved silicon in sediment pore waters although the evidence is conflicting. Volcanic ash has been offered as a source of silica for the silicification of wood (Murata, 1940). In a series of relatively low-temperature experiments involving albite and its decomposition products, Hemley, Meyer & Richter (1961) observed that montmorillonite and cristobalite form spontaneously from the feldspar. However, the experimental evidence of Heath & Jones (in Heath, 1974), suggests that volcanic ash, regardless of composition, is much less soluble in seawater than biogenous opal. Scheidegger (1973) reports, moreover, that volcanic glass shards and associated plagioclase crystals in sediments up to $3{\cdot}8 \times 10^6$ years old from the north-east Pacific appear to be completely unweathered. The shards lack hydration rinds and are not visibly devitrified. They are not unstable in marine sediments as is commonly assumed. On this evidence the amount of dissolved silicon supplied to sediment pore waters, at least over relatively recent geological time periods, appears to be minor.

The presence of sharp concentration gradients in dissolved silicon across the sediment/water interface and in the upper parts of the sediments implies a flux of silicon from the sediments into the overlying water. The exchange is both by diffusion and by advection wherever sedimentation rates, and hence compaction rates, are high. Recent estimates of the magnitude of this flux range from 0·6 to $7{\cdot}9 \times 10^{14}$ g SiO_2^{-1} (Fanning & Pilson, 1971; Bischoff & Sayles, 1972; Hurd, 1973; Heath, 1974), depending on the mean concentration gradient and the value of the diffusion coefficient selected. Thus, bottom sediments are a source of dissolved silicon for seawater at least as important as continental run-off.

DIAGENESIS OF SILICA IN MARINE SEDIMENTS

Early diagenetic reactions

It is clear from the discussion on dissolved silicon in pore waters that biogenous silica is a labile component of marine sediments. Measured concentrations commonly exceed equilibrium solubility values of a wide range of silicate minerals (Keller, Balgord & Reesman, 1963; Mackenzie & Garrels, 1965; Mackenzie *et al.*, 1967). On the other hand, silicon concentrations rarely approach the equilibrium solubility value with respect to amorphous silica (Alexander *et al.*, 1954; Krauskopf, 1956). The dissolved silicon concentrations in marine-sediment pore waters are therefore effectively buffered and this buffering mechanism is of considerable interest.

The dissolved silicon concentration profiles shown in Fig. 5 imply a change in the rate of solution of biogenous opal with depth in the sediment. Bischoff & Sayles (1972) and Hurd (1973) have examined a number of possible mechanisms involved. In the first place, if the crystallinity of the opal increased with age, lower equilibrium solubility levels of silica would lead to progressively lower dissolved silicon concentrations with burial. Differences in the crystallinity of Recent and fossil diatoms were reported by Rogall (1939) and it is known that the recrystallization of fossil radiolarians by heating is more difficult than that of modern forms (Heath, 1968). In addition, Heath & Jones (in Heath, 1974) have shown that Eocene radiolarians are much less soluble in seawater than amorphous silica.

The uptake of dissolved silicon released from the solution of biogenous opal by the formation of authigenic silicate minerals is a further intriguing possibility. Bischoff & Sayles (1972) and Hurd (1973) have pointed out that the concentrations of dissolved silicon in many sediment pore waters, using correct sampling temperatures, are quite close to the saturation value for equilibrium with sepiolite. Moreover, Hurd (1973) has observed that in fact sodic plagioclase, kaolinite, orthoclase, Mg and Na montmorillonite, leonhardite and talc would also be at equilibrium with the dissolved silicon concentrations observed in equatorial Pacific sediments. Therefore, reactions involving silicate synthesis in the presence of biogenous silica may be controlling dissolved silica concentrations in marine sediments by establishing new equilibria and by the reduction of opal solution rates by the presence of cations, such as Al, Fe and Ti, as observed by Lewin (1961).

Hurd (1973) has recently discovered some extremely fine-grained crystalline phases on the surfaces of radiolarian shells from some abyssal equatorial Pacific sediments. These have been interpreted as actively growing minerals on the opal surface which have the effect of extracting silicon from the pore water and at the same time reducing the solution rate of the opal. The identity of this material is of some considerable interest.

Recrystallization of biogenous silica

The recrystallization of opal and the formation of cherts and porcelanites in marine sediments is a reaction which is well documented by mineralogical and petrographic data obtained by the Deep Sea Drilling Project (Calvert, 1971a,b; Heath & Moberly, 1971; von der Borch, Gatehouse & Nesteroff, 1971; Pimm, Garrison & Boyce, 1971; von Rad & Rosch, 1972; Heath, 1973). Previously, information on the mechanism of silica recrystallization and the formation of chert was available only from land sequences and this information was conflicting.

An indication of the nature of the recrystallization products of biogenous silica in marine sediments was provided by a sample of so-called chert recovered in a sediment core from the northern equatorial Pacific in 1962 during Lusiad Expedition (Station LSDH88) of the Scripps Institution of Oceanography. The material was a very fine-grained, pale brown, porcelanous claystone from a sequence of radiolarian oozes. It has been established that the chert is middle Eocene in age (Heath & Moberly, 1971) similar to many of the cherts recovered by the Deep Sea Drilling Project to date (Fig. 6). The rock consists almost entirely of poorly-ordered cristobalite (Fig. 7a) and is not a chert in the strict mineralogical sense. It appears to be a slightly more ordered variety of biogenous opal which shows a broad reflection at about 4 Å (2θ CuKα=22°). Rocks containing abundant poorly ordered cristobalite in unconsolidated sediments have since been collected by conventional coring techniques at

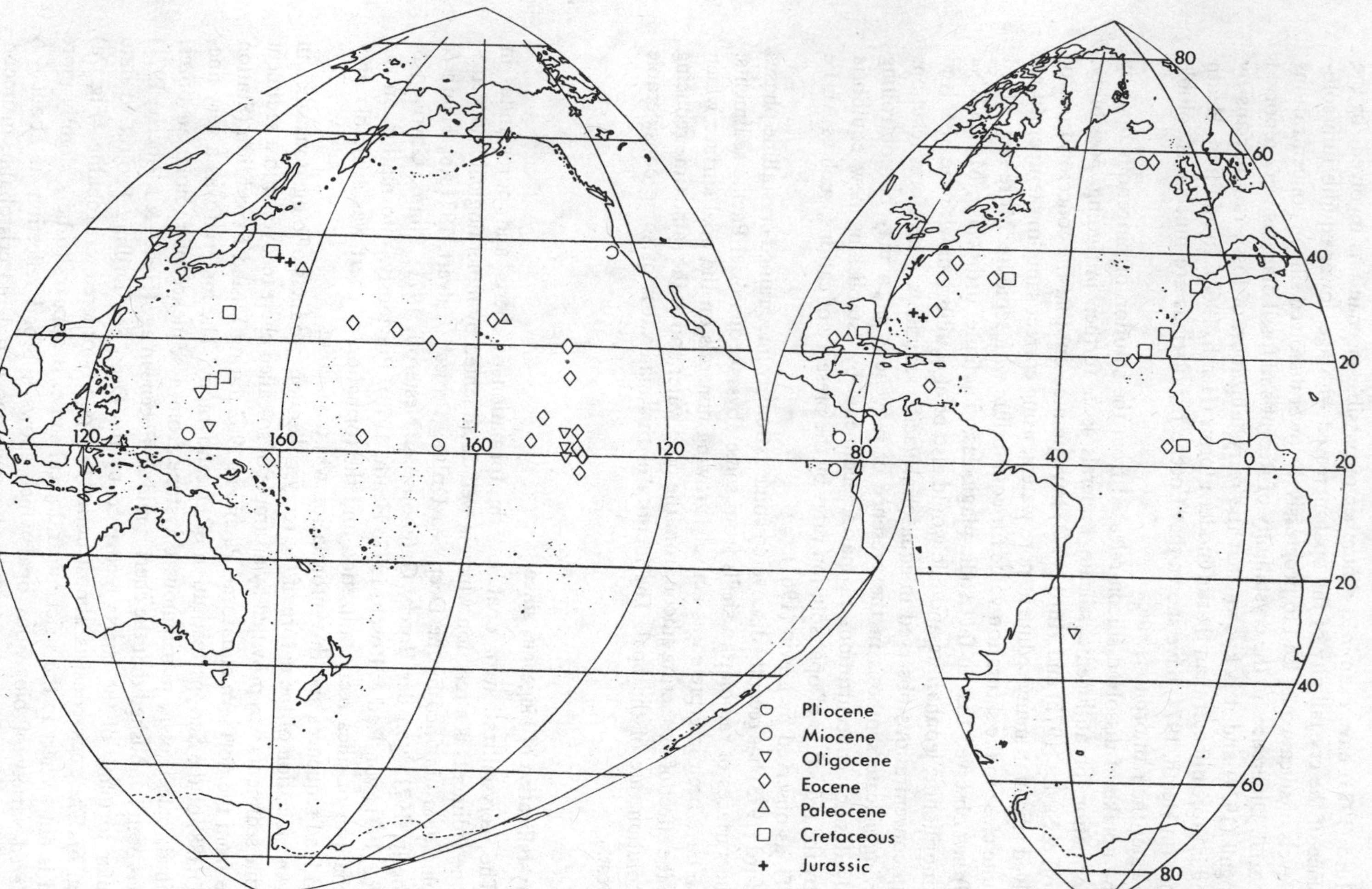

Fig. 6. Distribution and age of cherts and porcelanites collected by the Deep Sea Drilling Project, Legs 1–12, 14 and 16. Data from the Initial Reports.

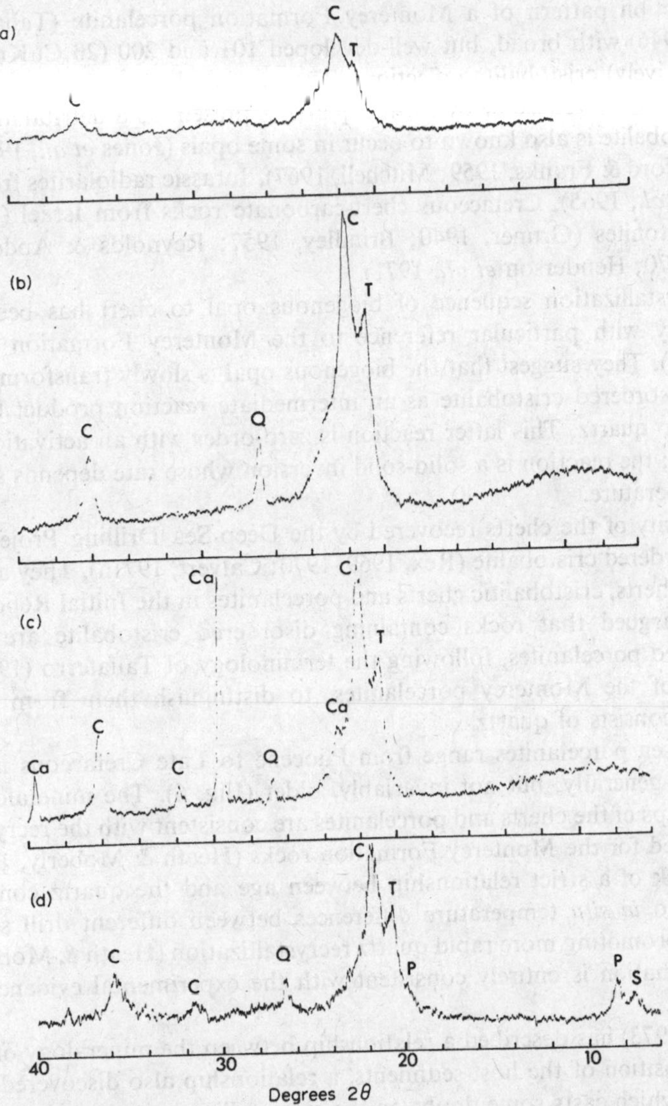

Fig. 7. X-ray diffraction patterns of porcelanites (a) Equatorial Pacific, Lusiad (Scripps Institution of Oceanography) Station 88. Age: Middle Eocene, Sample kindly made available by W. R. Riedel. (b) Monterey Formation, San Luis Obispo, California. Age: Miocene (see Ernst & Calvert, 1969). (c) Mishash Formation, Israel. Age: Upper Cretaceous (see Kolodny *et al.*, 1965). Sample kindly made available by Y. Kolodny. (d) Equatorial Atlantic, Deep Sea Drilling Project, Site 12B, Core 3, section 1, interval 119–120 cm (see Calvert, 1971b). Age: Eocene. C, cristobalite; T, tridymite; Q, quartz; Ca, calcite; P, palygorskite; S, sepiolite. Conditions: CuKα radiation, Ni filter, 36 kV, 20 mA, scintillation detector.

two other sites, in the central North Pacific (Heath & Moberly, 1971) and in the Southern Ocean (Wise, Buie & Weaver, 1972).

Subsequent to the discovery of cristobalitic rocks in Pacific sediments, it became apparent that disordered cristobalite was also widely distributed in the diatomaceous Monterey Formation of California (Ernst & Calvert, 1969). Figure 7b shows a typical

X-ray diffraction pattern of a Monterey Formation porcelanite (Taliaferro, 1934; Bramlette, 1946) with broad, but well-developed 101 and 200 (2θ CuKα = 21·7° and 36·0°, respectively) cristobalite reflections. The peak at 4·30 Å (2θ CuKα = 20·65°) is attributed to tridymite (see below) while quartz is present as a detrital impurity. Disordered cristobalite is also known to occur in some opals (Jones *et al.*, 1964), silicified wood (Swineford & Franks, 1959; Mitchell, 1967), Jurassic radiolarites from Hungary (Bardossy *et al.*, 1965), Cretaceous chert-carbonate rocks from Israel (Fig. 7c) and in many bentonites (Gruner, 1940; Brindley, 1957; Reynolds & Anderson, 1967; Reynolds, 1970; Henderson *et al.*, 1971).

The recrystallization sequence of biogenous opal to chert has been examined experimentally with particular reference to the Monterey Formation by Ernst & Calvert (1969). They suggest that the biogenous opal is slowly transformed into fine-grained or disordered cristobalite as an intermediate reaction product before being converted into quartz. This latter reaction is zero-order with an activation energy of 23 kcal mol^{-1}; the reaction is a solid-solid inversion whose rate depends solely on the reaction temperature.

The majority of the cherts recovered by the Deep Sea Drilling Project are composed of disordered cristobalite (Rex, 1969, 1970; Calvert, 1971a). They are variously described as cherts, cristobalitic cherts and porcelanites in the Initial Reports. Calvert (1917a) has argued that rocks containing disordered cristobalite are more conveniently called porcelanites, following the terminology of Taliaferro (1934) and the composition of the Monterey porcelanites, to distinguish them from chert *sensu stricto* which consists of quartz.

The deep-sea porcelanites range from Pliocene to Late Cretaceous in age while the cherts are generally, but not invariably, older (Fig. 6). The mineralogy and the age relationships of the cherts and porcelanites are consistent with the recrystallization scheme deduced for the Monterey Formation rocks (Heath & Moberly, 1971; Heath 1973). The lack of a strict relationship between age and the quartz content of the rocks is due to *in situ* temperature differences between different drill sites, higher temperatures promoting more rapid quartz recrystallization (Heath & Moberly, 1971). Such an explanation is entirely consistent with the experimental evidence (Ernst & Calvert, 1969).

Lancelot (1973) has described a relationship between the mineralogy of the cherts and the composition of the host sediments, a relationship also discovered by Greenwood (1973), which casts some doubt on the recrystallization scheme outlined above. Thus, cherts occurring in clayey sediments (zeolitic clays, clayey radiolarian oozes, marls and marly limestones) are composed of disordered cristobalite, while chert in carbonates consists predominantly of quartz. Furthermore, the cherts appear to be restricted to horizons where the sediments contain clinoptilolite rather than phillipsite, more K-feldspar than plagioclase and where palygorskite is the principal clay mineral. This association implies, according to Lancelot (1973), a more silicic environment for chert formation.

The lithological relationships are considered by Lancelot (1973) to be due to the preferential formation of a disordered cristobalite structure in environments where 'foreign cations' (see Flörke, 1955a) are abundant, the cations distorting the cristobalitic structure. Such cations would presumably be plentiful in clayey sediments but in low concentration in carbonate oozes where the silica polymorph precipitated would be quartz rather than cristobalite. Cherts containing cristobalite or quartz consequently range widely in age, from Tertiary to early Cretaceous, depending on the lithologies of the sediments and the availability of silica, and do not occur in an age sequence, quartzitic cherts being older than cristobalitic porcelanites.

Such an intriguing explanation for the different mineralogies of the deep-sea cherts requires confirmation in view of the reported compositions of cherts in the eastern Pacific (Heath, 1973) and the statement by Heath (in Greenwood, 1973) that many deep-sea cherts in calcareous sediments do in fact contain cristobalite. The explanation would not apply to the Monterey Formation cherts and porcelanites because both kinds of rocks occur in close association in diatomaceous sediments (Ernst & Calvert, 1969). However, the lithological control on the composition of chert would offer an explanation for the inhomogeneous distribution of porcelanite in clayey sequences recovered by the Deep Sea Drilling Project because of the low permeability of the sediments in restricting extensive migration of silica-rich interstitial solutions (Lancelot, 1973).

Nature of cristobalite in deep-sea porcelanite

The cristobalite in deep-sea porcelanite has a distinctive X-ray diffraction pattern, consisting of two rather broad reflections at 4·05 and 2·49 Å with a subsidiary reflection at 4·25–4·30 Å (Fig. 7). As pointed out previously, such patterns are also characteristic of a number of other low-temperature silica modifications. The complex of peaks in the 4·05–4·30 Å region have been identified in the past as reflections due to β-cristobalite, the high-temperature polymorph, in material formed at low temperatures (Greigg, 1932; Levin & Ott, 1933; Dwyer & Mellor, 1933; Buerger, 1951; Omure, Ookawara & Iwai, 1956; Buerger & Shoemaker, 1972).

The structure of cristobalite in opals, bentonites and glasses has been studied in considerable detail by Flörke (1955a,b, 1967). Accordingly, this phase consists of unidimensionally disordered low-temperature cristobalite; that is to say, it shows considerable stacking disorder which produces X-ray diffraction maxima attributable to tridymite and missing reflections of cristobalite at low angles. Flörke (1955a) refers to this material as *lussatite* (Laves, 1939; Braitsch, 1957), a usage which is further recommended by Jones & Segnit (1971). The extreme degree of stacking disorder in lussatite is confirmed by infrared absorption data which show spectra characteristic of more or less completely disordered low-cristobalite (Calvert, 1971a,b). Similar patterns in the long wave-length region are shown by common opals having X-ray diffraction patterns similar to those shown in Fig. 7 (Jones & Segnit, 1971).

Scanning electron microscopy reveals that the deep-sea porcelanites are composed of silica spheres, of the order of 10–15 μm diameter, that are themselves composed of fine bladed crystals, 300–500 Å thick (Weaver & Wise, 1972; Wise & Kelts, 1972). Microspheres of essentially similar morphology occur in Cretaceous silicified chalks from the Netherlands (Buurman & van der Plas, 1971). The bladed crystals in the porcelanites are identified as cristobalite by Weaver & Wise (1972) although the habit does not appear to be characteristic of the mineral.

Oehler (1973) has synthesized silica microspheres consisting of euhedral bladed crystals from silica gel which he refers to as tridymite rather than cristobalite. On the basis of their morphology, these crystals appear to be pseudohexagonal, similar to tridymite which commonly occurs as thin plates flattened on (0001). However, low-tridymite crystals, with which we are presumably dealing, invariably occur as inversion pseudomorphs after high tridymite (Frondel, 1962) so that the comparison is made with the high-temperature polymorph. Oehler (1973) has furthermore suggested by analogy with the synthetic product that the bladed crystals in the deep-sea porcelanites are hybrid crystals of cristobalite and tridymite with the morphology of tridymite dominating the structure.

Formation of lussatite

The formation of cristobalite at low temperatures takes place in the stability field of quartz and requires an explanation on energetic grounds.

The factors involved in the formation of opals and disordered cristobalite and tridymite at earth-surface conditions have been examined by Jones & Segnit (1972). They assume that low-temperature cristobalite and tridymite form by precipitation from hydrous solutions by polymerization of silicon molecules. This process unites SiO_4 tetrahedra with the elimination of water; the structure with the lowest energy so produced consists of six-membered unconstrained rings with the vertices of the tetrahedra pointing alternately up and down. Coalescence of the rings will produce extended sheets having the structure of both cristobalite and tridymite rather than quartz and it is not possible to produce a quartz structure without breaking Si-O bonds. Thus, as long as low-energy conditions persist or ionic mobilities are low, cristobalite-tridymite structures will form and may persist for geologically significant time periods.

In addition to these topological considerations, Jones & Segnit (1972) also suggest that of the two silica polymorphs produced under these conditions, cristobalite will be preferred over tridymite. This is because of the detailed arrangements of the silica tetrahedra in tridymite producing somewhat more energy in the rings.

This explanation accords with experimental work of Mizutani (1966) showing that cristobalite is the first phase formed under conditions which promote solution of amorphous silica. However, it is not consistent with the explanation of the formation of lussatite in deep-sea sediments by Lancelot (1973) which requires that the disordered structure is produced by the incorporation of cations in the precipitated silica phase.

Source of silica and the mode of formation of deep-sea porcelanites

The formation of deep-sea porcelanites by direct inorganic precipitation is suggested by Wise *et al.* (1972) on the basis of the micromorphology of the rocks discussed previously. This explanation is supported by the detailed arguments on the origin of lussatite by Jones & Segnit (1972).

The source of the silica in the deep-sea porcelanites and cherts is considered to be biogenous opal and volcanic debris. Heath & Moberly (1971), von der Borch *et al.* (1971), Wise *et al.* (1972) and Heath (1973) state that the porcelanites and cherts are formed by the diagenetic remobilization of opal, principally diatom and radiolarian debris. Such a mechanism is inferred by Peterson *et al.* (1970) on the basis of supposed sedimentary associations in drill cores from Leg 2 of the Deep Sea Drilling Project, and tacitly assumed by Ramsay (1971) and Herman (1972). Heath & Moberly (1971) observed that the cherts and porcelanites contain very little biogenous opal, but that the surrounding sediments often contain severely corroded siliceous shells. They inferred that the silica has migrated within the sediments over distances of the order of a few centimetres to metres.

Calvert (1971b) has pointed out that many of the porcelanites in the North Atlantic occur in sequences of zeolitic (clinoptilolite) and montmorillonitic clays (see also Heath, 1973) and also in sediments containing abundant sepiolite and palygorskite (Peterson *et al.*, 1970). Therefore, it is possible that volcanic debris or submarine hydrothermal activity, the presumed source or precursor of some of these associated minerals (Hay, 1966; Reynolds, 1970; Bowles *et al.*, 1971), are also sources of silica for porcelanite formation.

Griffin *et al.* (1972) have shown that Recent, rapidly accumulating sediments in the Lau Basin, western Pacific are composed almost entirely of volcanic debris and its alteration products. Among the diagenetic phases are abundant montmorillonite and disordered tridymite and cristobalite, the silica phases having reflections at about 2θ CuKα=21·6° and 21·9° (4·11 and 4·06 Å respectively) with the 4·11 Å peak the more intense. This material is most probably formed from the abundant volcanic debris present.

Clinoptilolite, montmorillonite and so-called low-cristobalite is a very common mineralogical association in altered pyroclastic deposits (Bramlette & Posnjak, 1933; Deffeyes, 1959; Hay, 1966; Reynolds & Anderson, 1967; Iijima & Utada, 1966; Reynolds, 1970). The cristobalite in these rocks appears to be identical with lussatite (Flörke, 1955a). X-ray diffraction patterns similar to those shown in Fig. 7 are illustrated by Reynolds (1970) from Tertiary pyroclastic claystones from Alabama, and Wermund & Moiola (1966) indicate that 'opal' in an Eocene quartzose sandstone from Mississippi and Alabama shows X-ray diffraction peaks at 4·07 and 4·29 Å which are characteristic of lussatite.

Gruner (1940) discounted a high-temperature volcanic origin for the cristobalite in bentonites but favoured its formation from the post-depositional alteration of volcanic ash, similar in fact to the origin of bentonitic montmorillonite. He further observed that the cristobalite is merely an intermediate stage in the alteration of ash, cristobalite recrystallizing to quartz with time. Thus, pre-Cretaceous bentonites do not contain cristobalite. The formation of cristobalite from the silica released from the alteration of volcanic ash and the concomitant formation of clinoptilolite, heulandite and montmorillonite is suggested by Reynolds (1970). The assemblage cristobalite–clinoptilolite–montmorillonite is also common in shallow-zone diagenetic products in the Neogene pyroclastic rocks of Japan (Iijima & Utada, 1966). The formation of cristobalite (lussatite) in many Tertiary volcanic sequences is therefore considered to be a result of the diagenetic redistribution of silica during the post-depositional alteration of volcanic debris.

Henderson *et al.* (1971, 1972) have been able to provide some more unequivocal evidence on this problem using the oxygen isotopic composition of the silica phases in bentonites and volcanic alteration products. Thus, cristobalite and quartz isolated from a Recent volcanic soil from New Zealand have a volcanic or hydrothermal origin. On the other hand, cristobalite from an Eocene bentonite from Texas has formed authigenically from the silica derived from biogenous opal and not from volcanic glass. They were also able to show that the cristobalite formed in meteoric waters whereas the associated montmorillonite formed in seawater. The application of this technique has therefore allowed some discrimination between the different reaction paths in volcanic debris and has shown that bentonitic cristobalite may have an origin both from biogenous and volcanically derived silica.

The formation of deep-sea porcelanites of the North Atlantic from volcanic debris has been specifically argued by Gibson & Towe (1971) and Mattson & Pessagno (1971). They suggested that Horizon A (Ewing *et al.*, 1966), which is correlated with the chert and porcelanite horizons in the North Atlantic (Gartner, 1970) and which is identified in drill cores as siliceous horizons, marks a widespread volcanic event in the Eocene with a source of pyroclastic debris in the Caribbean. Mattson & Pessagno (1971) further suggested that this horizon outcrops in Puerto Rico and the Dominican Republic as beds of green chert and altered vitric tuffs. Thus, the deposition of pyroclastic debris over the North Atlantic, derived from a Caribbean centre, is held to be the source of silica in the deep-sea cherts and porcelanites.

The siliceous rocks recovered in the North Atlantic by the Deep Sea Drilling Project are actually very variable in age and composition and have a wide range of lithological associations (Calvert, 1971b). They occur in nannoplankton chalks, marls, diatomaceous clays, radiolarian oozes, blue-green clays and clays containing zeolites, sepiolites and palygorskite. The porcelanites and cherts most probably had different modes of formation, including the solution and redeposition of silica from biogenous opal, by replacement of chalks, by the alteration of pyroclastic debris and by precipitation of material from hydrothermal sources.

At sites 8 and 9 in the North Atlantic, for instance, the porcelanites contain clinoptilolite and minor amounts of detrital plagioclase, quartz and mica, in addition to lussatite, and occur in unconsolidated sedimentary sections which have essentially the same composition (Rex, 1970). The porcelanites merely represent somewhat more silicified horizons of the cristobalite-zeolite sediments concerned (Calvert, 1971b). The same appears to be true for site 12 where the porcelanites occur in a thick sequence of volcanic ashes and clays containing abundant palygorskite and sepiolite and are themselves more coherent, silicified masses of the sedimentary section. On the basis of these compositions and associations, the porcelanites at these sites were probably formed from altered ash, or in the case of site 12, from hydrothermal solutions which were the sources of the ions necessary for the formation of the Mg silicates (see Bowles *et al.*, 1971).

Heath (1973) has recently argued that in spite of the mineralogical similarity of porcelanites and bentonites and altered volcanic debris on land, the silica in the porcelanites is in fact derived from biogenous silica. This is because of a complete absence of pseudomorphs or moulds of glass shards in the sections and the occasional presence of silicified siliceous ooze that is mineralogically indistinguishable from the porcelanites.

The source of silica in the deep-sea porcelanites and cherts is of some interest in determining the modes of supply and transformation of silica in marine sediments and in confirming the stratigraphic utility of cherty horizons in oceanic drill cores. Some more unequivocal criteria for establishing the origin of the silica is required, however, and the oxygen isotopic composition of the lussatite, following the work of Henderson *et al.* (1971, 1972), seems to be the most promising approach at the present time.

REFERENCES

Aleksina, I.A. (1960) The distribution of authigenic silica in mud deposits of the central part of the Caspian Sea. *Dokl. Akad. Nauk. S.S.S.R.* **130**, 624–626.

ALEXANDER, G.B., HESTON, W.M. & ILER, R.R. (1954) The solubility of amorphous silica in water. *J. Phys. Chem.* **58,** 453–455.

ARMSTRONG, F.A.J. (1965) Silicon. In: *Chemical Oceanography,* Vol. 1 (Ed. by J. P. Riley and G. Skirrow), pp. [illegible] Academic Press, London.

BARDOSSY, G., KONDA, I., RAPP-SHIK, S. & TOLNAI, V. (1965) Cristobalite in Bat-Kelloway radiolarites of the Bakony Mountains. In: *Problems of Geochemistry* (Ed. by N. I. Shitarov), pp. 564–587. Izdat. 'Nauka', Moscow. Israel Prog. Scientific Translations (1969).

BELL, R.J. & DEAN, P. (1972) The structure of vitreous silica: validity of the random network theory. *Phil. Mag.* **25,** 1381–1398.

BERGER, W.H. (1970) Biogenous deep-sea sediments: fractionation by deep-sea circulation. *Bull. geol. Soc. Am.* **81,** 1385–1401.

BISCHOFF, J.L. & KU, T.-L. (1970) Pore fluids of Recent marine sediments: I Oxidising sediments of 20° N, continental rise to Mid-Atlantic Ridge. *J. sedim. Petrol.* **40,** 960–972.

BISCHOFF, J.L. & KU, T.-L. (1971) Pore fluids of Recent marine sediments: II Anoxic sediments of 35° to 45° N. Gibraltar to Mid-Atlantic Ridge. *J. sedim. Petrol.* **41,** 1008–1017.

BISCHOFF, J.L. & SAYLES, F.L. (1972) Pore fluid and mineralogical studies of recent marine sediments: Bauer Depression of East Pacific Rise. *J. sedim. Petrol.* **42,** 711–724.

BOGOYAVLENSKIY, A.N. (1967) Distribution and migration of dissolved silica in the oceans. *Int. geol. Rev.* **9,** 133–153.

BONATTI, E. (1965) Palagonite, hyaloclastites and alteration of volcanic glass in the ocean. *Bull. volcan.* **28,** 3–15.

BOWLES, F.A., ANIGINO, E.E., HOSTERMAN, J.W. & GALLE, O.K. (1971) Precipitation of deep-sea palygorskite and sepiolite. *Earth Planet. Sci. Letts,* **11,** 324–332.

BRAITSCH, O. (1957) Über die natürlichen Faser-und Aggregationstypen beim SiO_2, ihre Verwachsungsformen, Richtungsstatistik und Doppelbrechung. *Heidelb. Beitr. Miner. Petrogr.* **5,** 331–372.

BRAMLETTE, M.N. (1946) The Monterey Formation of California and the origin of its siliceous rocks. *Prof. Pap. U.S. geol. Surv.* **213,** 57 pp.

BRAMLETTE, M.N. & POSNJAK, E. (1933) Zeolite alteration of pyroclastics. *Am. Mineral.* **18,** 167–171.

BRANDENBERGER, E. & FREY-WYSSLING, A. (1947) Über die Membransubstanzen von *Chlorochytridion tuberculatum W. Vischer. Experientia,* **3,** 492–493.

BRINDLEY, G.W. (1957) Fuller's earth from near Dry Branch, Georgia, a montmorillonite-cristobalite clay. *Clay Miner. Bull.* **3,** 167–169.

BROOKS, R.R., PRESLEY, B.J. & KAPLAN, I.R. (1968) Trace elements in the interstitial waters of marine sediments. *Geochim. cosmochim. Acta,* **32,** 397–414.

BRUEVICH, S.V. (1953) Concerning the geochemistry of silicon in the sea. *Izv. Akad. Nauk SSSR, Ser. Geol.* **4,** 67–69.

BUERGER, M.J. (1951) Crystallographic aspects of phase transformations. In: *Phase Transformations in Solids* (Ed. by R. Smoluchowski, J. E. Mayer and W. A. Weyl), pp. 183–209. J. Wiley, New York.

BUERGER, M.J. & SHOEMAKER, G.L. (1972) Thermal effect in opal below room temperature. *Proc. natn. Acad. Sci. U.S.A.* **69,** 3225–3227.

BURTON, J.D. & LISS, P.S. (1973) Processes of supply and removal of dissolved silicon in the oceans. *Geochim. cosmochim. Acta.* **37,** 1761–1773.

BUURMAN, P. & VAN DER PLAS, L. (1971) The genesis of Belgian and Dutch flints and cherts. *Geol. Mijnb.* **50,** 9–28.

CALVERT, S.E. (1966) Accumulation of diatomaceous silica in the sediments of the Gulf of California. *Bull. geol. Soc. Am.* **77,** 569–596.

CALVERT, S.E. (1968) Silica balance in the ocean and diagenesis. *Nature, Lond.* **219,** 919–920.

CALVERT, S.E. (1971a) Nature of silica phases in deep sea cherts of the North Atlantic. *Nature, Lond.* **234,** 133–134.

CALVERT, S.E. (1971b) Composition and origin of North Atlantic deep sea cherts. *Contr. Miner. Petrol.* **33,** 273–288.

CALVERT, S.E. & PRICE, N.B. (1971) Recent sediments of the South West African Shelf. In: *The Geology of the East Atlantic Continental Margin* (Ed. by F. M. Delany), pp. 173–185. Inst. Geol. Sci. Rept No. 70/16.

CARTZ, L. (1964) An X-ray diffraction study of the structure of silica glass. *Z. kristallogr. Miner.* **120,** 241–260.

CASEY, R.E. (1971) Radiolarians as indicators of past and present water masses. In: *The Micropalaeontology of Oceans* (Ed. by B. M. Funnell and W. R. Riedel), pp. 331–341. Cambridge University Press, London.

CHESTER, R. & ELDERFIELD, H. (1968) The infra-red determination of opal in siliceous deep-sea sediments. *Geochim. cosmochim. Acta,* **32,** 1128–1140.

COOMBS, J., SPANIS, C. & VOLCANI, B.E. (1967a) Studies on the biochemistry and fine structure of silica shell formation in diatoms. Photosynthesis and respiration in silicon-starvation synchrony of *Navicula pelliculosa. Plant Physiol.* **42,** 1607–1611.

COOMBS, J., DARLEY, W.M., HOLM-HANSEN, I. & VOLCANI, B.E. (1967b) Studies on the biochemistry and fine structure of silica shell formation in diatoms. Chemical composition of *Navicula pelliculosa* during silicon-starvation synchrony. *Plant Physiol.* **42,** 1601–1606.

COOMBS, J., HALICKI, P.J., HOLM-HANSEN, O. & VOLCANI, B.E. (1967c) Studies of the biochemistry and fine structure of silica shell formation in diatoms. Change in concentration of nucleoside triphosphates during synchronised division of *Cylindrotheca fusiformis* Reimann and Lewin. *Expl Cell. Res.* **47,** 302–314.

COOMBS, J., HALICKI, P.J., HOLM-HANSEN, O. & VOLCANI, B.E. (1967d) Studies on the biochemistry and fine structure of silica shell formation in diatoms. II Change in concentration of nucleoside triphosphates in silicon-starvation synchrony of *Navicula pelliculosa* (Bréb) Hilse. *Expl Cell. Res.* **47,** 315–328.

COOMBS, J. & VOLCANI, B.E. (1968) Studies on the biochemistry and fine structure of silica shell formation in diatoms. Chemical changes in the wall of *Navicula pelliculosa* during its formation. *Planta,* **82,** 280–292.

CRAIG, H., CHUNG, Y. & FIADEIRO, M. (1972) A benthic front in the South Pacific. *Earth Planet. Sci. Letts,* **16,** 50–65.

DARLEY, W.M. & VOLCANI, B.E. (1969) Role of silicon in diatom metabolism. A silicon requirement for deoxyribonucleic acid synthesis in the diatom *Cylindrotheca fusiformis* Reimann and Lewin. *Expl Cell Res.* **58,** 334–342.

DEACON, G.E.R. (1963) The Southern Ocean. In: *The Sea,* Vol. II (Ed. by M. N. Hill), pp. 281–296. Interscience, New York.

DEFFEYES, K.S. (1959) Zeolites in sedimentary rocks. *J. sedim. Petrol.* **29,** 602–609.

DESIKACHARY, T.V. (1957) Electron microscope studies on diatoms. *Jl R. Microsc. Soc.* **76,** 9–36.

DWYER, F.P. & MELLOR, D.P. (1933) A note on the occurrence of β-cristobalite in Australian opals. *J. Proc. R. Soc. N.S.W.* **66,** 378–382.

EDMOND, J.M. (1973) The silica budget of the Antarctic Circumpolar Current. *Nature, Lond.* **241,** 391–393.

EINSELE, W. & GRIM, J. (1938) Über den Kieselsäuregehalt planktischer Diatomeen und dessen Bedeutung für einige Fragen ihrer Okologie. *Z. Bot.* **32,** 545–590.

EISMA, D. & VAN DER GAAST, S.J. (1971) Determination of opal in marine sediments by X-ray diffraction. *Netherlands J. Sea Res.* **5,** 382–389.

EMERY, K.O. & RITTENBERG, S.C. (1952) Early diagenesis of California Basin sediments in relation to the origin of oil. *Bull. Am. Ass. Petrol. Geol.* **36,** 735–806.

ERIKSSON, E. (1962) Oceanic mixing. *Deep Sea Res.* **9,** 1–9.

ERNST, W.G. & CALVERT, S.E. (1969) An experimental study of the recrystallization of porcelanite and its bearing on the origin of some bedded cherts. *Am. J. Sci.* (Schairer Vol.), **267A,** 114–133.

EWING, J., WORZEL, J.L., EWING, M. & WINDISCH, C. (1966) Age of Horizon A and the oldest Atlantic sediments. *Science,* **154,** 1125–1132.

FANNING, K.A. & PILSON, M.E.Q. (1971) Interstitial silica and pH in marine sediments: some effects of sampling procedures. *Science,* **173,** 1228–1231.

FANNING, K.A. & SCHINK, D.R. (1969) Interaction of marine sediments with dissolved silica. *Limnol. Oceanogr.* **14,** 59–68.

FLÖRKE, O.W. (1955a) Zur Frage des 'Hoch'—Cristabalits in Opalen, Bentoniten und Gläsern. *Neues. Jb. Miner. Mh.* **1955,** 217–223.

FLÖRKE, O.W. (1955b) Strukturanomalien bei Tridymit und Cristabalit. *Ber. dt. keram. Ges.* **32,** 369–381.

FLÖRKE, D.W. (1967) The structures of $AlPO_4$ and SiO_2. In: *Science of Ceramics,* Vol. 3 (Ed. by G. H. Stewart), pp. 13–27. Academic Press, London.

FRONDEL, C. (1962) *The System of Mineralogy,* Vol. III, Silica Minerals, pp. 334. John Wiley and Sons, New York.

GARRELS, R.M. (1965) Silica: role in the buffering of natural waters. *Science,* **148,** 69.

GARRELS, R.M. & MACKENZIE, F.T. (1971) *Evolution of Sedimentary Rocks,* pp. 397. W. W. Norton and Co., N.Y.

GARRELS, R.M. & MACKENZIE, F.T. (1974) Chemical history of the oceans deduced from post-depositional changes in sedimentary rocks. In: *Studies in Paleo-Oceanography* (Ed. by W. W. Hay). *Spec. Publs Soc. Econ. Paleont. Miner., Tulsa,* **20** (in press).

GARTNER, S. (1970) Sea-floor spreading, carbonate dissolution level and the nature of Horizon A. *Science,* **169,** 1077–1079.

GESSNER, F. (1959) *Hydrobotanik,* Vol. II, Veb Deutsch., pp. 701. Verlag der Wissensch., Berlin.

GIBSON, T.G. & TOWE, K.M. (1971) Eocene volcanism and the origin of Horizon A. *Science,* **172,** 152–154.

GILBERT, J.Y. & ALIEN, W.E. (1943) The phytoplankton of the Gulf of California obtained by the E. W. SCRIPPS in 1939 and 1940. *J. mar. Res.* **5,** 89–110.

GOLDBERG, E.D. (1958) Determination of opal in marine sediments. *J. mar. Res.* **17,** 178–182.

GORDON, A.L. (1971) Antarctic circulation. *Trans. Am. geophys. Un.* **52,** 230–232.

GREENWOOD, R. (1973) Cristobalite: its relationship to chert formation in selected samples from the Deep Sea Drilling Project. *J. sedim. Petrol.* **43,** 700–708.

GREIGG, V.W. (1932) The existence of the high temperature form of cristobalite at room temperature and the crystallinity of opal. *J. Am. Chem. Soc.* **54,** 2846–2849.

GRIFFIN, J.J., KOIDE, M., HÖHNDORF, A., HAWKINS, J.W. & GOLDBERG, E.D. (1972) Sediments of the Lau Basin—rapidly accumulating volcanic deposits. *Deep Sea Res.* **19,** 139–148.

GRILL, E.V. (1970) A mathematical model for the marine dissolved silicate cycle. *Deep Sea Res.* **17,** 245–266.

GRUNER, J.W. (1940) Abundance and significance of cristobalite in bentonites and Fuller's Earth *Econ. Geol.* **35,** 867–875.

GUCLUER, S.M. & GROSS, M.G. (1964) Recent marine sediments in Saanich Inlet, a stagnant marine basin. *Limnol. Oceanogr.* **9,** 359–376.

HARDER, H. (1965) Experimente zur 'Ausfällung' der Kieselsäure. *Geochim. cosmochim. Acta,* **29,** 429–442.

HARRISS, R.C. (1966) Biological buffering of oceanic silica. *Nature, Lond.* **212,** 275–276.

HAY, R.L. (1966) Zeolites and zeolitic reactions in sedimentary rocks. *Spec. Pap. geol. Soc. Am.* **85,** 130 pp.

HEALEY, F.P., COOMBS, J. & VOLCANI, B.E. (1967) Changes in pigment content of the diatom *Navicula pelliculosa* (Bréb) Hilse in silicon-starvation synchrony. *Arch. Mikrobiol.* **59,** 131–142.

HEATH, G.R. (1968) *Mineralogy of Cenozoic deep-sea sediments from the equatorial Pacific Ocean,* pp. 168. PhD Dissertation, University of California, San Diego.

HEATH, G.R. (1973) Cherts from the eastern Pacific, Leg 16, Deep Sea Drilling Project. In: *Initial Reports of the Deep Sea Drilling Project,* Vol. XVI (Tj. van Andel, G. R. Heath *et al.*), pp. 609–613. U.S. Government Printing Office, Washington.

HEATH, G.R. (1974) Dissolved silica and deep-sea sediments. In: *Studies in Paleo-Oceanography* (Ed. by W. W. Hay). *Spec. Publs Soc. Econ. Paleont. Miner., Tulsa,* **20** (in press).

HEATH, G.R. & MOBERLY, R. (1971) Cherts from the western Pacific, Leg 7, Deep Sea Drilling Project. In: *Initial Reports of the Deep Sea Drilling Project,* Vol. VII (E. L. Winterer *et al.*), pp. 991–1008. U.S. Government Printing Office, Washington.

HELMCKE, J.G. (1954) Die Feinstruktur der Kieselsäure und ihre physiologische Bedeutung in Diatomeenschalen. *Naturwissenschaften,* **11,** 254–255.

HEMLEY, J.J., MEYER, C. & RICHTER, D.H. (1961) Some alteration reactions in the system Na_2O—Al_2O_3—SiO_2—H_2O. *Prof. Pap. U.S. geol. Surv.* **424D,** 338–340.

HENDERSON, J.H., JACKSON, M.L., SYERS, J.K., CLAYTON, R.N. & REX, R.W. (1971) Cristobalite authigenic origin in relation to montmorillonite and quartz origin in bentonites. *Clays Clay Miner.* **19,** 229–238.

HENDERSON, J.H., CLAYTON, R.N., JACKSON, M.L., SYERS, J.K., REX, R.W., BROWN, J.L. & SACHS, I.B. (1972) Cristobalite and quartz isolation from soils and sediments by hydrofluosilicic acid treatment and heavy liquid separation. *Proc. Soil Sci. Soc. Am.* **36,** 830–835.

HERMAN, Y. (1972) Origin of deep-sea cherts in the North Atlantic. *Nature, Lond.* **238,** 392–393.

Hughes, J.C. & Lund, J.W.G. (1962) The rate of growth of *Asterionella formosa* Hass. in relation to its ecology. *Arch. Mikrobiol.* **42**, 117–129.

Hull, W.Q., Keel, H., Kenny, J. & Gamson, B.W. (1953) Diatomaceous earth. *Ind. Engng Chem.* **45**, 256–269.

Hurd, D.C. (1973) Interactions of biogenic opal, sediment, and seawater in the central equatorial Pacific. *Geochim. cosmochim. Acta,* **37**, 2257–2282.

Iijima, A. & Utada, M. (1966) Zeolites in sedimentary rocks, with reference to the depositional environments and zonal distribution. *Sedimentology,* **7**, 327–357.

Jørgensen, E.G. (1953) Silicate assimilation by diatoms. *Physiologia Pl.* **6**, 301–315.

Jørgensen, E.G. (1955) Variations in the silica content of diatoms. *Physiologia Pl.* **8**, 840–845.

Jones, L.H.P. & Milne, A.A. (1963) Studies of silica in the oat plant. I Chemical and physical properties of the silica. *Pl. Soil,* **17**, 207–220.

Jones, J.B. & Segnit, E.R. (1971) The nature of opal. I. Nomenclature and constituent phases. *J. geol. Soc. Aust.* **18**, 57–68.

Jones, J.B. & Segnit, E.R. (1972) Genesis of cristobalite and tridymite at low temperatures. *J. geol. Soc. Aust.* **18**, 419–422.

Jones, J.B., Sanders, J.V. & Segnit, E.R. (1964) Structure of opal. *Nature, Lond.* **204**, 990–991.

Kamatani, A. (1971) Physical and chemical characteristics of biogenous silica. *Mar. Biol.* **8**, 89–95.

Kanaya, T. & Koizumi, I. (1966) Interpretation of diatom thanatocoenoses from the North Pacific applied to a study of core V20–130. *Sci. Repts Tohoku Univ.* 2nd Ser. (Geol.) **37**, 89–130.

Keller, W.D., Balgord, W.D. & Reesman, A.L. (1963) Dissolved products of artificially pulverised silicate minerals and rocks. *J. sedim. Petrol.* **33**, 191–204.

Kolodny, Y., Nathan, Y. & Sass, E. (1965) Porcellanite in the Mishash Formation, Negev, southern Israel. *J. sedim. Petrol.* **35**, 454–463.

Knauss, J.A. (1962) On some aspects of the deep circulation of the Pacific. *J. geophys. Res.* **67**, 3943–3954.

Krauskopf, K.B. (1956) Dissolution and precipitation of silica at low temperatures. *Geochim. cosmochim. Acta,* **10**, 1–26.

Lancelot, Y. (1973) Chert and silica diagenesis in sediments from the central Pacific. In: *Initial Reports of the Deep Sea Drilling Project,* Vol. XVII (E. L. Winterer, J. I. Ewing *et al.*), pp. 377–405. U.S. Government Printing Office, Washington.

Lanning, F.C., Ponnaiya, B.W.X. & Crumpton, C.F. (1958) The chemical nature of silica in plants. *Pl. Physiol.* **33**, 339–343.

Laves, F. (1939) Über der Einfluss von Spannungen auf die Regelung von Quartz-und Cristabolit-Kriställchen im Chalcedon, Quarzin und Lussatit. *Naturwissenschaften,* **27**, 705–707.

Levin, J. & Ott, E. (1933) X-ray study of opals, silica glass and silica gel. *Z. Kristallogr.* **85**, 305–318.

Lewin, J.C. (1954) Silicon metabolism in diatoms. I Evidence for the role of reduced sulphur compounds in Si utilisation. *J. gen. Physiol.* **37**, 589–599.

Lewin, J.C. (1955a) Silicon metabolism in diatoms. II Sources of silicon for growth of *Navicula pelliculosa. Pl. Physiol.* **30**, 129–134.

Lewin, J.C. (1955b) Silicon metabolism in diatoms. III Respiration and silicon uptake in *Navicula pelliculosa. J. gen. Physiol.* **39**, 1–10.

Lewin, J.C. (1957) Silicon metabolism in diatoms. IV Growth and frustule formation in *Navicula pelliculosa. Can. J. Microbiol.* **3**, 427–433.

Lewin, J.C. (1961) The dissolution of silica from diatom walls. *Geochim. cosmochim. Acta,* **21**, 182–198.

Lewin, J.C. (1962) Silicification. In: *Physiology and Biochemistry of Algae* (Ed. by R. A. Lewin), pp. 445–455. Academic Press, London.

Lewin, J.C., Lewin, R.A. & Philpott, D.E. (1958) Observations on *Phaeodactylum tricornitum. J. Gen. Microbiol.* **18**, 418–426.

Lisitzin, A.P. (1964) Distribution and chemical composition of suspension in waters of the Indian Ocean. *Resultaty Issledovaniy po Programme MCG, X Section, Okeanologii,* **10**, 135 pp.

Lisitzin, A.P. (1967) Basic relationships in distribution of modern siliceous sediments and their connection with climatic zonation. *Int. Geol. Rev.* **9**, 631–652.

Lisitzin, A.P. (1971) Distribution of siliceous microfossils in suspension and in bottom sediments. In: *The Micropalaeontology of Oceans* (Ed. by B. M. Funnell and W. R. Riedel), pp. 173–195. Cambridge University Press, London.

LISITZIN, A.P. (1972) Sedimentation in the World Ocean (Ed. by K. S. Rodolfo). *Spec. Publs Soc. Econ. Paleont. Miner., Tulsa,* **17,** 218 pp.

LISITZIN, A.P., BELVAYEV, Y.I., BOGDNOV, Y.A. & BOGOYAVLENSKIY, A.N. (1967) Distribution relationships and forms of silicon suspended in waters of the world ocean. *Int. Geol. Rev.* **9,** 604–623.

LIVINGSTON, D.A. (1963) Chemical composition of rivers and lakes. *Prof. Pap. U.S. geol. Surv.* **440-G,** 64 pp.

LOWENSTAM, H.A. (1971) Opal precipitation by marine gastopods (Mollusca). *Science,* **171,** 487–490.

LUND, J.W.G. (1950) Studies on *Asterionella formosa* Hass. II. Nutrient depletion and the spring maximum. *J. Ecol.* **38,** 15–35.

MACKENZIE, F.T. & GARRELS, R.M. (1965) Silicates: reactivity with sea water. *Science,* **150,** 57–58.

MACKENZIE, F.T. & GARRELS, R.M. (1966a) Chemical mass balance between rivers and oceans. *Am. J. Sci.* **264,** 507–525.

MACKENZIE, F.T. & GARRELS, R.M. (1966b) Silica-bicarbonate balance in the ocean and early diagenesis. *J. sedim. Petrol.* **36,** 1075–1084.

MACKENZIE, F.T., GARRELS, R.M., BRICKER, D.P. & BICKLEY, F. (1967) Silica in sea water: control by silica minerals. *Science,* **155,** 1404–1405.

MARTIN, J.H. & KNAUER, G.A. (1973) The elemental composition of plankton. *Geochim. cosmochim. Acta,* **37,** 1639–1653.

MATTSON, P.H. & PESSAGNO, E.A. (1971) Caribbean Eocene volcanism and the extent of Horizon A. *Science,* **174,** 138–139.

MENARD, H.W. (1964) *Marine Geology of the Pacific,* 271 pp. McGraw-Hill, New York.

METCALF, W.G. (1969) Dissolved silicate in the deep North Atlantic. *Deep Sea Res.* Suppl. **16,** 139–145.

MITCHELL, R.S. (1967) Tridymite pseudomorphs after wood in Virginian Lower Cretaceous sediments. *Science,* **158,** 905–906.

MIZUTANI, S. (1966) Transformation of silica under hydrothermal conditions. *J. Earth Sci.* **14,** 56–88.

MURATA, K.J. (1940) Volcanic ash as a source of silica for the silicification by wood. *Am. J. Sci.* **238,** 586–596.

NAYUDU, Y.R. (1962) A new hypothesis for origin of guyots and sediment terraces. In: *The Crust of the Pacific Basin* (Ed. by G. A. MacDonald and H. Kuno). *Geophys. Monogr.* **6,** 171–180.

OEHLER, J.H. (1973) Tridymite-like crystals in cristobalitic 'cherts'. *Nature, Phys. Sci.* **241,** 64–65.

OLAUSSON, E. (1967) Climatological, geoeconomical and palaeo-oceanographical aspects of carbonate deposition. In: *Progress in Oceanography* (Ed. by M. Sears), Vol. 4, pp. 245–265. Pergamon Press, London.

OMURE, M., OOKAWARA, S. & IWAI, S.L. (1956) Über die Entstehung des Hoch-Cristobalits in Roseki-Schmotte. *Naturwissenschaften,* **43,** 495.

PETERSON, M.N.A. & GOLDBERG, E.D. (1962) Feldspar distributions in South Pacific pelagic sediments. *J. geophys. Res.* **67,** 3477–3492.

PETERSON, M.N.A. & GRIFFIN, J.J. (1964) Volcanism and clay minerals in the south-eastern Pacific. *J. Mar. Res.* **22,** 13–21.

PETERSON, M.N.A. *et al.* (1970) *Initial Reports of the Deep Sea Drilling Project,* Vol. II, 501 pp. U.S. Government Printing Office, Washington.

PETRUSHEVSKAYA, M.A. (1971) Spumellarian and nasselarian Radiolaria in the plankton and bottom sediments of the Central Pacific. In: *The Micropalaeontology of Oceans* (Ed. by B. M. Funnell and W. R. Riedel), pp. 309–317. Cambridge University Press, London.

PIMM, A.C., GARRISON, R.E. & BOYCE, R.E. (1971) Sedimentology synthesis: lithology, chemistry and physical properties of sediments in the north-western Pacific Ocean. In: *Initial Reports of the Deep Sea Drilling Project,* Vol. VI (A. G. Fischer *et al.*), pp. 1131–1252. U.S. Government Printing Office, Washington.

RAMSAY, A.T.S. (1971) Occurrence of biogenous siliceous sediments in the Atlantic Ocean. *Nature, Lond.* **233,** 115–117.

REDFIELD, A.C., KETCHUM, B.H. & RICHARDS, F.A. (1963) The influence of organisms on the composition of sea-water. In: *The Sea,* Vol. II (Ed. by M. N. Hill), pp. 26–77. Interscience, New York.

REID, J.L. (1962) On circulation, phosphate-phosphorus content and zooplankton volumes in the upper part of the Pacific Ocean. *Limnol. Oceanogr.* **7,** 287–306.

REID, J.L. & LYNN, R.J. (1971) On the influence of the Norwegian-Greenland and Weddell seas on the bottom waters of the Indian and Pacific Oceans. *Deep Sea Res.* **18**, 1063–1088.

REIMANN, B.E.F. (1964) Deposition of silica inside a diatom cell. *Expl Cell Res.* **34**, 605–608.

REIMANN, B.E.F., LEWIN, J.C. & VOLCANI, B.E. (1965) Studies on the biochemistry and fine structure of silica shell formation in diatoms. I. The structure of the cell wall of *Cylindrotheca fusiformis* Reimann and Lewin. *J. Cell Biol.* **24**, 39–55.

REIMANN, B.E.F., LEWIN, J.C. & VOLCANI, B.E. (1966) Studies on the biochemistry and fine structure of silica shell formation in diatoms. II. The structure of the cell wall of *Navicula pelliculosa* (Breb) Hilse. *J. Phycol.* **2**, 74–84.

REYNOLDS, R.C. & ANDERSON, D.M. (1967) Cristobalite and clinoptilolite in bentonite beds of the Colville Group, northern Alaska. *J. sedim. Petrol.* **37**, 966–969.

REYNOLDS, W.R. (1970) Mineralogy and stratigraphy of Lower Tertiary clays and claystones of Alabama. *J. sedim. Petrol.* **40**, 829–838.

REX, R.W. (1969) X-ray mineralogy studies—Leg 1. In: *Initial Reports of the Deep Sea Drilling Project*, Vol. I (M. Ewing *et al.*), pp. 354–367. U.S. Government Printing Office, Washington.

REX, R.W. (1970) X-ray mineralogy studies—Leg 2. In: *Initial Reports of the Deep Sea Drilling Project*, Vol. II (M. N. A. Peterson *et al.*), pp. 329–346. U.S. Government Printing Office, Washington.

RIEDEL, W.R. (1963) The preserved record: Paleontology of pelagic sediments. In: *The Sea*, Vol. III (Ed. by M. N. Hill), pp. 866–887. Interscience, New York.

ROGALL, E. (1949) Über den Feinbau der Kieselmembran der Diatomeen. *Planta*, **29**, 279–291.

ROUND, F.E. (1968) The phytoplankton of the Gulf of California, Part II. The distribution of phytoplanktonic diatoms in cores. *J. exp. Mar. Biol. & Ecol.* **2**, 64–86.

SCHEIDEGGER, K.F. (1973) Volcanic ash layers in deep sea sediments and their petrological significance. *Earth Planet. Sci. Letts*, **17**, 397–407.

SCHINK, D.R., FANNING, K.A. & PILSON, M.E.Q. (1973) Distribution of dissolved silica in interstitial waters of Atlantic sediments. *Trans. Am. geophys. Un.* **54**, 336.

SCHRADER, H.J. (1971) Fecal pellets: role in sedimentation of pelagic diatoms. *Science*, **174**, 55–57.

SIEVER, R. (1957) The silica budget in the sedimentary cycle. *Am. Miner.* **42**, 821–841.

SIEVER, R. (1968) Establishment of equilibrium between clays and sea water. *Earth Planet. Sci. Letts*, **5**, 106–110.

SIEVER, R., BECK, K.C. & BERNER, R.A. (1965) Composition of interstitial waters of modern sediments. *J. Geol.* **73**, 39–73.

SILLÉN, L.G. (1961) The physical chemistry of sea water. In: *Oceanography* (Ed. by M. Sears), pp. 549–581. Am. Ass. Adv. Sci. Publ. No. 67.

SPENCER, D.W. (1972) Geosecs II, the 1970 North Atlantic Station: hydrographic features, oxygen and nutrients. *Earth Planet Sci. Letts*, **16**, 91–102.

STERLING, C. (1967) Crystalline silica in plants. *Am. J. Bot.* **54**, 840–844.

STOERMER, E.F. & PANKRATZ, H.S. (1964) Fine structure of the diatom *Amphipleura pellucida* I. Wall Structure. *Am. J. Bot.* **51**, 986–990.

STOERMER, E.F., PANKRATZ, H.S. & DRUM, R.W. (1964) The fine structure of *Mastogloia grevillei* Wm. Smith. *Protoplasma*, **59**, 1–13.

STOERMER, E.F., PANKRATZ, H.S. & BOWEN, C.C. (1965) Fine structure of the diatom *Amphipleura pellucida* II. Cytoplasmic fine structure and frustule formation. *Am. J. Bot.* **52**, 1067–1078.

STOMMEL, H. & ARONS, A.B. (1960) On the abyssal circulation of the world ocean—II. An idealised model of the circulation pattern and amplitude in oceanic basins. *Deep Sea Res.* **6**, 217–233.

SVERDRUP, H.U. (1938) On the process of upwelling. *J. Mar. Res.* **1**, 155–164.

SWINFORD, A. & FRANKS, P.C. (1959) Opal in Ogallala Formation of Kansas. In: *Silica in Sediments* (Ed. by H. A. Ireland). *Spec. Publs Soc. Econ. Paleont. Miner., Tulsa*, **7**, 111–120.

TALIAFERRO, N.L. (1934) Contraction phenomena in cherts. *Bull. geol. Soc. Am.* **45**, 189–232.

TUREKIAN, K.K. (1965) Some aspects of the geochemistry of marine sediments. In: *Chemical Oceanography*, Vol. 2 (Ed. by J. P. Riley and G. Skirrow), pp. 81–126. Academic Press, London.

VON DER BORCH, C.C., GALEHOUSE, J. & NESTEROFF, W.D. (1971) Silicified limestone-chert sequences cored during leg 8 of the Deep Sea Drilling Project: A petrologic study. In: *Initial Reports of the Deep Sea Drilling Project*, Vol. VIII (J. I. Tracey, Jr *et al.*), pp. 819–828. U.S. Government Printing Office, Washington.

VON RAD, U. & RÖSCH, H. (1972) Mineralogy and origin of clay minerals, silica and authigenic silicates in Leg 14 sediments. In: *Initial Reports of the Deep Sea Drilling Project,* Vol. XIV (D. E. Hayes, A. C. Pimm *et al.*), pp. 727–751. U.S. Government Printing Office, Washington.

VINOGRADOV, A. P. (1953) The elementary chemical composition of marine organisms. *Mem. Sears Fdn. mar. Res.* II. 647 pp. Yale University Press, New Haven.

WARREN, B.E. & BISCOE, J. (1938) The structure of silica glass by X-ray diffraction. *J. Am. Ceram. Soc.* **21,** 49–54.

WEAVER, F.M. & WISE, S.W., JR (1972) Ultramorphology of deep sea cristobaltic chert. *Nature, Phys. Sci.* **237,** 56–57.

WERMUND, E.G. & MOIOLA, R.J. (1966) Opal zeolites and clays in an Eocene neritic bar sand. *J. sedim. Petrol.* **36,** 248–253.

WISE, S.W., JR & KELTS, K.R. (1972) Inferred diagenetic history of a weakly silicified deep sea chalk. *Trans. Gulf-Cst Ass. geol. Socs,* **22,** 177–203.

WISE, S.W., JR, BUIE, B.F. & WEAVER, F.M. (1972) Chemically precipitated sedimentary cristobalite and the origin of chert. *Eclog. geol. Helv.* **65,** 157–163.

WOOSTER, W.S. & REID, J.L. (1963) Eastern boundary currents. In: *The Sea,* Vol. II (Ed. by M. N. Hill), pp. 253–280. Interscience, New York.

Spec. Publs int. Ass. Sediment. (1974) **1**, 301–326

Chertification of oceanic sediments

SHERWOOD W. WISE, JR *and* FRED M. WEAVER

Department of Geology, Florida State University, Tallahassee, Florida, U.S.A.

ABSTRACT

Classical theories of marine chert formation formulated by land geologists who worked with older (Palaeozoic and Mesozoic) rock sequences have been considerably altered or updated by modern studies of younger (Tertiary) sediments of uplifted marginal basins and of present-day ocean basins where long sediment cores have encountered widespread cherts in incipient stages of development. Scanning electron microscopy, electron-microprobe, and X-ray analyses have enabled silica and carbonate migrational pathways during chertification to be traced in detail; nevertheless, questions such as silica sources for chertification (volcanic or biogenous) remain controversial. Both nodular and bedded cherts can form exclusively from biogenous silica as shown by a Southern Ocean Pliocene cristobalite 'chert' which grades into diatom ooze (ELTANIN Core 47–15). Unidimensionally disordered alpha-cristobalite, a metastable form of silica, is characteristic of incipient cherts, and forms the outer growth margins of developing chert nodules where the following complete transition can be observed: biogenous silica → cristobalite → true quartz chert. Such examples have elucidated not only modes of formation of deep-sea chert, but also of shelf and near-shore equivalents such as the cristobalite-rich Lower Tertiary sequences of the south-eastern U.S. Coastal Plain. Our studies indicate that chert formation can best be explained by a 'maturation' theory rather than the recently proposed 'quartz precipitation' theory, and that biogenous silica is by far the most important immediate silica source for oceanic cherts.

INTRODUCTION

Although chert is one of the more common sedimentary rocks on earth and perhaps the first to be exploited by man in the manufacture of tools and weaponry, the geological origin of marine chert has long been a baffling and controversial problem. Fundamental to this problem is the fact that marine cherts have not been observed in the process of formation in Recent sediments. Geologists, therefore, have had to infer their probable modes of origin from studies of well-aged and altered cherts present in ancient sediments exposed on land. With little direct evidence in the way of fossils or original texture preserved in many of the older rocks, geologists have had to plumb their imaginations to envisage what the formative stages must have been for those ancient rocks. Extrapolations from laboratory and experimental studies coupled with field observations of rocks tens or hundreds of millions of years old did enable classical geologists to formulate numerous interesting and varied theories on chert formation. Many of these theories, in one version or another, still pervade the thinking of modern earth scientists. For instance, in regard to nodular chert formation, Tarr (1917) strongly advocated a theory of primary precipitation of silica from sea water. He envisaged silica concentrations in the sea building up periodically to a low threshold value, whereupon the silica would coagulate into spherical blobs of gel, which would eventually harden into chert nodules. The idea of direct precipitation of silica from sea water is still current in the literature (e.g. Reynolds, 1970), and modifications of the gel theory are still very much alive (e.g. Kneller *et al.*, 1968; Berger & von Rad, 1972). These theories persist despite the fact that throughout the past three decades of modern ocean-basin exploration, no discernible accumulations of silica gel have been found on the sea floor or within the underlying sediment.

In opposition to Tarr's theory of chert-nodule formation, Barton (1918) and van Tuyl (1918) advocated penecontemporaneous diagenetic replacement of the carbonate host rock by silica. Support for this theory has been provided by Newell *et al.* (1953) who also suggested a biogenous source for the silica replacement after observing desilicified sponge spicules adjacent to chert nodules (see also Lowenstam, 1948). A biogenous silica source for deep-sea nodular cherts is now generally accepted (Davies & Supko, 1973). The almost exclusive association of these nodules with biogenous carbonate ooze or rock strongly suggests that siliceous microfossils disseminated throughout the host rock constitute the immediate source of silica.

Little agreement exists on the origin of bedded cherts. Davies & Supko (1973, p. 381) state simply that 'the source of silica in deep-sea bedded cherts is unknown.' This is not to say, however, that many opinions have not been expressed on the subject. Davis (1918) studied the famous iron-stained radiolarian bedded cherts of the Franciscan mélange in California and concluded that silica was supplied directly to the sea by emanations accompanying submarine lava flows. It formed layers of silica gel on the ocean floor which entrapped radiolarian tests, then solidified into chert. Davis extended his theory of primary precipitation to explain the origin of the Miocene Monterey chert of California. Bramlette (1946), however, presented convincing evidence to show that quartz chert in the Monterey is the end product of a series of diagenetic alterations of the diatomite which is the predominant rock type in that formation. More importantly, Bramlette showed that the biogenic opal of the diatomite is first converted to an opaline-rich claystone called porcelanite by Taliaferro (1934). The opaline component has since been identified as disordered alpha-cristobalite (a metastable polymorph of silica) by Ernst & Calvert (1969) who also reproduced the diagenetic reaction experimentally by subjecting Monterey porcelanite to high temperature and pressure.

From these studies emerged a 'maturation' theory which suggests that chert can be produced by the reaction: biogenous silica $\rightarrow$ disordered cristobalite $\rightarrow$ true quartz chert*. Significantly, a similar transition of rock types was observed during initial studies of geologically young cherts recovered by the Deep Sea Drilling Project

* Strictly speaking, opaline or cristobalite-rich rocks are not true chert, although the term 'chert' in the broad sense is often applied to all such rocks.

(Heath & Moberly, 1971). According to Heath & Moberly (1971), the rate of the reaction would be determined by kinetics, requiring either an introduction of heat into the system or the passage of a sufficiently long period of time.

Despite the impressive work by Bramlette (1946) and his successors, mineralogists have been hesitant to accept wholeheartedly a biogenous origin for bedded cherts, even when such cherts are often replete with the tests of siliceous microfossils. In fact, historically there seems to have existed a near conspiratorial effort to reject the obvious in order to invoke exotic or non-uniformitarian solutions to the problem. For example, although radiolarians and siliceous sponge spicules are common or abundant in certain beds of the Stanley (Mississippian) Formation of the Ouachita area (USA), Goldstein & Hendricks (1953) attribute occurrences of novaculite and chert in that unit to submarine alteration of volcanic ash. This conclusion was drawn despite the fact that in their careful petrographic work, they report no volcanic ash shards in any of the siliceous rocks themselves. After studying the Rex chert (Permian) of south-eastern Idaho, Mansfield (1927) and Keller (1941) reject an organic silica source even though sponge spicules are common in certain beds. They follow Tarr (1917, 1926) in believing that the bulk of the chert is a direct precipitate from sea water.

Geologists have long recognized that volcanic glass is a possible source of silica for bedded chert because its devitrification may place silica into solution. The diagenetic alteration of zeolite to montmorillonite, and montmorillonite to illite or kaolinite, might also constitute a source of free silica. The latter is supported by the decrease in abundance of montmorillonite in older sedimentary rocks (Siever, 1962).

Today, proponents of the biogenous theory of bedded chert formation are most often challenged by advocates of a volcanic origin for the chert in question. For instance, in lieu of the biogenic origin suggested by shipboard scientists on Leg 2 of the Deep Sea Drilling Project (Peterson *et al.*, 1970, p. 419), Gibson & Towe (1971) and Mattson & Pessagno (1971) have proposed a volcanic origin for the extensive Eocene Horizon A and A″ cherts encountered by Deep Sea Drilling in the Caribbean and North Atlantic as well as for near-shore equivalents of those units deposited along adjacent continental shelves. The principal evidence advanced by advocates of the volcanic school is the detection in the cherty sediments of presumed volcanic alteration products such as montmorillonite, palygorskite, sepiolite, alpha-cristobalite, and clinoptilolite (a zeolite). Although some of these diagenetic minerals may have other origins, the co-occurrence of clinoptilolite, disordered cristobalite, and montmorillonite has been cited as *prima facie* evidence of extensive volcanic ash deposition (Reynolds, 1970; Gibson & Towe, 1971).

Numerous studies have shown that montmorillonite, clinoptilolite, and disordered cristobalite are common alteration products of rhyolitic volcanic glass (Bramlette & Posnjak, 1933; Urashima, 1959; Deffeyes, 1959; Hathaway & Sachs, 1965; Hay, 1966; Wang, 1967). Not yet demonstrated, however, is whether significant volumes of chert or porcelanite are produced as a result of the devitrification of volcanic ash. The unwary might judge from the literature that the quantity of silica released is nearly infinite; however, the opposite appears to be true. Bentonites which contain appreciable amounts of cristobalite are usually found concentrated within massive, volcanic tuff sequences (e.g. Bramlette & Posnjak, 1933; Urashima, 1959; Wang, 1967), not within pelagic sediments of the marine realm. Little if any chert is concentrated in such deposits because any cristobalite present tends to remain well disseminated within the clay matrix of the bentonite until leached from the rock. We examined six deep-sea ashes from DSDP Leg 6 (Sites 45·1, 53·1, 52·0), and no detectable opaline or cherty material was observed except where siliceous microfossils were also present. Abundant montmorillonite was present in the ash layers. Berger & von Rad (1972) describe a thin bed of white altered ash in DSDP Core 138–6–2 which contains abundant montmorillonite with traces of palygorskite, but no cristobalite or chert within the bentonite or the surrounding rock. In his DSDP Initial Report, Lancelot (1973, p. 378) notes that 'volcanogenic sediments such as tuff and volcanic siltstones, very frequent at Leg 17 sites, are conspicuously devoid of chert'. To date, the only reasonably well-established example from Deep Sea Drilling of a cristobalite-rich bed associated with volcanogenic material is a tuffaceous mudstone in DSDP sample 7–61·0–1–1, 83–89. X-ray study by Zemmels & Cook (1973) shows this cristobalite to be the variety, 'opal-C', which Jones & Segnit (1971) consider characteristic of volcanogenic cristobalite. Nevertheless, the amount of opaline silica produced seems modest compared to the amount of bentonite and associated volcanogenic material which accompanies an opal-C occurrence. The latter point is telling. If a chert is derived from a volcanogenic source, it should most likely be associated with sufficient bentonite or tuffaceous sediment to indicate its source.

Occurrences of bentonites derived from rhyolitic ash in carbonate environments further suggest that the contribution of free silica from the devitrification of volcanic glass is somewhat meagre. Zones of silica-replaced carbonate occur directly beneath bentonite beds within the Middle Ordovician dolomites and limestones of the Southern Appalachians (Hergenroder, 1973). The silica does not occur as nodular chert, but merely as a silicified layer about one-tenth the thickness of the overlying bentonite (Hergenroder, personal communication, 1973). This thickness ratio indicates that most of the original rhyolitic ash converted to clay minerals rather than to free silica. Thus, in the deep sea, one should expect to find a considerable amount of bentonite associated with any volcanogenic chert.

Similar volume problems do not exist when biogenic opal is converted to chert because all of the parent material (except for expelled water) is made available as free silica. Of course, one feature common to nearly all cherts cored during the exploratory activities of the Deep Sea Drilling Project is the presence of siliceous microfossils, either unaltered or, more commonly, as moulds or ghosts. These are well illustrated by Calvert (1971, Fig. 3 from Site 9) in his study of North Atlantic DSDP cherts. Nevertheless, Calvert (1971, p. 284) states: 'from the wealth of evidence in favour of the derivation of clinoptilolite and cristobalite assemblages from volcanic glass . . . it is suggested that the cherts and porcelanites at site 9 were derived from the diagenetic alteration of volcanic materials'. With regard to porcelanites from sites 6 and 7, Calvert (1971, p. 283) concludes that 'the source of the amorphous silica in the North Atlantic cores cannot be identified with the information available; opaline microfossils in the sediment would probably be insufficient, however'.

Because some detectable quantity of siliceous microfossils and diagenetic minerals are present in most deep-sea sediments, the debate over the origin of bedded cherts may be for all time endless. Nevertheless, in an attempt to clarify certain points in this debate, we have investigated the formation of bedded cherts and siliceous claystones, with special reference to those which have been heralded as exclusively volcanic in origin. We present our results in light of the discussion of the literature given above. Following this presentation, we will discuss the formation of nodular chert.

To distinguish nodular chert from bedded chert, we follow the definitions of

Dunbar & Rodgers (1957) and Heath & Moberly (1971) who characterized nodular cherts as those which occur in carbonate host rocks, whereas bedded cherts occur in siliceous sequences. Mineralogy of the silica phases is not considered in these definitions.

BEDDED CHERT

With regard to the origin of bedded cherts, we find most illuminating the occurrence of a pure white monomineralic opaline chert cored on the Kerguelen Plateau in the Southern Ocean (ELTANIN Core 47–15; see Wise, Buie & Weaver, 1972; Wise & Weaver, 1972; Weaver & Wise, 1973b). The chert consists entirely of chemically precipitated authigenic cristobalite in the form of uniform-sized (5–10 μm) bladed microspherulites called lepispheres (Wise & Kelts, 1972). Identifiable via light and scanning electron microscopy (Fig. 1), the porous, spherulitic chert grades into the overlying white Pliocene diatom ooze so as to leave no doubt that the chert-forming silica was derived from the dissolution of the biogenous opal (Weaver & Wise, 1973b). X-ray analysis (Wise *et al.*, 1972) shows the opaline chert to be composed of unidimensionally disordered alpha-cristobalite (Flörke, 1955). The disordering in the silica tetrahedral layers of this mineral (Flörke, 1955) may result from the small crystal size, the incorporation of cations in the crystal lattice (Flörke, 1955), and/or the irregular mode of growth of the mineral along the serrate edges of the crystallites (Wise & Weaver, 1973). Different names given to this substance by other investigators are listed in Table 1.

A number of points about this chert are clear.

(1) The biogenous silica-cristobalite conversion is a dissolution-reprecipitation reaction, therefore the migration of silica over distances of centimetres or metres is entirely feasible (Wise *et al.*, 1972).

(2) There are no volcanic alteration products in either the chert or the diatom ooze from which it was derived. Therefore conversion of relatively unstable biogenous silica to opaline chert may be effected without the aid, influence or presence of volcanic materials in the sediment.

(3) Said conversion may occur at a very early stage of diagenesis, as indicated by the Pliocene age of the deposit.

(4) Once begun, the conversion may go very nearly to completion; thus, only rare fragments of diatom or radiolarian tests can be found within the opaline chert. Those illustrated in Fig. 1b occur in a chert pellet sieved from the transition zone near the base of the diatom unit. Therefore, the presence of only a few rare traces of siliceous microfossils in an opaline or quartz chert may well represent all that remains of a former diatomaceous or radiolarian-rich parent ooze. How often such tell-tale traces of siliceous fossils are summarily dismissed as casual trespassers into a supposedly volcanogenic sedimentary environment!

(4) No amorphous gel resolvable via electron microscopy or X-ray diffraction can be detected in the cristobalitic chert. Similar observations in young cherts from other ocean basins have prompted the conclusion that, in all probability, no gel phase exists in the conversion of biogenous opal to marine cristobalitic chert (Weaver & Wise, 1972).

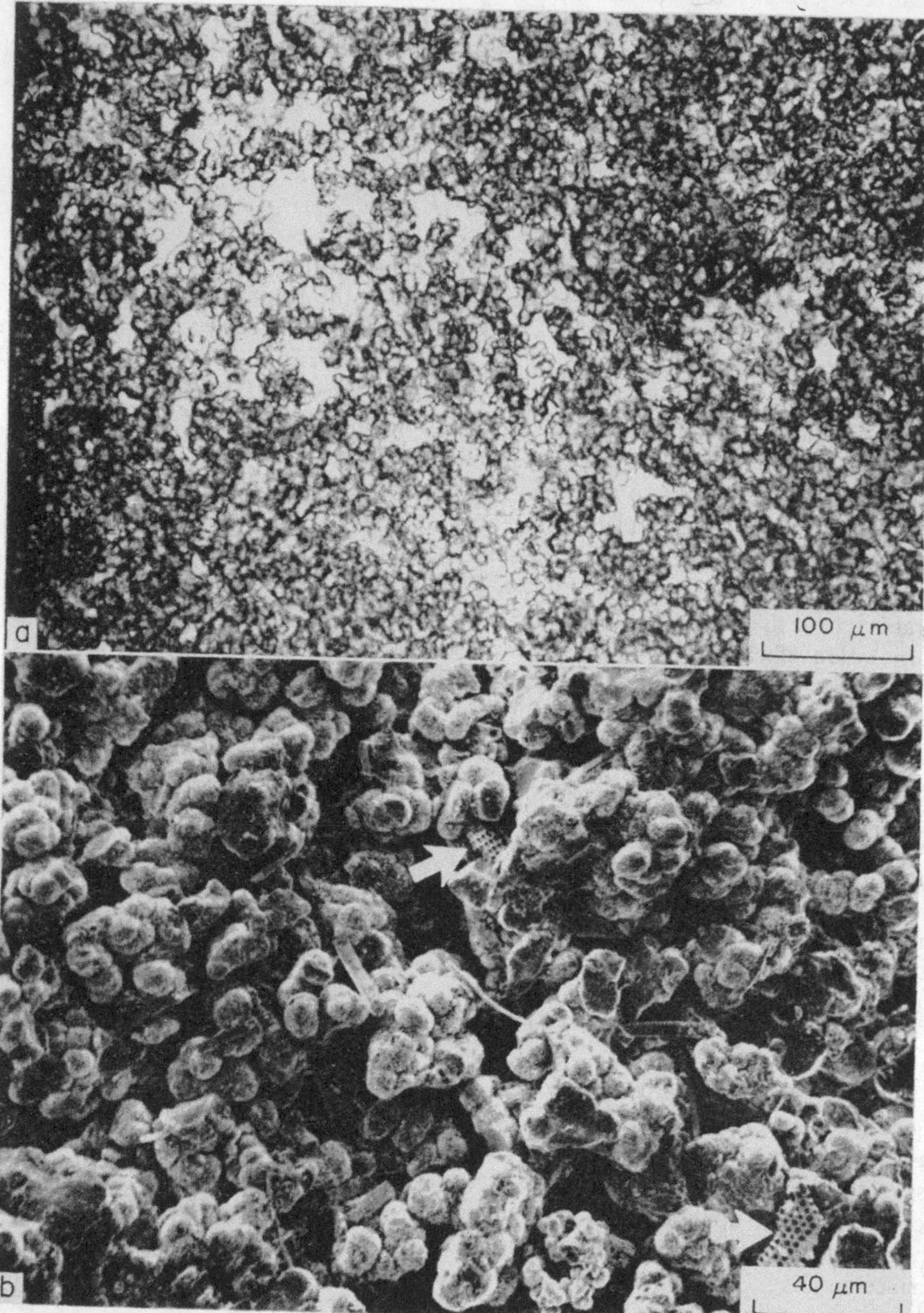

Fig. 1. Pliocene cristobalite bedded chert from the Kerguelen Plateau (Southern Ocean) recovered in ELTANIN Core 47–15. (a) Light micrograph of a thin section from this material reveals the spherulitic nature of the cristobalite matrix. (b) Scanning electron micrograph of a fracture section through this chert. Arrows point to numerous skeletal fragments of partially dissolved siliceous microfossils incorporated within a very porous cristobalitic matrix.

Table 1. Different terminologies used in marine opaline chert mineralogy (modified from Lancelot, 1973)

Terms used in this report	Equivalent terms used in the literature	Mineralogical structure
Unidimensionally disordered alpha-cristobalite or disordered cristobalite*	Cristobalite, alpha-cristobalite, or low cristobalite (DSDP X-ray reports; Reynolds, 1970) Disordered cristobalite (Lancelot, 1973) Lussatite‡ (Berger & von Rad, 1972) Opal-CT (Jones & Segnit, 1971) Opal (Frondel, 1962; Berger & von Rad, 1972†) Opal-cristobalite (Gibson & Towe, 1971) Unidimensionally disordered cristobalite (Flörke, 1955) Type 2 opal (Segnit, Anderson & Jones, 1970)	Disordered stacking of 'low' cristobalite and 'low'-tridymite layers*

* See X-ray diagram in Wise *et al.* (1972, Fig. 1).

† Opal described by these authors appears intermediate between amorphous silica and disordered cristobalite (highly disordered cristobalite).

‡ Lussatite was described by Mallard (1890) as a fibrous variety of cristobalite.

(5) Because the sedimentary sequence in ELTANIN Core 47–15 contains practically no calcium carbonate, we may assume that the opaline chert formed in interstitial waters of relatively low pH. Under certain natural conditions, therefore, the conversion of biogenous opal to cristobalite may occur readily at low or moderate pH.

(6) Essentially no detectable clay minerals are present in the Kerguelen chert or in the overlying pure white ooze.

(7) The opaline chert and the diatom ooze are highly porous and permeable.

The last two points bear special discussion in view of the special emphasis Lancelot (1973) places on the role of clay minerals and permeability in the formation of deep-sea chert. Lancelot (1973, p. 394) notes the well-known role that foreign cations are thought to play in the formation of disordered cristobalite (see Flörke, 1955; Millot, 1960, 1964), and suggests that clay minerals are a likely source for easily exchangeable cations. Thus Lancelot (1973, p. 394) suggests that clay minerals are 'responsible for the diagenetic precipitation of disordered cristobalite in clayey sediments, while only quartz can precipitate in a carbonate environment' (which lacks such a cation source). Lancelot (1973, p. 394) goes on to say that: 'Disordered cristobalite is confined to sediments in which clay minerals represent a major obstacle to circulation and create a micro-environment in which pore spaces are small enough to keep the ratio of silica to metallic cations at a relatively low value. Whenever the permeability increases, such as in a large void, chalcedony and quartz can precipitate because of a sharp increase of this ratio.'

According to Lancelot's formula, as we understand it, the deficiency of clay minerals and the high permeability of the Kerguelen diatomite should have caused the precipitation of quartz chert rather than the disordered cristobalite which actually formed.

We conclude that Lancelot's suggestions as to the role clay minerals and permeability play in early chert diagenesis are not applicable in this case. We further believe that disordered cristobalite may form regardless of lithology as long as there is an adequate supply of biogenous silica available to serve as a silica source.

One point is clear. The Kerguelen chert is identical to the porcelanite of the Monterey Formation (Bramlette, 1946) except for the clay content of the latter. The silica source for both of these bedded cherts has been clearly demonstrated, and certainly should not be labelled 'unknown'.

The Monterey Formation of California, so fascinating to western geologists, was deposited in marginal Tertiary basins a few hundred to over 3000 m deep (Ingle, 1973). On the other hand, coastal plain geologists of the southeastern United States have long been intrigued by occurrences of opaline claystones in shallow-shelf Eocene strata from Alabama to South Carolina. These hard, siliceous but light-weight, highly porous rocks are not unlike Monterey porcelanites in appearance (Hastings & McVay, 1963), and are composed principally of disordered cristobalite (Heron, Robinson & Johnson, 1965; Heron, 1969; Reynolds, 1970; Carver, 1972). They also contain variable amounts of montmorillonite and at some localities, traces or generous quantities of clinoptilolite (mostly confined to intercalated clay-rich beds). Although early workers (Smith, Johnson & Langdon, 1894) reported identifiable siliceous microfossils (diatoms, radiolarians, sponge spicules) in the 40 m thick opaline claystones of Alabama, most subsequent investigators (Grim, 1936; Reynolds, 1970) dismissed these as vagrant forms entombed in a rapidly deposited sediment. That sediment was interpreted to have comprised vast quantities of volcanic ash which supposedly blanketed brackish and near-shore coastal areas at various times during the Early and Middle Eocene. This interpretation was accepted by workers in South Carolina where opaline claystones had long been considered to be devoid of siliceous microfossils (Heron, 1969). Gibson & Towe (1971) and Mattson & Pessagno (1971) noted that the Alabama and South Carolina opaline claystones are time equivalent to the Lower-Middle Eocene Horizon A and A″ cherts drilled by DSDP Legs 1, 2 and 4 in the North Atlantic and Caribbean. In order to explain a synchronous deposition of opaline-rich sediments of similar mineralogy over a wide geographical range of environments ranging from deep sea to nearshore and brackish, these authors embellished on Heron's (1969) suggestion of volcanic activity to the south (Caribbean region) by invoking the production of appreciable quantities of ash for dispersal by atmospheric and water currents throughout the Gulf of Mexico, Eastern Seaboard, and North Atlantic Ocean. Such ash, presumed to be rhyolitic or dacitic in nature, would have been responsible for the formation of opaline claystones on the continental shelves, and opaline cherts in the deep sea.

With further investigation, however, a number of flaws have appeared in the volcanic theory. Weaver & Wise (1973a), Wise & Weaver (1973) and Wise *et al.* (1974) studied fracture surfaces of opaline claystones via scanning electron microscopy and discovered moulds of siliceous microfossils in all of the presumed unfossiliferous volcanogenic materials from the Eastern Coastal Plain (South Carolina and Georgia). These fossils could readily be identified as marine diatoms, sponge spicules, and radiolarians (example, Fig. 2; compare with Fig. 3). These fossils, plus the general similarity of the material to the porcelanite of the Monterey Formation and the Kerguelen chert, suggest that many more siliceous microfossils may have been present in the original deposit than are now preserved as moulds. Secondly, the presence of the

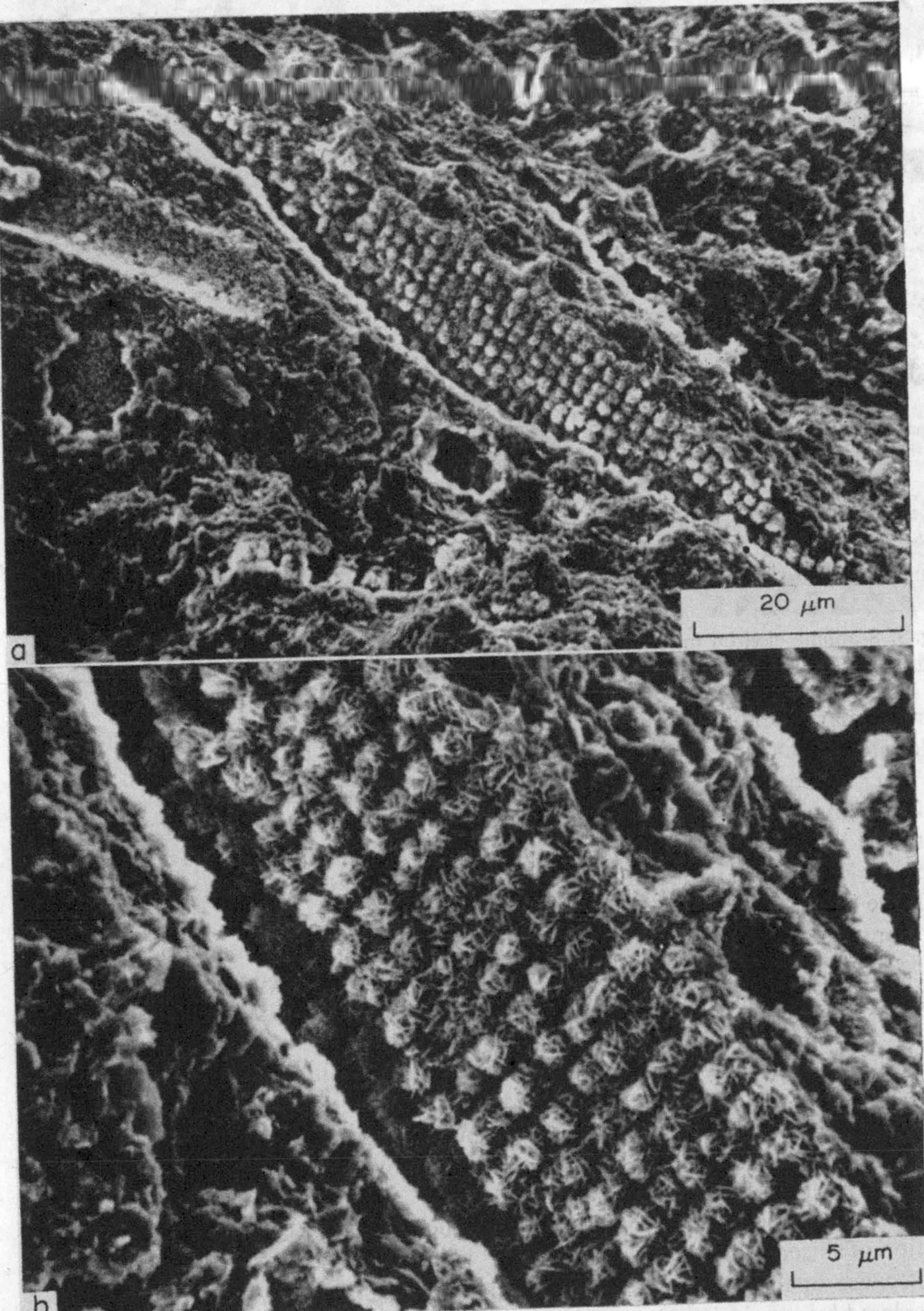

Fig. 2. Microfossil moulds in the Upper Eocene Twiggs Clay member of the Barnwell Formation, southeastern coastal plain, U.S.A. (a) Pennate diatom (*Cymatosera* sp.). (b) A high magnification scanning electron micrograph of this diatom mould reveals parallel rows of alpha-cristobalite spherulites approximately 2 μm in diameter which reflect the relict structure of the diatom frustule surface.

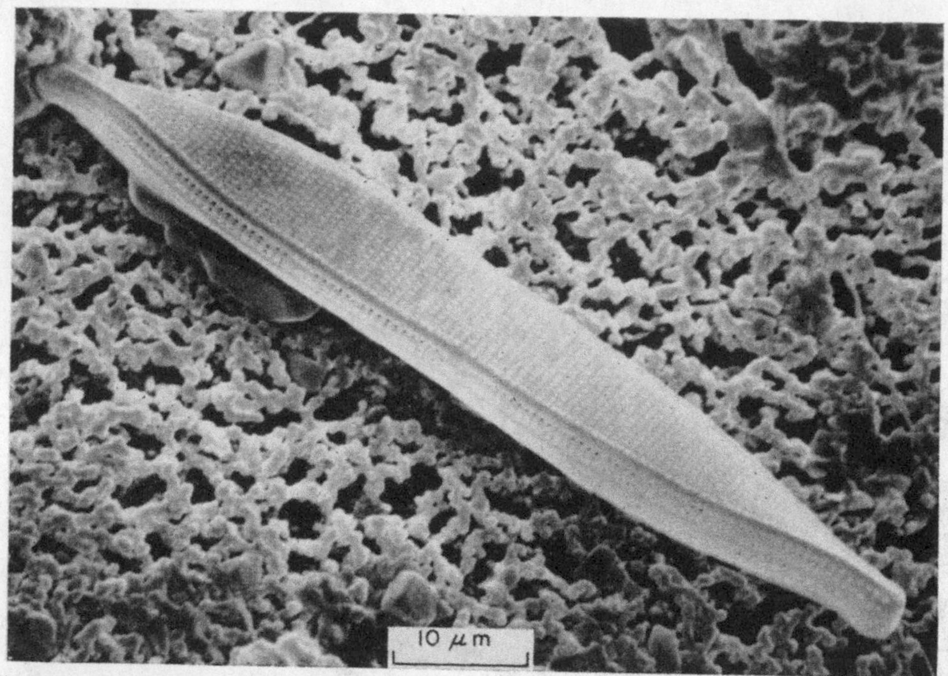

Fig. 3. Scanning electron micrograph of a Recent pennate diatom, *Nityschia* sp.; for comparison with Fig. 2 (A. W. Collier and W. I. Miller, unpublished).

normal marine radiolarian fauna and a restudy of stratigraphic relationships in the field shows that those opaline claystones previously thought to have been deposited in brackish, back-bay environments are actually normal marine shelf sediments. Thirdly, one prominent 10 m thick opaline claystone in Georgia (Twiggs Clay mbr, Barnwell Fm), not discussed by Gibson and Towe, is of Late Eocene age and was not deposited during the period of rhyolitic volcanism in the Caribbean defined by Gibson & Towe (1971), and Mattson & Pessagno (1971).

Considering the above, Wise & Weaver (1973) concluded that the opaline claystones of the southeastern Coastal Plain which they studied are actually highly altered diatomite deposits, not ashes, and that their early diagenesis followed the pattern inferred for the opaline cherts of the Kerguelen Plateau (Wise *et al.*, 1972). The presence of diatom-rich sediments on the continental shelves of the eastern United States is not unknown (e.g., the Miocene Calvert diatomite of Maryland), and such lithologies deposited during the Eocene can reasonably be explained by the distribution of ocean current patterns postulated by Ramsay (1971). These current patterns would have made nutrient-rich waters available to plankton in the Gulf of Mexico, Caribbean, and western North Atlantic areas, thus stimulating high production of siliceous organisms.

Because opaline claystones of the southeastern United States were for so long considered by a majority of geologists to be volcanic in origin, we now find it difficult to suppress the suspicion that other marine cherts considered to be of volcanic origin may, if carefully scrutinized, also be shown to have been biogenically derived.

NODULAR CHERT

Nodular cherts are generally considered to comprise those types which form in carbonate host rock (Dunbar and Rodgers, 1957; Heath & Moberly, 1971). Well-formed nodules in relatively young deep-sea chalks or limestones typically consist of two parts: an inner quartz core surrounded by an outer rind (up to several millimetres thick) composed of isotropic silica. The rind may consist totally of disordered cristobalite or may consist of a mixture of cristobalite and cryptocrystalline chalcedony. The proportions of cristobalite and cryptocrystalline chalcedony in the latter situation are difficult to assess by X-ray analysis because of the common occurrence of fibrous chalcedony in microfossil tests of the outer zone. A third zone may be distinguished surrounding the nodule in which cristobalite lepispheres are deposited within the interstices of the carbonate host rock. Foraminiferal tests in this region may also be filled or partially filled with cristobalite lepispheres or fibrous chalcedony.

Based on the above observations, two basic theories of nodular chert formation are currently under debate. One is a 'maturation' theory advocated by Heath & Moberly (1971), Weaver & Wise (1972), and Berger & von Rad (1972). The other is a 'quartz precipitation' theory advanced by Lancelot (1973).

The maturation theory holds that chert nodule formation begins with precipitation in the host rock of disordered cristobalite which, through time, will convert to chalcedonic quartz. Quartz conversion begins in the centre of the nodule and conceivably might continue while more cristobalite is precipitated at the periphery of the nodule. Heath & Moberly (1971) recognize four stages of development in chert-nodule formation (see also Davies & Supko, 1973).

(1) Infilling of empty foraminiferal chambers with chalcedony or chalcedonic quartz.

(2) Replacement of the groundmass of the carbonate with fine-grained cristobalite.

(3) Replacement of foraminiferal tests by chalcedony or chalcedonic quartz.

(4) Final infilling of all the voids and pore spaces in the rock with silica; inversion of the cristobalite matrix to quartz.

The quartz precipitation theory considers primary quartz precipitation to be the driving mechanism behind nodule formation with disordered cristobalite a by-product of the process. According to Lancelot (1973, p. 397), the sequence is as follows.

(1) Quartz precipitates directly in the carbonate matrix where clay minerals are rare and dispersed.

(2) The nodule develops outwards by accretion, and in the process all the clay minerals and dissolved cations that cannot be accommodated in the quartz structure are excluded and move along a 'quartzification front'. This can be achieved only because of the high permeability of the sediment, allowing relatively free circulation of the interstitial waters.

(3) At the periphery of the quartz nodule the concentration of clay minerals and foreign cations increases, while that of dissolved silica decreases because of limited supply. These conditions favour precipitation of disordered cristobalite that makes the rim commonly observed around the nodule.

In support of his hypothesis, Lancelot (1973) cites the tendency for nodules in relatively young carbonates to be quartz-rich whereas some porcelanites in clayey sequences resist conversion to quartz in sediments as old as Early Cretaceous (Zemmels & Cook [1973], however, do note an increase in the ratio of tridymite/cristobalite layers in these older porcelanites, indicating a possible re-ordering of the structure with time). Lancelot (1973) also emphasizes his belief that an abundance of clay minerals acting as a cation exchange reservoir and a permeability barrier, favour disordered cristobalite precipitation whereas the paucity of these minerals (as would be the case in a carbonate host rock) favours direct precipitation of quartz.

We find Lancelot's hypothesis difficult to apply in a number of instances. First, Wise & Hsü (1971) and Wise & Kelts (1972) observed early silicification phenomena in a highly permeable, white Oligocene nannofossil chalk which contained a negligible amount of clay minerals. The first and only formed silica precipitate, which constitutes 5–10% of the rock, was disordered cristobalite in the form of lepispheres, not quartz as Lancelot has predicted. Other significant occurrences of disordered critobalite in chalks and silicified limestones have been documented by Pimm, Garrison & Boyce (1971, Pls 27 and 28) and Heath (1973). Secondly, our study of the Kerguelen chert (Wise *et al.*, 1972) shows unequivocably that disordered cristobalite can form in a highly permeable siliceous sedimentary environment free of clay minerals. Again, the opposite of Lancelot's prediction proves to be the case.

We do not question the accuracy of the empirical data Lancelot (1973) has gathered to support his hypothesis, but we do point out that strong correlations of silica mineralogy to host rock lithology are not evident in all situations examined by us. In some cases, it may be possible to explain apparent discrepancies in Lancelot's correlations by postulating cation sources other than clay mineral reservoirs. For instance, the devitrification of volcanic glass might provide a cation source in a chalk otherwise free of impurities. However, it is not yet evident that a special cation source is necessary for the formation of disordered cristobalite in marine sediments. Perhaps any relationship between the carbonate matrix and a predominance of quartz in nodular chert could be a function of the carbonate replacement process or the chemistry of the carbonate itself. Another factor which should be critically important in determining the predominant silica phase present is the thermal history of the chert in question (Heath & Moberly, 1971; Heath, 1973; Heath, personal communication, 1973). For example, high heat flow may account for quartz present in relatively young (Miocene) chert nodules of the Eastern Pacific (see Heath, 1973). At present, each case will have to be evaluated on its own merits, and careful compilation of data will be necessary to test further Lancelot's interesting hypothesis.

We are impressed by Heath's documentation of high cristobalite-to-quartz ratios (as high as 450) in the geologically young (Miocene) cherts recovered from chalks and limestones of the Panama Basin during Leg 16 of the Deep Sea Drilling Project. Heath (1973, p. 609) also finds that the 'oldest cherts (Campanian at DSDP 163) contain roughly equal proportions of quartz and cristobalite, suggesting that the stabler polymorph (quartz) increases with time at the expense of cristobalite'. Inasmuch as these observations accord well with our own (see also Heath & Moberly, 1971; Berger & von Rad, 1972), but apparently are not well explained by Lancelot's hypothesis as presently stated (Lancelot, 1973), we will relate our observations of chert-nodule formation in terms of the maturation theory. For illustration, we present a series of scanning electron micrographs of a fracture surface (Fig. 4a) through a Middle Eocene cristobalitic chert nodule and the adjacent chalk host rock. The sample was cored in the Central Pacific Basin (DSDP 7–64·1–11CC). The chert ('S', smooth portion of rock fragment, right side of Fig. 4a) is light grey in colour, and is surrounded by a whitish, weakly silicified nannofossil chalk ('C', rough textured portion of fragment, left side of Fig. 4a) which grades into a somewhat lithified nannofossil chalk (not shown). An enlargement of Fig. 4a reveals an irregular silicification front separating the dense chert (right) from the chalk matrix. A close-up view of the chalk

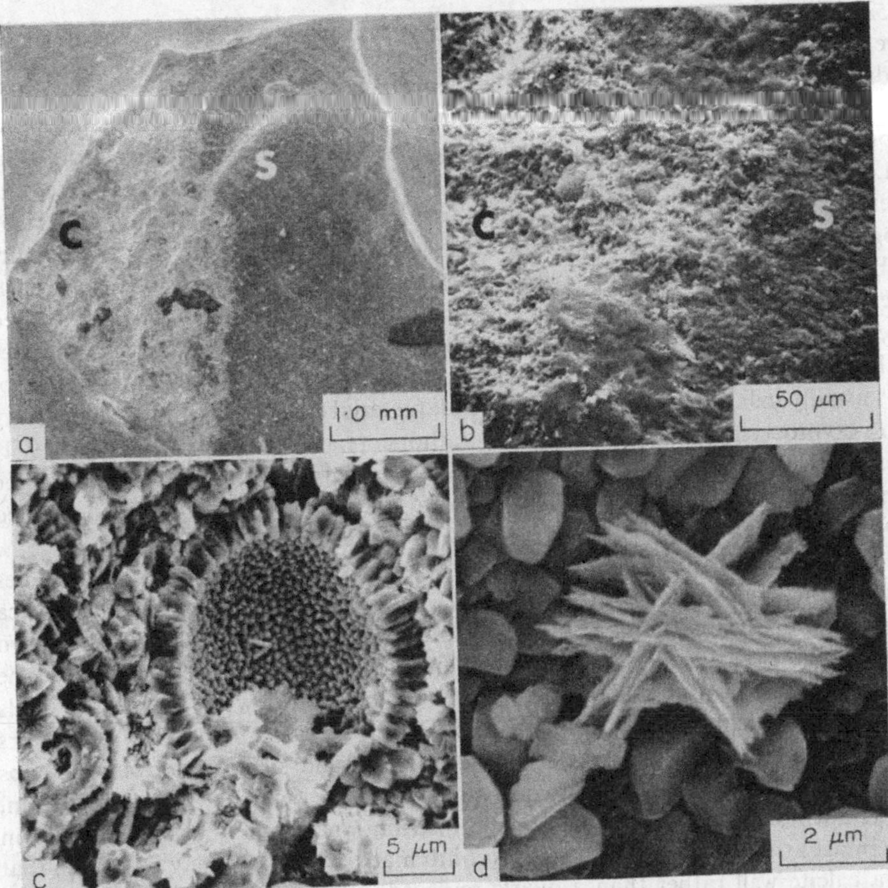

Fig. 4. Scanning electron micrographs of Middle Eocene (NP 16) nodular chert and partially silicified nannofossil chalk from the Central Pacific Basin (DSDP sample 7–64.1–11CC). (a) Fracture surface through the chert nodule(s) and silicified chalk matrix(c). (b) Detail view of the irregular silicification front along the periphery of the chert nodule. (c) Partially silicified calcareous nannofossil and foraminiferal matrix material, calcisphere in centre. (d) Interpenetrating crystallites of alpha-cristobalite nucleating around calcite crystals.

portion reveals a coccolith matrix and occasional spherical-shaped chambers. Termed 'calcispheres' by sedimentary petrographers, these spherical chambers, composed of doubly terminated radiating prisms, are apparently not foraminiferal tests, but rather some type of calcareous algal remains (Hans M. Bolli, personal communication, 1973), although the exact taxonomic affinity of these objects is difficult to assess. Perch-Nielsen (1971, Pl. 53, Fig. 2) illustrates an isolated example of a similar object from Eocene sediment. Note the small wedge-shaped particle in the centre of the calcisphere in Fig. 4c. This wedge is composed of two blade-shaped crystals of disordered cristobalite, and represents the beginning of silica deposition within the chamber. Note that there is no sign of an amorphous gel anywhere within the chamber. Instead, the initial cristobalite is deposited as ragged-edge blades which grow in discrete clusters (Fig. 4d), not as fibres growing out of a gel as recently suggested (Berger & von Rad, 1972, p. 853). These clusters develop into lepispheres which, in

Fig. 5a, fill the greater portion of a recrystallized foraminiferal chamber. In this instance, the wall structure of the foraminiferal test has recrystallized into granular, anhydral calcite crystals. The cristobalite lepispheres, having grown firmly against the chamber wall, form a cast of the interior surface of the chamber which appears as a pockmarked surface when the wall is separated from the filling by fracture (Fig. 5b). Lepispheres are characteristic free growth forms of disordered cristobalite, and are readily visible in optical thin sections (Fig. 5c) as well as in scanning electron micrographs. The continued deposition of silica, however, eventually fills entire chambers in the chalk zone with tightly intergrown spherulites which form solid casts of the interior chamber surfaces, and leave little indication of the original bladed nature of the material (note bladed structure remaining, however, within the free growth area at the bottom centre of Fig. 5d). At this stage of development, light micrographs usually indicate that a portion of the fill material consists of fibrous chalcedony and/or cryptocrystalline quartz. Some authors (e.g. Heath & Moberly, 1971; Lancelot, 1973) consider this chalcedony to be a primary precipitate; however, some of it may be an inversion product which grows at the expense of cristobalite deposited previously within the chambers.

A major problem in interpreting scanning electron micrographs such as these lies in distinguishing between the various mineral phases present, particularly where silica has partially replaced carbonate. A helpful tool in discriminating between these two minerals is the electron microprobe attachment of the scanning electron microscope which provides elemental analysis over the entire field of view or of select crystals mere microns in diameter. In dealing with silica fills of calcispheres or foraminifers (Fig. 6a, b), the electron microprobe allows calcium and silicon distribution maps (Fig. 6 c, d respectively) to be plotted of the subject area (Wise, Weaver & Güven, 1973). The white dot pattern of the calcium distribution map (Fig. 6c) of the area in Fig. 6b images the calcisphere wall and coccolith matrix. Thus one may be certain that this is a calcite wall rather than a quartz-replaced, recrystallized radiolarian test as has been suggested (Schlanger *et al.*, 1973, p. 415, Fig. 11). On the other hand, cristobalite fillings have been misidentified previously as aragonite (see discussion by Weaver & Wise, 1972, p. 57). However, in Fig. 6d, the silicon distribution map positively identifies the material filling the chamber as silica. Silica has also replaced portions of the chalk matrix (right portion of Fig. 6b, d). Spot probe analyses on the prisms of the calcite chambers in Fig. 5 indicate that none have been replaced by silica through solid–solid conversion involving an intermediary mineral phase. Instead, calcite wall replacement by silica apparently occurs through a molecular dissolution-reprecipitation reaction.

Aside from the normal calcite lithification experienced by the chalk matrix during early diagenesis (such as secondary calcite overgrowth; e.g. Wise & Hsü, 1971, Wise & Kelts, 1972, or Wise, 1973) or during compaction of the sediment (e.g. Schlanger *et al.*, 1973), a certain amount of calcite cementation may result directly from growth of the chert nodule. As the silicification front (Fig. 4a) moves through the chalk host rock, considerable amounts of chalk matrix are displaced by dissolution. Some of this calcite is lost to the rock, but some apparently is reprecipitated beyond the growth front as overgrowths on chalk components or as minute euhedral calcite crystals. In this manner, the chalk matrix immediately surrounding the chert nodule becomes even more tightly cemented. Figure 7a shows the outside surface of a calcisphere located near the silicification front which, on close examination (Fig. 7b), exhibits tightly

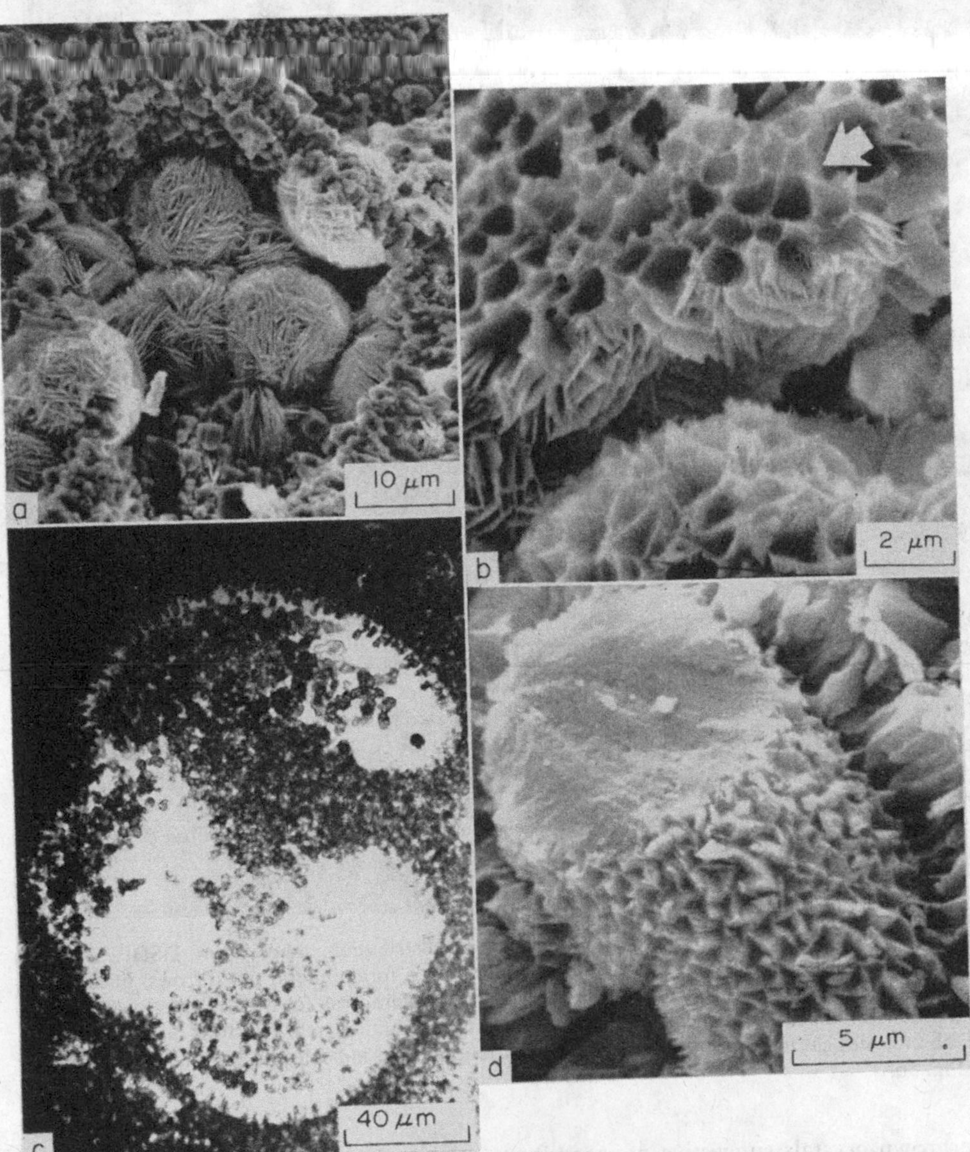

Fig. 5. Progressive sequence of silica filling in microfossil tests in Middle Eocene (NP 16) nodular chert and partially silicified chalk matrix (DSDP sample 7–64.1–11CC). (a) Cluster of cristobalite microspherulites (lepispheres) filling a recrystallized foraminiferal test chamber (scanning electron micrograph). (b) Cristobalitic spherulites detached from a foraminiferal or calcisphere test wall reveals moulds of euhedral terminations of the calcite wall crystals (arrow) (scanning electron micrograph). (c) Light micrograph of foraminiferal test partially filled with cristobalite microspherulites. (d) Fracture through calcisphere reveals solid plug of tightly intergrown silica microspherulites; bladed free growth surface preserved only at lower centre portion of micrograph.

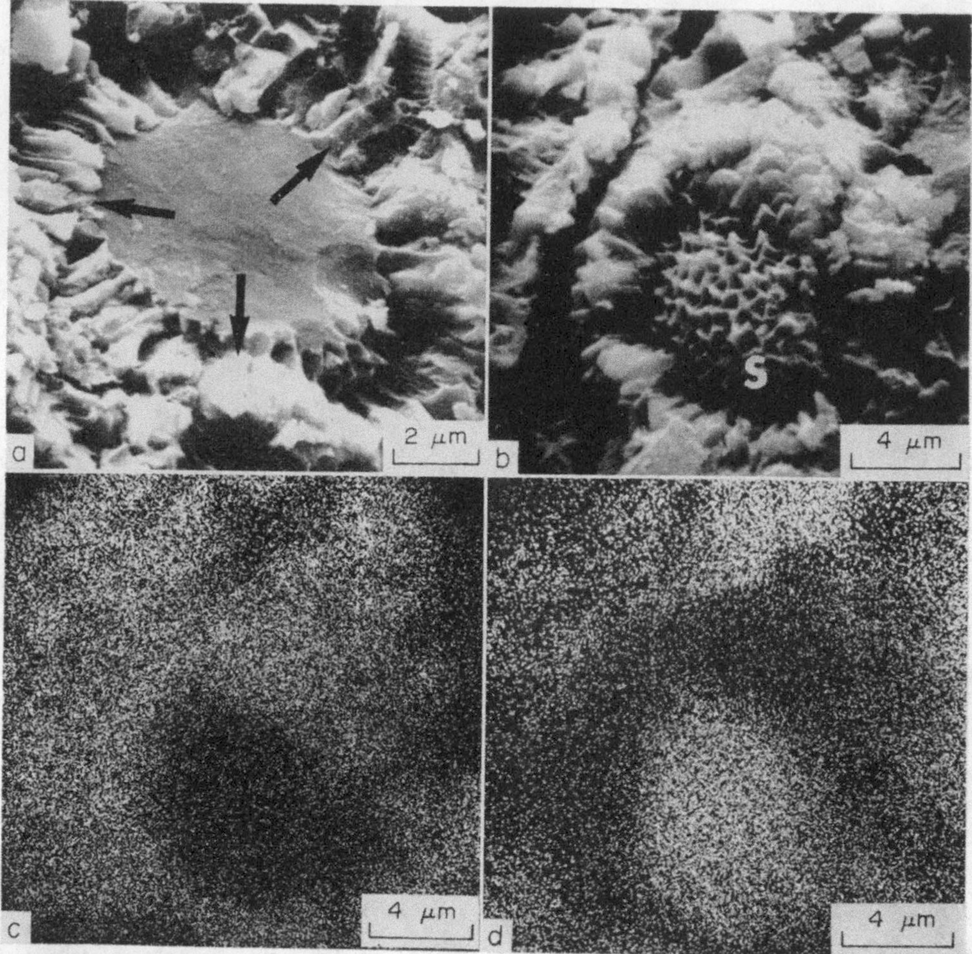

Fig. 6. Scanning electron micrographs and elemental distribution maps from DSDP sample 7–64.1–11CC. (a) Fracture surface through a calcisphere or a foraminiferal test (arrows) filled with dense quartz silica. (b) Calcisphere or recrystallized foraminiferal test (c) with euhedral wall prism terminations moulded by a silica filling(s). (c) Calcium distribution map (electron microprobe) of Fig. 5b indicates calcite distribution. (d) Silicon distribution map indicates a silica-filled test interior for Fig. 5b.

intergrown crystals suggestive of secondary overgrowth. More explicit, however, are euhedral calcite crystals which precipitated within the interstices of the chalk, some of which envelope portions of the coccolith matrix (Fig. 7c). Such crystals are common close to the silicification front (Fig. 7d). Also abundant within the chalk matrix near the silicification front are cristobalite lepispheres (Fig. 7d). As more lepispheres are deposited, more of the chalk matrix is dissolved and driven out by diffusion through the interstitial waters. At the boundary of the silicification front (Fig. 8a), little calcite remains, and the fracture surface takes on a somewhat smooth appearance except for occasional punctuations formed by small holes (centre of micrograph). Close examination (Fig. 8b) shows these to be hollow centres of silica microspherulites. The significance and genesis of these forms is not clear. Beyond this narrow zone the chert

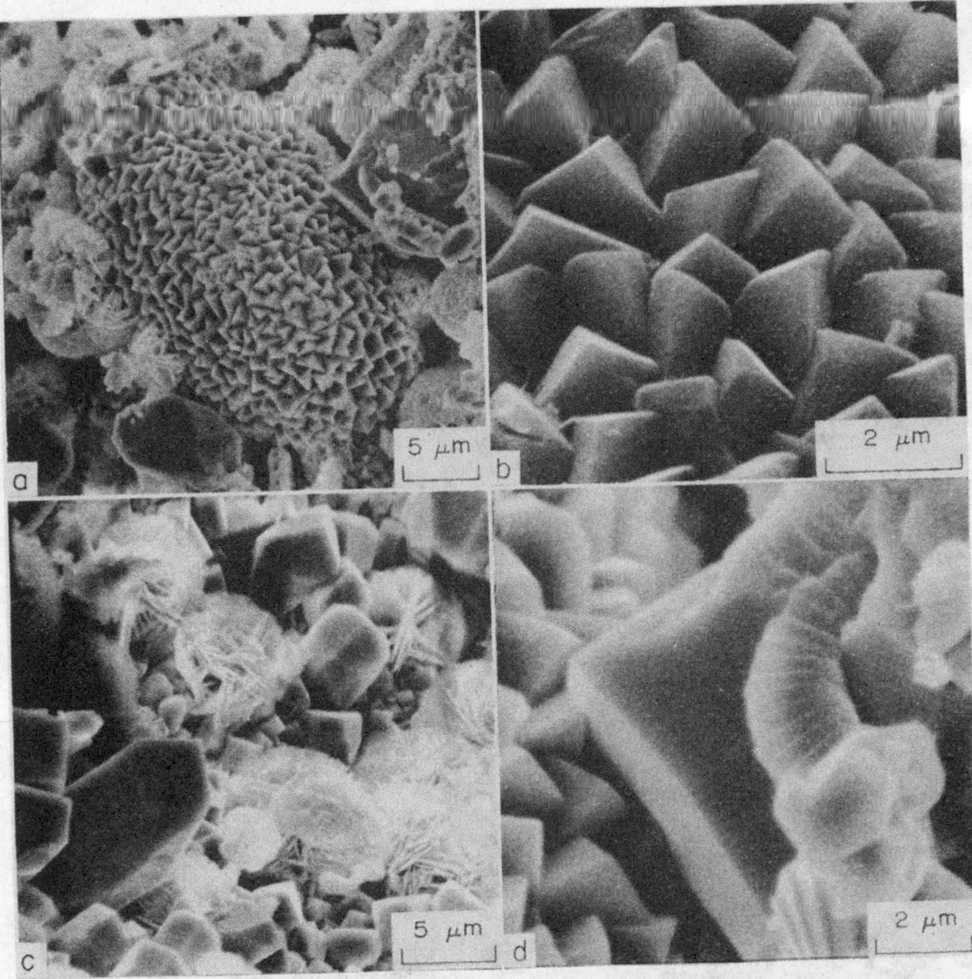

Fig. 7. Scanning electron micrographs of Middle Eocene silicified chalk matrix (DSDP sample 7–64.1–11CC). (a) Apparent secondary calcite overgrowths on the outer surface of a calcisphere (recrystallized foraminifer test?). (b) Detail of Fig. 7a. (c) Large euhedral calcite authigenic calcite crystal which has partially enveloped a coccolith plate. (d) Reprecipitated euhedral crystals of calcite and alpha-cristobalite spherulites within the chalk matrix adjacent to the silicification front.

matrix is dense, and exhibits a rough conchoidal fracture (Fig. 8c). Some siliceous microfossils within dense nodular (or bedded) cherts may be preserved intact or in various stages of dissolution (e.g. Fig. 8d).

Because fractures through all tightly silicified portions of chert nodules exhibit smooth, featureless surfaces, light microscopy is most practical for studying silica phases within these nodules. As stated previously, the outer cortex of well-developed nodules is composed of an isotropic groundmass which may contain varying mixtures of disordered cristobalite and cryptocrystalline quartz. The true proportions are impossible to estimate petrographically, and can only be estimated by quantitative X-ray analysis, although such analyses are usually complicated by the presence of fibrous chalcedony fillings within microfossil tests which cannot be isolated from the

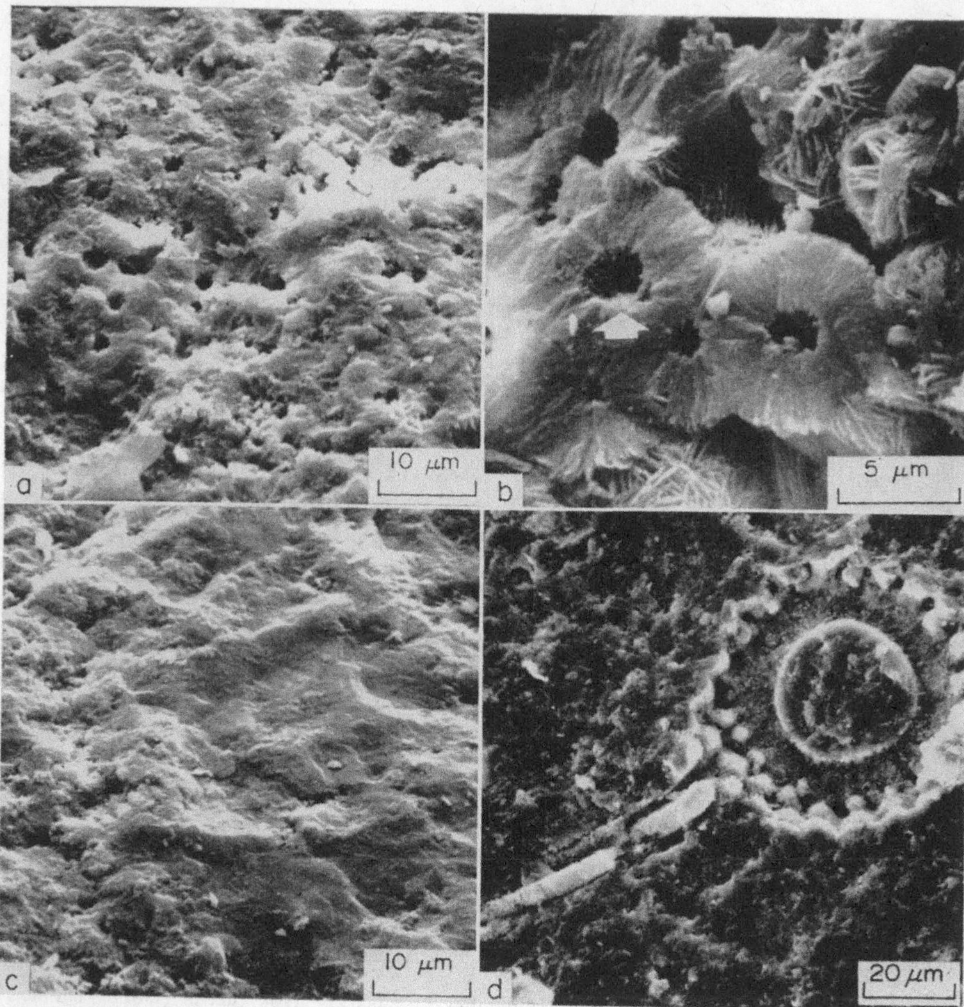

Fig. 8. Scanning electron micrographs of the transitional zone between silicified chalk matrix and chert nodule from DSDP sample 7–64.1–11CC. (a) Transitional contact between the silicified chalk matrix and the chert nodule. (b) High concentration of cristobalitic spherulites within the transitional zone surrounding the peripheral edges of the chert nodule. The arrow points to the hollow central area of one of the spherulites. (c) Dense chert nodule matrix (isotropic silica). (d) Partially dissolved siliceous microfossil in nodular chert.

sample. Percentages may range from nearly pure cristobalite to a predominance of quartz. The quartz content generally increases gradationally toward the centre of the nodule. Some nodules (such as DSDP sample 7–64·1–11CC) do not have a well-defined quartz interior, although in others, the quartz inversion front is quite sharply defined. An excellent example of the latter is DSDP sample 8–70–2–0, Piece #2, in which the isotropic 'transition' zone forming the outer rind of the nodule is 5 mm wide. Figure 9a illustrates, under crossed nicols, the quartz inversion front in that nodule (isotropic silica on the left, chalcedonic quartz exhibiting a salt and pepper-like birefringent pattern on the right). The isotropic silica in this nodule is composed of disordered cristobalite apparently mixed with an appreciable amount of finely dispersed cryptocrystalline quartz. Fibrous chalcedony is present within microfossil tests; therefore, exact percentages of cristobalite/chalcedony within the isotropic matrix cannot be ascertained. The proportion of cristobalite present, however, is sufficient to maintain the isotropic optical properties of this portion of the nodule. To the right of the inversion front, however, no opaline silica is present. The inversion, therefore, is quite complete.

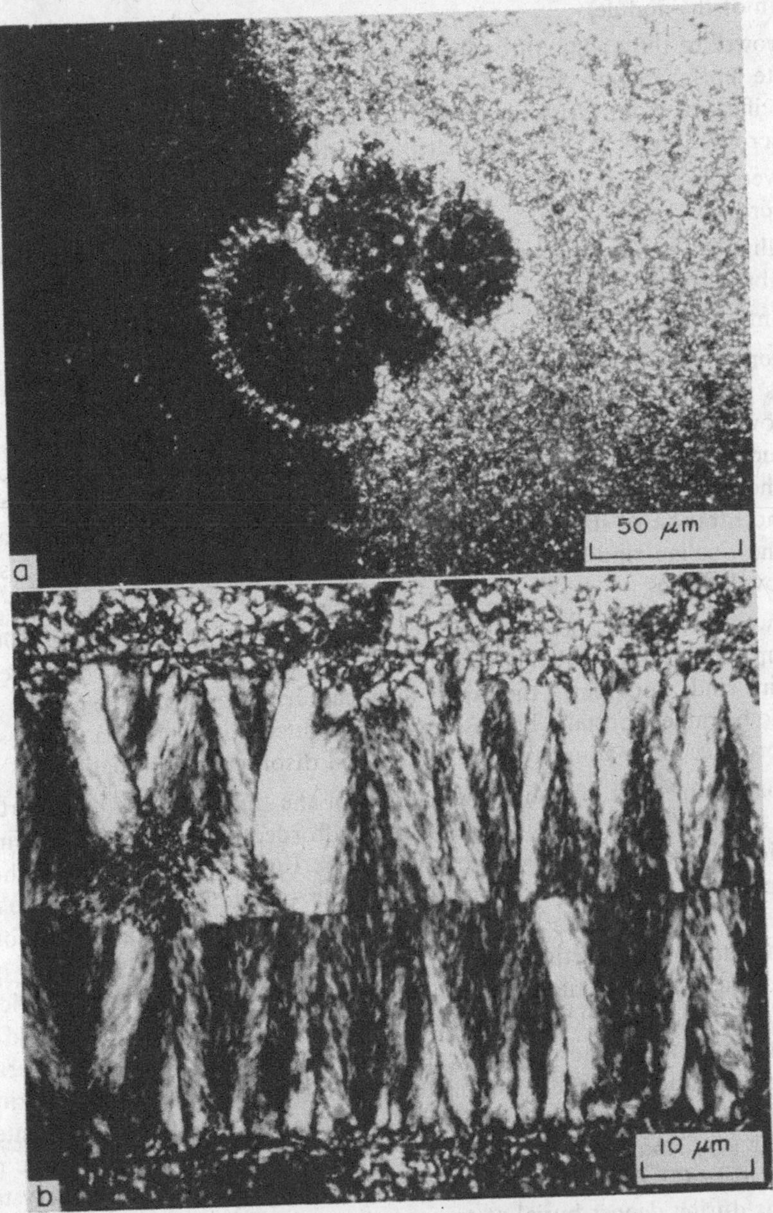

Fig. 9. (a) Interior of chert nodule showing boundary between isotropic silica of outer rind or cortex (left) and chalcedony of the central core (right); DSDP sample 72–7–0, piece #2, light micrograph, crossed nicols. (b) Fibrous chalcedony vein filling brecciated radiolarian chert; DSDP sample 70B–2–1, piece #19, light micrograph, crossed nicols.

In summary, our interpretation of chert nodule formation is as follows.

(1) Mobilization of silica within interstitial fluids.

(2) Precipitation of silica as authigenic disordered cristobalite lepispheres within the interstices of the chalk or limestone matrix and within microfossil tests.

(3) Development of a dense opaline nodule nucleus with concomitant dissolution and expulsion of host rock carbonate (part of the calcite may be reprecipitated beyond the margin of the nodule).

(4) Growth of the nodule through accretion of opaline silica at the margins with cristobalite lepispheres precipitated within the surrounding chalk matrix in advance of the silicification front. Cristobalite (and fibrous chalcedony???) may be precipitated within microfossil tests in the chalk matrix or within the nodule itself.

(5) Inversion of cristobalite to quartz beginning in the centre of the nodule and within microfossil tests.

(6) Filling of remaining pore space and microfossil tests within the nodule by primary fibrous chalcedony and cryptocrystalline quartz.

(7) Continuation of 4, 5 and 6 until silica supply is exhausted or shut off.

An alternative explanation for our observations is provided by Lancelot's (1973) hypothesis which would suggest the following events. As the quartz nucleus of the nodule grows, clay minerals and impurities which cannot be accommodated in the quartz structure would be pushed aside and driven in advance of the 'quartzification'. Then, as the silica supply declines, the ratio of silica/cations would drop and, because of the concentration of impurities, disordered cristobalite would be precipitated to produce the opaline rind of the nodule. With further decline in the silica supply, nodule growth ceases.

If one were to apply Lancelot's hypothesis to the case of DSDP sample 8–70–2–0, the build-up of impurities would have to be considerable in order to produce the 5 mm thick opaline rind. Apparently, the supply of silica during active nodule growth would also have to be rather constant, because any drastic fluctuations in silica supply should probably produce alternating layers of quartz and disordered cristobalite.

One possible method of testing the validity of the 'maturation' and 'quartz precipitation' theories for a given situation may be afforded by isotopic palaeotemperature techniques. Preliminary studies by Knauth & Epstein (1973) indicate that the palaeotemperature of crystallization of silica phases of deep-sea cherts can be determined and that such data can be correlated with depth of burial (and therefore the relative time) at which a particular silica phase was precipitated. Thus, if the quartz nucleus of a given chert nodule yields a higher temperature than the cristobalitic rind, a maturation process would be indicated since the quartz inversion would have occurred at a later time (and greater depth of burial, therefore, higher temperature) than the precipitation of the cristobalite (assuming that isotopic equilibrium is achieved). Interestingly enough, in their initial studies of cherts from DSDP sites 61, 64, 70B, and 149, Knauth & Epstein (1973, p. 695) find that oxygen isotope ratios 'suggest that opal-CT forms before significant burial while granular microcrystalline quartz forms during deeper burial at somewhat warmer temperatures. The temperature at which opal-CT forms may thus be approximately equal to the temperature of the overlying bottom water'. Further studies of this type on all forms of quartz in deep-sea cherts would be of considerable interest.

MICROBRECCIATION AND VEIN FILLING

A final event in chert formation may be microbrecciation and vein filling, phenomena common to both nodular and bedded cherts (e.g. Pimm *et al.*, 1971; Heath & Moberly, 1971; von der Borch, Galehouse & Nesteroff, 1971; Laughton *et al.*, 1972). Veins may show cross-cutting relationships, and are usually filled by length-fast chalcedony (Fig. 9b), often emplaced in two or more successive generations. Cristobalite and mosaic quartz fillings are also observed. Similar vein fillings are not uncommon in Palaeozoic bedded cherts, such as novaculite deposits (E. F. McBride, personal communication, 1973). Microbrecciation or repeated fracturing may be due to compaction by overlying sediments, but is likely to be aided by differential solution within the sedimentary sequence. Both chert and chalk diagenesis involve dissolution and reprecipitation of organic sediment. If not effected on a uniform scale, this could certainly produce differential compaction within the sediment.

DISCUSSION AND CONCLUSIONS

Most students of silica diagenesis who have participated in the Deep Sea Drilling Project have been impressed by the magnitude of the biogenous contribution to deep-sea chert formation (e.g. Fischer *et al.*, 1970; Heath & Moberly, 1971; Wise & Weaver, 1972; Berger & von Rad, 1972; Hay, 1973; Heath, 1973; Lancelot, 1973). In view of the pronounced undersaturation of sea water with respect to silica, these workers recognize that any primary input of dissolved silica into the oceans is immediately utilized by siliceous organisms for skeletal secretion, and that large amounts of amorphous biogenic silica are delivered to the sea floor for burial in deep-sea sediments, particularly in zones of high planktonic productivity or sponge bank environments. As explained by Lancelot (1973, p. 387), 'interstitial water is generally undersaturated with respect to amorphous silica and supersaturated with respect to the crystalline forms. This implies that amorphous silica can be selectively dissolved and reprecipitated as any of the crystalline forms'. Krauskopf (1956, 1959) has shown that the role of pH in such reactions is negligible at values typical of interstitial fluids of most deep-sea sediments. We show that biogenous opal can be readily converted into opaline chert regardless of host rock lithology.

Nevertheless, a strong historical bias toward exotic theories of chert formation still pervades the literature. For instance, regional volcanism is invoked to explain the Horizon A cherts of the North Atlantic. On close examination, however, supposedly pure volcanic deposits on land associated with these deep-sea cherts are found to be altered diatomites instead of ashes. Thus, volcanic glass as an immediate source of significant quantities of ocean chert has become suspect. With the recent discovery, during Deep Sea Drilling operations, of widespread Eocene chert through portions of the Pacific, Indian and Southern Oceans as well as the Caribbean and North Atlantic, the idea of regional volcanism as the immediate silica source for all these chert deposits becomes untenable. Suggestions that the dissolution of volcanic glass may provide a source of nutrient to stimulate planktonic blooms (Gibson & Towe, 1971) provide a possible area of compromise between the volcanic and biogenic schools of thought, in that dissolution of volcanic glass could be a cause for increased biogenic productivity as well as an *ultimate* silica source (along with terrestrial run-off, etc.). In such a case, biogenic supply would still remain the *immediate* silica source for chert formation. This theory, however, needs support of independent evidence. Far more attractive at present is the suggestion based on documented study (Berger, 1970) that biogenous silica deposits form where high planktonic productivity is stimulated by nutrient-laden waters brought to the surface by favourable ocean circulation patterns. Ramsay's (1971) palaeocurrent model, proposed as an explanation for the biogenic silica source for Horizon A, is plausible, and could be modified to explain Indo-Pacific Eocene chert deposits. In the latter region, ocean circulation patterns were apparently in a state of flux during mid-Tertiary times, culminating with radical changes in velocity and direction near the end of the Eocene as a result of continental separation (Frakes & Kemp, 1972; Kennett *et al.*, 1972); thus, possibilities for overturn and upwelling of nutrient-rich waters in various sectors of those oceans were definitely enhanced during this period of earth history.

The classical bias against biogenic silica sources for chertification remains particularly strong with regard to bedded chert formation. For instance, Davies & Supko (1973, p. 389) write that:

> The formation of the bedded cherts poses more of a problem. Bedded cherts seem to be found only in association with siliceous ooze or abyssal clay, rather than calcareous ooze. Further, the source of the silica is unknown but it is apparently not organic since no recognizable biogenous silica remains are found associated with these cherts.

A detailed rebuttal of the above statement would be redundant since the very definite association of biogenous silica with most deep-sea bedded cherts and porcelanites has already been discussed in this article, and should not be repeated here. We do note, however, that the latest palaeogeographic and electron microscopic work done on the classic bedded cherts of the Franciscan Formation of California (Pessagno, 1973) indicate an undoubted biogenic silica source for those units. In addition, the significance of a few traces of siliceous microfossils as indicators of wholesale dissolution of an original diatomaceous sediment during the formation of bedded chert (see page 305) cannot be overemphasized.

Numerous instances have been encountered during Deep Sea Drilling in which a few ghosts or moulds of siliceous microfossils in a chert provide the only 'fingerprints' to indicate a biogenous silica source. When such clues are combined with mineralogical and petrographic data, a biogenous source for the cherts in question may seem quite reasonable. Lancelot (1973) reports such an example south of the Hawaiian Islands at DSDP Site 164 where numerous poorly fossiliferous (Lancelot, 1973, personal communication) Upper Cretaceous chert beds are intercalated with non-fossiliferous zeolitic brown clays. X-ray determinations show the chert to be the 'opal-CT' variety of Jones & Segnit (1971). The absence of 'opal-C' or obvious signs of wholesale bentonite or tuff deposition (neither are reported at Site 164) argue against a volcanic origin for the chert in question. However, a biogenous silica source is especially appealing in view of the fact that Site 164 is marginal to the equatorial belt of high planktonic productivity, and probably experienced biogenous opal deposition during the Late Cretaceous (particularly since plate motion should have positioned the site closer to the equatorial belt during the Mesozoic than at present).

Our basic conclusions may be listed as follows.

(1) The vast majority of oceanic cherts deposited since Mid-Palaeozoic times (and

probably since the beginning of that era) are biogenic in origin. This applies to the immediate source of silica precipitated *in situ* during early diagenesis.

(2) Opal variety (opal C or opal-CT), presence or absence of bentonite beds as well as volcanic alteration minerals, sediment [illegible], proximity to planktonic productivity zones, and contained fossils are factors which should all be considered when determining the immediate silica source of an oceanic chert.

(3) The occurrence of a few traces of siliceous microfossils in a chert sequence may well indicate that many more such fossils were present in the original sediment than are presently preserved, the majority having been dissolved during diagenesis.

(4) Exotic theories of chertification (e.g. direct precipitation of silica from sea waters, silica emanations from volcanic sources, vast ash falls) need not be invoked if traces of siliceous microfossils are present in the chert unit.

(5) Appreciable chert deposition begins with the precipitation of disordered cristobalite, regardless of host rock lithology.

(6) The maturation theory of chert nodule and bedded chert formation is considered valid by us. The 'quartz precipitation' theory of chert nodule formation provides an interesting alternative explanation for some observations, and should be studied further.

(7) Within the limits of resolution of optical and electronic equipment presently available, no amorphous gel phase has yet been detected in incipient oceanic chert; no such gel phase, therefore, is thought to exist during the formation of these rocks.

(8) Oceanic cherts may form wherever abundant biogenous opal is deposited, be it along shallow continental shelves and margins, submerged banks (e.g. Rockall Plateau [Laughton *et al.*, 1972, p. 411]), or in the deep sea.

ACKNOWLEDGMENTS AND RESPONSIBILITIES

Scanning electron micrographs were taken primarily by F. M. Weaver, who also performed the X-ray and petrographic analyses; the paper was written by S. W. Wise. Professors B. Vincent Hall and William W. Hay (University of Illinois) and Pasquale P. Graziadei (Florida State University) kindly made available Cambridge Mark IIA scanning electron microscopes at their institutions; Mrs Olive Stayton, and Messrs Paul F. Ciesielski, Norman Peterson, Ronald D. Parker, William I. Miller, III, and Frank H. Wind provided valuable technical assistance. We thank Professor Albert W. Collier and Mr Miller (FSU) for use of their unpublished micrograph of a Recent diatom, and Messrs Ciesielski and Miller for taking the micrographs in Fig. 2. Messrs Andrew M. Gombos and David W. McCollum (FSU) provided helpful instruction on diatom test morphology. Professor Lyman D. Toulmin (FSU) guided us to many important localities in the field, and Professor S. Duncan Heron (Duke University) provided numerous samples of South Carolina opaline claystone. We are grateful to Dr Yves Lancelot (Lamont-Doherty Geological Observatory) for critical and helpful review of the manuscript, and for challenging discussion of points yet under debate. Dr G. Ross Heath (Oregon State University) contributed helpful discussion.

We especially thank the Deep Sea Drilling Project for supplying samples through the courtesy of the U.S. National Science Foundation. Support provided by a Sigma Xi research grant to F. M. Weaver, FSU Faculty Research Grants to S. W. Wise, and by the Donors of the Petroleum Research Fund administrated by the American

Chemical Society. National Science Foundation Grant GB-35952 to S. W. Wise provided, in part, the optical equipment.

This study was initiated at the Swiss Federal Institute of Technology (Zürich) in collaboration with Professor Kenneth J. Hsü during the tenure of an NSF Post-doctoral Fellowship held by S. W. Wise. Professor Hans M. Bolli (host professor) kindly provided laboratory facilities, and Dr Hans-Ude Nissen made available the scanning electron microscope at the ETH Laboratory for Electron Microscopy, Hönneggberg.

REFERENCES

Barton, D.C. (1918) Notes on the Mississippian chert of the St. Louis area. *J. Geol.* **26,** 361–374.

Berger, W.H. (1970) Biogenous deep-sea sediments: fractionation by deep-sea circulation. *Bull. geol. Soc. Am.* **81,** 1385–1402.

Berger, W.H. & von Rad, U. (1972) Cretaceous and Cenozoic sediments from the Atlantic Ocean. In: *Initial Reports of the Deep Sea Drilling Project,* Vol. XIV (D. E. Hayes and A. C. Pimm *et al.*), pp. 787–954. U.S. Government Printing Office, Washington.

Bramlette, M.N. (1946) The Monterey Formation of California and the origin of its siliceous rocks. *Prof. Pap. U.S. geol. Surv.* **212,** 57 pp.

Bramlette, M.N. & Posnjak, E. (1933) Zeolitic alteration of pyroclastics. *J. Miner. Soc. Am.* **18,** 167–171.

Calvert, S.E. (1971) Nature of silica phases in deep sea cherts of the North Atlantic. *Nature, Phys. Sci.* **234,** 133–134.

Carver, R.E. (1972) Absorption characteristics of opaline clays from the Eocene of Georgia. *Spec. Publs Fla Dept Nat. Resourc. Bur. Geol.* **17,** 91. pp.

Davies, T.A. & Supko, P.R. (1973) Oceanic sediments and their diagenesis: some examples from deep-sea drilling. *J. sedim. Petrol.* **43,** 381–390.

Davis, E.F. (1918) The radiolarian cherts of the Franciscan group. *Univ. Calif. Publs Bull. Dep. Geol.* **11,** 235–432.

Deffeyes, K.S. (1959) Zeolites in sedimentary rocks. *J. sedim. Petrol.* **29,** 602–609.

Dunbar C.O. & Rodgers, J. (1957) *Principles of Stratigraphy,* pp. 356. Wiley and Sons, New York.

Ernst, W.G. & Calvert, S.E. (1969) An experimental study of the recrystallization of porcelanite and its bearing on the origin of some bedded cherts. *Am. J. Sci.* **267-A,** 114–133.

Fischer, A.G., Heezen, B.C., Boyce, R.E., Bukry, D., Douglas, R.G., Garrison, R.E., Kling S.A., Krasheninnirov, V., Lisitzin, A.P. & Pimm, A.C. (1970) Geologic history of the western North Pacific. *Science,* **168,** 1210–1214.

Flörke, O.W. (1955) Structuranomalien bei Tridimyt und Cristobalit. *Ber. dt. keram. Ges.* **10,** 217–223.

Frakes, L.A. & Kemp, E.M. (1972) Influence of continental positions on Early Tertiary climates. *Nature, Lond.* **240,** 97–100.

Frondel, Clifford (1962) *The System of Mineralogy, Vol. III, the silica minerals,* pp. 334. John Wiley and Sons, New York.

Gibson, T.G. & Towe, K.M. (1971) Eocene volcanism and the origin of Horizon A. *Science,* **172,** 152–154.

Goldstein, A., Jr & Hendricks, T.A. (1953) Siliceous sediments of Ouachita facies in Oklahoma. *Bull. geol. Soc. Am.* **64,** 421–441.

Grim, R.E. (1936) The Eocene sediments of Mississippi. *Bull. Miss. St. geol. Surv.* **30,** 240 pp.

Hastings, E.L. & McVay, T.N. (1963) Tallahatta diatomite, Choctaw and Clarke Counties, Alabama. *Invest. Rep. U.S. Bur. Mines,* **6271,** 1–12.

Hathaway, J.C. & Sachs, P.L. (1965) Sepiolite and clinoptilolite from the Mid-Atlantic Ridge. *Am. Miner.* **50,** 852–867.

Hay, R.L. (1966) Zeolites and zeolitic reactions in sedimentary rocks. *Spec. Pap. geol. Soc. Am.* **85,** 130 pp.

Hay, W.W. (1973) Significance of paleontologic results of Deep Sea Drilling Project, Legs 1–9. *Bull. Am. Ass. Petrol. Geol.* **57,** 55–62.

HEATH, G.R. (1973) Cherts from the Eastern Pacific, Leg 16, Deep Sea Drilling. In: *Initial Reports of the Deep Sea Drilling Project,* Vol. XVI (Tj. van Andel and G. R. Heath *et al.*), pp. 609–613. U.S. Government Printing Office, Washington.

HEATH, G.R. & MOBERLY, R. (1971) Cherts from the Western Pacific, Leg 7, Deep Sea Drilling Project. In: *Initial Reports of the Deep Sea Drilling Project,* Vol. VII (E. L. Winterer *et al.*), pp. 991–1007. U.S. Government Printing Office, Washington.

HERGENRODER, J.D. (1973) Stratigraphy of the Middle Ordovician bentonites in the Southern Appalachians. *Prog. Abs. geol. Soc. Am.* **5,** 403.

HERON, S.D. (1969) Mineralogy of the Black Mingo mudrocks. *S.C. St. Dev. Bd, Div. Geol., Geol. Notes,* **13,** 27–41.

HERON, S.D., ROBINSON, G.C. & JOHNSON, H.S. (1965) Clays and opal claystones of South Carolina Coastal Plain. *Bull. S.C. St. Dev. Bd, Div. Geol.* **31,** 64 pp.

INGLE, J.C. (1973) Pliocene-Miocene sedimentary environments and biofacies, Southeastern Los Angeles Basin—San Joaquin Hills area, Orange County, California. In: *Miocene Sedimentary Environment and Biofacies, Southeastern Los Angeles Bay,* pp. 1–6. SEPM Trip 1 Guidebook, 1973 Annual Meeting A.A.P.G., S.E.P.M., S.E.G.

JONES, J.B. & SEGNIT, E.R. (1971) The nature of opal I. Nomenclature and constituent phases. *J. geol. Soc. Aust.* **18,** 57–68.

KELLER, W.D. (1941) Petrography and origin of the Rex chert. *Bull. geol. Soc. Am.* **52,** 1279–1297.

KENNETT, J.P., BURNS, R.E., ANDREWS, J.E., CHURKIN, M., JR, DAVIES, T.A., DUMITRICA, P., EDWARDS, A.R., GALEHOUSE, J.S., PACKHAM, G.H. & VAN DER LINGEN, G.J. (1972) Australian-Antarctic continental drift, paleocirculation changes and Oligocene deep-sea erosion. *Nature, Phys. Sci.* **239,** 51–55.

KNAUTH, L.P. & EPSTEIN, S. (1973) Hydrogen and oxygen isotope ratios in cherts from the JOIDES Deep Sea Drilling Project. *Prog. Abs. geol. Soc. Am.* **5,** 694–695.

KNELLER, W.A. *et al.* (1968) *The properties and recognition of deleterious cherts which occur in aggregate used by Ohio concrete producers,* pp. 201. Research Foundation, University of Toledo, Toledo, Ohio.

KRAUSKOPF, K.B. (1956) Dissolution and precipitation of silica at low temperatures. *Geochim. cosmochim. Acta,* **10,** 1–27.

KRAUSKOPF, K.B. (1959) The geochemistry of silica in sedimentary environments. In: *Silica in Sediments* (Ed. by H. A. Ireland). *Spec. Publs Soc. Econ. Paleont. Miner., Tulsa,* **7,** 4–20.

LANCELOT, Y. (1973) Chert and silica diagenesis in sediments from the Central Pacific. In: *Initial Reports of the Deep Sea Drilling Project,* Vol. XVII (E. L. Winterer and J. I. Ewing *et al.*), pp. 377–405. U.S. Government Printing Office, Washington.

LAUGHTON, A.S. *et al.* (1972) Sites 116 and 117. In: *Initial Reports of the Deep Sea Drilling Project,* Vol. XII (A. S. Loughton *et al.*), pp. 395–671. U.S. Government Printing Office, Washington.

LOWENSTAM, H.A. (1948) Biostratigraphic studies of the Niagaran inter-reef formation in northeastern Illinois. *Pap. Illinois St. Mus. Sci.* **4,** 146 pp.

MALLARD, E. (1890) Sur la lussalite, nouvelle variété minéral cristallisée de silice. *Bull. Soc. fr. Minér.* **13,** 63–66.

MANSFIELD, G.R. (1927) Geography, geology, and mineral resources of part of southeastern Idaho. *Prof. Pap. U.S. geol. Surv.* **152,** 409 pp.

MATTSON, P.H. & PESSAGNO, E.A. (1971) Caribbean Eocene volcanism and the extent of Horizon A. *Science,* **174,** 138–139.

MILLOT, G. (1960) Silice, silex, silicifications et croissance des cristaux. *Bull. Serv. Carte géol. Alsa.-Lorr.* **13,** 129–146.

MILLOT, G. (1964) *Géologie des Argiles,* Masson et Cie, Paris, 499 pp.

NEWELL, N.D. *et al.* (1953) *The Permian reef complex of the Guadalupe Mountains region, Texas and New Mexico: a study in Paleoecology,* pp. 236. Freeman, San Francisco.

PERCH-NIELSEN, K. (1971) Elektronenmicroskopische Untersuchungen an Coccolithen und verwandten Formen aus dem Eozän von Dänemark. *K. danske Vidensk. Selsk. Biol. Skr.* **18,** 76 pp.

PESSAGNO, E.A. (1973) Age and geologic significance of radiolarian cherts in the California Coast Ranges. *Geology,* **1,** 153–156.

PETERSON, M.N.A. *et al.* (1970) Cruise leg summary and discussion. In: *Initial Reports of the Deep Sea Drilling Project,* Vol. II (M. N. A. Peterson *et al.*), pp. 413–427. U.S. Government Printing Office, Washington.

PIMM, A.C., GARRISON, R.E. & BOYCE, R.E. (1971) Sedimentology synthesis: lithology, chemistry, and physical properties of sediments in the northwestern Pacific Ocean. In: *Initial Reports of the Deep Sea Drilling Project*, Vol. VI (A. G. Fischer *et al.*). pp. 1131–1252. U.S. Government Printing Office, Washington.

RAMSAY, A.T.S. (1971) Occurrence of biogenic siliceous sediments in the Atlantic Ocean. *Nature, Lond.* **233,** 115–117.

REYNOLDS, W.R. (1970) Mineralogy and stratigraphy of Lower Tertiary clays and claystones of Alabama. *J. sedim. Petrol.* **40,** 829–838.

SCHLANGER, S.O., DOUGLAS, R.G., LANCELOT, Y., MOORE, T.C., JR & ROTH, P.H. (1973) Fossil preservation and diagenesis of pelagic carbonates from the Magellan Rise, Central North Pacific Ocean. In: *Initial Reports of the Deep Sea Drilling Project*, Vol. XVII (E. L. Winterer and J. I. Ewing *et al.*), pp. 407–427. U.S. Government Printing Office, Washington.

SEGNIT, E.R., ANDERSON, C.A. & JONES, J.B. (1970) A scanning microscope study of morphology of opal. *Search,* **1,** 349–350.

SIEVER, R. (1962) Silica solubility, 0°–200°C., and the diagenesis of siliceous sediments. *J. Geol.* **70,** 127–150.

SMITH, E.A., JOHNSON, L.C. & LANGDON, D.W. (1894) Report of the geology of the Coastal Plain of Alabama. *Spec. Rep. geol. Surv. Ala.* **6,** 759 pp.

TALIAFERRO, N.L. (1934) Contraction phenomena in cherts. *Bull. geol. Soc. Am.* **45,** 189–232.

TARR, W.A. (1917) Origin of the chert in the Burlington limestone. *Am. J. Sci.* **44,** 409–452.

TARR, W.A. (1926) The origin of chert and flint. *Univ. Mo. Stud.* **1,** 1–54.

URASHIMA, Y. (1959) On the silica minerals of the Abuta mine in Hokkaido Japan. *J. Earth Sci. Hokkaido Univ.* ser. IV, **10,** 235–254.

VAN TUYL, F.M. (1918) The origin of chert. *Am. J. Sci.* **45,** 449–456.

VON DER BORCH, C.C., GALEHOUSE, J. & NESTEROFF, W.D. (1971) Silicified limestone-chert sequences cored during Leg 8 of the Deep Sea Drilling Project: a petrologic study. In: *Initial Reports of the Deep Sea Drilling Project*, Vol. VIII (J. I. Tracey, Jr *et al.*), pp. 819–827. U.S. Government Printing Office, Washington.

WANG, Y. (1967) Studies on rock-forming minerals from Taiwan (3) Low-temperature cristobalite from altered andesites, Tatun volcanic region. *Proc. geol. Soc. China,* #10, 125–129.

WEAVER, F.M. & WISE, S.W., JR (1972) Ultramorphology of deep sea cristobalitic chert. *Nature, Lond.* **237,** 56–57.

WEAVER, F.M. & WISE, S.W., JR (1973a) Chemically precipitated sedimentary cristobalite in Tertiary Atlantic and Gulf Coastal Plain sediments. *Prog. Abs. geol. Soc. Am.* **5,** 449.

WEAVER, F.M. & WISE, S.W., JR (1973b) Early diagenesis of a deep-sea bedded chert. *U.S. Antarctic J.* **8,** 298–300.

WISE, S.W., JR (1973) Calcareous nannofossils from cores recovered by leg 18, Deep Sea Drilling Project: biostratigraphy and observations of diagenesis. In: *Initial Reports of the Deep Sea Drilling Project*, Vol. XVIII (L. D. Kulm and R. von Huene *et al.*), pp. 565–615. U.S. Government Printing Office, Washington.

WISE, S.W. JR, BUIE, B.F. & WEAVER, F.M. (1972) Chemically precipitated cristobalite and the origin of chert. *Eclog. geol. Helv.* **65,** 157–163.

WISE, S.W., JR, CIESIELSKI, P.F., SCHMIDT, W. & WEAVER, F.M. (1974) Altered Upper Eocene diatomite in the coastal plain of Georgia. *Prog. Abs. geol. Soc. Am.* **6,** 414–415.

WISE, S.W., JR & HSÜ, K.J. (1971) Genesis and lithification of a deep sea chalk. *Eclog. geol. Helv.* **64,** 273–278.

WISE, S.W., JR & KELTS, K.R. (1972) Inferred diagenetic history of a weakly silicified deep sea chalk. *Trans. Gulf-Cst Ass. geol. Socs,* **22,** 177–203.

WISE, S.W., JR & WEAVER, F.M. (1972) Origin of deep sea cristobalite chert. *Prog. Abs. geol. Soc. Am.* **4,** 116.

WISE, S.W., JR & WEAVER, F.M. (1973) Origin of cristobalite-rich Tertiary sediments in the Atlantic and Gulf Coastal Plain. *Trans. Gulf-Cst Ass. geol. Socs,* **23,** 305–323.

WISE, S.W., JR, WEAVER, F.M. & GÜVEN, N. (1973) Early silica diagenesis in volcanic and sedimentary rocks: devitrification and replacement phenomena. *31st Ann. Proc. Electron Microsc. Soc. Am.* 206–207.

ZEMMELS, I. & COOK, H.E. (1973) X-ray mineralogy of sediments from the Central Pacific Ocean. In: *Initial Reports of the Deep Sea Drilling Project*, Vol. XVII (E. L. Winterer and J. I. Ewing *et al.*), pp. 517–559. U.S. Government Printing Office, Washington.

Spec. Publs int. Ass. Sediment. (1974) **1**, 327–347

Petrography and diagenesis of deep-sea cherts from the central Atlantic

ULRICH VON RAD *and* HEINRICH RÖSCH

Bundesanstalt für Bodenforschung (Federal Geological Survey), Hannover, Germany

ABSTRACT

Chert layers of Cenomanian to Early Miocene age and covered by 100–700 m of sediments were recovered by Leg 14 of the Deep Sea Drilling Project. Four types of silicified sediments can be distinguished by optical, X-ray, infrared absorption, scanning and transmission electron microscope analysis.

(I) Clay-rich lussatite porcelanite (Maastrichtian–Lower Miocene).

(II) Bedded, carbonaceous, more or less zeolitic lussatite-chalcedony-quartz porcelanite (Upper Cretaceous–Palaeocene).

(III) Faintly bedded, dolomitic and pyritic palygorskite-lussatite (-quartz) porcelanite (mostly Palaeocene).

(IV) Homogenized, ferruginous quartz (-lussatite) chert (Middle–Upper Cretaceous).

Remobilization of dissolved, mainly biogenous opal–A and precipitation of lussatite blades, randomly crystallizing in microspherules, are the first steps of chertification. The following diagenetic sequence is suggested. (1) Biogenous opal–A in siliceous oozes and ± zeolitic (marl) muds → (2) 'precursor stage' (semi-consolidated porcelaneous mudstone with well-preserved, tangentially welded siliceous organisms, incipient mass polarization and precipitation of lussatite spherules, palygorskite and/or clinoptilolite) → (3) 'immature' lussatite porcelanite (type I) → (4) type II or (the palygorskite-rich) type III porcelanite → (5) 'mature' quartz chert (type IV) with almost all siliceous fossils obliterated. This maturation involves a gradual decrease of the lussatite : quartz ratio from 28 (type I) via 5–14 (II–III) to about zero (IV).

In the sites investigated, the maximum age of well-preserved skeletal opal–A ranges from 20 to 40 million years. Starting from the deposition of suitable sediments, it takes at least 20–50 million years to produce the porcelanites of types I–III and at least 70–90 million years to produce quartz cherts (type IV).

INTRODUCTION

One of the most exciting sedimentological results of the Deep Sea Drilling Project was the discovery of a great variety of heterogeneous cherts, associated with practically all hemi- and eupelagic lithofacies of Tithonian to Miocene age. In some drilling sites one can study an almost continuous series of original sediments, not yet silicified, grading into the various stages of porcelanite and finally into quartz chert. Thus, it is now possible to interpret the important early diagenetic history of silicified sediments which is generally obscured in the older mature cherts exposed on land.

These are the main questions discussed in the recent literature on chert genesis.

(1) What is the source of the free silica (biogenous *v.* volcanic origin)? This topic is discussed in some detail in the accompanying papers of Wise & Weaver (1974) and Calvert (1974).

(2) What are the processes and factors responsible for the formation of lussatite porcelanites and quartz cherts? Two opposing theories have been formulated.

(a) The 'maturation theory', stating that the transformation of biogenous opal-A ⟶ lussatite porcelanite ⟶ quartz chert is mainly a time-dependent (ageing) process (Heath & Moberly, 1971; Berger & von Rad, 1972; Weaver & Wise, 1972; Wise & Weaver, 1974) and (b) the 'quartz precipitation theory' of Lancelot (1973), suggesting that environmental conditions (mainly the clay content and permeability of the host rock and the influence of foreign cations) determine whether lussatite porcelanite or quartz chert is formed at any given time (see p. 344).

The objective of this paper is to contribute to the verification of the maturation theory by a detailed petrographic description and diagenetic interpretation of the four chert types which were distinguished during preliminary investigations of Leg 14 sediments (Berger & von Rad, 1972, pp. 846–854; von Rad & Rösch, 1972). In this paper *chert* is used as general rock term for silicified sediments. This term includes porous, fine-grained lussatite-bearing *porcelanites* and dense true *cherts s. str.* consisting of microcrystalline quartz (see Table 1, and Bramlette, 1946; Ernst & Calvert, 1969; Calvert, 1971a).

PETROGRAPHY OF CHERT TYPES

Occurrence and age of Leg 14 cherts

Layers of silicified rocks were found in the sediments of Sites 135 and 137–140 off north-west Africa and in Site 144, all of Leg 14 (Berger & von Rad, 1972; Hayes, Pimm *et al.*, 1972). The cherts of Sites 12 and 13 (Peterson *et al.*, 1970; Maxwell *et al.*, 1970) are also included in the fence diagram of Fig. 1.

Except for the Lower Miocene porcelanites of Site 139, all silicified rocks were found in sediments of pre-Oligocene age (Figs 1 and 2). Most cherts have an age ranging from Santonian to Early Eocene with a concentration near the Cretaceous/Palaeocene boundary, not in the Eocene as in so many other Deep Sea Drilling sites. The oldest cherts of Legs 14 and 3 occur near the Lower/Upper Cretaceous boundary (Albian–Turonian) and have analogues in the Cretaceous cherts of the northwest Atlantic (Beall & Fischer, 1969; Lancelot, Hathaway & Hollister, 1972) and of the Gulf of Mexico (Beall *et al.*, 1973).

Chert classification

By means of optical, X-ray diffraction, and electron microscope analysis, four major chert types were distinguished, mainly on the basis of the proportion and mineralogy of the various SiO_2 phases in the matrix, the preservation and filling of siliceous organisms, and the general lithology of the host rock. These are the four chert types, the occurrence, age, associated lithofacies and composition of which are shown in Fig. 2 and discussed by Berger & von Rad (1972, pp. 846–850).

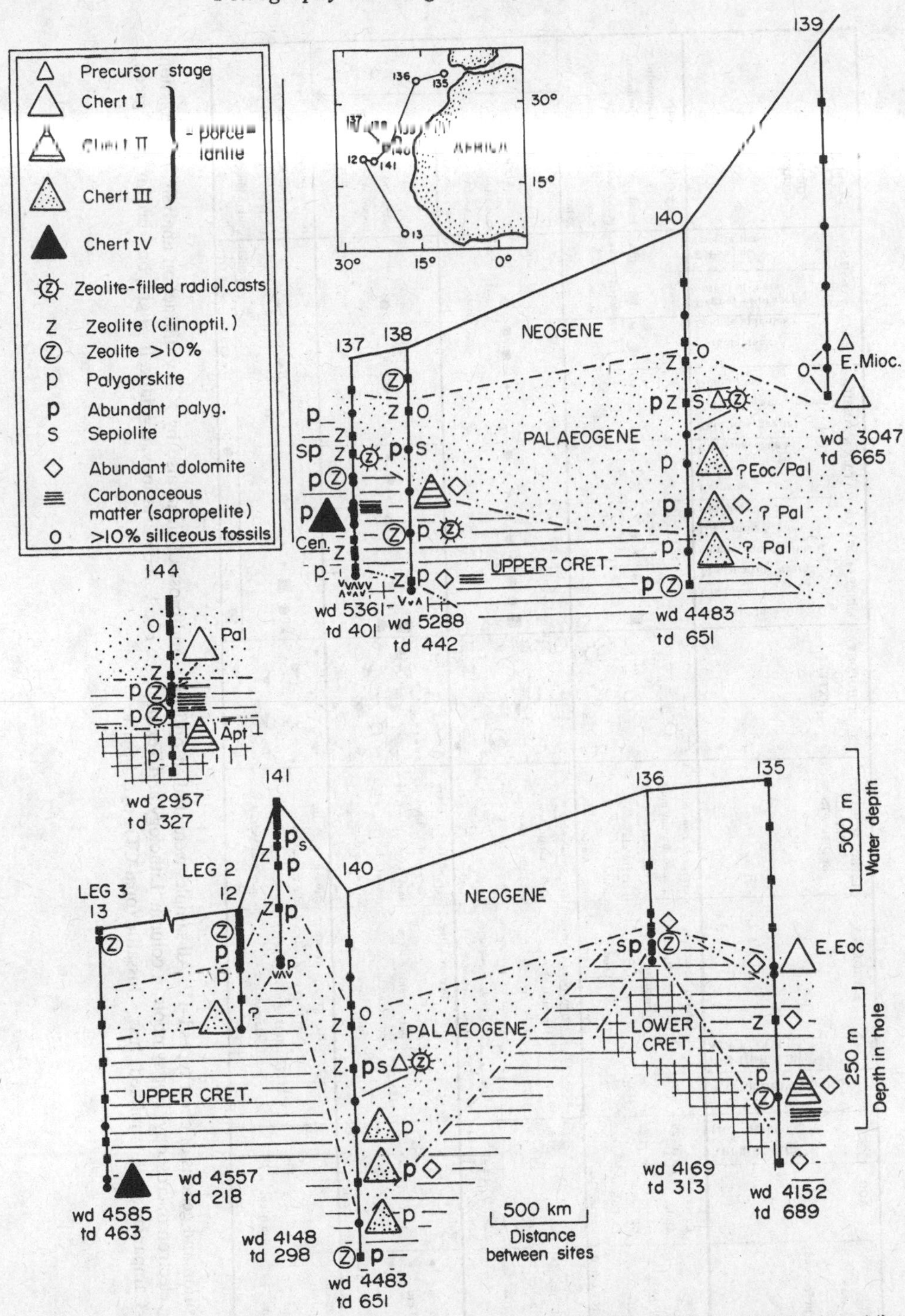

Fig. 1. Distribution of chert types and associated silicates in Sites 135–141, 144 (DSDP, Leg 14) and Sites 12, 13 (Legs 2 and 3). Chert type numbers I–IV indicate increasing maturity (see Fig. 2, Table 2). wd = Water depth (m), td = total depth of hole (note different vertical scales for wd and td). Fence diagram graph adapted from Berger & von Rad (1972, Figs 15, 23, 39).

Eoc = Eocene, E = Early, Pal = Palaeocene, O = abundant (10–30%) siliceous organisms (mainly diatoms and radiolarians) with mostly well-preserved opaline skeletons.

CHERT TYPE	SAMPLE NUMBER (▲ quantitative X-ray analysis (Table 2); * (*) chert layers preserved in situ)	AGE	APPROX. AGE (10^6 years)	WATER DEPTH (m)	DEPTH BELOW SEA FLOOR (m)	LITHOLOGY	LITHOFACIES OF ASSOCIATED SEDIMENTS	LITHOLOGY AND TEXTURE
I	139-7-3	Early Mioc.	16-23	3047	700	brown, homogen. rad.-spiculitic porcelanite	ol.-grey hemipel. qtzose diatom mud, terrig. silt & sand	
	139-7-5, 121 cm							
	▲ 139-7-6, 47-50 cm							
	135-5 cc	E. Eoc.		4152	340	brown	pelagic clay	
	144-2-1	Lt. Paleoc.	50-72	2957	110	green- } rad.	zeolitic	
	▲ 144-3-2, 105-107 cm *	Maastr./Cp.			165	grey } mudst.	marl / mud	
II	▲ 138-4 cc			5288	255	d.grey porc.(-breccia)	grey mudstone	
	▲ 135-8-1, 55-58 cm (*)	Late Cret.	65 - 100	4152	565	laminated dark-grey spiculitic mudstone	dark, laminated, zeolitic carbonac. shale; silty ls	
	135-8-1, 125-135 cm (*)							
	135-8 cc							
	136-bit	? Cret.	?	4196	?	bn. rad. mudstone		
	144A-5cc	Santon.	80-84	2957	185	dark carbonac. mudst.	black carb. shale+ls.	
III	▲ 140-5cc	Paleogene			435	light- to dark olive grey dolomitic palygorskite porcelanite, faintly bedded	pyr. shale; silt layers ol.-grn dol. mudst., dark-grey silicified mudst. layers, a few to 10 cm thick	
	▲ 140-6-2, 84-86 cm *							
	140-6-2, 86-90 cm *							
	▲ 140-6-2, 95 cm (*)	? Paleoc.	~ 55-63	4483	515			
	▲ 140-6-2, 125 cm							
	140-7 cc	Lt. Cret.			582		mudstone / pyr. shale	
IV	137-7-1, 117-120 cm				?	lt.-grey to brown highly recrystallized ferruginous chert, ± laminated	olive-grey shale, pyritic carbonac. shale nanno marl/chalk ooze	
	137-8-1, 52 cm	Cenom.	95 - 105	5361	255			
	▲ 137-12-2, 116 cm				270			
	137-16-3, 130 cm	Alb-Cen.			380			

CHERT TYPE	Opal-A + lussatite	Chalcedony	Microcryst. quartz	Terrig. qtz & feldsp.	Clay min. & mica	Palygorskite	Fe oxides	Pyrite	Carbonaceous m.	Zeolite & barite	Calc. (forams, nanno)	Dolomite	Radiolarians	Sponge spicules	Degree of preservation of tests	Filling of fossils (casts)	DESCRIPTIVE ROCK NAME AND GENERAL TRENDS
I	■		•	+	●			•	•	+	• f	+	●	•	■ ?opal luss. ghosts	unf. luss.	clay-rich opal-A - lussatite porcelanite homogeneous; early diagenetic 'immature' high lussatite: quartz ratio
II	■	●	●	•	●	•	•	•	●			•	●	●	▲ ?opal luss. qtz	unf. qtz chalc.	bedded silicified carbonac. spicul. & radiol. mudstone (lussatite porcelanite)
III	■ ▲	•	●	•		▲	•	●	•			●	•	+	● pyr. ?opal ghosts	qtz	dolomitic & pyritic palygorskite - lussatite porcelanite
IV	•	●	■	+	•		●				• n	•	(●)	(+)	• ghosts pyrite	qtz + chalc	homogeneous ferruginous, recryst. quartz chert; late diagen. 'mature'; very low luss.: qtz ratio

■ > 40 % ▲ 20 - 50 % ● 10 - 20 % • < 10 % + < 3 %

Fig. 2. Composition and genesis of chert types I–IV. All samples were studied optically, and most of them also by X-ray diffraction analysis, transmission and/or scanning electron microscopy. For symbols in column 'Lithology and Texture', see Fig. 8. Abbreviations: d=dark, lt=light, bn=brown, grn=green, ol=olive; Cp=Campanian; unf.=unfilled; luss.=lussatite (opal-CT); pyr.=pyrite.

Type I: Immature, brown, clay-rich, opal-A-lussatite porcelanite

Age of associated sediments: Maastrichtian–Miocene (72–23 million years); high lussatite/quartz ratio; relatively well-preserved siliceous skeletons (opal–A or ?lussatite); rare fossil ghosts, and lumina rarely quartz- or chalcedony-filled (see Fig. 3d).

Type II: Dark-grey, bedded, silicified, carbonaceous, spiculitic or radiolarian mudstone=lussatite (-quartz-chalcedony) porcelanite

Upper Cretaceous–Palaeocene (90–65 million years); associated with black, zeolitic, carbonaceous mudstones; matrix consisting of lussatite, mixed with various amounts of carbonaceous matter (tiny bituminous spherules), palygorskite, clay minerals and terrigenous minerals (Fig. 3e); poor preservation of siliceous skeletons: ?radiolarian cross-sections either with open pores or filled by fibrous-spherulitic chalcedony.

Type III: Faintly bedded, partly homogenized, light grey to olive green, dolomitic and pyritic palygorskite-lussatite porcelanite

?Palaeocene (55–65 million years); matrix consisting of fine-grained fibrous palygorskite (Fig. 7d) with cryptocrystalline silica (30–50% lussatite); radiolarians preserved as 'ghosts' or pyrite pseudomorphs, exceptionally also as 'aged' opal–A (?) in dark ferruginous layers (Fig. 3f).

Type IV: Dark-grey, mature, homogenized, ferruginous, recrystallized quartz chert.

Associated with Upper Cretaceous carbonaceous nannofossil marl oozes (137–12) and black carbonaceous shales (137–7); matrix coloured by disseminated iron oxides; 'pin-point extinction' typical of crypto- to micro-crystalline quartz (Carozzi, 1960, p. 310; Folk & Weaver, 1952); siliceous fossils mostly destroyed (Fig. 3c), diffuse outlines of 'ghosts', filled by chalcedonic or microcrystalline quartz.

In summary, these four types range from poorly crystallized, 'immature' lussatite porcelanites to highly crystallized, 'mature' quartz cherts, implying a general decrease of opal–A+lussatite and increase of chalcedonic and microcrystalline quartz from types I to IV. The 'lussatite : quartz ratio' varies from 28 (type I) to 0·05 in type IV (see Table 2 and Fig. 6). Most cherts contain variable admixtures of detrital quartz (optically determined), feldspar, pyrite, carbonates, clay minerals, palygorskite, and zeolite. The percentages of well-preserved radiolarians and sponge spicules decrease from type I to types II/III to a few percent, and to zero in type IV. The lumina of these fossils vary from 'open pores' in type I, to mainly quartz- and chalcedony-filled ones (II/III), and fossil 'ghosts' in chert type IV.

MINERALOGY

Silica modifications

Many authors have stated that in general the formation of chert is a two-stage process (e.g. Heath & Moberly, 1971; Ernst & Calvert, 1969; Calvert, 1971a, b; Wise, Buie & Weaver, 1972).

The first step represents the early diagenetic remobilization of hydrous, X-ray amorphous, biogenic or volcanogenic opal–A and its low-temperature alteration product, lussatite. The latter was probably chemically precipitated by polymerization of silicon molecules from hydrous pore solutions as an authigenic mineral (Weaver & Wise, 1972). Lussatite (a term originally used by Mallard, 1890; see Jones & Segnit,

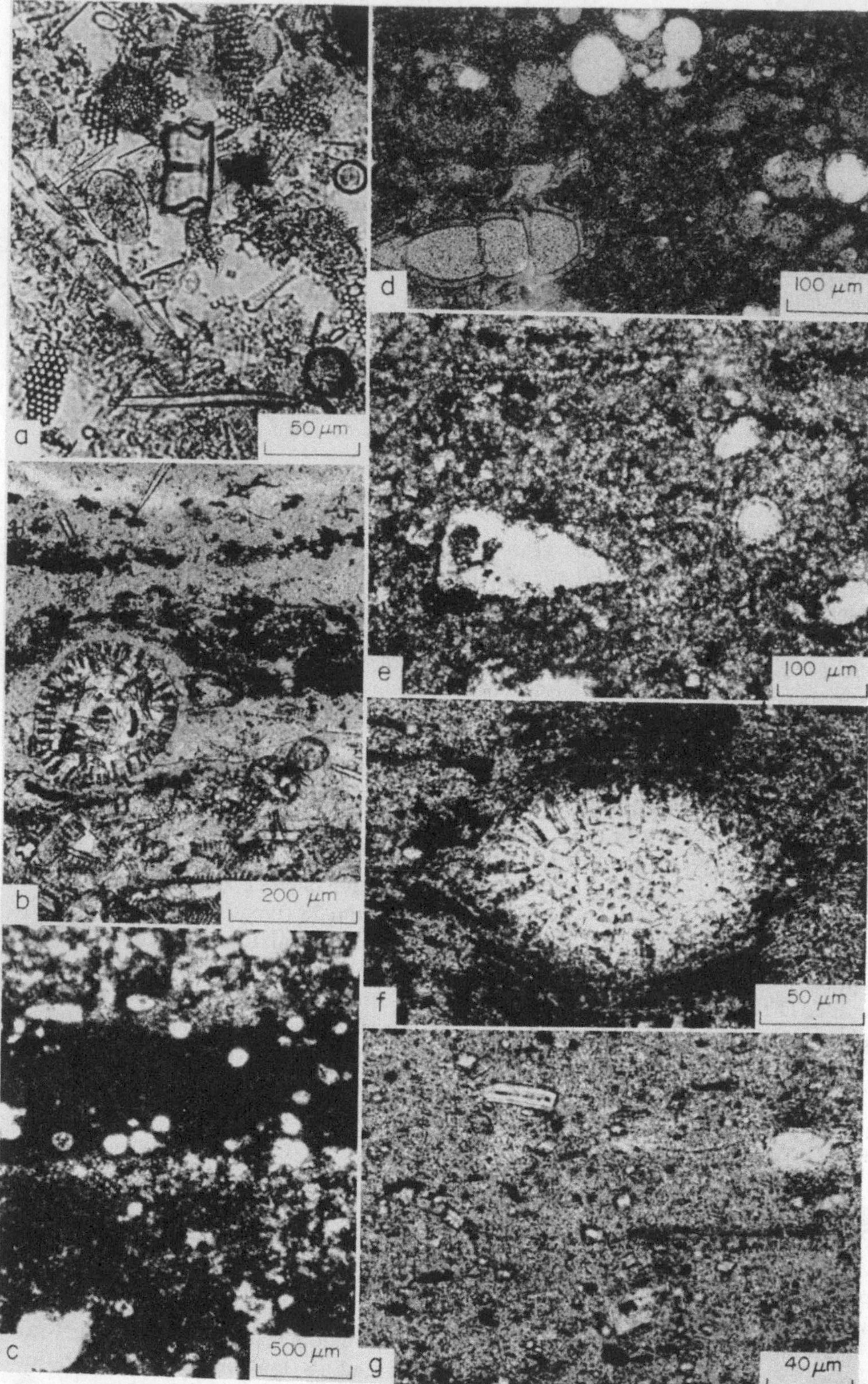

Fig. 3. (a) DSDP (Leg) 14–(Site) 139–(Core) 5–(Section) 1, 140 cm (smear slide; Lower to Middle Miocene). Relatively pure *diatom ooze* with about 40% diatoms, sponge spicules, silicoflagellates, very few radiolarians, all highly fragmented. Clay (and silt) fraction contains calcareous nanno-plankton and clay minerals; the sand fraction contains quartz. All skeletal fragments consist of

continued opposite

1971) is a stacking low-temperature cristobalite with an irregular interstratified structure of cristobalite layers (analogous to cubic close packing) alternating with tridymite layers (analogous to hexagonal close packing; Flörke, 1955, 1962). The term lussatite is here used in the same sense as 'opal–CT' (Jones & Segnit, 1971), 'opal cristobalite' (Braitsch, 1957), or 'low-temperature cristobalite' (Flörke, 1962).

The second transition stage is the late diagenetic alteration of lussatite into chalcedony and/or micro- (>1 μm) to cryptocrystalline (<1 μm) quartz. The exact mechanism of this inversion is still controversial: Ernst & Calvert (1969), Heath & Moberly (1971) and Weaver & Wise (1972) suggest slow solid-solid zero order conversion under large activation energies. Mizutani (1966), however, and Wise *et al.* (1972) do not exclude dissolution of cristobalite and chemical redeposition of quartz as a possible alternative mechanism. On the basis of structural and thermodynamical considerations, Jones & Segnit (1972) suggest that the formation of the metastable cristobalite-tridymite structure will be favoured under low energy/temperature conditions (which are typical for the marine environment) and/or low ionic mobilities.

Cherts, porcelanites and other silica-bearing sediments investigated in this paper contain the four silica modifications or varieties opal–A, lussatite, chalcedony, and quartz, the mineralogical and genetic characteristics of which are summarized in Table 1. Opal–C, the natural hydrous α-cristobalite with only minor tridymitic stacking disorder, is not present in our material; nor does it seem to exist in any deep-sea sediment, unless associated with lava flows (Güven & Grim, 1972). The somewhat misleading term 'α-cristobalite', though still frequently used in the recent literature (e.g. Rex, 1970; Calvert, 1971; Weaver & Wise, 1972), should be changed into 'opal–CT' or 'lussatite'.

Unfortunately, the pure silica phases hardly ever occur in deep-sea sediments. This fact, together with inadequate experimental conditions, seem to cause a certain

(unaltered) opal–A. (b) DSDP 14–140–3–2, 74–78 cm (thin section of Araldite-impregnated undisturbed specimen; Middle Eocene). The carbonate-free matrix of the *siliceous mudstone* consists of chlorite, kaolinite, and illite, with small amounts of lussatite, palygorskite, and clinoptilolite. Abundant siliceous fragments (mainly diatoms, some radiolarians and sponge spicules) are fragmented and 'tangentially welded', but still preserved as opal–A, although matrix shows first signs of diagenetic alteration (formation of palygorskite, oriented birefringence). This siliceous mudstone is a precursor of the porcelanite of sample 140–6–2 (Fig. 3g). (c) DSDP 14–137–12–2, 116 cm. Cenomanian highly recrystallized, distinctly bedded, 'dusty' *quartz chert* (type IV). Note ferruginous dark laminae and diffuse outlines ('ghosts') of siliceous fossils (?radiolarians), partly filled by chalcedony, partly by microcrystalline quartz. (d) DSDP 14–144–3–2, 105–107 cm (thin section of Araldite-impregnated semi-indurated sediment; Maastrichtian): foraminiferal and radiolarian silicified mudstone (=*type I porcelanite*). Foraminifera, with preserved calcitic tests partly filled by silica, and 'ghosts' of ?radiolarians are seen in a cryptocrystalline lussatite (58%) matrix mixed with amorphous matter and clay minerals. Tiny crystals of euhedral clinoptilolite and authigenic barite crystals are growing in some of the open cavities and fossil lumina. (e) DSDP 14–135–8–1, 55–58 cm. Upper Cretaceous, dark grey, carbonaceous, laminated, silicified radiolarian-spiculitic mudstone (*type II porcelanite*). Longitudinal and axial cross-sections of sponge spicules and one nassellarian radiolarian, the skeletons of which are recrystallized into lussatite or microcrystalline quartz. (f) DSDP 14–140–6–2, 84·5–91·5 cm (?Palaeocene). Dolomitic and pyritic *palygorskite-lussatite porcelanite* (=*type III*) with relatively well-preserved radiolarian cross-section. Local preservation of siliceous skeletons as opal–A (or ?lussatite replacing opal–A), especially in dark (?carbonaceous) micro-environments. (g) Same sample as (f): light-coloured ('white') porcelanite layer showing rounded terrigenous quartz grains (right), dark biotite flakes, well-preserved sponge spicule (top, ?opal–A) and many tiny dolomite rhombs, floating in a cryptocrystalline lussatite-palygorskite matrix.

disagreement between observations concerning natural cherts and laboratory syntheses (Oehler, 1973).

Analytical procedures

In addition to the preliminary investigations of von Rad & Rösch (1972) twelve typical chert samples were analysed by X-ray diffraction (+infrared absorption, differential thermal analysis and scanning electron microscopy). The X-ray analyses (Table 2) were carried out using the 'method of transparent specimens' (Bergin, 1964; Rösch, 1968). Copper radiation was monochromatized by a LiF crystal between sample and detector. The measurement of mass attenuation coefficients is automatically taken into consideration by the special geometrical arrangement of this method. An 'X-ray transparent specimen' is put upon a crystalline carrier material. Thus, the intensity of a suitable Bragg reflection coming from the external standard will be partly absorbed by the overlying sample to be analysed.

The standard samples were carefully selected according to the condition of the best agreement in peak intensity sequence and half width (grain-size similarity). The column 'amorphous clay minerals' represents the remainder to total up to 100% in a sample evidently containing a considerable amount of amorphous scattering in the *d*-range of the main clay mineral peaks.

The standards for lussatite and amorphous silica were pure samples, kindly provided by Professor Flörke, Bochum. The following standard samples were used: Opal–A=potch opal, Coober Pedy, Australia; lussatite, fairly disordered=rose opal, Lake Co., Oregon; lussatite, high-grade disordered=opal, Lake Eyre, Australia.

The accuracy of the mentioned analytical method equalled that of other phase analysis procedures, with a relative error between 6 and 15%, provided that the appropriate standards were used, the limits of detection were considered, and the lines to be analysed did not coincide with peaks of other minerals.

We also tried to determine the amount of opaline material by heat-transformation of opal–A into opal–C (or α-cristobalite) and inferring the amount of amorphous silica from the measured opal–C content (Goldberg, 1958; Calvert, 1966). For reasons unknown this method failed completely, as did the method of directly measuring peak heights or integrated intensities of the opal 'bulge' (Eisma & van der Gaast, 1971).

Precise Guinier photographs were made in order to determine the degree of disorder in the lussatite component (Flörke, 1962; Jones & Segnit, 1971). All samples containing considerable amounts of lussatite proved to be of high-grade stacking disorder indicating low temperatures of genesis. The α-cristobalite double peak at d=2·486 and 2·469 Å is not split into two lines.

Because this phenomenon could be caused by the extremely fine grain size of the lussatite particles, it should not be over-estimated. The crystallite size normal to (101), determined from the Scherrer equation (Klug & Alexander, 1954), proved to range from 10 to 250 Å. Regarding the electron micrographs, at least this magnitude has to be expected. But, similar to the conclusions of Heath & Moberly (1971), the crystal size of lussatite plotted against geological age confirmed that there is no simple relationship between the two variables.

Infrared analyses of selected samples, kindly run by Professor Jones, Adelaide, generally confirmed the X-ray results but did not yield further information.

Table 1. Characteristics of silica varieties of deep sea cherts*

	Mineralogy and crystal structure	X-ray pattern	SEM	Occurrence, paragenesis	Genesis
Opal (='Opal-A', Jones & Segnit, 1971)	Highly disordered nearly amorphous natural hydrous silica. Crystallite size ⊥ (101) : 11–15 A. Discontinuous framework of SiO_4 tetrahedrons + water	Prominent very diffuse band between 6 and 2·8 A with maximum at about 4·1 A; no cristobalite/tridymite reflections; coinciding with the lussatite peaks	Close-packed aggregates of silica spheres of uniform size (0·1–2 μm)	Tests of siliceous organisms (rads, sponge spic., diatoms, etc.) (Also in precious opal, 'potch', diatomites, geysirites, glass-clear hyalites)	Original organic skeletons <30–50 million years
Lussatite (='Opal-CT', Jones & Segnit, 1971)	Obviously crystalline, natural hydrous silica = unidimensionally highly disordered low- (α-) Cristobalite consisting of irregularly stacked low-cristobalite or 'k'-layers (cubic close packing) and low-tridymite or 'h'-layers (hexagonal close packing), water and foreign ions. Cryptocrystalline spherulitic variety (see SEM), crystallite size ⊥ (101) : 10–250 A	Broadened, but well-defined peaks of low-cristobalite (mainly $d_{(101)}$ at 4·1 A and $d_{(200)}$ at 2·5 A with tridymite peaks, e.g. at 4·25–4·3 A)	Closely packed silica spheres of irregular size and packing order (3–8, max 20 μmφ), consisting of a network of minute, irregularly oriented blades	Porcelanites of low to intermediate maturity (types I–III) (most common opals, bentonitic clays, silica glass, porcelanites)	Early diagenetic origin at low temperatures from skeletal opal, volcanic glass etc. Authigenic precipitation from pore solutions
Chalcedony	= disordered quartz, fibrous ⊥ [0001] = ∥ [1120] length-fast, crystals mostly ≫ 1 μm	Sharp quartz peaks, the double peak at d=1·375 and 1·372 A is split in the case of macro- to micro-crystalline quartz, and not split in the case of cryptocrystalline quartz		Filling of fossil lumina (especially type II) and in the matrix of mature quartz-cherts (type IV)	
Quartz	= quartz (a) cryptocrystalline (< 1 μm), no orientation (b) 'quarzine' (< 1μm), fibres ∥ [0001] = length-slow (c) microcrystalline quartz (> 1μm)			In the matrix of mature cherts (type III–IV); recrystallizing siliceous tests and filling of fossil lumina	'Late' diagenetic origin in sediments older than about 70–90 million years

* In part after Flörke (1955, 1962); Mizutani (1966); Calvert (1971); Heath & Moberly (1971); Weaver & Wise (1972).

Table 2. Mineralogical composition of selected opaline and chert samples

No. in Figs 4 and 6	Sample no. DSDP 14-site-core-barrel, depth (cm)	Age	Approx. age range (10^6 years)	Probable age (10^6 years)	Chert type	Mineral composition (wt %)															Lussatite/quartz
						Quartz	Lussatite (=opal–CT)	Feldspar	Chlorite (+kaolinite)	Illite	Palygorskite	Clinoptilolite	Calcite	Siderite	Dolomite	Montmorillonite	Haematite	Pyrite	Amorphous silica (opal–A)	Amorphous clay minerals	
4	139–5 cc	Early Miocene	22·5–16	18	P	3	6	3	4	2									82		2
6	140–3–2, 68–72	Middle Eocene	49 –43	46	P	8	5	(5)	(20)	(10)	(10)	(5)		(2)	(2)	(10)			20*		0·6
5	139–7–6, 47–50	Early Miocene	22·5–16	23	I	3	83		2		3	3	6								28
12	144–3–2, 105–107	Campan-Maastr.	80 –65	72	I	3	88			1		5	3								29
3	138–4 cc A	Turon-Maastr.	100 –53·5	67	II	6	80	2			5	3			4						13
11	144–3 cc	Campan-Maastr.	80 –65	72	II	7	58	4	2	3		8				3				~15	8
1	135–8–1, 55–58	Late Cretaceous	100 –65	93	II	13	81	1		1	2			2							6
7	140–5 cc	Palaeocene	65 –53·5	55	III	5	68	4			14			3				6			14
8	140–6–2, 84·5–85	Palaeocene	65 –53·5	60	III	4	49	2	2		35			2	6						12
9	140–6–2, 95–98	Palaeocene	65 –53·5	60	III	6	29	5	2		36	4		4	14						5
10	140–6–2, 125	Palaeocene	65 –53·5	60	III	7	51	6			18			5	6	3		4			7
2	137–12–2, 116	Cenomanian	100 –91	98	IV	87	4	2				2	3				2				0·05

'Probable age'=hypothetical age taken from Berger & von Rad (1972, Figs 57–58).
Chert types I–IV are lithofacies types defined by mineralogy *and* general petrography (e.g. presence of carbonaceous material, nature and preservation of siliceous fossils: see Fig. 2).
P=Precursor stage of porcelanite; ( )=percentages semiquantitatively estimated from X-ray traces.

* Skeletal opal estimated from thin section.

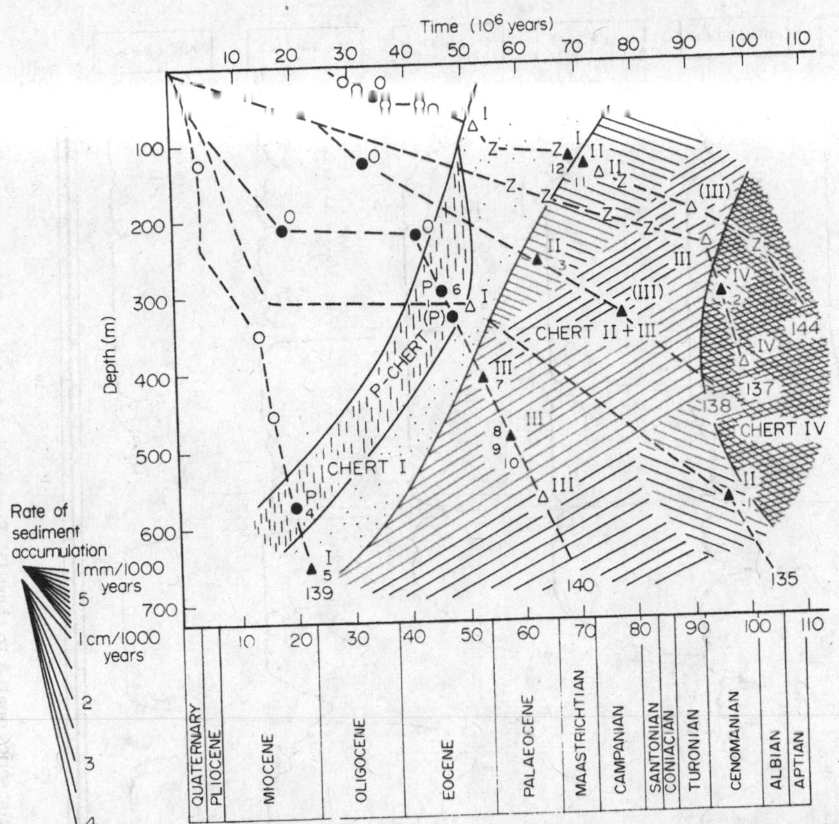

Fig. 4. Distribution of chert types in Sites 135, 137–140, 144 with respect to age and depth in hole. Rates of sediment accumulation from Berger & von Rad (1972, Fig. 6). ○, Cores containing unaltered opaline tests of siliceous fossils (●=ditto, X-rayed samples, see von Rad & Rösch, 1972, Table 1–2). ▲, Chert samples analysed for this paper, index numbers 1–12 refer to Table 2, roman numbers see Fig. 2. P-chert=precursor stage of porcelanites; Z=zeolite common; (P), (III)=according to analyses of Pow-Foong Fan & Rex (1972).

Notes on X-ray mineralogical results

One of the main analytical problems of the X-ray analysis of porcelanites is the determination of opal–A in the presence of lussatite. Unless the content of amorphous silica exceeds a considerable value (see no. 4, Table 2), opal–A cannot be resolved, but is included in the lussatite peak area. Thus, the lussatite : quartz ratio (Table 2) should be viewed with caution.

Figure 5 shows selected diffractometer traces characteristic of the different chert types. The sample termed 'precursor' is a 'normal' unconsolidated clay with 5% lussatite, an estimated portion of about 20% opal–A and 8% quartz.

This sample together with no. 4 (Table 2) seems to take a transitional (precursor) position between unaltered skeletons of siliceous fossils consisting almost entirely of amorphous silica (opal–A) and the lussatite-bearing porcelanites. The hypothetical area of this 'p-chert' in both the time *v.* depth-in-hole diagram (Fig. 4) and in the age *v.* lussatite : quartz graph (Fig. 6) has, in fact, an exceptional position (low age and low lussatite : quartz ratio) and caused us to separate the 'precursor' field in these diagrams. As a further peculiarity of these samples, the quartz double peak at

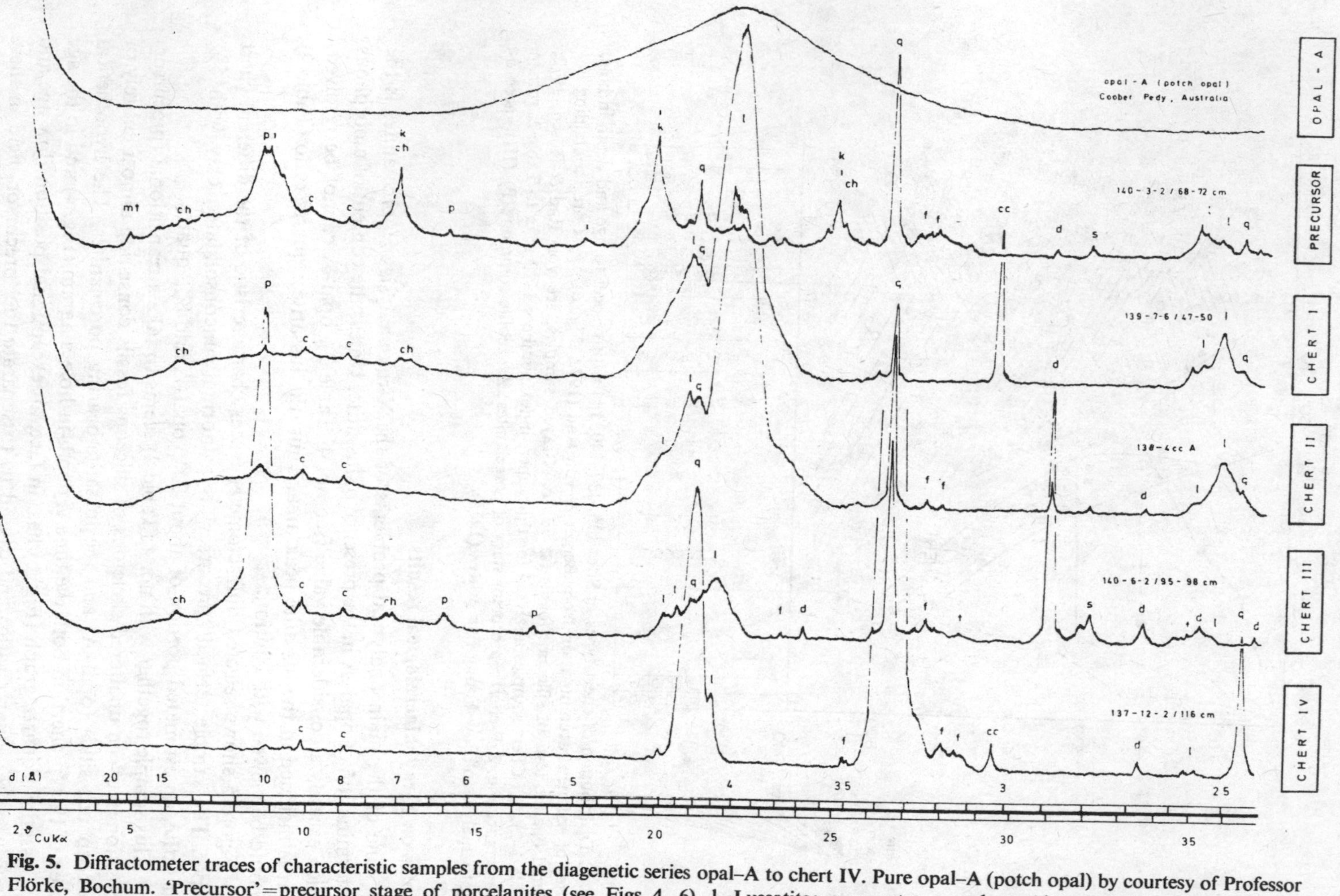

Fig. 5. Diffractometer traces of characteristic samples from the diagenetic series opal–A to chert IV. Pure opal–A (potch opal) by courtesy of Professor Flörke, Bochum. 'Precursor' = precursor stage of porcelanites (see Figs 4, 6). l, Lussatite; q, quartz; p, palygorskite; f, feldspar; d, dolomite; c, clinoptilolite; s, siderite; cc, calcite; mt, montmorillonite; k, kaolinite; ch, chlorite; i, illite (-mica).

d=1·375 and 1·372 Å on Guinier photographs is well divided in contrast to all other samples investigated. This generally indicates a coarser grain size and detrital origin of the quartz.

It should be noted that the wt-ratio lussatite : quartz (Fig. 6, Table 2) represents minimum values, since small amounts of detrital (i.e. non-diagenetic) quartz are included in the 'quartz' percentage, especially in the hemipelagic cherts. The lack of disintegration of the quartz double peak (except for nos. 4 and 6) suggests, however, that the analysed quartz is mainly of non-detrital origin.

The X-ray mineralogical characteristics (Table 2) tend to support the classification scheme of Fig. 2 which originally was based on optical investigations only. Type I and IV cherts, as well as the so-called 'precursor type' and the original siliceous oozes, are restricted to well-defined areas in the plots of Fig. 4 and Fig. 6. Chert type III is distinguished from type II by its high palygorskite and dolomite content.

Discussion of scanning electron micrographs

Even optical examinations at high magnitude demonstrate that the originally opaline tests of radiolarians were transformed into an almost isotropic mass of spherulitic silica. The cavities of former radiolarians are lined by tiny spherules, 3–10 μm in diameter (Fig. 7a; see also Beall & Fischer, 1969, their Fig. 19C) which

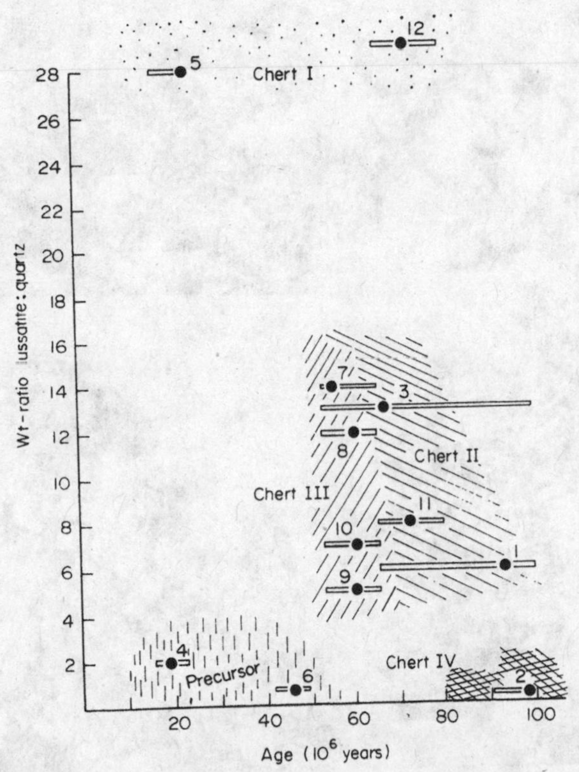

Fig. 6. Lussatite : quartz ratio as a function of geological age. Length of horizontal bars represent maximum age range of investigated samples (numbers 1–12, see Table 2). Assumed age (heavy dot) taken from Berger & von Rad (1972, Figs 57, 58). Types II and III are not distinguished by their lussatite : quartz ratio, but mainly by their different lithofacies (see Fig. 2, Table 2).

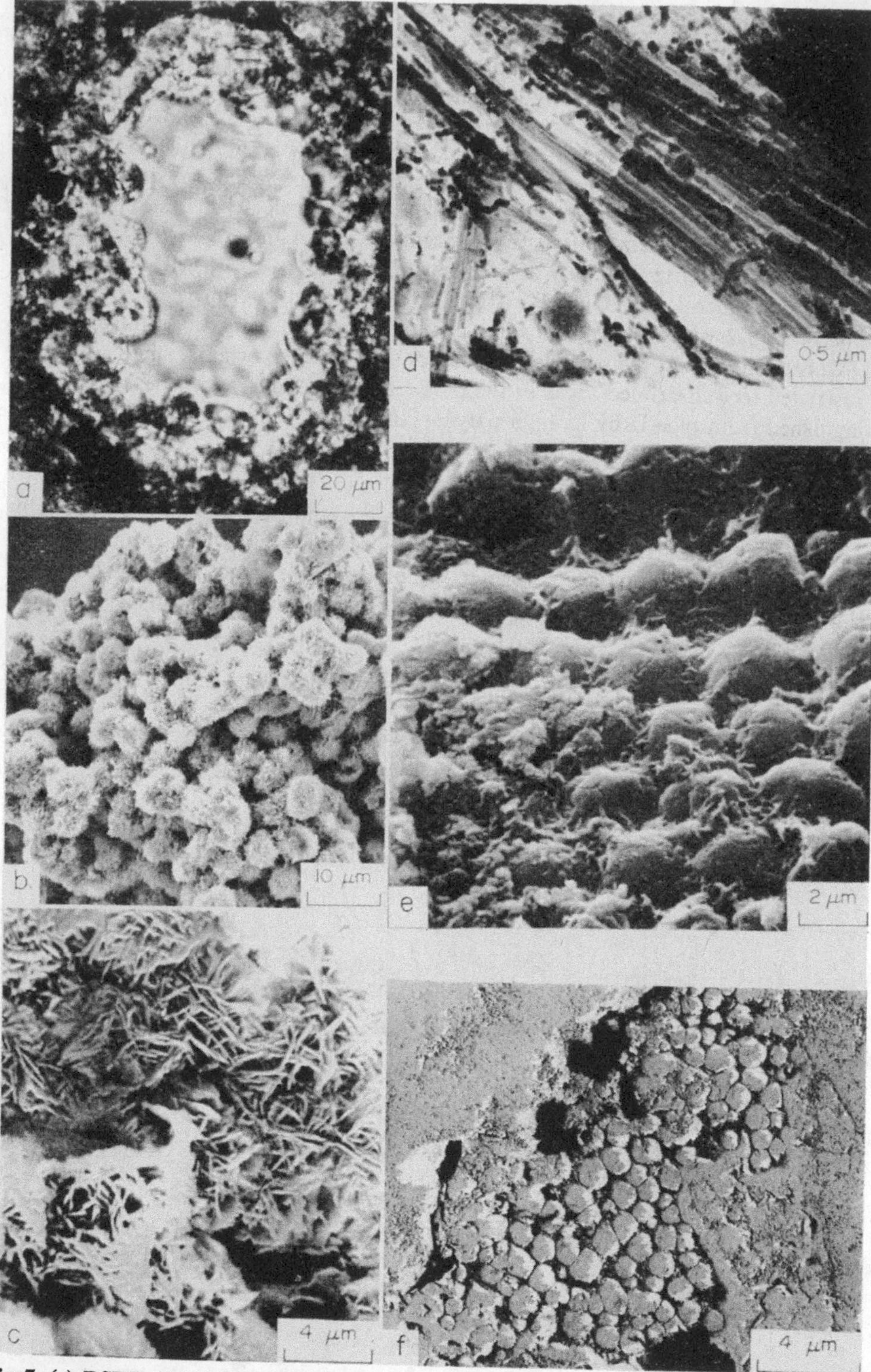

Fig. 7. (a) DSDP 14–135–8–1, 55–58 cm (Upper Cretaceous type II porcelanite). Radiolarian (?) cross-section ('ghost'). Originally opaline test is recrystallized to lussatite spherules, 5–15 μm in diameter, that grow into the open pore space of the unfilled lumen. Lussatite spherules are present also in the cryptocrystalline matrix, but difficult to resolve under the microscope. (b) DSDP 14–140–5 cc (scanning electron micrograph, 10 KV, by Mrs Scheuermann, Hannover). Palaeogene

continued opposite

can be further resolved by the scanning electron microscope (Fig. 7). Clusters of closely packed silica spherules from deep-sea cherts, identified as lussatite (Fig. 7b), were illustrated for the first time by Weaver & Wise (1972), Wise *et al.* (1972), von Rad & Rösch (1972, Plate 2), and Berger & von Rad (1972, Plate 14).

The diameter of those spherules ranges from about 3 to 15 μm (average 5 μm). Some transmission electron micrographs illustrate the 'cobblestone pattern' typical of 'cristobalite' (Fig. 7f). These microspherules are about an order of magnitude smaller (0·2–2 μm ϕ) than the ones described above. More detailed investigations will clarify whether or not those minute spherules indeed represent an incipient gelatinous phase which is not yet completely crystallized into lussatite (cf. also the tiny spherules making up larger lepispheres in Lancelot, 1973, Fig. 8). Wise *et al.* (1972), Wise & Kelts (1972), and Oehler (1973) published excellent scanning electron micrographs which show that the minute 'spheres' consist of thin, ?hexagonal, tridymite-like crystals of 'α-cristobalite' (=lussatite) radiating outwards from the centre. Thus, the spherules are rather 'spheres of blades' ('lepispheres', Wise & Kelts, 1972). According to our observations the blades are approximately 0·05–0·1 μm thick and a few microns wide and long. They appear to be randomly oriented within each lepisphere (Fig. 7c). The typical gel-spheres of truly amorphous natural opal (opal–A) have a similar diameter (about 3 μm) to the lussatite lepispheres (Fig. 7e; scanning electron micrograph by courtesy of Dr H.-U. Nissen, Zürich). Those spherules are, however, characterized by uniform size, high regularity of packing, and smooth surface.

ORIGIN AND DIAGENESIS

General remarks

A tentative genetic interpretation of the chert types distinguished in Leg 14 sediments and of other deep-sea cherts is given in Fig. 8. The porcelanites of types I–III belong to the group of '*bedded cherts*', commonly found in non-carbonate sediment sequences (Heath & Moberly, 1971). Some porcelanite layers are preserved *in situ*; they are interbedded with sediments that are only slightly less silicified and consolidated, but otherwise of very similar colour and composition (see Fig. 2). Most of

---

type-III lussatite porcelanite consisting mainly of irregular clusters of weakly cemented lussatite 'lepispheres', about 5 μm in diameter. Lussatite was precipitated in open pores during early diagenesis (possibly with amorphous silica gel spheres as precursor?). Note high remaining porosity in these porcelanites. (c) Same sample as (a) scanning electron micrograph, showing detail of 'lepisphere' with randomly oriented, thin (<0·05 μm) lussatite blades (1–4 μm wide and long). According to Oehler (1973) the blades appear to be hexagonal (tridymite-like) crystals. (d) Same sample as (a) (broken surface, transmission electron micrograph, Triafol-SiO-two stage replica, shadow-casting with Pt at 30° and 90°, Mrs Scheuermann). Clusters of palygorskite needles (diameter 0·05–0·1 μm) in porcelanite. (e) 'Black opal' (Australia; sample no. 94129a, scanning electron micrograph no. 6–107/2 by courtesy of H.-U. Nissen and R. Wessicken, Zürich, 1970). Natural opal–A precipitated by flocculation of an aqueous colloid and consisting of close-packed aggregates of silica spheres of uniform size (diameter about 3 μm). Note regular packing and smooth surface (in contrast to (b)). Small blades and fibres of ?lussatite bridging those spherules might indicate first stage of crystallization of the amorphous silica. (f) DSDP 14–135–8–1, 55–58 cm (transmission electron micrograph of polished section; two stage replica as in (d); same sample as (a). Minute silica spherules and polygonal grains (0·2–2 μm), showing 'cobblestone pattern' typical for lussatite.

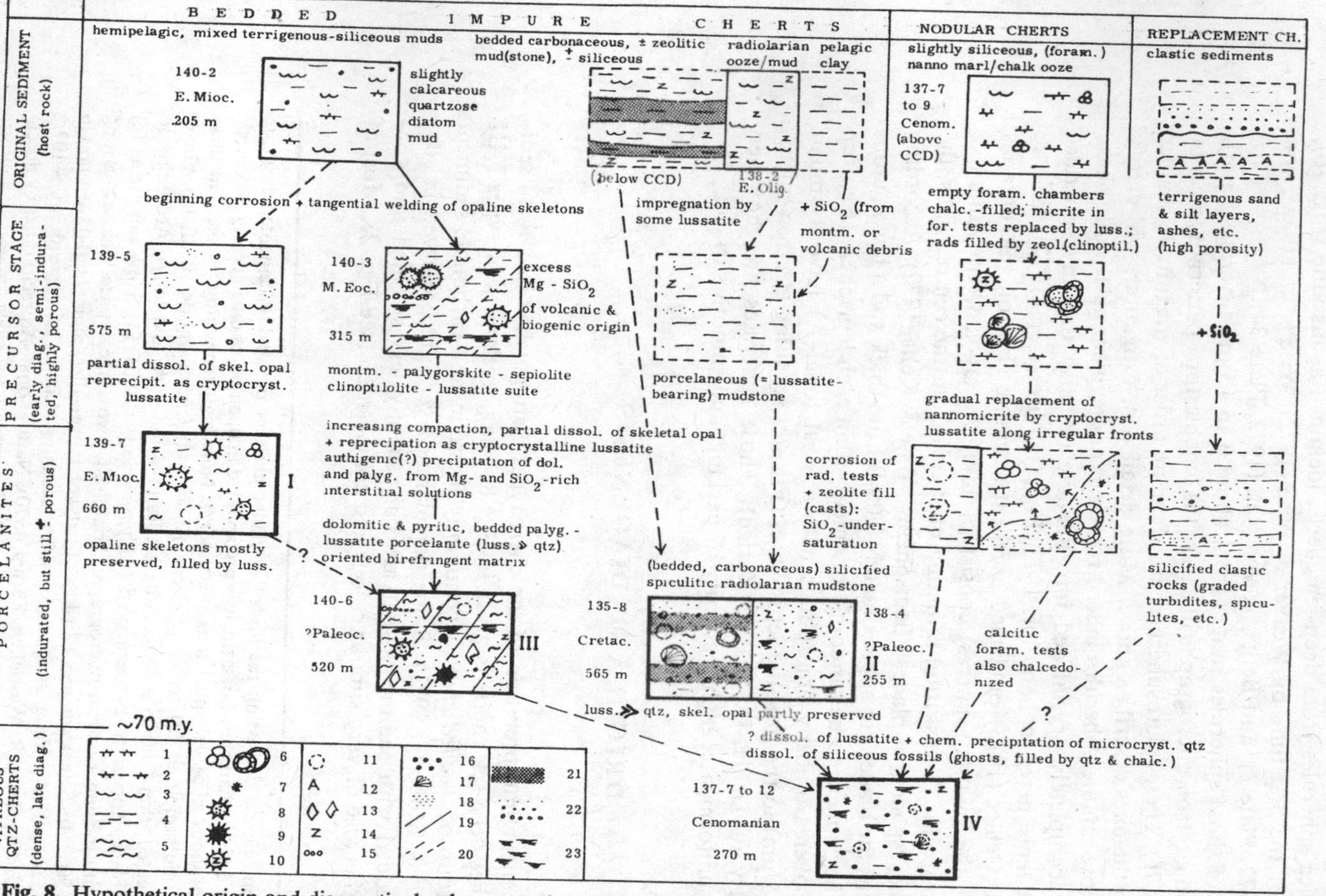

Fig. 8. Hypothetical origin and diagenetic development of common deep-sea chert types. The lithological sketches show the sequence original sediments → 'precursor stage' → lussatite porcelanite → vitreous quartz chert. Generalized results of Leg 14 (cherts I–IV) are supplemented by those from Legs 1–4 (Beall & Fischer, 1969; Peterson *et al.*, 1970; Calvert, 1971; Wise *et al.*, 1972) and from Legs 5–7 and 10 (von der Borch, Nesteroff & Galehouse, 1971; Pimm, Garrison & Boyce, 1971; Heath & Moberly, 1971). Sketches in broken boxes and broken arrows indicate hypothetical sediments and transformations, inferred but not observed in Leg 14 sediments. Symbols: 1=nannofossil chalk ooze, 2=nannofossil marl ooze, 3=siliceous organisms (diatoms, radiolarians, sponge spicules, etc.), 4=pelagic clay (montmorillonite, kaolinite, illite, etc.), 5=palygorskite (-sepiolite), 6=preserved Foraminifera (with calcitic tests), 7=relict coccoliths, 8=opaline skeletons (radiolarians), 9=pyritized radiolarians, 10=zeolite-filled radiolarians, 11=(radiolarian?) ghosts, 12=ash layer, 13=(corroded) dolomite (and siderite) rhombs, 14=zeolite (clinoptilolite), 15=detrital quartz, 16=non-detrital, microcrystalline quartz, 17=chalcedony, 18=lussatite (=opal–CT), 19=mass polarization (probably due to oriented palygorskite), 20=ditto incipient, 21=

the well-indurated cherts (especially the type IV cherts and the type II porcelanites) were only found as displaced fragments. Typical *nodular cherts,* interbedded with calcareous oozes, are rare in Leg 14 sediments: the quartz cherts of type IV (Site 137) might fit into that group. *Replacement cherts,* i.e. silicified terrigenous sand and silt layers, ashes, etc., as described by Beall & Fischer (1969) and Calvert (1971a), appear to be absent in our material.

Original sediments and the source of free silica

Type I porcelanites were found associated with hemipelagic, mixed terrigenous-siliceous muds, e.g. quartzose diatom mud, and with zeolite-bearing siliceous nannofossil marly oozes (Figs 1 and 8); type II and IV rocks are typically interbedded with black, carbonaceous, ± zeolitic shales, radiolarian oozes and greenish pelagic clays, whereas the typical association of the type III porcelanites is a hemipelagic, olive-green, dolomite-rich palygorskite mudstone. All facies interbedded with chert are characterized by olive-green to black colours (reducing conditions) and the presence of siliceous fossils and/or zeolite or palygorskite (Fig. 1), indicating high silica availability. On the other hand, pelagic brown clay was not found associated with chert, regardless of the presence of volcanogenic components. Also the volcanogenic sequence of Site 136 is essentially free of chert (Berger & von Rad, 1972).

Sediments with more than 30% siliceous fossils (diatom muds and radiolarian oozes) are quite rare (Fig. 3a). This abundance is only reached in Mid-Eocene, Oligocene, and Lower Miocene sediments, especially in the high fertility area off north-west Africa (Fig. 1), where the eastern boundary current and upwelling causes a high plankton production and silica supply. The best preserved siliceous skeletons occur in Lower Miocene continental-rise sediments (139–5, 140–2), where high supply of siliceous skeletons, rapid burial, and relatively silica-rich bottom and pore waters prevent silica leaching (Berger & von Rad, 1972). The maximum age of siliceous skeletons preserved as amorphous opal–A ranges apparently from 20 to 40 million years (Fig. 4), depending mainly on the sedimentation rates.

In cases where chertification is at an early stage, the *original sediment* is not too difficult to reconstruct, as in the foraminiferal and radiolarian mudstones of chert type I. The original unsilicified sediments of the cherts of type II probably consisted of hemipelagic, laminated, zeolitic and partly carbonaceous, siliceous and carbonate-free silty mudstones and pelagic clays (Fig. 8). The original sediments from which cherts III and IV were probably derived are discussed in the following section.

Although suggestive, the presence of siliceous fossils within cherts does not present a sufficient proof for the *source of free silica.* A reasonably high burial rate (high fertility) and silica-rich bottom-waters prevent the dissolution of siliceous skeletons at the sediment-water interface and are the first prerequisite for the later formation of a 'biogenous chert'. If siliceous skeletons were displaced together with terrigenous and calcareous material into water depth beyond the calcite compensation depth (CCD; about 4600 m, at Site 140), the selective dissolution of calcite would also favour the concentration of the opaline tests. The supply of volcanic debris would likewise reduce opal dissolution; the devitrification of volcanic glass increases the SiO_2 content of the pore water.

The second possible source of silica for chertification is the diagenetic alteration and devitrification of volcanic glass finely dispersed in pelagic clays. In these sediments silica is freed to form clinoptilolite, smectites, and presumably also opal–CT.

If excess magnesium is available, palygorskite and sepiolite can also be precipitated (Nayudu, 1964; Peterson & Griffin, 1964; Bonatti, 1965; Hay, 1966; Bonatti & Joensuu, 1968; Hathaway & Sachs, 1965; Peterson *et al.*, 1970; Hathaway, McFarlin & Ross, 1970).

The question of the origin of silica from sources that have vanished is, of course, difficult to decide without, at least, negative incidental evidence. The presence of chert in sequences which lack any volcanogenic components (e.g. Sites 139, 144), the preservation of the more resistant sponge spicules, and the frequency of ghosts of siliceous fossils in almost all chert types, favour biogenous opal as the most important source for the silica of Leg 14 cherts (see also Wise *et al.*, 1972; Heath, 1973).

Early diagenesis: mobilization of silica and the formation of lussatite-porcelanites

The processes involved in the mobilization of the silica and its migration during early diagenesis are poorly understood. Presumably, opaline skeletons (Fig. 3a) dissolve and/or volcanic material releases silica during devitrification. The silica-rich interstitial solutions migrate along bedding planes or fractures and vertically—in the order of cm to m?—from areas where silica is released to nearby more permeable lenses or layers, e.g. sand- or silt-rich sediments (Heath & Moberly, 1971, p. 996).

At an early stage of diagenesis, minute 'lepispheres' of lussatite are precipitated from silica-rich pore solutions and partially fill the interstices of semi-indurated porous mudstones (Fig. 8) or carbonate rocks (Wise & Kelts, 1972). X-ray analyses show that lussatite is quite common in un- to semi-consolidated deep-sea sediments of Leg 14 (von Rad & Rösch, 1972; Pow-foong Fan & Rex, 1972). It is commonly associated with clinoptilolite, especially in hemipelagic, olive-black carbonaceous marls, in black carbonaceous, zeolitic shales and in pelagic, greenish grey, zeolitic clays (Fig. 1). Most of these sediments contain little or none of the montmorillonite that is supposed to form before clinoptilolite and amorphous silica is an intermediate alteration product of rhyolitic-dacitic ashes (Hay, 1966, p. 82; Calvert, 1971a).

The semi-indurated, yellowish, mid-Eocene 'porcelaneous' mudstone (5% lussatite) of core 140–3–2 might indeed represent a '*precursor*' *stage* relating the original Lower Miocene diatom mud (140–2) to the ?Palaeocene palygorskite-lussatite porcelanite of core 140–6 (see Fig. 8). Compared with the porcelanite, this mudstone is about 10–15 million years younger and is buried by about 200 m less sediment. Age and overburden were apparently insufficient for chertification. The carbonate-free matrix consists of a mixture of chlorite, kaolinite, illite and palygorskite with small amounts of lussatite (5%), clinoptilolite and montmorillonite, and traces of dolomite and siderite. Bedding is produced by lenses and layers of dark carbonaceous and ferruginous material. The cohesion of these siliceous sediments suggests partial mobilization of silica, providing silica-silica bondings (Fig. 3b) and a 'welding' of opaline skeletons (Heath & Moberly, 1971, p. 989). Although not quite silicified, the matrix shows an oriented birefringence, i.e. mass extinction parallel to the bedding under crossed nicols. This effect is probably produced by a preferred orientation of palygorskite fibres, either by crystallization under directional stress (e.g. compaction) which produces a foliation parallel to the original bedding plane (Carozzi, 1960, p. 317), or by authigenic crystal growth from a saturated solution (Bowles *et al.*, 1971, Fig. 7). This mudstone contains about 20% siliceous organisms with well-preserved although mostly fragmented skeletons, largely radiolarians and sponge spicules, but traces of the more easily soluble diatoms were also seen. In contrast, in site 140–6, diagenetic processes presumably obliterated most of the original fossil record (Fig. 3g).

Because of the typical association of clinoptilolite-cristobalite-palygorskite, Calvert (1971a) suggested that the silica was derived from volcanic sources. The palygorskite could have been diagenetically formed together with clinoptilolite + amorphous opal or lussatite from the alteration of smectites and/or altered volcanic ash (Bonatti & Joensuu, 1968; Hathaway & Sachs, 1965). As an alternative, the direct precipitation of authigenic palygorskite and lussatite from hydrothermal solutions, supersaturated with magnesium and silica and reacting with sea water, was suggested by Bowles *et al.* (1971). Because the palygorskite in the > 120 m thick Palaeocene-Eocene palygorskite-rich section of Site 140 is mostly associated with montmorillonite (von Rad & Rösch, 1972; Pow-foong Fan & Rex, 1972) a diagenetic origin of the palygorskite starting from a montmorillonitic parent material is suggested for our type III porcelanites (see also Hayes & Pimm, 1972, p. 969). Although these porcelanites contain traces of siliceous fossils, it is possible that biogenous opal was insufficient as silica source. In view of the associated clay minerals, an additional 'volcanogenic' silica source cannot be ruled out.

In the Upper Cretaceous (to Palaeogene) sections of Sites 137, 138 and 140, radiolarians are generally poorly preserved with lussatite lepispheres and euhedral clinoptilolite crystals precipitated from pore waters on the internal walls of the skeletons (Berger & von Rad, 1972, pp. 800–806, Plates 12 and 18). Hurd (1973) has observed similar authigenic silicates actively growing on the surface of biogenic opal. By extracting silicon from the pore water these silicates reduce the solution rate of opal (see also Calvert, 1974).

Berger & von Rad (1972, pp. 880–881) suggested that the initial preservation of the Cenomanian radiolarians and their zeolitic fill was favoured by higher-than-average sedimentation rates on the ridge flank above the lysocline and supported by relatively high silica concentrations in less oxygenated bottom waters. During a later stage, these zeolite-filled radiolarian tests were redeposited into deeper water below the CCD, where the calcareous nanno- and micro-plankton were dissolved. In this environment, the alkalinity of the pore water was apparently increased and the displaced siliceous fossils were exposed to silica-undersaturated bottom waters under extremely low rates of sedimentation. This could explain the final dissolution of the alkaline tests and the observed corrosion of their zeolitic fill (Berger & von Rad, 1972, Plate 18; see also Fig. 8). It is interesting to note that the zeolite-filled radiolarians or zeolite aggregates forming casts of dissolved radiolarians can generally be found somewhat higher in the sedimentary column than the youngest cherts of each Site (Fig. 1, Site 137, 140). Given more time and deeper burial conditions, they will probably be completely dissolved, thus freeing silica for the formation of porcelanites.

The relatively high proportions of solution-resistant 'sponge' spicules noted in most porcelanites (especially type II) indicate that most of the original fossil assemblage within the cherts is dissolved. The destruction may have occurred long before chertification, with silica escaping into overlying bottom waters; or in connection with diagenetic mobilization of silica or recrystallization (Berger & von Rad, 1972).

Reducing conditions, indicated by the presence of pyrite, pyritized radiolarians and by the preservation of carbonaceous matter and laminations, were apparently typical of the deepest parts of the restricted Late Cretaceous Proto-Atlantic. The presence of organic-rich layers (negative Eh and pH) appears to have favoured the precipitation of both authigenic clinoptilolite and lussatite in the laminated carbonaceous mudstones of Sites 135 and 144 (see Fig. 1) to form the porcelanites of types I and II (see also Hathaway *et al.*, 1970).

Most of the cherts (I–III) recovered in Leg 14 sediments have only reached the lussatite-porcelanite stage with partly preserved or recrystallized siliceous organisms. The transformation of skeletal opal to lussatite by dissolution and reprecipitation appears to be a mainly time-dependent process. Apparently, it takes at least 20–50 million years to produce a porcelanite from the suitable original sediments (Figs 4 and 6). In this calculation it is of course an open question at what time after deposition the incipient stages of chertification began.

Chalcedony-cemented porcelanite breccias are common. The fracturing took place at an unknown time after complete consolidation and chertification of the rock and was probably caused by differential compaction or submarine slumping (cf. Lancelot, 1973). It was followed by repeated ingressions of silica-precipitating pore solutions, cementing the angular fragments by one or more generations of length-fast chalcedony (Berger & von Rad, 1972, Plate 31).

Late diagenesis: recrystallization of lussatite porcelanites to quartz cherts

The late diagenetic, slow conversion of cryptocrystalline lussatite into chalcedony and microcrystalline quartz has been discussed earlier. Quartz is the only stable phase in the dense, mature, vitreous cherts, where almost all fossils are obliterated and all porosity is lost (see Fig. 8). Some of the 'mature', recrystallized cherts of type IV contain a few percent of relict coccoliths. This suggests that the original sediment was a nannofossil marly ooze similar to the associated sediments in that core, but with admixtures of siliceous organisms.

Several factors, such as time, pH, eH, foreign cations (pore water chemistry), *in situ*-temperature, partial pressures, and permeability of the sediment seem to control this diagenetic process. Under normal deep-sea conditions, *time* appears to be the single most important parameter. According to our observations, about 70–90 million years have passed from the deposition of the original sediments to the completion of mature quartz cherts (see Figs 4, 6, 8). The minimum age of the sediments associated with other quartz cherts from the Atlantic is likewise Senonian (Beall & Fischer, 1969; Wise *et al.*, 1972; Calvert, 1971a; Beall *et al.*, 1973).

The *thickness of overlying sediments*, however, might also play an important role by raising the *in situ*-temperatures. According to Heath (1973) high concentrations of biogenous silica and *in situ*-temperatures close to 30°C are necessary to form cherts in less than 30 million years. The Lower Miocene (type I) porcelanite of Site 139 which is overlain by 700 m of hemipelagic sediment (Figs 1–2) might be an example of this process. The Cretaceous and Palaeogene cherts were formed under much shallower burial conditions (100–500 m; Fig. 2).

Based on Millot (1960), Lancelot (1973) has suggested that bedded lussatite porcelanites will form with the aid of foreign cations preferably in clay-rich sediments (zeolitic clays, clayey radiolarian oozes, marls, and marly limestones). On the other hand (nodular) quartz cherts are supposed to be restricted to clay-free carbonate environments and may form by direct precipitation during early diagenesis. Most of the points made by Lancelot in favour of his theory cannot be confirmed by our results. Practically all of our Leg 14 cherts—i.e. porcelanites *and* quartz cherts—are associated with clayey or marly sediments and contain appreciable amounts of clay minerals, zeolites, pyrite, dolomite, carbonaceous and opaque matter (Fig. 2, Table 2). No quartz cherts younger than Late Cretaceous were discovered, although skeletal opal is common in the Neogene carbonate oozes of Site 139 and 140 (Fig. 4). Also the suggestion that cherts are restricted to sediments with high K-feldspar/plagioclase, palygorskite/montmorillonite, and clinoptilolite/phillipsite ratios is not generally supported by our data. Relatively high K-feldspar contents are only typical for pyroclastic sediments devoid of chert; except for Site 136, clinoptilolite is the *only* zeolite in our sediments, occurring only in rocks older than about 20 million years and buried by at least 100 m (von Rad & Rösch, 1972). High palygorskite contents are only characteristic of our type III porcelanites. For a more detailed discussion of Lancelot's 'quartz precipitation hypothesis' see Wise & Weaver (1974).

Concluding, it should be emphasized that the chert classification and the genetic interpretations discussed in this paper are still tentative working hypotheses. More thorough investigations from a greater variety of deep-sea cherts from all oceans and facies, refined analytical methods, and more experimental data are needed to provide greater insight into all stages of the diagenesis of silica in marine sediments.

ACKNOWLEDGMENTS

Many of the environmental interpretations presented in this paper were developed in discussions with Dr W. H. Berger (La Jolla) during the preparation of a joint paper on the Leg 14 sediments (Berger & von Rad, 1972). Dr W. H. Berger, Dr M. Kastner (La Jolla) and Dr O. W. Flörke (Bochum) kindly criticized the manuscript and suggested several improvements. We are very grateful to Dr Flörke for several discussions on the mineralogy of silica phases in cherts and for contributing standard samples of opal–A and lussatite. Dr J. B. Jones (Adelaide) kindly made a few scanning electron micrographs and infrared analyses from our porcelanites. Mrs Scheuermann (Hannover) provided transmission and scanning electron photomicrographs. We are also indebted to Drs Eckhardt, Kürsten, von Stackelberg and Venzlaff (all of the Bundesanstalt für Bodenforschung, Hannover) for their support of our investigations of Deep Sea Drilling Project (DSDP) material. All studied samples are from Leg 14 of the DSDP which is financed by the National Science Foundation (USA).

REFERENCES

BEALL, A.O. JR & FISCHER, A.G. (1969) Sedimentology. In: *Initial Reports of the Deep Sea Drilling Project*, Vol. I (M. Ewing *et al.*), pp. 521–593. U.S. Government Printing Office, Washington.

BEALL, A.O., LAURY, R., DICKINSON, K. & PUSEY, W.C., III (1973) Sedimentology. In: *Initial Reports of the Deep Sea Drilling Project*, Vol. X (J. L. Worzel, W. Bryant *et al.*), pp. 699–729. U.S. Government Printing Office, Washington.

BERGER, W.H. & VON RAD, U. (1972) Cretaceous and Cenozoic sediments from the Atlantic Ocean. In: *Initial Reports of the Deep Sea Drilling Project*, Vol. XIV (D. E. Hayes, A. C. Pimm *et al.*), pp. 787–954. U.S. Government Printing Office, Washington.

BERGIN, R. (1964) Quantitative diffractometric analysis of X-ray transparent specimens. *J. scient. Instrum.* **41,** 558–560.

BONATTI, E. (1965) Palagonite, hyaloclastites and alteration of volcanic glass in the ocean. *Bull. volcan.* **28,** 1–13.

BONATTI, E. & JOENSUU, O. (1968) Palygorskite from Atlantic deep sea sediments. *Am. Miner.* **53,** 975–983.

BOWLES, F.A., ANGINO, E.E., HOSTERMAN, J.W. & GALLE, O.K. (1971) Precipitation of deep-sea palygorskite and sepiolite. *Earth Planet. Sci. Letts,* **11,** 324–332.

BRAITSCH, O. (1957) Über die natürlichen Faser- und Aggregationstypen beim SiO_2 und ihre Verwachsungsformen, Richtungsstatistik und Doppelbrechung. *Heidelb. Beitr. Miner. Petrogr.* **5,** 331–372.

BRAMLETTE, M.N. (1946) The Monterey Formation of California and the origin of its siliceous rocks. *Prof. Pap. U.S. geol. Surv.* **212,** 1–55.

CALVERT, S.E. (1966) Accumulation of diatomaceous silica in the sediments of the Gulf of California. *Bull. geol. Soc. Am.* **77,** 569–596.

CALVERT, S.E. (1971a) Composition and origin of North Atlantic deep sea cherts. *Contr. Miner. Petrol.* **33,** 273–288.

CALVERT, S.E. (1971b) Nature of silica phases in deep sea cherts of the North Atlantic. *Nature Phys. Sci.* **234,** 133–134.

CALVERT, S.E. (1974) Deposition and diagenesis of silica in marine sediments. In: *Pelagic Sediments: on Land and under the Sea* (Ed. by K. J. Hsü and H. C. Jenkyns). *Spec. Publs int. Ass. Sediment.* **1,** 273–299.

CAROZZI, A.V. (1960) *Microscopic Sedimentary Petrography,* pp. 485. J. Wiley & Sons, New York.

EISMA, D. & VAN DER GAAST, S.J. (1971) Determination of opal in marine sediments by X-ray diffraction. *Neth. J. Sea Res.* **5,** 382–389.

ERNST, W.G. & CALVERT, S.E. (1969) An experimental study of the recrystallization of porcelanite and its bearing on the origin of some bedded cherts. *J. Am. Sci.* **267-A,** 114–133.

FLÖRKE, O.W. (1955) Zur Frage des 'Hoch'—Cristobalits in Opalen, Bentoniten und Gläsern. *Neues Jb. Miner. Mh.* **10,** 217–233.

FLÖRKE, O.W. (1962) Untersuchungen an amorphem und mikrokristallinem SiO_2. *Chemie Erde,* **22,** 91–110.

FLÖRKE, O.W. (1967) Die Modifikationen von SiO_2. *Fortschr. Miner.* **44,** 181–230.

FOLK, R.L. & WEAVER, C.E. (1952) A study of the texture and composition of chert. *Am. J. Sci.* **250,** 498–510.

GOLDBERG, E.G. (1968) Determination of opal in marine sediments. *J. mar. Res.* **17,** 178–182.

GÜVEN, N. & GRIM, R.E. (1972) X-ray diffraction and electron optical studies on smectite and α-cristobalite associations. *Clays Clay Miner.* **20,** 89–92.

HATHAWAY, J.C., MCFARLIN, P.F. & ROSS, D.A. (1970) Mineralogy and origin of sediments from drill holes on the continental margin off Florida. *Prof. Pap. U.S. geol. Surv.* **581-E,** 26 pp.

HATHAWAY, J.C. & SACHS, P.L. (1965) Sepiolite and clinoptilolite from the Mid-Atlantic Ridge. *Am. Miner.* **50,** 852–867.

HAY, R.L. (1966) Zeolites and zeolitic reactions in sedimentary rocks. *Spec. Pap. geol. Soc. Am.* **85,** 130 pp.

HAYES, D.E., PIMM, A.C. *et al.* (1972) *Initial Reports of the Deep Sea Drilling Project,* Vol. XIV, pp. 975. U.S. Government Printing Office, Washington.

HEATH, G.R. (1973) Cherts from the Eastern Pacific. In: *Initial Reports of the Deep Sea Drilling Project,* Vol. XVI (Tj. H. van Andel, G. R. Heath *et al.*), pp. 609–613. U.S. Government Printing Office, Washington.

HEATH, G.R. & MOBERLY, R. JR (1971) Cherts from the Western Pacific, Leg 7, Deep Sea Drilling Project. In: *Initial Reports of the Deep Sea Drilling Project,* Vol. VII (E. L. Winterer *et al.*), pp. 991–1007. U.S. Government Printing Office, Washington.

HURD, D.C. (1973) Interactions of biogenic opal, sediment, and seawater in the central equatorial Pacific. *Geochim. cosmochim. Acta,* **37,** 2257–2282.

JONES, J.B., SANDERS, J.V. & SEGNIT, E.R. (1964) Structure of opal. *Nature, Lond.* **204** (4962), 990–991.

JONES, J.B. & SEGNIT, E.R. (1971) The Nature of opal. I. Nomenclature and constituent phases. *J. geol Soc. Aust.* **18,** 57–68.

JONES, J.B. & SEGNIT, E.R. (1972) Genesis of cristobalite and tridymite at low temperatures. *J. geol. Soc. Aust.* **18,** 419–422.

KLUG, H.P. & ALEXANDER, L.E. (1954) *X-ray Diffraction Procedures,* pp. 716. J. Wiley & Sons, New York.

LANCELOT, Y. (1973) Chert and silica diagenesis in sediments from the Central Pacific. In: *Initial Reports of the Deep Sea Drilling Project*, Vol. XVII (E. L. Winterer, J. I. Ewing *et al.*), pp. 377–506. U.S. Government Printing Office, Washington.

LANCELOT, Y., HATHAWAY, J.C. & HOLLISTER, C.D. (1972) Lithology of sediments from the North-west Atlantic, Leg 11, Deep Sea Drilling Project. In: *Initial Reports of the Deep Sea Drilling Project*, Vol. XI (C. D. Hollister, J. I. Ewing *et al.*), pp. 901–949.

MALLARD, M.E. (1890) Sur la lussatite, nouvelle variété minérale cristallisée de silice. *Bull. Soc. fr. Miner. Cristallogr.* **13**, 63–66.

MAXWELL, A.E. *et al.* (1970) *Initial Reports of the Deep Sea Drilling Project*, Vol. III, pp. 806. U.S. Government Printing Office, Washington.

MILLOT, G. (1960) Silice, silex, silicifications et croissance des cristaux. *Bull. Serv. Carte géol. Als. Lorr.* **13**, 129–146.

MIZUTANI, S. (1966) Transformation of silica under hydro-thermal conditions. *J. Earth Sci.* **14,** 56–88.

NAYUDU, Y.R. (1964) Palagonite tuffs (hyaloclastites) and the products of posteruptive processes. *Bull. volcan.* **27**, 1–20.

OEHLER, J.H. (1973) Tridymite-like crystals in cristobalitic 'cherts'; *Nature, Phys. Sci.* **241,** 64–65.

PETERSON, M.N.A., EDGAR, N.T., VON DER BORCH, C.C. & REX, R.W. (1970) Cruise Leg summary and discussion. In: *Initial Reports of the Deep Sea Drilling Project*, Vol. II (M. N. A. Peterson *et al.*), pp. 413–427. U.S. Government Printing Office, Washington.

PETERSON, M.N.A. & GRIFFIN, J.J. (1964) Volcanism and clay minerals in the southeastern Pacific. *J. Mar. Res.* **22,** 13–21.

PIMM, A.C., GARRISON, R.E. & BOYCE, R.E. (1971) Sedimentology synthesis: lithology, chemistry and physical properties of sediments in the northwestern Pacific Ocean. In: *Initial Reports of the Deep Sea Drilling Project*, Vol. VI (A. G. Fischer *et al.*), pp. 1131–1252. U.S. Government Printing Office, Washington.

POW-FOONG FAN & REX, R.W. (1972) X-ray mineralogy studies–Leg 14. In: *Initial Reports of the Deep Sea Drilling Project*, Vol. XIV (D. E. Hayes, A. C. Pimm *et al.*) pp. 677–725. U.S. Government Printing Office, Washington.

REX, R.W. (1970) X-ray mineralogy studies–Leg 2. In: *Initial Reports of the Deep Sea Drilling Project*, Vol. II (M. N. A. Peterson *et al.*), pp. 329–346. U.S. Government Printing Office, Washington.

RÖSCH, H. (1968) *Quantitative Phasenanalyse mit transparenten Diffraktometer-Proben.* Lecture, German Mineralogical Society, Köln.

VON DER BORCH, C.C., NESTEROFF, W.D. & GALEHOUSE, J.S. (1971) Iron-rich sediments cored during Leg 8 of the Deep Sea Drilling Project. In: *Initial Reports of the Deep Sea Drilling Project*, Vol. VIII (J. I. Tracey Jr *et al.*), pp. 829–835. U.S. Government Printing Office, Washington.

VON RAD, U. & RÖSCH, H. (1972) Mineralogy and origin of clay minerals, silica and authigenic silicates in Leg 14 sediments. In: *Initial Reports of the Deep Sea Drilling Project*, Vol. XIV (D. E. Hayes, A. C. Pimm *et al.*), pp. 727–751. U.S Government Printing Office, Washington.

WEAVER, F.M. & WISE, S.W. JR (1972) Ultramorphology of deep sea cristobalitic chert. *Nature Phys. Sci.* **237,** 56–57.

WISE, S.W. JR, BUIE, B.F. & WEAVER, F.M. (1972) Chemically precipitated sedimentary cristobalite and the origin of chert. *Eclog. geol. Helv.* **65,** 157–163.

WISE, S.W. JR & KELTS, K.R. (1972) Inferred diagenetic history of a weakly silicified deep sea chalk. *Trans. Gulf-Cst Ass. geol. Socs* **22,** 172–203.

WISE, S.W., JR & WEAVER, F.M. (1974) Chertification of oceanic sediments. In: *Pelagic Sediments: on Land and under the Sea* (Ed. by K. J. Hsü & H. C. Jenkyns). *Spec. Publs int. Ass. Sediment.* **1,** 301–326.

Spec. Publs int. Ass. Sediment. (1974) **1,** 349

Formation of deep-sea chert: role of the sedimentary environment

YVES LANCELOT

Lamont-Doherty Geological Observatory, Palisades, New York 10964, *U.S.A.*

ABSTRACT

The mineralogy of chert recovered in the central Pacific at several Deep Sea Drilling sites appears strongly influenced by the lithology of host sediments. Chert in clayey sediments is porcelanitic, while quartzose chert nodules are found exclusively in calcareous oozes and rocks. This composition is related to the amount of clay minerals in the sediment. Poor permeability and abundant foreign cations in clay-rich layers lead to precipitation of disordered cristobalite while good circulation and lack of exchangeable cations in carbonates favour direct precipitation of quartz. Chert formation results from a transfer from the highly soluble amorphous silica of biogenic origin into much less soluble crystalline forms. This transfer may be controlled by silico-aluminates present in the sediments. In most of the post-Lower Tertiary sediments these minerals might have been able to take up enough silica from interstitial waters to prevent chert formation.

The reference for the full text of this article is as follows.

LANCELOT, Y. (1973) Chert and silica diagenesis in sediments from the Central Pacific. In: *Initial Reports of the Deep Sea Drilling Project*, Vol. XVII (E. L. Winterer, J. I. Ewing *et al.*), pp. 377–405. U.S. Government Printing Office, Washington.

Spec. Publs int. Ass. Sediment. (1974) **1,** 351–366

Siliceous turbidites: bedded cherts as redeposited, ocean ridge-derived sediments

EUAN G. NISBET *and* ILFRYN PRICE

Department of Geology, University of Cambridge

ABSTRACT

The Neraida cherts of the Othris Mountains, Greece, were deposited during the initial opening phase of a small Mesozoic ocean. They are bedded radiolarites with many features similar to those of turbidites. Typical chert beds contain several intervals: top and bottom structureless intervals and a central parallel laminated zone enclosing a band with an anastomosing texture. Lithological components are clearly graded and give consistent facing directions. The clay content resembles products of submarine weathering of basalt. The cherts could have been deposited from low-density turbidity flows, possibly from a mid-ocean ridge. A model is suggested for the diagenetic development of such turbidite-derived cherts.

INTRODUCTION

The origin of bedded cherts has long been controversial. Debate has centred on the mode of deposition and the biogenic or volcanogenic origin of the silica (see Grunau, 1965, for review of the problems). Of the wide variety of cherts in the Othris Mountains, eastern central Greece, one, the Neraida Chert Member of the Agrilia Formation (stratigraphy by Smith *et al.*, in preparation) throws some light on these problems as the cherts display many apparently primary sedimentary structures. The type section of this unit, discussed in this paper, is 1 km south of the village of Neraida on the road to the town of Stylis, 15 km east of Lamia, central Greece.

The chert member forms the lowest unit of a sedimentary sequence. It overlies spilitized pillow lavas of the Agrilia Formation and is overlain by distal calcareous turbidites and pelagic carbonates. The volcanics have been interpreted as continental margin basalts (Hynes, 1972), whilst trace element studies show they have oceanic affinities (Bickle & Nisbet, 1972). It is probable that they formed during early ocean spreading. A so-called 'Tethyan' ophiolite complex, containing very similar cherts, tectonically overlies the Neraida chert-bearing sequence (Hynes *et al.*, 1972).

Thus the Neraida Chert Member is interpreted as having been deposited directly on top of basalt, at the extreme distal end of a continental slope, as the first sediments in a developing ocean basin.

The exposure is complex (Fig. 1). In both the southern and northern parts 7·5 m of cherts underlie grey bedded micrites and in the central zone 5 m of similar cherts

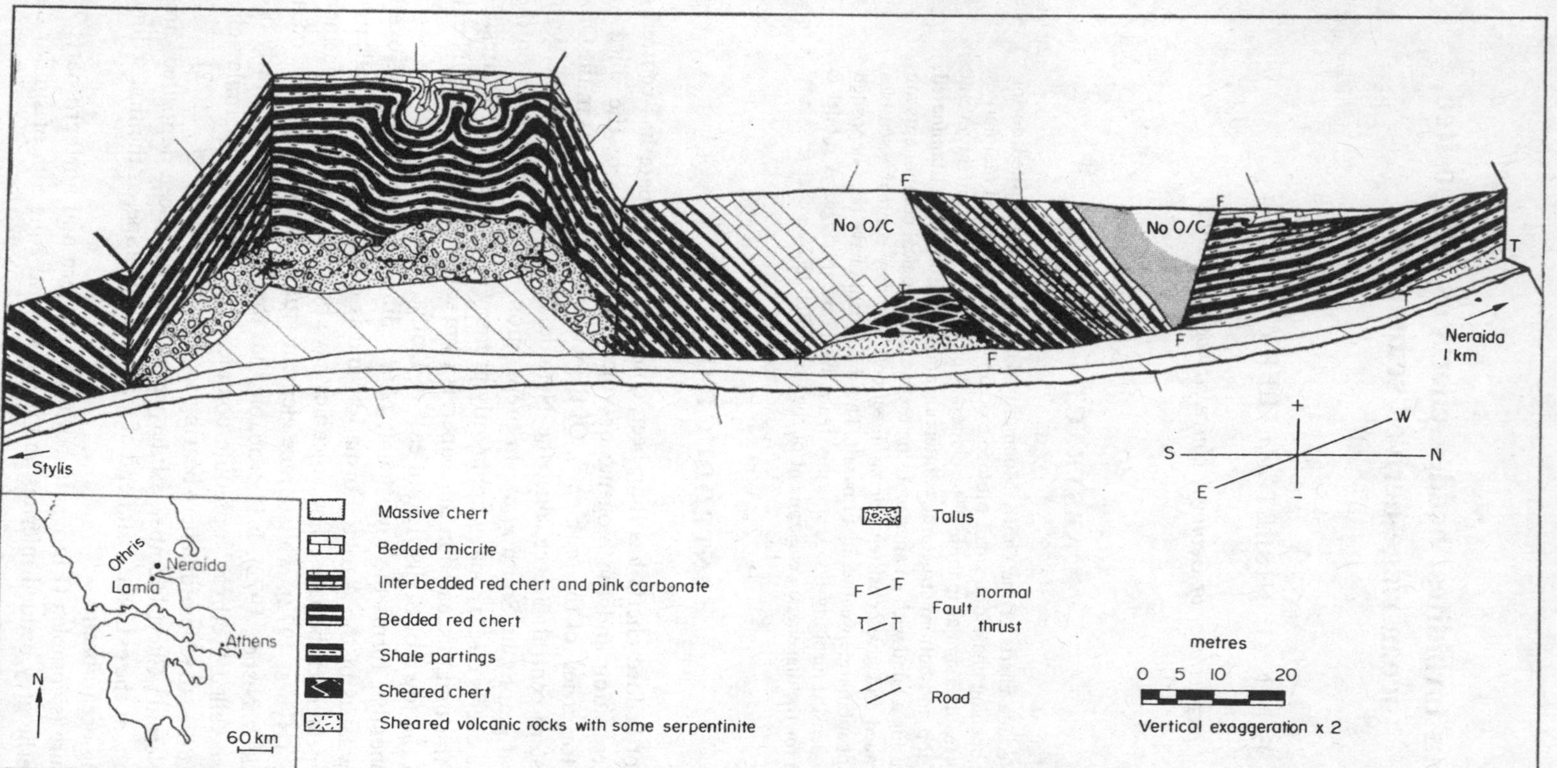

Fig. 1. The type outcrop of the Neraida Chert Member; a cliff and quarry section following a road. The cherts and cherts with carbonates in a faulted block (mid-section) are not discussed here.

Fig. 2. The cherts in outcrop. (a) Low angle thrusts in the section. (b) Detail: note the grouping of beds of similar thickness (hammer handle=30 cm).

occur, in a faulted block, underlying cherts interbedded with pink carbonates. The southern end of the outcrop is strongly folded, and low-angle bedding plane thrusts are present (Fig. 2a).

PETROLOGY

Field appearance

In outcrop the Neraida chert resembles many 'Tethyan' radiolarites. Regular chert beds, 1–6 cm thick, are separated by thin shaly partings (Fig. 2). Under close inspection in the field, most chert beds are either featureless or show zones of fine parallel lamination (Fig. 3a, c). Such zones are usually confined to the central part of a bed in which the top and bottom intervals are structureless. In rare instances a

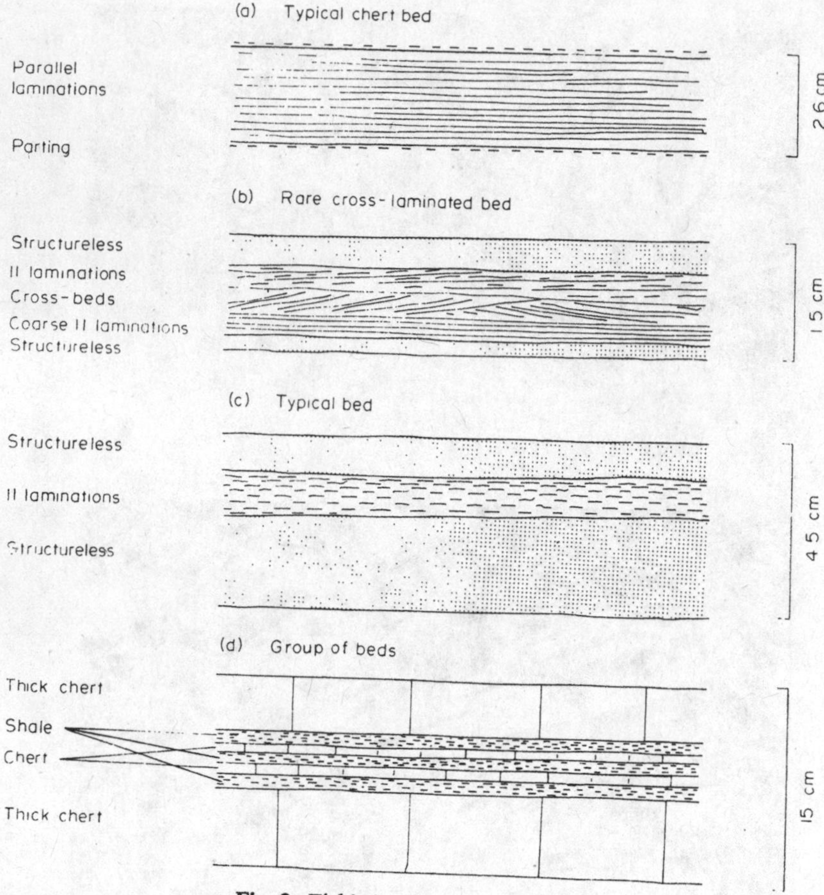

Fig. 3. Field sketches of chert beds.

central interval may show cross-lamination (Fig. 3b). Rounded radiolarian outlines are sometimes visible in the structureless parts of a bed. There is a tendency towards grouping of beds of similar thickness (Fig. 2b).

Petrography

Five distinct components can be recognized. Textures are probably primary. Large areas of microcrystalline quartz are absent and the ornament is often preserved on radiolarian skeletons; both features imply little recrystallization of silica (Thurston, 1972). Petrological and structural distinctions become less apparent in more re-crystallized samples.

The components are the following.

(1) Red clay. We use this term for apparently amorphous red brown material. It cannot be resolved optically but shows a suggestion of graininess in both acetate peels and thin sections. Electron microprobe and electron microscope studies indicate a granular nature and a high iron content, and the red clay may be either granular iron oxide (D. R. Thurston, personal communication) or iron oxide sorbed onto clay micelles (Carroll, 1958).

(2) Silica-siltite (Fig. 4c). This, the main siliceous component, consists of silt- and fine sand-grade, approximately ellipsoidal aggregates of cryptocrystalline and microcrystalline quartz and chalcedony in a matrix of red clay. Larger aggregates are distinct but smaller particles merge into a siliceous red clay.

(3) Quartz debris (Fig. 4a). Minute individual grains of quartz and cryptocrystalline silica are often visible as dispersed fragments in the red clay. They have a maximum diameter of about 0·05 mm, one quarter that of radiolarian tests. Broken biogenic material (including spicules) is placed in this category. Similar fragments to those in red clay can be found in silica-siltite but their recognition is more difficult.

(4) Radiolarian tests. These are often well preserved with spines and internal structures visible. The degree of preservation does vary and almost invariably the better preserved specimens are found in red clay. Recrystallized tests are infilled by either microcrystalline quartz or chalcedony. Rarely, tests are infilled by red clay.

(5) Iron-rich stringers (Fig. 4a, b). Thin dark brown stringers forming a flattened mesh texture are common in red clay and sometimes occur in silica-siltite. Studies of iron, aluminium and silica distribution with a scanning electron microprobe indicate a strong concentration of iron and aluminium in the stringers in comparison to both silica-siltite and red clay. Some stringers cut Radiolaria and are probably microstylolites. Similar material is present in jagged vertical stylolites. Most stringers, however, wrap round the microfossils and it is possible that some were deposited in periods of weakened current with iron oxide sorbed onto light clay micelles.

Mineralogy

X-ray diffraction traces of silica-siltite, red clay and parting samples (Fig. 5) show quartz and haematite to be the dominant minerals, though the latter is less abundant in silica-siltite. Pure samples of any lithological component are difficult to obtain and all samples are inevitably contaminated by other components. Parting samples usually contain red clay and the distinction between a parting and red clay with iron-rich stringers may be trivial (see section on texture).

The calcite content of the cherts is negligible and is accounted for by the presence of small calcite filled fractures. In the partings calcite is more variable and seems to increase up section towards the overlying carbonates.

Most significant is a set of peaks, whose relative intensity varies consistently from sample to sample, at 4·5Å, 2·5Å and 1·49–1·51Å. This set with a weak peak at about 3·5Å corresponds to peaks recorded by Matthews (1971) from palagonites in weathered submarine basalt samples. Together with chlorite they form the principal clay component of both red clay and parting samples. These are generally similar in composition but the partings also have a strong peak at 4·98Å probably indicating the presence of a micaceous clay mineral absent from red clay. The chlorite has not been positively identified but could also have an igneous origin (Griffin, Windom & Goldberg, 1968).

SEDIMENTARY LAYERING

Various inter-relationships of the five component lithologies, coupled with different sedimentary structures, produce distinct layering (Figs 4, 6) within the chert beds. A bed is taken as the interval between two partings. In its simplest form this layering

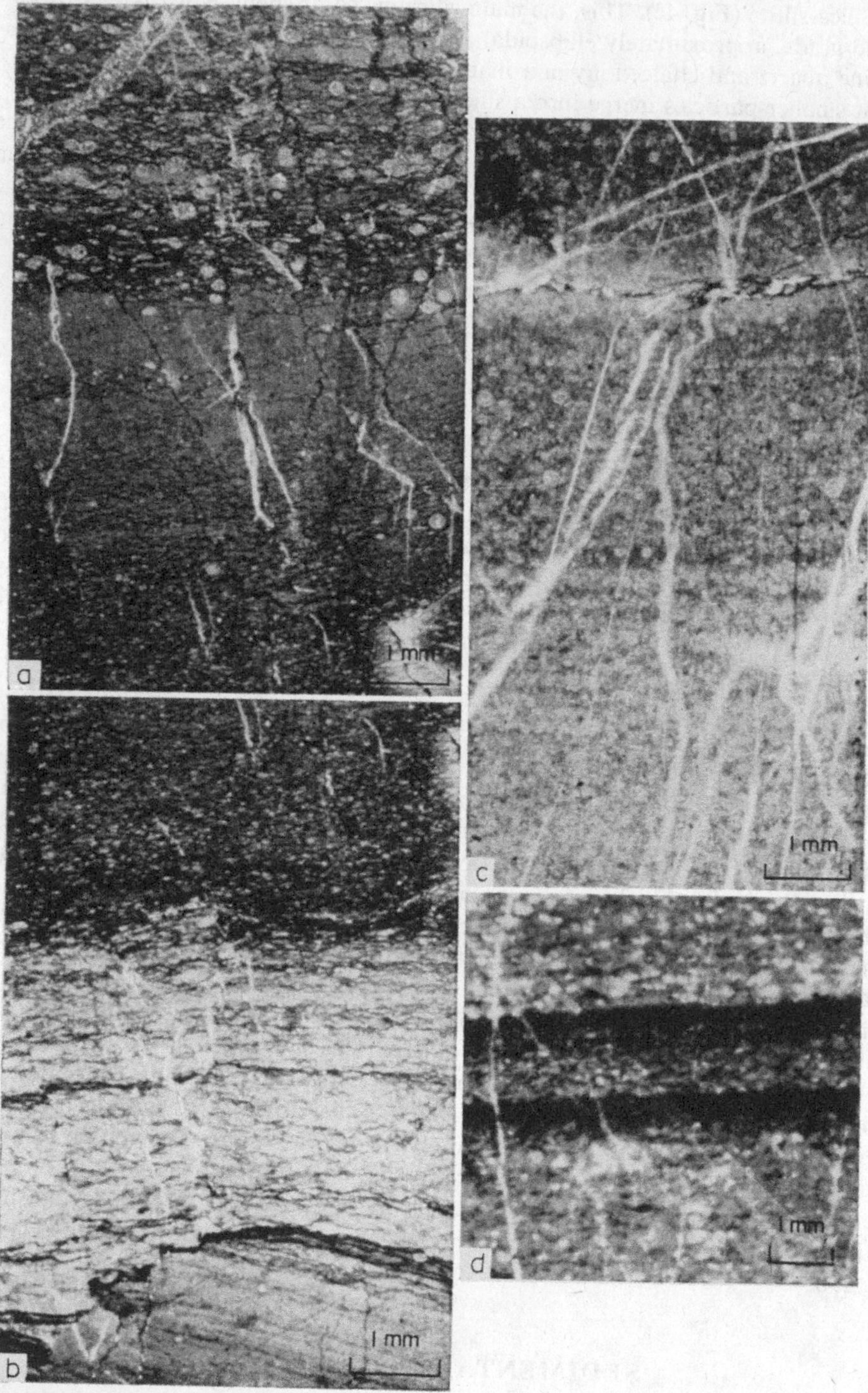

Fig. 4. Photomicrographs. (a) Red clay, Radiolaria and graded detrital quartz. (b) Silica-siltite, iron-rich stringers and base of a set. (c) Lamination developing in silica-siltite. (d) A complete graded chert set: silica-siltite white, red clay dark. All these photographs are the right way up. (a) and (b) overlap to show a complete interval (one set).

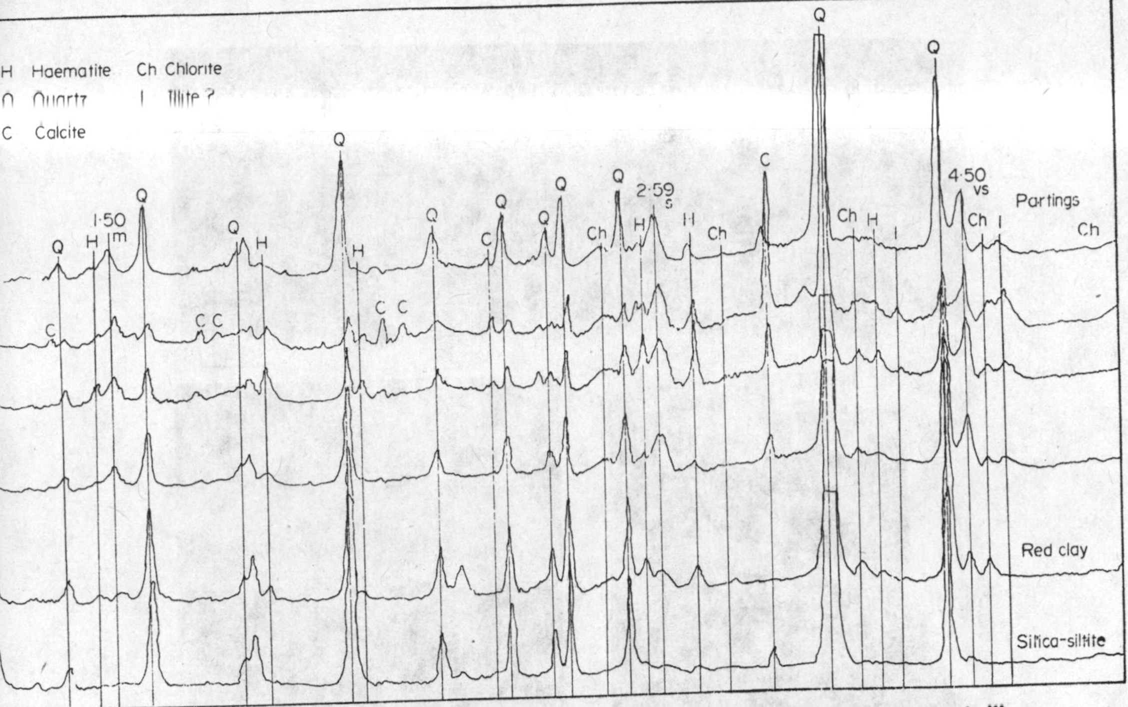

Fig. 5. X-ray diffraction traces (40 kV, 35 A, CuK) of four parting samples, red clay and silica-siltite. The numbered peaks correspond to those listed by Matthews (1971) for submarine palagonites. Intensities and *d*-spacings are shown (m, medium, s, strong, vs, very strong).

reduces to a single graded interval with silica-siltite grading upwards into red clay (Fig. 4d). Such intervals have sharp upper and lower contacts which can be marked by possible erosional and load cast (Fig. 6c) structures. We term such an interval a *set* for discussion purposes. A set so defined has sharp upper and lower contacts which need not be partings between beds.

Three distinct intervals are usually present in a set (Fig. 6b).

(1) Basal interval. The structureless basal interval consists of either silica-siltite or red clay with abundant Radiolaria and detrital material. Silica-siltite is usually predominant.

(2) Laminated interval. The laminated interval occurs above the basal interval. Laminations can be marked either by darker red clay laminae or, more commonly, by a lamination of silica-siltite and red clay. Small, discrete lenses of pure red clay are frequent in this interval. Radiolaria are less common than in the basal interval.

(3) Top interval. The upper part of a set is once again structureless. It is composed of red clay, with or without detrital quartz, and occasional Radiolaria. The uppermost part of a set usually consists of almost pure red clay. Iron-rich stringers can occur at the contact of two sets.

The position of the iron-rich stringers varies. They are not found at the extreme base of any set, but appear mainly in the laminated interval. The lowermost stringers are usually parallel to the bedding and to each other, but upwards they become curved and often cut out other stringers. They become parallel again and less common or absent in the top interval. A central mesh-structured interval (the anastomosing

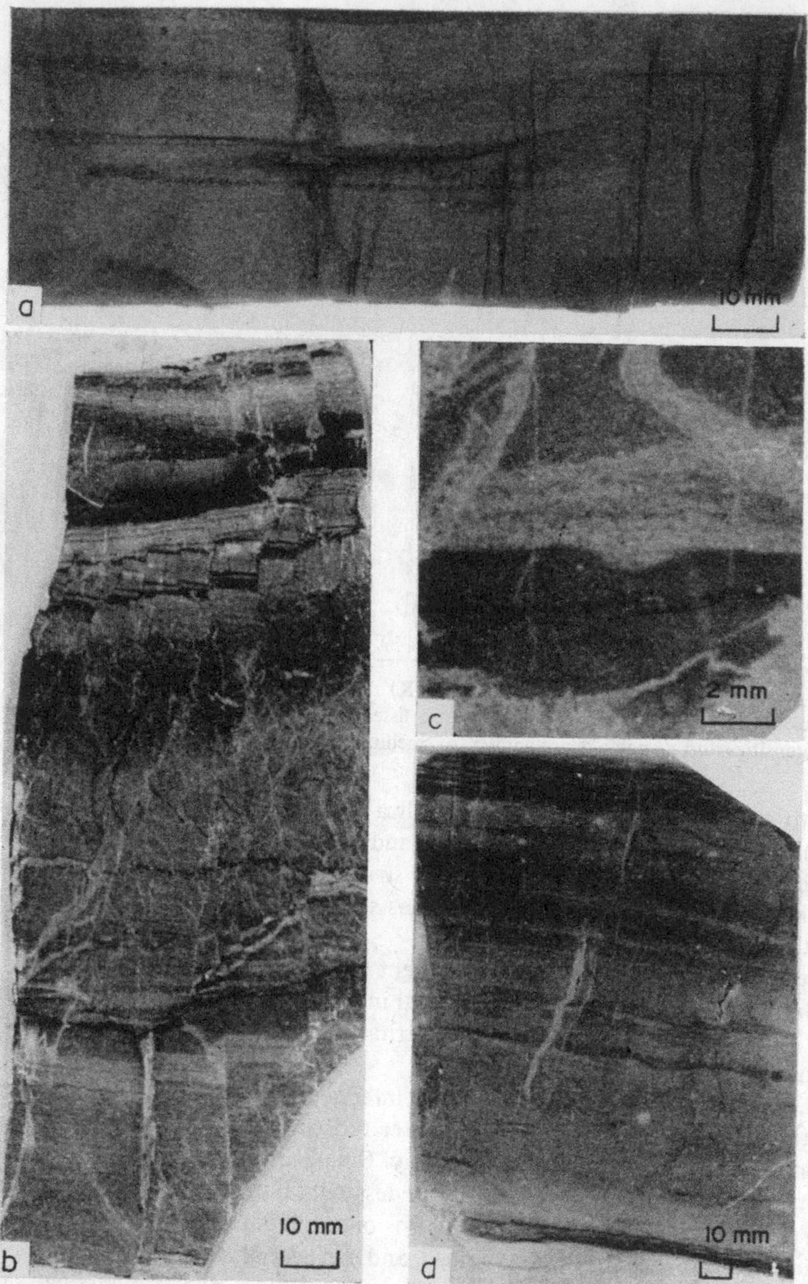

Fig. 6. Sedimentary structures in the cherts. (a) X-radiograph: note preferential concentration of Radiolaria which appear as dark specks and the overall laminated appearance. (b–d) Acetate peels: positive prints. (b) One chert bed—apparently homogeneous in outcrop. (c) Detail of the base of a set showing a possible erosional bottom structure. (d) Alternation of silica-siltite and red clay laminae with anastomosing iron-rich stringers at the top of the specimen.

interval) is present in most sets, marked by curved and intersecting stringers. It varies in thickness and can show radiolarian concentrations. Even though many iron-rich stringers may be micro-stylolites, they could still reflect a pre-existing texture. Shale partings are extremely difficult to sample. The sharp contact between sets is often marked by a very thin interval of iron-rich stringers, thicker and more closely spaced than those in the central anastomosing intervals. Acetate-peel comparisons suggest that interbed partings are composed of similar material. Low-angle cross bedding characterizes the contacts of some sets (Fig. 4b) and some contacts show complex, probably penecontemporaneous deformational and load structure (Fig. 6b).

GRADING

Each lithological component in the rock is graded, both in abundance and, for coarser components, in size. Over the central laminated interval silica-siltite gradually grades out and is replaced by red clay. Detrital quartz is similarly graded, both in size and in abundance and is only occasionally present in the top interval. Radiolaria are markedly concentrated at the base of sets and are less common in the central interval. They grade out in abundance upwards. Inversely, red clay becomes commoner upwards.

Both grading and cross-cutting contacts establish a consistent facing direction for the cherts. They indicate that the succession is the right way up as is also inferred from the regional stratigraphy.

BED THICKNESS

Approximately 700 measurements of bed and parting thickness were taken along lines perpendicular to bedding over the outcrop. This represents roughly three measurements per bed in the 7·5 m thickness of the Neraida Chert Member. Folding, though locally intense, appears to have little effect on bed thickness as deformation is mainly by slip along partings. Measurements of thickness variation of a number of individual beds round fold hinges revealed little change even close to the hinge. Readings considered in the analysis presented below are taken, where possible, from areas of low deformation.

A second independent group of readings was taken at three sites in the lower part of the section. Results of both groups are displayed in Figs 7 and 8. Over the whole section (Fig. 7) virtually all beds are between 1 and 7 cm thick with a wide distribution in this range. Beds were taken as the interval between two partings or visible fractures, however thin. Figure 7a groups the readings in 0·2 cm intervals to reduce observer error in measurement. The distribution is not uniform. Readings from the more limited stratigraphic range (Fig. 8) indicate a bimodal distribution. Readings in this range were grouped into 0·5 cm intervals to allow for slight lateral thickness variation when sampling limited intervals. Minor faults (Fig. 2b) make it impossible to measure beds along one line through the complete section, but the outcrop gives a qualitative impression of a general upwards thickening of all beds. Set thickness measurements made on acetate peels of twenty samples from the lower part of the section show, in contrast, a unimodal distribution (Fig. 8) indicating that sets, as

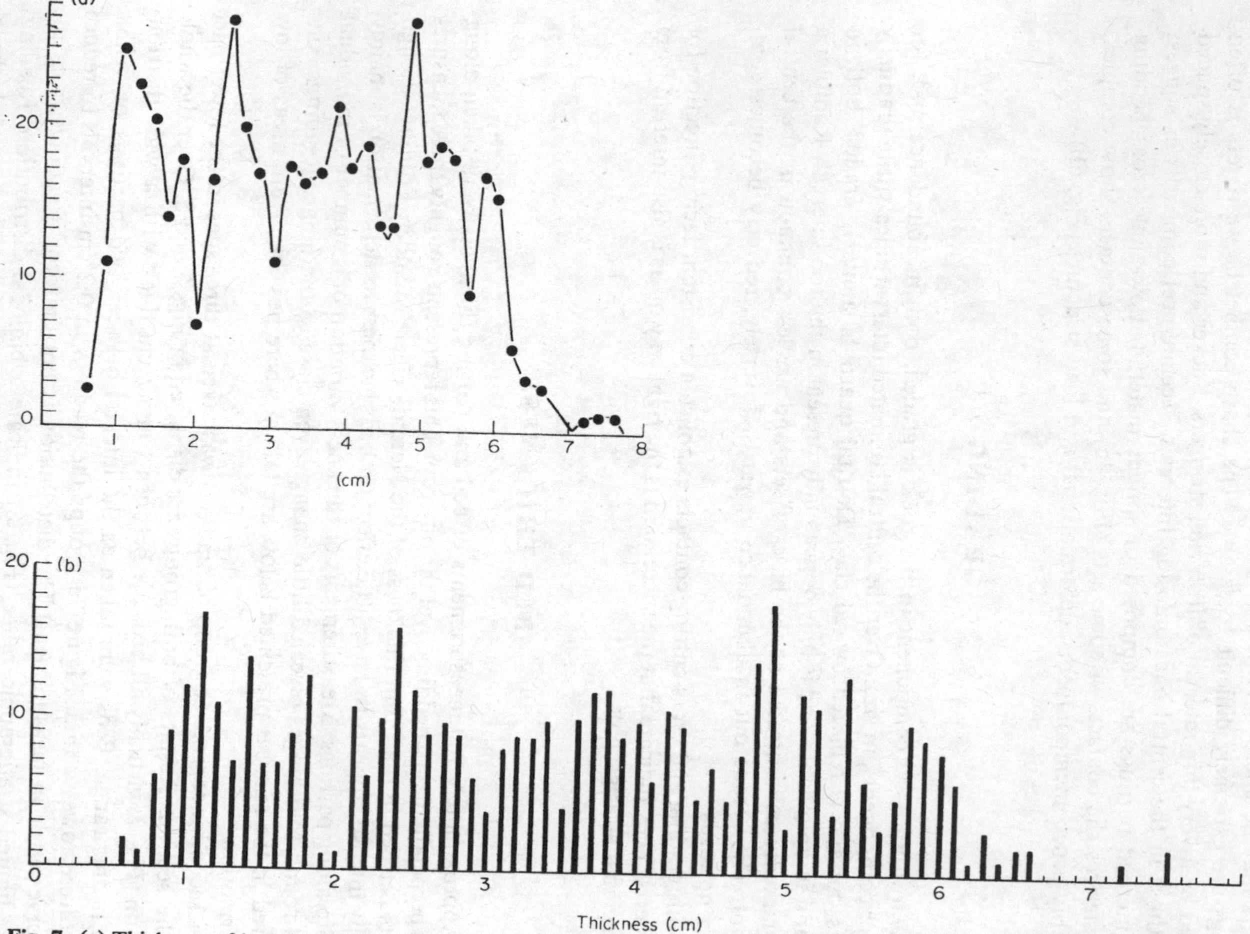

Fig. 7. (a) Thickness of beds in the Neraida chert, grouped in 0·2 cm intervals. (b) Thickness of beds in the Neraida chert.

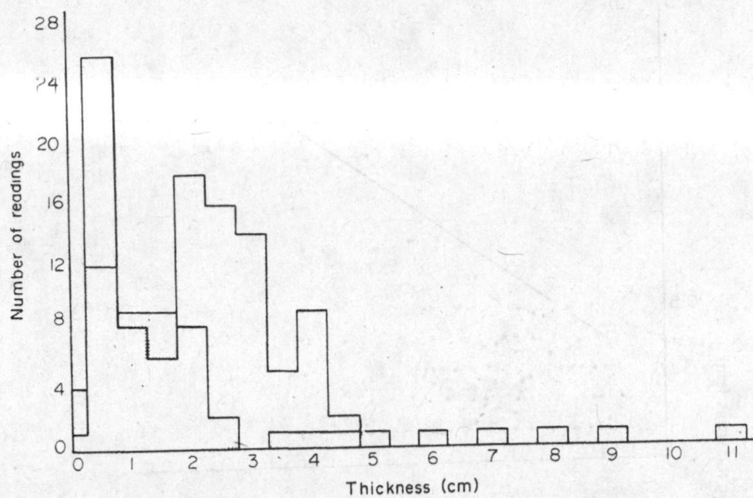

Fig. 8. Comparison of bed (unshaded) and set (shaded) thickness of cherts over a limited interval of the section.

defined above, are a more realistic indication of the sedimentary processes which formed the cherts than are the interbed partings.

Figure 9a, b shows the relationship between parting and bed thickness. From these it is apparent that there is a relationship between the maximum thickness of each shaly band and its chert bed, but no minimum relationship. If the shale were diagenetically derived from the chert (cf. Davis, 1918) and the silica/clay ratio remained constant throughout the section, a linear relationship between the two would be expected with thick partings consistently separating thick beds. This is not found and partings may be barely recognizable even between the thickest beds in the outcrop.

The observed relationship is easily explained if the fine interval were partly eroded between the deposition of two chert beds. Gentle ocean bottom winnowing or sub-turbidite erosion could produce this effect. Locally, absence of parting may result from deformation but measurements of parting thickness were confined to the less deformed parts of the exposure.

INTERPRETATION

The cherts of the Neraida Member are current-deposited sediments. All the evidence presented above supports this conclusion. Internal structures, grading, erosional contacts and partings imply that the chert is a turbidity-current deposit, with a set as the fundamental unit deposited from a single current (cf. Fagan, 1962). Partings occur where either an inter-turbidite interval or the finest sediment from one turbidity flow is preserved. The textural sequence observed supports a turbidite model with a massive graded basal interval, a central laminated interval and a fine massive upper interval. The central anastomosing interval, with iron-rich stringers cutting out each other, may correspond to a cross-laminated interval, although lithological components may not be primary.

The provenance of the various lithological components is uncertain. Radiolaria

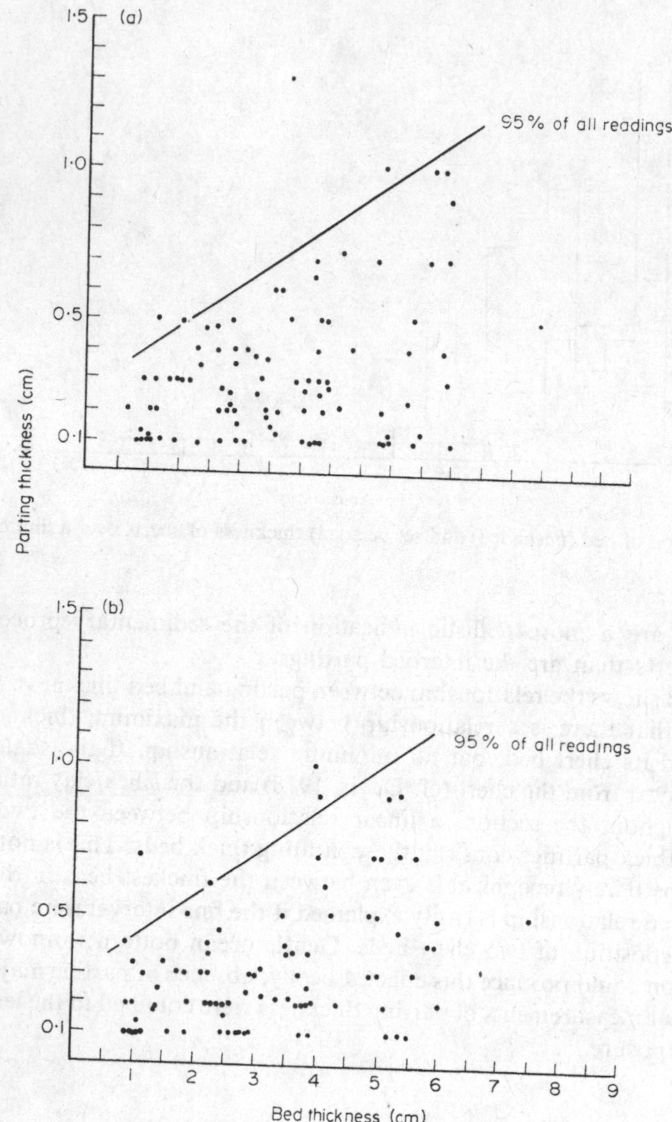

Fig. 9. (a) Relationship between thickness of top parting and bed thickness. (b) Relationship between thickness of lower parting and bed thickness.

are typical pelagic components but their concentration in definite layers implies relative enrichment of the source area of the turbidite over the depositional basin. Concentration of debris from a relatively wide area in a small sediment trap of limited extent would achieve this effect. Greater radiolarian production in the source area and gathering from pelagic deposits by basal erosion during transport might also contribute. The quartz debris may be submarine volcanic ejecta, biogenic debris, aeolian pelagic sediment (Radczewski, 1939) or some combination of these. Palagonitic deposits of submarine volcanism are a likely source of the red clay. Two alternative origins for silica-siltite are possible (e.g. Berger & von Rad, 1972). A biogenic component is certainly present but it could be supplemented by siliceous volcanic debris. Both are compatible with the formation of a silica-rich deposit prior to turbidite transport.

To generate siliceous turbidites the source area would need sufficient relief to sustain a turbidity flow and an irregular topography to provide suitable sediment traps. Submarine volcanism as a sediment source and perhaps seismic activity to trigger flows are also likely.

IMPLICATIONS

Feasibility

Siliceous turbidites, if they occur, must be either extremely distal, as the material deposited is very fine, or derive from a source area lacking coarse sediment. Since the cherts are wholly siliceous, and show no sign of ever having had a calcareous component, it is probable that the current was at all times a very low-density current and never had an important coarse calcareous fraction. It is likely, however, that a small part of the current was once moderately coarse siliceous debris which would have been significant in transport (Riddell, 1969). The order of magnitude of the current density in the area of deposition can be calculated, given some assumptions. The depositional site of the cherts (following section and Fig. 10) was probably a 'pond' between an ocean ridge and a continental shelf. This allows the assumption that the current deposited material in this pond over much of its ponded length. Assuming a conservative thickness of 10–25 m for the turbidity flow, a maximum set thickness of 5–7 cm and a density of 2·5 g/cm^3 for the suspended material, a density increment over seawater ($\delta\rho$) of the order of 0·005–0·02 g/cm^3 is obtained for the current, with a probable head velocity of between 5 and 20 cm/s (using the equations developed by Middleton, 1966). This is consistent with the small amount of basal erosion observed and the frequent preservation of shaly partings. Circulating currents in the head of the flow, with velocities of perhaps 10–40 cm/s, could adequately account for the basal erosion seen.

Middleton (1966) has pointed out that, in a low-density flow, particles first deposited on the bottom come to rest rapidly, with little movement by traction. The interval above the base is deposited particle by particle or layer by layer with considerable traction, followed by rapid deposition from suspension as supply is depleted and, finally, very slow deposition of the finest sediment from the tail of the current. In a low-density current it would perhaps be expected that the middle interval would be different from that normally found in turbidites, because of the short supply of sediment, particularly coarse material. The 'anastomosing' interval in the Neraida cherts (Figs 4c, 6e) may represent this interval with parallel laminations above and below it and a structureless interval top and bottom.

Riddell (1969) has shown the importance of fine material in turbidity currents and demonstrated how they can be boosted by a small coarse component. It is probable that, with a suitable gradient, fine turbidity flows are stable and reach a constant velocity. With an adequate slope to maintain the flow low-density currents should be stable over extremely long distances, at low velocities.

Environmental aspects

The stratigraphic position of the Neraida Chert Member indicates that it was probably deposited soon after the cessation of volcanic activity on a continental margin and was contemporaneous with active ocean-floor spreading. On the continental side of the depositional site a carbonate platform was developing (Hynes *et al.*, 1972) and there is no evidence of a suitable source of siliceous material in that direction. Nor is there evidence of any original carbonate component to these cherts. A differing source area is thus implied. An obvious possibility is that the material in the cherts originally accumulated on the flanks of a mid-ocean ridge. Thurston (1972) has shown the similarity of metal-cation ratios in ophiolite-associated cherts and the underlying mafic material. The similarity of the clay mineral assemblage in the cherts to that in submarine palagonites supports this hypothesis. An active ridge would fulfil all the requirements of a source area, including the possible supply of a coarser siliceous fraction, volcanic or seismic 'triggers' and high initial slopes.

Petrological observations in the overthrust ophiolites (Hynes, 1972 and our own results) suggest a slow to moderate spreading rate, probably of the order of 1 cm per year or slightly more. Following Sclater & Harrison's (1971) model for ridge elevation, the crest of the ridge was probably about 2 km below sea level. Since both the continental-margin volcanics and the ophiolite complex in Othris have chert directly overlying volcanics, it is probable that the carbonate compensation depth at the time was above the ridge crest; i.e. less than 2 km. Assuming a deposition rate of 3 m per million years for the Neraida cherts (estimate from data in Berger & von Rad, 1972) and 2–3 million years for the cessation of continental-margin volcanism after initiation of the ridge, it is likely that the cherts were formed 40–50 km from the ridge crest at a depth of about 3 km or slightly more, in an ocean perhaps 150 km wide. These are probably maximum estimates since the elevation model is affected by the proximity of the continental margin.

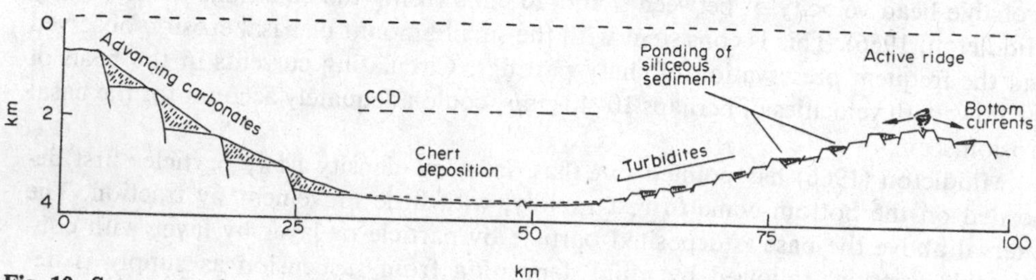

Fig. 10. Summary of the depositional mode suggested for the Neraida chert. CCD=carbonate compensation depth. Ridge elevations from Sclater & Harrison (1971).

The slope from the source on the ridge flanks to the site of deposition was probably about 1–2°; adequate to maintain indefinitely a low-density current of fine material travelling at low velocity with little erosional coupling at its base. Deposition would only occur at the gradient reversal as the current reached the continental margin (Fig. 10). With little erosion or sediment loss during transport, deposits from such a current would be of roughly constant thickness independent of distance travelled.

ROLE OF DIAGENESIS

The critical evidence for siliceous turbidites is based on structures considered to be primary. We consider the model could apply to other cherts lying on oceanic or continental-margin crust. Some assessment of the possible diagenetic alteration of such cherts is necessary.

Two diagenetic 'trends' can be seen in the Neraida cherts. Some recrystallization has occurred in silica-siltite. Radiolaria and biogenic debris are best preserved in red clay suggesting that the clay hinders recrystallization of the silica in some fashion. Secondly there is a degree of stylolitization involved in the formation of iron-rich stringers. These develop at red clay/silica-siltite contacts. Between sets this lithological change is abrupt and the stringers are densely packed compared to those in the anastomosing interval where the lithological change is gradational. Continued solution along the stringers would tend to sharpen the separation of silica and clay and would in particular disrupt the laminated interval of a set.

A combination of this process and increased recrystallization of silica could produce a rhythmic alternation of beds of microcrystalline chert and red clay, with no obvious sedimentary link between them. It is possible that the rhythmic bedding of many radiolarian cherts could be a result of diagenetic enhancement of the silica/clay alternation produced by siliceous turbidites.

ACKNOWLEDGMENTS

The basic concepts in this paper came after a field discussion with Dr A. G. Smith and Mr E. T. C. Spooner. We should like to thank Dr Smith, Dr P. F. Friend, Dr R. E. Garrison, Dr H. C. Jenkyns, Dr R. W. O'B. Knox and Dr G. V. Middleton for critically reading the manuscript and for helpful comments; and Dr E. M. Moores, Dr D. R. Thurston and Mr A. H. F. Robertson for very useful discussions. The work was supported by the Natural Environment Research Council, Sidney Sussex College, Cambridge and the Sedgwick Museum.

REFERENCES

BERGER, W.H. & VON RAD, U. (1973) Cretaceous and Cenozoic sediments from the Atlantic Ocean. In: *Initial Reports of the Deep Sea Drilling Project*, Vol. XIV (D. E. Hayes, A. C. Pimm *et al.*), pp. 787–954. U.S. Government Printing Office, Washington.

BICKLE, M.J. & NISBET, E.G. (1972) Oceanic affinities of some Alpine mafic rocks based on their Ti-Zr-Y contents. *J. geol. Soc.* **128,** 267–271.

CARROLL, D. (1958) Role of clay minerals in the transportation of iron. *Geochim. cosmochim. Acta,* **14,** 1–28.

DAVIS, E.F. (1918) The radiolarian cherts of the Franciscan group. *Univ. Calif. Publs Bull. Dep. Geol.* **11,** 235–432.

FAGAN, J.J. (1962) Carboniferous cherts, turbidites and volcanic rocks in Northern Independence Range, Nevada. *Bull. geol. Soc. Am.* **73,** 595–612.

GRIFFIN, J.J., WINDOM, H. & GOLDBERG, E.D. (1968) The distribution of clay minerals in the world oceans. *Deep Sea Res.* **15,** 433–459.

GRUNAU, H.R. (1965) Radiolarian cherts and associated rocks in space and time. *Eclog. geol. Helv.* **58,** 157–208.

HYNES, A.J. (1972) *The geology of part of the western Othris Mountains, Greece,* pp. 196. Unpublished Ph.D. Thesis, University of Cambridge.

HYNES, A.J., NISBET, E.G., SMITH, A.G., WELLAND, M.J.P. & REX, D.C. (1972) Spreading and emplacement ages of some ophiolites in the Othris region (Eastern Central Greece). *Z. dt. geol. Ges.* **123,** 455–468.

MATTHEWS, D.H. (1971) Altered basalts from Swallow Bank, an abyssal hill in the NE Atlantic and from a nearby seamount. *Phil. Trans. R. Soc.* **268,** 551–572.

MIDDLETON, G.V. (1966) Experiments on density and turbidity currents, I: *Can. J. Earth Sci.* **3,** 523–546.

RADCZEWSKI, O.E. (1939) Eolian deposits in marine sediments. In: *Recent Marine Sediments* (Ed. by P. D. Trask). *Spec. Publs Soc. econ. Paleont. Miner., Tulsa,* **4,** 496–502.

RIDDELL, J.F. (1969) A laboratory study of suspension-effect density currents. *Can. J. Earth Sci.* **6,** 231–246.

SCLATER, J.G. & HARRISON, C.G.A. (1971) Elevation of mid ocean ridges and the evolution of the South West Indian Ridge. *Nature, Lond.* **230,** 175–177.

SMITH, A.G., HYNES, A.J., NISBET, E.G., PRICE, I. & WELLAND, M.J.P. (In preparation.) The stratigraphy of the Othris Mountains, eastern central Greece.

THURSTON, D.R. (1972) Studies on bedded cherts. *Contr. Miner. Petrol.* **36,** 324–334.

Spec. Publs int. Ass. Sediment. (1974) **1**, 367–399

Radiolarian cherts, pelagic limestones, and igneous rocks in eugeosynclinal assemblages

ROBERT E. GARRISON

Earth Sciences Board, University of California, Santa Cruz, U.S.A.

ABSTRACT

Among the many unresolved problems concerning eugeosynclinal rocks, none has proved more durable than the controversy over the role of submarine volcanism in the genesis of the thin-bedded cherts and pelagic limestones that commonly occur interbedded with, and above, pillow lavas and volcaniclastic rocks. The composition of the pelagic rocks, their relationships to eugeosynclinal igneous rocks, and data from modern ocean basins all argue against a direct connection between volcanism and formation of pelagic sediments, such as volcanically induced chemical precipitation or plankton blooms. Nonetheless, the persistent joint occurrence of bedded cherts and ophiolites, particularly in the Tethyan region, remains a puzzle.

The palaeobathymetric interpretation of bedded radiolarian cherts is equivocal because the calcite compensation depth (CCD) has apparently fluctuated in time and space. But the appearance of the calcareous microplankton (coccolithophorids, planktonic Foraminifera) in mid to late Mesozoic time marked a significant change in the distribution of biogenic pelagic sediments. Whereas, prior to mid Mesozoic time, pelagic sedimentation in regions of high plankton productivity was largely radiolarian ooze regardless of water depth, radiolarian oozes most commonly have accumulated at abyssal depths below the CCD from the Cretaceous onward, and pelagic calcareous oozes of organic origin have been deposited in oceanic settings since the Jurassic. Measurements of vesicle size and abundance in eugeosynclinal pillow basalts may provide an independent palaeobathymetric evaluation; but ancient water depths obtained in this manner should be applied to the associated pelagic sediments with some reserve because the sea floor may have changed elevation substantially following basalt emplacement but prior to sedimentation.

Sediment ponding, disruption of sediment bodies by volcanic erosion and intrusion on the sea floor, and tectonic shearing, all of which induce lenticularity, determine the configuration of pelagic sedimentary bodies in eugeosynclinal assemblages. Large submarine volcanic structures like ridges and seamounts strongly influence the regional distribution of pelagic facies, for example, by elevating local areas of the sea floor above the CCD.

Varied depositional-tectonic situations can be recognized in eugeosynclinal assemblages by determination of the geometry and size of pelagic sedimentary bodies, their relations to contiguous igneous rocks, and the vertical succession of sedimentary and igneous facies.

INTRODUCTION

Since Steinmann (1905, 1927) emphasized the frequent spatial connection between radiolarian cherts and mafic and ultramafic igneous rocks, the significance of this stratigraphic association has been repeatedly debated. Moreover, it is the basis for several generalizations about eugeosynclinal rocks,* the most recurrent of which is the hypothesis that submarine volcanism is in some way responsible for the formation of many deposits of thin-bedded cherts or limestones, either by chemical precipitation or by its more indirect influences on plankton productivity. A more recent deduction is that ophiolite complexes on land are segments of oceanic crust and upper mantle which, not unexpectedly, are often overlain by lithified oceanic deposits like radiolarian cherts.

Like the concept of an eugeosyncline itself, this latter generality about ophiolites is useful to a first approximation. But it is of limited utility to geologists who wish to unravel the geology and to reconstruct the history of specific eugeosynclinal assemblages of rocks, which are among the most difficult groups of rocks to map and understand. Their structural complexity, complicated and variable metamorphism, lenticular stratigraphic units and rapid facies changes, and their lack of marker beds and paucity of guide fossils often combine to hinder detailed reconstruction of palaeogeography and sequential history, approaches which are so fruitful in less intricate geological situations.

An additional difficulty is that we understand relatively little about the genesis and environmental significance of eugeosynclinal sedimentary rocks. We may, in fact, comprehend more about the origin of the volcanic rocks interbedded with them, yet many sedimentologists are reluctant to study terrains that consist largely of volcanic rocks.

By focusing attention upon some critical details of the association between igneous rocks and pelagic sediments—in mountain belts and in the ocean basins—this paper seeks to encourage collection of more detailed information from eugeosynclinal sequences, as well as the development of models which may guide geologists in understanding these perplexing rocks. This discussion is limited to only two kinds of pelagic sedimentary rocks: radiolarian cherts and nannofossil-rich limestones. The circumstances of their occurrences and the origins of these rocks are considered first; then some broad aspects of relationships between igneous and pelagic sedimentary rocks are examined; and finally these relationships are generalized and extended into the models illustrated by Fig. 6.

RADIOLARIAN CHERTS

The 'chert problem'

Whether the common spatial association of bedded cherts and igneous rocks indicate that the cherts formed due to igneous activity has been a contentious question for over half a century (see reviews in Davis, 1918; Kobayshi & Kimura, 1944; Grunau, 1965; Aubouin, 1965, pp. 115–126; McBride & Thompson, 1970, pp. 74–75).

* Used here in the broad sense suggested by Hsü (1972) to designate packages of rocks found at converging plate margins; these packages may include ophiolite as well as island-arc complexes.

One view is that the cherts were abyssal radiolarian oozes, unrelated to igneous activity and sedimented below the calcite compensation depth (CCD); this is the *bathymetric hypothesis* of Aubouin (1965, p. 115).

Alternatively, the cherts are considered as direct or indirect products of submarine volcanism. This view, called by Aubouin (1965, pp. 123–124) the *volcanic-sedimentary hypothesis,* has several variants, among which are the following: (1) Silica, supplied either directly from subaqueous magmas or leaching of congealed lava on the sea floor, precipitates inorganically. (2) Silica released during subaqueous volcanism or during submarine weathering of volcanic ash enhances the productivity and/or preservation of siliceous microplankton. (3) The CCD is elevated to relatively shallow depths during periods of intense volcanic activity. The resiliency of these ideas is illustrated by a partial list of those who have argued for some variety of the volcanic-sedimentary hypothesis within the past decade: Bailey, Irwin & Jones (1964); Cooper (1965); Elderfield *et al.* (1972); Gibson & Towe (1971); Grunau (1965); Maxwell (1969); Rheinhardt (1969); Herman (1972); Thurston (1972); Mattson, Pessagno & Helsley (1972); Bernoulli (1972); and Kanmera (1968, 1974).

More recent views have linked the deposition of siliceous biogenic oozes to regions of upwelling and high plankton productivity, these being in turn controlled by geography and circulation patterns (e.g. Calvert, 1966; Ramsay, 1973).

Petrology

Typical radiolarian cherts consist of poorly preserved, quartz-filled radiolarian moulds set in a very fine-grained matrix of poorly resolvable microcrystalline quartz and iron oxides (Fig. 1a, b). In some cases the matrix may contain abundant clay minerals or microcrystalline calcite, producing a somewhat friable, radiolarian-rich rock called 'radiolarite' by some workers (e.g. Audley-Charles, 1965). The percentage and preservation of radiolarian moulds is quite variable, due apparently in some instances to partial or complete metamorphic recrystallization of the rock. Other petrological variants are produced by secondary calcitization of the radiolarian tests (Fig. 1c); by current winnowing leading to concentration of radiolarian remains (Fig. 1c); and by addition of substantial amounts of fine-grained, altered volcanic debris (Fig. 1d; see also Diersche, 1973).

Contacts with igneous rocks

The precise field relationships between eugeosynclinal bedded cherts and spatially associated igneous rocks commonly are unclear. One reason is because tectonic shearing tends to be concentrated along contacts between massive igneous rocks and thinly bedded cherts, thereby obliterating primary relationships and distorting thicknesses (Grunau, 1946; Dietrich, 1967; Lemoine, Steen & Vuagnat, 1970; Allemann & Peters, 1972; Glennie *et al.*, 1973).

However, the majority of detailed descriptions of unsheared contacts indicate that most cherts rest depositionally upon subaqueous extrusive rocks, the latter being either pillow lavas, massive flow rocks, or hyaloclastites and breccias. This seems to be the most common situation in Mesozoic eugeosynclinal rocks of the California Coast Ranges (Fig. 2; Bailey *et al.*, 1964; Bailey, Blake & Jones, 1970; see also Kanmera, 1968), and can also be well documented for the radiolarian cherts and radiolarites which lie above the Vourinos and Troodos ophiolitic complexes of Greece and Cyprus

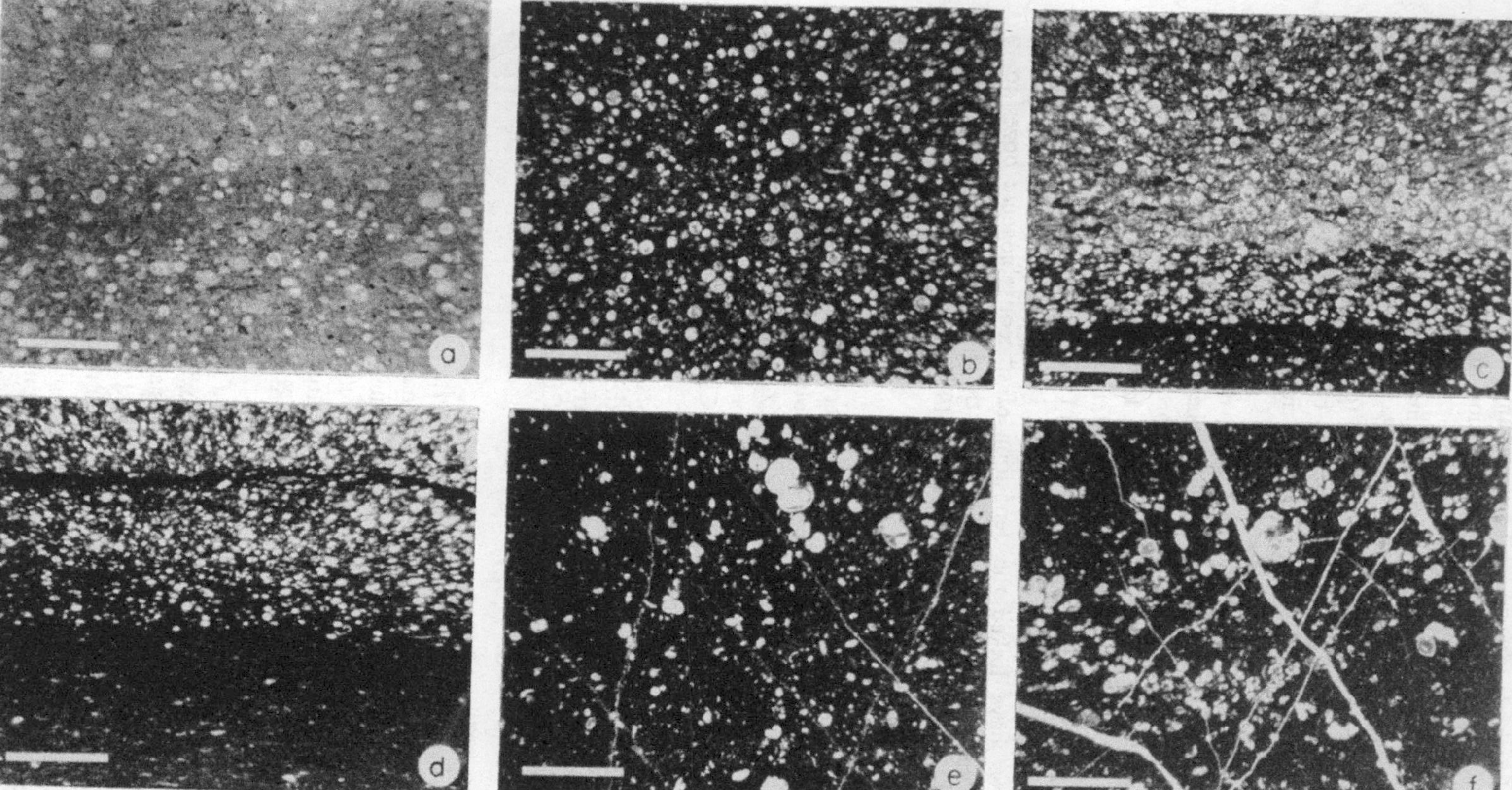

Fig. 1. Photomicrographs of pelagic sedimentary rocks from eugeosynclinal assemblages; scale bars represent 1 mm. (a) Highly siliceous radiolarian chert, Franciscan Formation, Jurassic or Cretaceous, San Luis Obispo County, California. (b) Dense concentration of radiolarian skeletons in Jurassic radiolarite, Vara Supergroup, near Rochetta di Vara, northern Italy. (c) Dense concentration of calcitized radiolarian tests at base of probable turbiditic layer (cf. Fig. 3) in redeposited Jurassic radiolarite, Vara Supergroup, Rochetta di Vara, northern Italy. (d) Upper half of the photomicrograph is the radiolarian-rich bottom part of a redeposited radiolarite. Bottom half of the photomicrograph is a very fine-grained, altered tuff. Vara Supergroup, Rochetta di Vara, northern Italy. (e) Pelagic limestone with tests of planktonic Foraminifera in micritic matrix, Eocene, Olympic Peninsula, Washington. (f) Pelagic limestone (Calera type) of Cretaceous age from the Franciscan Formation, San Mateo County, California. Note abundant Foraminifera tests in micritic matrix, and calcite veins.

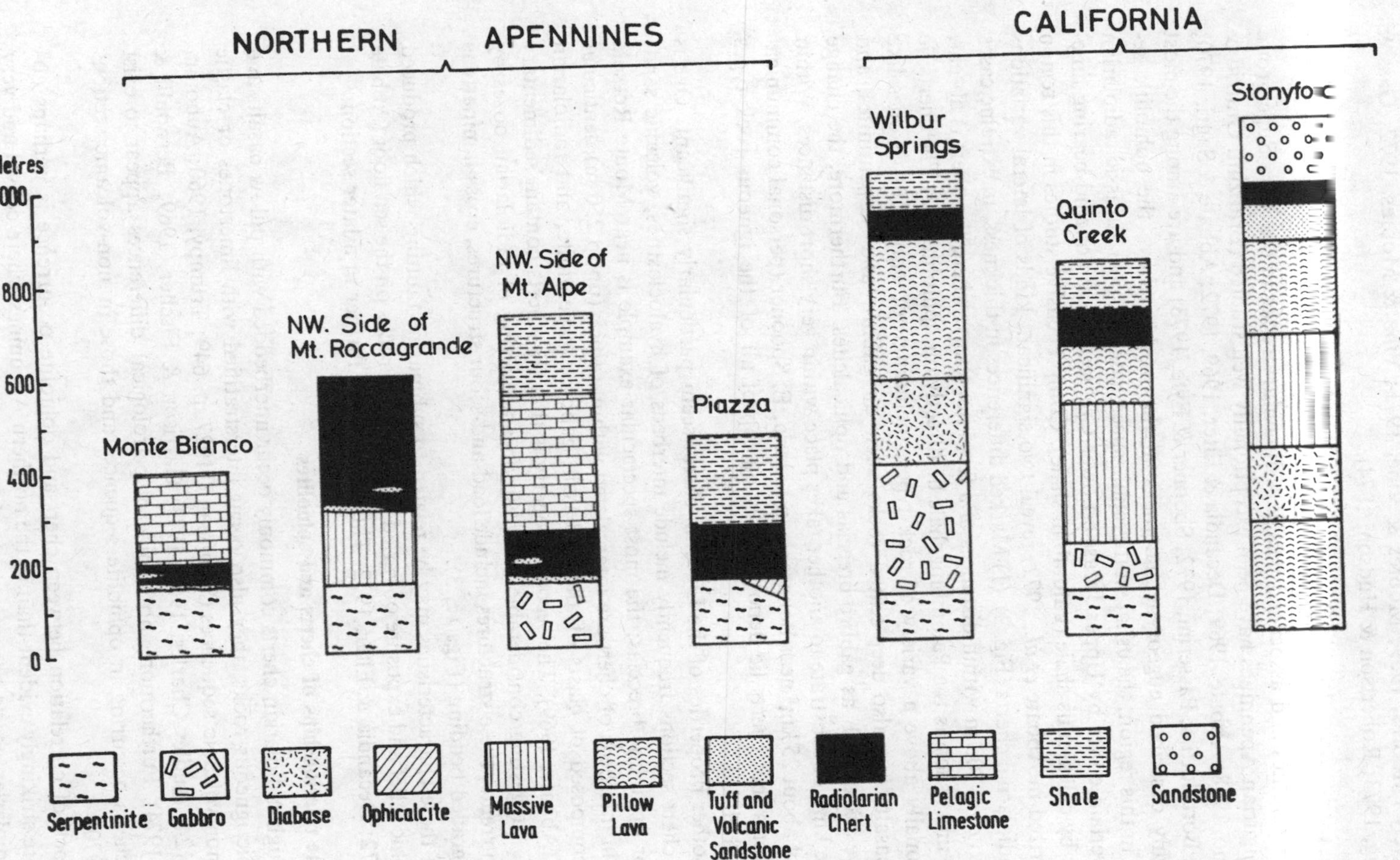

Fig. 2. Typical relationships between pelagic sedimentary rocks and basic igneous rocks in unfaulted Mesozoic eugeosynclinal sequences of northern Italy (Abbate *et al.*, 1972) and the California Coast Ranges (Bailey *et al.*, 1970).

respectively (Moores, 1969; Moores & Vine, 1971; Vine & Moores, 1972; Gass & Smewing, 1973; Robertson & Hudson, 1974).

Relationships in the northern Apennines

The ophiolites and associated pelagic sedimentary rocks in the Vara Supergroup of the Ligurian Apennines have been exceptionally well studied (Franzini, Gratziu & Schiaffino, 1968; Abbate, 1969; Decandia & Elter, 1969, 1972; Abbate & Sagri, 1970; Abbate, Bortolotti & Passerini, 1972; Spooner & Fyfe, 1973) and are among the most thoroughly described eugeosynclinal rocks in the world. Above the ophiolitic sequences in this region, the usual stratigraphic succession is Upper Jurassic radiolarian chert overlain either by Upper Jurassic-Lower Cretaceous calpionellid-bearing limestone or by calcareous shale (Palombini shale). Comprehensive studies in this region (summarized in Abbate *et al.*, 1972) reveal two significant kinds of lateral variations in the radiolarian cherts (Fig. 2). (1) Marked differences in thickness, in extreme cases from 0 to over 300 m within distances of a few kilometres; where the chert is absent, younger units such as the Palombini shale lie directly on ophiolites. (2) The cherts lie depositionally above a variety of rock types, including not only effusive ones like pillow basalts but also deep-seated rocks such as gabbro and serpentinites, and derivative rocks such as gabbro breccias and ophicalcites. Furthermore, the change from one kind of substrate to another takes place within very short distances. Within an area of about 25 km^2 near Bonassola in Liguria, E. Spooner (personal communication) reports that cherts lie depositionally on top of all of the igneous rock types mentioned above.

Two other properties of the Ligurian cherts seem particularly significant. One is that the chert sections frequently include interbeds of hyaloclastites, volcanic sandstones, or ophiolitic breccias; the most spectacular example is near Monte Rossola where thin interbeds of chert are intercalated within a thick (up to 250 m), lenticular breccia composed of clasts of altered basalt, gabbro, serpentinite, and radiolarian siltstone (Abbate, 1969). The second property is that the cherts contain sedimentary structures suggesting considerable redeposition and sorting of radiolarian oozes by bottom currents; these structures include load-and-scour structures, cross-laminations and fine-graded bedding (Figs 1c, 3).

All of these characteristics may be related to submarine faulting which produced topographic relief and exposure of varied basic igneous rocks on the sea floor (Abbate *et al.*, 1972; Decandia & Elter, 1972. This is discussed further in a later section.

Space-time relationships of cherts and ophiolites

Although radiolarian cherts commonly occur interbedded with pillow basalts and other basic igneous rocks, they also occur interstratified with limestones or clastic rocks in non-volcanic sequences (Steinmann, 1927, p. 649; Trümpy, 1960; Aubouin, 1965, p. 124; Audley-Charles, 1965, 1973; Garrison & Fischer, 1969; Bernoulli & Jenkyns, 1974). Furthermore, no significant petrological differences appear to exist between cherts occurring in ophiolite sequences and those in non-volcanic sections (Parea, 1970).

This lack of correlation between chert and ophiolite occurrence is perhaps nowhere more strikingly evident than in the northern Apennines where coeval and very similar radiolarian cherts occur not only above ophiolites in the Ligurian region, but also interbedded with pelagic limestones in the non-volcanic, miogeosynclinal Tuscan

Fig. 3. Polished face of a laminated, turbiditic radiolarite of Jurassic age; from section near Rochetta di Vara, northern Italy. Light-coloured layers consist of densely packed, calcitized radiolarian skeletons (cf. Fig. 1c); note scour and load structures at base of light-coloured turbiditic layer just below middle of specimen. Bright white patches are secondarily discoloured zones. Scale bar represents 5 cm.

and Umbrian sequences to the east (Abbate & Sagri, 1970; Bortolotti *et al.*, 1970). Likewise, although most bedded cherts in the Franciscan and related Mesozoic rocks in California are spatially associated with basaltic rocks, some also occur interbedded with sandstones in non-volcanic sections (Davis, 1918; Blake & Jones, 1974).

Thus, clearly the genesis of radiolarian cherts is not due to *local* inorganic precipitation of silica released by submarine volcanism. If the latter is involved, it must be effective over wide regions beyond the actual eruptive sites (cf. Maxwell, 1969, p. 493).

Sedimentation rates

Where bedded cherts lie depositionally between fossiliferous strata and can be dated accurately by biostratigraphic methods, average rates of sedimentation can be estimated based on time-thickness relations. In most cases the chert sections are relatively thin yet appear to represent long time intervals, and thus have low average sedimentation rates. Garrison & Fischer (1969) estimated 0·7–1·0 mm/10^3 years as the average accumulation rate for a Jurassic radiolarian chert section in the Eastern Alps; and D. Bernoulli (quoted in Schlager & Schlager, 1973, p. 84) has calculated 4 mm/10^3 years as the average rate for a number of Tethyan radiolarian cherts. McBride & Thompson (1970, pp. 89–90) give rates of 1·0–5·3 mm/10^3 years as the best estimates for uncompacted bedded cherts in the Palaeozoic Caballos Novaculite of Texas.

A typically thick sections of radiolarian cherts, such as the 300-m section at

Roccheta di Vara in the northern Apennines (Abbate, 1969, p. 985), may indicate much higher average rates of sediment accumulation, perhaps brought about by substantial sediment redeposition into local depressions. Many of the Ligurian chert sections have laminated and graded beds (Figs 1c, 3) suggesting local redeposition of radiolarian oozes by currents (cf. Nisbet & Price, 1974) and some also have thin layers of ophiolitic sandstone and hyaloclastite which locally increase stratigraphic thicknesses.

The long duration of deposition and slow rates of sediment accumulation for radiolarian cherts are difficult to reconcile with any form of the sedimentary-volcanic hypothesis because the effects of igneous activity (e.g. siliceous microplankton blooms) would have to be operative not only over vast geographical regions; they also would have to persist through millions of years of geological time, during which other types of sedimentation would be excluded.

Observations in modern oceans

The effects of submarine eruptions on the chemistry of sea water have been studied in only a few cases, but all these suggest that the silica content of sea water becomes increased locally. Stefansson (1966) monitored seawater chemistry around the Surtsey eruptive site from November 1963 to November 1964. He found a detectable increase in silica content in waters within 30 km of the eruption and a three-fold increase of silica concentration immediately around the volcano. He attributed the excess silica primarily to the leaching effect of sea water on hot erupted material.

Elderfield (1972) gave chemical analyses of lake and sea waters affected by fumarolic activity preceding the eruption of February 1969 on Deception Island, Antarctica. Compared to average sea water, these waters showed increases in silicon and manganese, especially water from enclosed lakes wherein silicon concentrations up to 50,700 mg/l and manganese up to 2420 mg/l were noted. According to Elderfield (1972, p. 195), it is unclear whether these enrichments '. . . represent a primary volcanic supply . . . or high-temperature leaching of pyroclasts'.

Zelenov (1964) reported analyses of hot fumarolic waters from the submarine Banu Wuhu eruption in Indonesia. Iron and manganese hydroxides apparently precipitated as the water cooled, and his measurements showed silica concentrations in the fumarolic waters up to ten times greater than those in the surrounding sea water.

Whether the effects noted during these short term studies have any relevance to longer time spans, and whether these silica enrichments would eventually lead to inorganic precipitation or to siliceous plankton blooms if the volcanism persisted, are unknown. In the case of the violent Surtsey eruption, the volume of tephra produced during the subaerial eruptive phases undoubtedly would have completely diluted any deposits formed by chemical precipitation or by plankton blooms.

Perhaps the most significant observation in this regard from the present ocean basins, however, is the negative one that no extensive deposits of undoubted chemically precipitated silica have yet been found,* despite substantial sampling of oceanic sediments of varied age and location during the Deep Sea Drilling Project; and also despite intensive collection and study of metal-rich ridge crest sediments from areas

* Elderfield (1972, p. 195) mentions possible siliceous inorganic precipitates in lakes on Deception Island, but as yet this material has not been thoroughly examined (H. Elderfield, personal communication).

of repeated submarine volcanism (e.g. Boström, 1969). Also of possible relevance here are Calvert's (1968) tabulations suggesting that the amount of silica supplied annually to the oceans by submarine volcanism is insignificant compared to that delivered in solution by streams.

Conclusions

Nearly a decade ago Grunau (1965, p. 199) remarked that radiolarian cherts were not fully understood, a statement still applicable today. But during the intervening period, the oceanic sedimentary column has been diligently sampled by the Deep Sea Drilling Project, and the theory of sea-floor spreading has gained wide acceptance. Both these developments have rendered the volcanic-sedimentary hypothesis less tenable as an explanation for the widespread bedded cherts of the geological past. Sea-floor spreading requires virtually continuous igneous activity at spreading ridges, but clearly this type of continuity has not been accompanied by continuous inorganic precipitation of silica or by blooms of siliceous plankton (Ramsay, 1973, p. 204).

If a connection between sea-floor igneous activity and the formation of bedded siliceous deposits does exist, it may be an indirect one, such as the preferential preservation of siliceous microfossil tests in the chemical milieu created by the alteration of basic igneous rocks (e.g. Thurston, 1972, p. 333). Possibly the preservation of silica in radiolarian-rich sediments which lie on basalts is aided by diagenetic or hydrothermal processes which supply silica to pore waters. Hart (1970, 1973) and Fein (1973) have demonstrated that low-temperature submarine weathering of tholeiitic ridge basalts releases silica along with other elements, a process probably enhanced through hydrothermal leaching of the basalts as heated sea water circulates within them (Corliss, 1971; Piper, 1973).

PELAGIC LIMESTONES IN EUGEOSYNCLINAL ASSEMBLAGES

In the Western Cordillera of North America, fine-grained limestones interbedded with pillow lavas have been considered, like the radiolarian cherts, as chemical precipitates formed by submarine volcanism. A widely quoted exposition of this idea was given by Kania (1929) who maintained that heating and agitation of sea water above a submarine vent would expel CO_2, causing precipitation of $CaCO_3$ from sea water. These precipitates would tend to be concentrated around the vent and would become further localized by a tendency for the precipitates to flow into depressions on the surface of the lava flows, thus accounting for lensoid limestone bodies which Kania had observed within eugeosynclinal sequences in British Columbia and elsewhere.

Kania's ideas were adapted by Park (1946) to explain the origin of red Eocene limestones interbedded with submarine basalts in the Crescent Formation of the Olympic Peninsula, Washington. Likewise, Bailey *et al.* (1964, p. 77) utilized Kania's model in postulating a largely inorganic origin for Cretaceous limestones in the Franciscan Formation of California, which are similarly interbedded with volcanic rocks. The limestones from both localities, however, contain abundant planktonic Foraminifera (Fig. 1e, f); and electron-microscopic study shows that the fine-grained carbonate in them—presumably the product of inorganic precipitation in the Kania

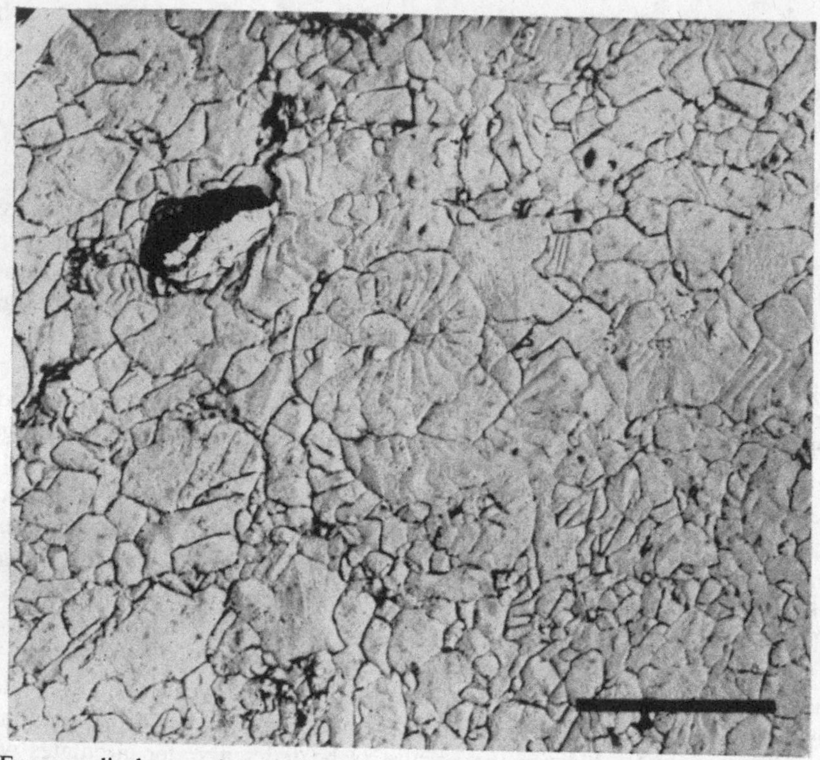

Fig. 4. Eugeosynclinal nannoplankton limestone (Laytonville type) of Cretaceous age, partly recrystallized. From the Franciscan Formation, Sonoma County, California. Scale bar represents 5 μm.

model—actually consists largely of coccoliths and other calcareous nannoplankton in varied stages of recrystallization (Fig. 4; Garrison & Bailey, 1967; Garrison, 1967; Wachs, 1973a, b). These limestones therefore probably originated as pelagic nannoplankton-foraminiferal oozes which were slowly deposited on a substratum of volcanic rocks between episodes of submarine volcanism. Although compositionally similar to the widespread chalk oozes of deep-sea regions, commonly they have become very well indurated and sometimes partly metamorphosed during intense tectonic deformation of geosynclinal sequences.

Nannoplankton limestones of this kind are common in Mesozoic and Cenozoic geosynclinal sequences, both volcanic and non-volcanic, but the presence of these minute fossils in well-indurated limestones can usually be determined only with the electron microscope (Bramlette, 1958; Fischer, Honjo & Garrison, 1967; Flügel, Franz & Ott, 1968; Honjo, 1969). Calcareous nannofossils in rock-forming quantities, however, are known only in Jurassic and younger limestones (see below). On the other hand, eugeosynclinal limestone lenses of the kind discussed by Kania (1929) are found also in pre-Jurassic sequences and are said by him to be especially abundant in Archean terrains.

These pre-Jurassic eugeosynclinal limestones thus require intensive investigation to establish their origin. Until they are more thoroughly understood, however, the possibility of significant inorganic carbonate precipitation due to submarine volcanism cannot be discounted. Bonatti (1966) seems to have established the feasibility of this mechanism to produce at least small amounts of calcite and dolomite in modern deep-sea sediments from the Pacific. But it remains to be demonstrated that submarine volcanic processes can generate chemically precipitated carbonates on the scale needed to form eugeosynclinal limestone bodies.

GEOLOGICAL HISTORY OF MICROPLANKTON AND THE RECORD OF PELAGIC SEDIMENTATION

A peculiar characteristic of the fossil record is that most of the important sediment-forming groups of planktonic organisms first appear relatively late in geological time (Bramlette, 1958; Lipps, 1970). Although the Radiolaria are known from the Cambrian onwards, the other siliceous microplankton are comparative newcomers with both diatoms and silicoflagellates first appearing in rocks of mid-Cretaceous age. Even more striking is the record of the calcareous microplankton. Although problematic coccoliths are reported in Upper Carboniferous rocks (Gartner & Gentile, 1972), calcareous nannoplankton first became significant rock-formers in the Jurassic Period (Fischer *et al.*, 1967) and planktonic Foraminifera are important sedimentary components starting in Cretaceous time.

These asymmetries of plankton evolution critically influenced the distribution of pelagic facies through geological time. In pre-Jurassic time, biogenic pelagic sediments must have been dominantly radiolarian oozes which had low rates of sedimentation, whereas black shales may have been the typical pelagic deposit in areas of less prolific plankton productivity (Ramsay, 1973, p. 231). Sedimentation rates of siliceous oozes may have increased significantly beginning in mid-Cretaceous when diatoms and silicoflagellates became parts of plankton communities. Chalks and nannoplankton limestones are limited to the post-Triassic part of the geological record.

PALAEOBATHYMETRIC SIGNIFICANCE OF PELAGIC SEDIMENTARY ROCKS

Assuming that radiolarian cherts are lithified radiolarian oozes, the palaeobathymetric significance of pre-Jurassic bedded cherts must be fundamentally different from that of post-Jurassic ones. Rather than accumulating at sub-CCD depths below regions of high plankton productivity, as at present, the earlier radiolarian oozes could have been deposited beneath fertile regions in relatively shallow as well as in deep water, due to the absence of calcareous microplankton. This would explain, for example, the evidence presented by Kanmera (1972, 1974) and Danner (1967, 1970) pointing to shallow-water depositional sites for eugeosynclinal bedded cherts of Triassic and Palaeozoic age in Japan and Washington State; it might account in part also for the Palaeozoic peritidal cherts postulated by Folk (1973).

In post-Jurassic time, following the onset of vigorous calcite production by the coccolithophorids and planktonic Foraminifera, radiolarian oozes must have been sedimented most commonly below the CCD. But the latter has apparently fluctuated in time and in space, at least during Cenozoic and the later part of Mesozoic time (Hay, 1970, 1973; Berger, 1972, 1973; Ramsay, 1973), and the depth of the lysocline (Berger, 1970) probably has responded in similar fashion. Hay (1970) believes that both levels may have been elevated as high as the photic zone in the Atlantic at the end of Cretaceous time. Thus, even in the post-Jurassic record, the possibility that radiolarian cherts formed at relatively moderate depths (e.g. a few hundred metres) cannot be completely excluded.

The Jurassic Period itself, however, was a transitional time during which the coccolithophorids developed, and it is the palaeobathymetry of the Jurassic radiolarian cherts, especially the Tethyan ones, which have engendered the most controversy. One stream of thought has regarded them as sub-CCD deposits formed at abyssal depths (Cayeux, 1924; Steinmann, 1925; Trumpy, 1960; Aubouin, 1965; Garrison & Fischer, 1969). But others have doubted whether the formation of these sediments was as depth dependent as modern radiolarian oozes seem to be (Hallam, 1971; Abbate *et al.*, 1972; Bernoulli, 1972; Bernoulli & Jenkyns, 1974). One view is that, like their Palaeozoic counterparts, the Jurassic siliceous oozes may have been deposited in relatively shallow as well as deep water because the manufacture of biogenic calcite for pelagic sediments had not yet attained the high rates characteristic of later geological time. Another view relates deposition of the Tethyan cherts to Jurassic rifting of a continental margin and creation of small ocean basins where a number of factors favoured deposition and preferential preservation of radiolarian oozes at varied depths (A. G. Smith, personal communication); among these factors might have been the effects in semi-confined basins of vigorous upwelling (like that described in the Gulf of California by Calvert, 1966) or intense volcanism (Bernoulli, 1972, p. 820). But a totally satisfactory explanation for the Tethyan ophiolite-sediment sequences is not yet at hand.

BATHYMETRIC INDICATORS IN PILLOW LAVAS

Because the hydrostatic pressure of water in the oceans acts to counterbalance the internal vapour pressure of subaqueous lava (Rittman, 1958; McBirney, 1963), there exists a relationship between the size and density of vesicles in pillow basalts and pillow breccias and the water depth at which they were emplaced. Moore (1965) and Jones (1969) have sought to quantify this relationship empirically, thereby providing a potentially useful assessment of palaeobathymetry in eugeosynclinal assemblages—one, moreover, which is independent of the customary qualitative estimates based on sedimentological criteria.

Utilizing this method, Matthews & Wachs (1973) have suggested that Cretaceous nannoplankton-foraminiferal limestones associated with pillow lavas and pillow breccias in the Franciscan Formation were deposited in both relatively shallow water (200–300 m) and relatively deep water (greater than 3500 m) environments. In similar fashion, Kanmera (1972) estimated that radiolarian cherts and pelagic limestones occurring with submarine basalts in the Triassic Sambosan Group of Japan were deposited at depths less than 500 m.

Although promising for palaeobathymetric evaluation of pillow basalts, perhaps this method should be applied only with some caution to the associated pelagic sediments. The latter cannot always be assumed to be deposited at the same depths where the basalts were emplaced. Considerable discrepancies between the depths of basalt emplacement and of overlying pelagic deposition are in fact likely on tops of ridges and seamounts which may subside rapidly during sea-floor spreading (Menard,

1969b). Kaneoka (1972) described a striking example of this in the north-western Pacific where highly vesicular basalts were dredged from seamounts at depths below 2000 m. Judging by their vesicularity, however, they should have been erupted at water depths of only a few hundred metres, and Kanoeka attributed this discrepancy to more than 2000 m of subsidence since the time of eruption.

A qualitative indication of water depth may be provided by variolitic structure, whose presence in pillow lavas is linked by Furnes (1973) to eruption in deep water.

SOME PHYSICAL AND AGE RELATIONSHIPS BETWEEN PELAGIC SEDIMENTS AND IGNEOUS ROCKS

Precise field descriptions of the relationships between pelagic sedimentary rocks and submarine volcanic rocks are surprisingly few, especially considering the importance attributed by many workers to the ophiolite-chert association and to sub-sea volcanism in the genesis of cherts and limestones. Some kinds of relationships (such as sheared versus depositional contacts of ophiolites and cherts) were treated in a previous section, and discussed below are five additional aspects, arranged more or less in order of increasing size, which bear on the origin and significance of assemblages of igneous and pelagic sedimentary rocks.

Inclusions of pelagic sediments in basaltic rocks

On a small scale, at the level of hand specimen or outcrop investigation, inclusions of pelagic sedimentary rocks can be observed in massive flow rocks or hypabyssal intrusive bodies in eugeosynclinal terrains (Fig. 5d). Davis (1918, pp. 270–274) described variously sized blocks of radiolarian chert included in basalts and gabbroic intrusive rocks of the Franciscan Formation. Similar inclusions of limestone are recorded by Park (1946) and Garrison (1973) in Eocene basalts and diabases on the Olympic Peninsula of Washington State. Some limestone xenoliths occur as angular blocks, probably plucked and engulfed from consolidated bodies of sediment; but in other outcrops they may show evidence of plastic deformation as if they had been in a semi-consolidated state at the time they were incorporated in the magma. As noted by Davis (1918, p. 273), incorporation of completely unconsolidated ooze into magma would tend to disperse and obliterate the sediment.

Nayudu (1971), who described inclusions of pelagic carbonate sediments in basalts and basaltic breccias dredged from Pacific seamounts, discussed the significance of this kind of relationship. The most important implication is that the sediment is unequivocally older than the igneous rocks, thus fossils in the sediment can establish only the maximum possible age of igneous activity on the sea floor.

Pelagic sediment between and within basalt pillows

Sedimentary rocks between and within pillows of basalt are present in at least small amounts in most eugeosynclinal volcanic rocks. Arrangements of this type have been recorded for radiolarian cherts (Dewey & Flett, 1911; Davis, 1918, p. 272) and for pelagic limestones (Park, 1946; Garrison, 1972; Varne, Gee & Quilty, 1969; Varne & Rubenach, 1972), as well as for argillites and skeletal limestones (Park, 1918; Snyder & Fraser, 1963; Lewis, 1972). Long regarded as supporting evidence for volcanically induced precipitation of sediment, this kind of intimate juxtaposition can

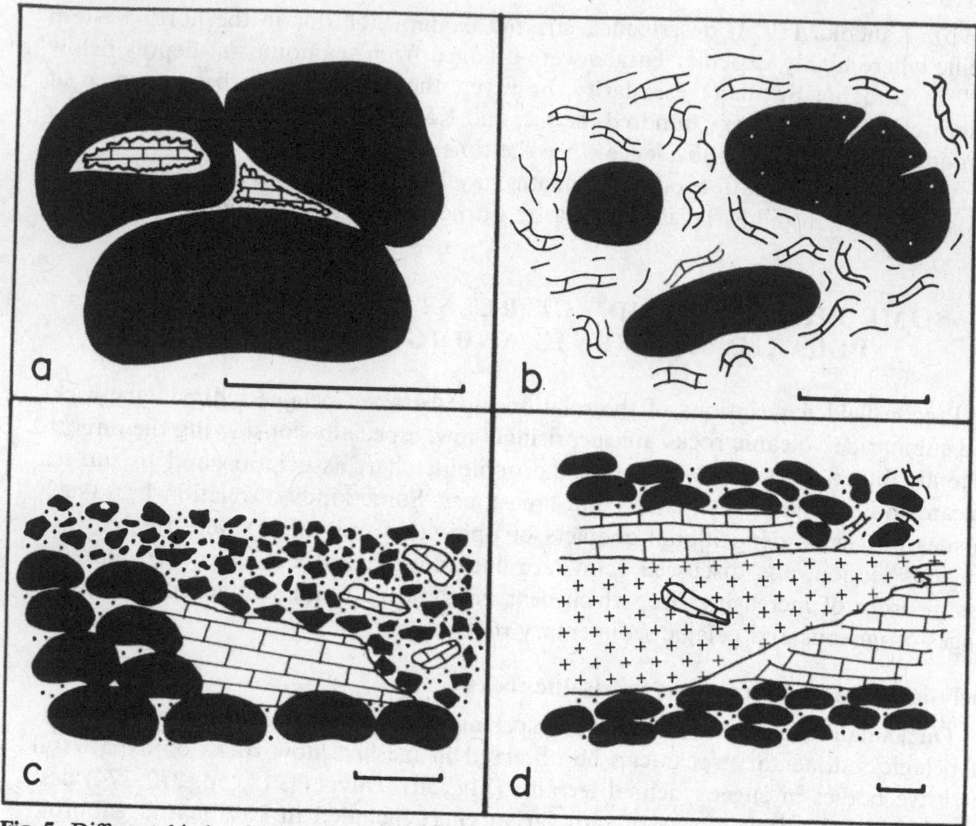

Fig. 5. Different kinds of relationships between pelagic sediments and igneous rocks, producing sediment lenses of varied size (Garrison, 1973). Scale in each drawing represents 50 cm. (a) Inter-pillow and intrapillow limestone produced by sediment infiltration into primary void space between and within pillows (*infiltrated mode*). Former void space commonly lined by drusy secondary calcite. (b) Disrupted interpillow sediment formed by intrusion of lava pillows into soft ooze (*intrusive mode* of interpillow sediment). (c) A small limestone lens formed by a combination of: (1) primary sediment ponding in a small depression on the sea floor, and (2) subsequent partial erosion by a volcanic breccia on the sea floor. (d) Disruption of a carbonate sediment body by intrusion of a diabase sill, followed by later intrusion of a pillow lava into the calcareous sediment (upper right).

be generated in two very different ways, neither requiring inorganic precipitation (Garrison, 1972).

In one mode, here termed the *intrusive mode,* pillowed lava intruded into pre-existing unconsolidated sediment so that the sediment was squeezed between the pillows and injected into cracks and other voids within the pillows (Fig. 5b). Sediment flowage during this process disrupted primary sedimentary fabrics, and the resulting sedimentary rocks tend to be structureless and massive; in addition they may or may not show evidence of thermal metamorphism. Intrusion of more consolidated sediment by pillowed lava resulted in the formation of angular blocks of sedimentary rock between pillows, and these rarely show signs of thermal metamorphism.

In contrast to the above are small sediment bodies formed by filtration of pelagic sediment into voids between pre-existing pillows on the sea floor (Fig. 5a), a process defined here as the *infiltrated mode.* Frequently laminated or cross-laminated, these sediment bodies often also contain an admixture of fragmental volcanic debris derived from submarine erosion of adjacent pillows. Limestone-filled cavities of this sort commonly contain single or multiple crusts of void-filling fibrous calcite, which provide decisive evidence (1) for the presence of former void space, and (2) that the sediment is younger than the adjacent volcanic rocks, and that fossils in the sediment record the minimum possible age of the volcanism.

High- and low-temperature alteration of pelagic sediment at contacts with igneous rocks

Geologists traditionally regard thermal alteration of sediments at their contacts with igneous rocks as definitive evidence of intrusive relationships. But recent work suggests some uncertainties about this criterion both in eugeosynclinal sections on land and in pelagic sediments from modern oceans.

Davis (1918, pp. 270–274) discussed the intrusion of cherts by mafic igneous rocks in the Franciscan Formation, and described a variety of effects which he attributed to thermal metamorphism; these include changes in colour and texture, due to recrystallization, as well as apparent fusion and assimilation in some localities. At other localities, however, he noted that included blocks of chert and shale were only slightly altered. My own studies on the Olympic Peninsula (Garrison, 1972, 1973) revealed a similar variability in the thermal alteration of intruded pelagic limestones, although in general the limestones seem to show less marked effects of thermal metamorphism than the cherts. Possibly this reflects the lower melting point of opaline silica and quartz compared to calcite, as well as the contrasting effects of increased temperature on the solubilities of the two materials.

Among the more obvious thermal effects noted in both cherts and limestones (Davis, 1918, pp. 270–274; Garrison, 1972, p. 321) are alterations in colour, especially of reddish rocks. Sometimes there is an increase in the intensity of colouration, but more often the colours become variegated or bleached. Thermally altered limestones are sometimes partly silicified or chloritized, they may contain secondary pyrite or dolomite, and both cherts and limestones are occasionally recrystallized to coarser grained rocks. Recent experimental work by Adelseck, Geehan & Roth (1973) casts light on how nannoplankton oozes can become thermally recrystallized at elevated temperatures by simultaneous dissolution of less resistant species and precipitation of secondary calcite overgrowths on more robust ones. Davis (1918, pp. 272–274) noted the common development of spherulites in cherts near igneous contacts and the less common occurrence of mamillary structure in chert xenoliths within intrusive rocks.

To complicate matters, however, on the Olympic Peninsula, limestone xenoliths showing the kinds of alteration described above may sit side by side with xenoliths which are virtually unaltered, and limestones at the contacts of some intrusive bodies are similarly unaltered. These vagaries in contact metamorphism doubtless reflect differences in the cooling histories of the igneous magmas as well as in the physical condition of the sediments at the time of intrusion.

A further complication is that contact zones of altered sediment apparently can be formed by low-temperature diagenesis of basaltic glass which causes palagonitization and precipitation of secondary calcite and zeolites (Nayudu, 1964; Hay & Iijima, 1968). Lithified zones of sediment produced in this manner at the contacts of basalts with calcareous pelagic sediments may mimic zones of contact metamorphism, even though the contact is depositional, but the stable isotope composition of the carbonates indicates that high-temperature alteration has not occurred (Thompson, 1972; Lloyd & Hsü, 1972; Garrison, Hein & Anderson, 1973).

These complications of course may make it difficult to decide whether a given basalt-sediment contact is intrusive or depositional, especially when the contact is viewed in a small core from the ocean floor.

Lenses of pelagic sedimentary rocks in eugeosynclinal terrains

Kania (1929) noted the widespread occurrence of highly lenticular limestone bodies in eugeosynclinal terrains; he attributed this to local inorganic precipitation of calcareous sediment due to submarine volcanism, followed by flow of the sediment into depressions on the sea floor. Somewhat similar conclusions were reached by Davis (1918, pp. 376 and 383) to explain lenses of radiolarian chert in the Franciscan Formation.

Clearly these inorganic mechanisms cannot apply to biogenic deposits like the Eocene limestone on the Olympic Peninsula, and elsewhere (Garrison, 1973, pp. 590–591). I have listed three processes contributing to their lenticularity. (1) Sediment redeposition and ponding on an uneven sea floor of volcanic rocks. (2) Tectonic shearing during deformation. (3) Post-depositional igneous activity on or near the sea floor, including intrusion and erosion by subaqueous volcanic breccia flows (Fig. 5c, d). I felt the latter was most important in producing lenticular limestone bodies on the Olympic Peninsula. Greenwood (1956) and Carlisle (1963) also emphasized the importance of sub-sea volcanic breccia flows in eroding and disrupting sediment.

Herm (1967), in describing marine limestones interbedded with Cretaceous andesites in Chile, stressed the effects of submarine topographic relief on a sea floor of volcanic rocks in producing rapid lateral changes in both thickness and facies of sediments. The main effects of this relief in his view were on the oxygen content and degree of agitation in bottom waters, important in this context since the limestones he examined consist chiefly of benthonic rather than planktonic organisms. Similar highly lenticular, shallow-water limestones are interbedded with Permian eugeosynclinal rocks in northern California; Demiren & Harbaugh (1965) interpreted them as products of deposition on an elevated platform in an otherwise detrital sedimentary setting.

For pelagic sediments the effects of local submarine topography in restricting the lateral extent of individual sediment bodies are very evident in studies of modern oceanic ridge regions, particularly rifted ridges with high local relief like the Mid-Atlantic Ridge (Ewing & Ewing, 1964; Siever & Kastner, 1967; van Andel, 1968; van Andel & Bowin, 1968; van Andel & Komar, 1969; Aumento & Loncarevic, 1969; Keen & Manchester, 1970; van Andel & Heath, 1970; van Andel *et al.*, 1973). According to van Andel & Bowin (1968, pp. 1288–1289) the thick 'valley fill' in depressions on the flanks of ridges, when compared to the thin or non-existent sediment cover on the surrounding hills, indicates local sediment transport from surrounding 'drainage areas' by turbidity currents and slumping. Variations in thickness of sediment ponds (100–500 m) were attributed by these authors mainly to drainage areas of different sizes. Most of the ponded sediment sampled on the flanks of the Mid-Atlantic Ridge is redeposited pelagic ooze, but a small portion is locally derived from mafic igneous and metamorphosed igneous rocks exposed on the sea floor (Siever & Kastner, 1967; Bonatti, Honnorez & Gartner, 1973).

In contrast to ponded sediments which accumulate in depressions, other lenticular bodies of pelagic oozes form sediment caps on the higher parts of irregular topography (Houtz & Markl, 1972, p. 152; Talwani, Windisch & Langseth, 1971, p. 488), in some cases apparently because adjacent depressions may be current-scoured. Sediment caps also occur on the tops of some seamounts (Karig, Peterson & Short, 1970) where they may form isolated bodies of calcareous pelagic ooze if the surrounding sea floor lies below the calcite compensation depth (Lonsdale, Normark & Newman, 1972, p. 306). This type of perch for lenticular accumulation of pelagic oozes may be especially important in areas of intensive detrital sedimentation, for example, near continental margins or island-arc systems, where pelagic sedimentation normally is completely masked by terrigenous sediment (see below).

There is thus a tendency for bodies of pelagic sediments in areas of active tectonism and submarine volcanism to accumulate in lenticular bodies, not because of local precipitation, but rather due to various topographic effects. This primary lenticularity is doubtless enhanced by subsequent volcanic processes (erosion, intrusion, etc.) and by tectonic shearing during deformation. Intense shearing during the formation of mélanges may explain the extreme lenticularity of many pelagic rocks, such as the radiolarian cherts in parts of the Franciscan Formation (Hsü, 1971).

Influences of oceanic ridges on the regional distribution of pelagic sediments

As indicated in the previous section, sub-sea topographic prominences like ridges and seamounts form natural sites for the deposition of pelagic oozes on volcanic rocks by locally elevating the sea floor above areas of detrital sedimentation or calcite dissolution. Models of this sort have been suggested for pelagic rocks in the Franciscan Formation by Bailey *et al.* (1964, p. 76), Garrison & Bailey (1967), Chipping (1971) and Hsü (1971).

In addition to these local effects which induce lenticularity, submarine volcanic prominences also strongly influence pelagic sedimentation on a regional scale, particularly in the case of active oceanic ridges. Perhaps nowhere is this more clearly illustrated than in the segment of the southern Indian Ocean between Australia and Antarctica. The Southeast Indian Ridge lies about midway between these two continents, and, except for scattered, small sediment ponds, the crestal region is virtually free of sediment (Houtz & Markl, 1972) because the spreading rate has exceeded the rate of pelagic sedimentation by several orders of magnitude, at least since Late Pleistocene time (Conolly & Payne, 1972). During this interval pelagic sediments have been accumulating on the flanks of the ridge at rates of 1·0–10·0 mm/10^3 years. Wedges of pelagic sediment thicken away from the ridge in both northerly and southerly directions until they become masked through the more rapid deposition of terrigenous sediments on the topographically lower abyssal plains flanking Australia and Antarctica (Conolly & Payne, 1972; Hayes *et al.*, 1973). A further influence on the regional sedimentation pattern is oceanographic : pelagic sedimentation during Late Cenozoic time has been mainly calcareous nannoplankton ooze in relatively warm waters north of the ridge, siliceous diatom ooze in cooler Antarctic water south of it.

In addition to the shielding effect provided by the elevation of oceanic ridges, their changes in elevation during sea–floor spreading also markedly affect the character of pelagic deposition. Sclater and his colleagues, expanding on work by Menard (1969a), have documented the consistent relationships between depth and age of oceanic crust (see Sclater & Detrick, 1973, and references therein); they have shown that as newly formed oceanic crust moves laterally away from crestal margins, it subsides at exponentially decreasing rates. Pelagic sediments in crestal regions, which have average depths of about 2700–2800 m, will be dominantly calcareous oozes, but as the seafloor subsides below first the lysocline and then the CCD, non-calcareous sedimentation will ensue; depending on plankton productivity in the overlying waters, this sediment will be either pelagic red clay or siliceous biogenic ooze.

This interplay of lateral plate motion and subsidence with oceanographic variables and the resulting patterns of sedimentation are well documented in recent work reported by Fischer & Gealy (1969), Heezen *et al.* (1973), Winterer (1973), Berger (1973), and Ewing *et al.* (1973). In addition, van Andel *et al.* (1973, p. 1538) have shown how, other things being equal, rates of pelagic sedimentation are strongly influenced by position relative to the Mid-Atlantic Ridge. In the uneven topography of the crestal region, downslope movement of redeposited sediment produces local rates of around 40 mm/10^3 years in sediment ponds; at a distance of 400 km from the ridge crest, where local relief has become subdued, this is halved to 20 mm/10^3 years, which is apparently the normal regional rate of pelagic deposition; and at 1000 km from the crest, the rate is reduced to around 6 mm/10^3 years as calcium carbonate is progressively subtracted from the sediment during subsidence of the seafloor through the lysocline and the CCD.

Since most crestal regions of ridges today lie above the CCD, calcareous ooze normally accumulates above mafic igneous rocks there. Assuming many of the ophiolite sequences of mountain ranges were formed at oceanic ridges, it may at first seem puzzling why basaltic rocks in them are so commonly overlain by radiolarian cherts rather than pelagic limestones; and why in many Tethyan sequences, particularly in the northern Apennines (Abbate *et al.*, 1972) and parts of the Alps (Lemoine *et al.*, 1970), the cherts give way stratigraphically upward to Upper Jurassic or Lower Cretaceous pelagic limestones (Fig. 1), rather than the reverse situation which would be expected as oceanic crust subsides below the CCD (Fischer & Gealy, 1969; Heezen *et al.*, 1973). As previously discussed, important elements may have been the absence of significant calcareous microplankton in pre-Jurassic time (when the dominant biogenic pelagic sediment must have been radiolarian ooze), as well as the progressive depression of the CCD brought about by proliferation of the coccolithophorids during Late Jurassic time (Garrison & Fischer, 1969, p. 48).

The calcareous microplankton, however, were presumably well developed by Late Cretaceous time. Therefore the occurrence above the Upper Pillow Lavas in the Troodos Massif of Cyprus of, first, Campanian radiolarites and then Maastrichtian chalks above the radiolarites is puzzling if these lavas are ridge basalts (Moores & Vine, 1971; Vine & Moores, 1972; Robertson & Hudson, 1973, 1974), unless the ridge was much deeper than the average ridge today and became uplifted in the Maastrichtian (on the other hand, Abbate *et al.*, 1972, summarizes evidence suggesting some of the Tethyan ridges must have been considerably shallower than those today); or unless the CCD was much shallower in the Tethyan region during Campanian time and dropped markedly in the Maastrichtian. Alternatively, the stratigraphic succession is somewhat less puzzling if the Upper Pillow Lavas represent off-ridge volcanism (Gass & Smewing, 1973; Robertson & Hudson, 1974) below the CCD, with uplift of the sea floor or depression of the CCD by Maastrichtian time. The situation on Cyprus regarding the succession of pelagic sediments is, however, but one part of the Tethyan enigma in this regard.

MODELS FOR ASSOCIATIONS OF IGNEOUS ROCKS AND PELAGIC SEDIMENTS

The suggestion made by Dietz (1963) and Hess (1965) that ophiolite complexes in mountain belts are slices of old oceanic crust and mantle which were formed at spreading ridges has by now been widely accepted. Less generalized models proposed in the last few years have sought to compare different kinds of ophiolite complexes with more specific tectonic settings in the present ocean basins. Examples include attempts to distinguish between ophiolite complexes formed at fast versus slow spreading ridges (Moores & Vine, 1971, pp. 462–465); or between those formed at spreading ridges in major ocean basins or at ridges in small marginal seas (Dewey, 1971; Ramsay, 1973). Among criteria employed for this purpose are geochemical properties of the basic igneous rocks (see references in Pearce & Cann, 1973) and the internal structure of the ophiolite complex (Moores & Vine, 1971; Ramsay, 1973).

Increased knowledge of the ocean floors, however, continues to reveal unexpected complications and in some instances serves to blur previously established categories. Thus, for example, Cann (1968) demonstrated little or no correlation between ridge structures, topography, and rate of spreading; and recent work by Hawkins (1973) suggests that the crust beneath marginal basins may be nearly identical in petrology, magnetic properties and structure to that in the major ocean basins. Moreover, as pointed out by Abbate *et al.* (1972, p. 273), it may not be possible to construct uniformitarian models which will apply to all ancient ophiolite complexes.

Bearing these limitations in mind, spatial associations of pelagic sediments and oceanic igneous rocks should nonetheless be divisible into a finite number of categories. Some of the more obvious of these are illustrated in Fig. 6 and are discussed below, with emphasis on sedimentary and volcanological characteristics by which the different settings might be recognized in ancient eugeosynclinal assemblages. This descriptive listing is neither comprehensive nor does it correspond exactly with those subdivisions favoured by igneous petrologists or marine geologists; it may serve, however, as a guide to the kinds of palaeogeographical reconstructions which might be attempted in eugeosynclinal rocks by using current levels of knowledge.

Rifted ridges (Fig. 6a)

The frequency of faulting in the crestal regions of this type of ridge exposes lower crustal or upper mantle rocks on the sea floor, and produces considerable relief (maximum of about 500–1500 m; van Andel & Bowin, 1968), which leads to sediment ponding on the flanks of the ridge (Menard, 1967; van Andel, 1968; Melson, Thompson & van Andel, 1968; Atwater & Mudie, 1968; McBirney, 1971; Schneider, 1972). As a consequence, pelagic sediment will lie depositionally above a variety of igneous rocks, ranging from effusive pillow lavas to deep-seated gabbros, serpentinites or amphibolites. Pelagic sedimentary bodies will tend to have a primary lenticularity, and thickness variations will be pronounced at right angles to the elongation of sediment ponds, less so parallel to it. These bodies of sediment, some quite large, will contain numerous layers of pelagic sediment which were redeposited from surrounding higher areas by turbidity currents and slumping, and which have a characteristic assemblage of sedimentary structures, including graded bedding, flute casts, and

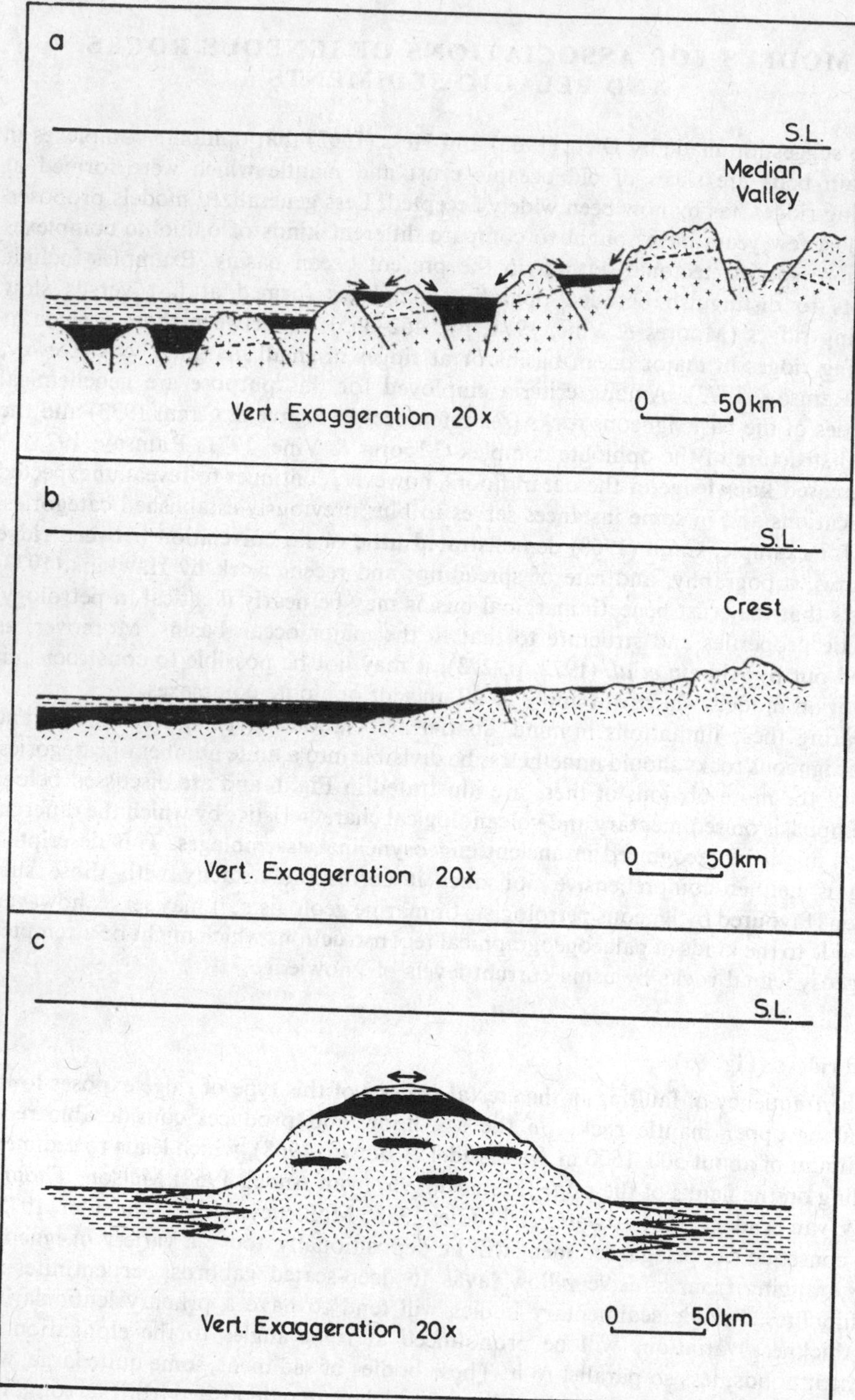
a
S.L.
Median
Valley
Vert. Exaggeration 20x
0 50km
b
S.L.
Crest
0 50km
Vert. Exaggeration 20x
c
S.L.
0 50km
Vert. Exaggeration 20x

Fig. 6.

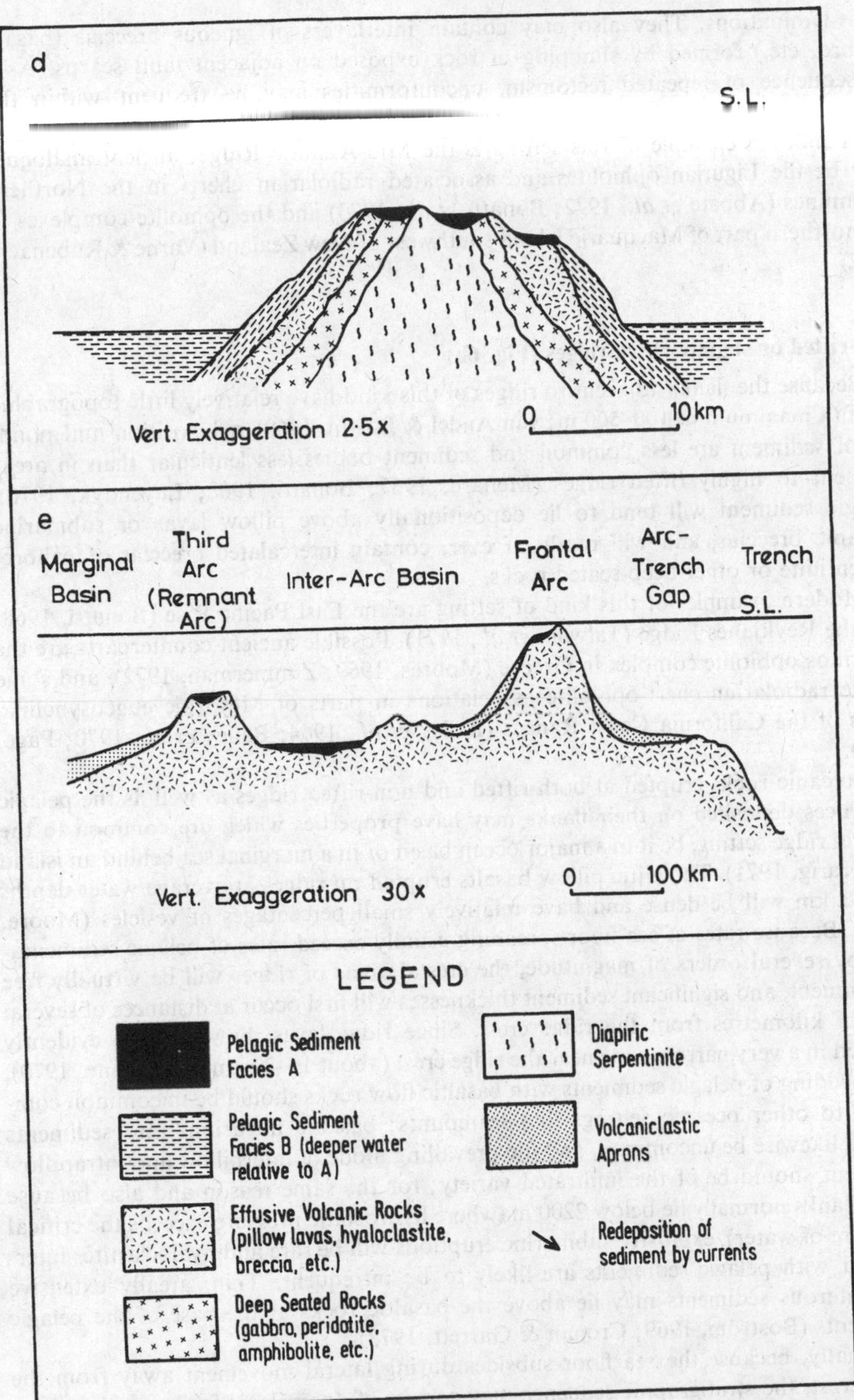

Fig. 6. Depositional-tectonic models of relationships between pelagic sediments and oceanic igneous rocks. See text for discussion. (a) Rifted spreading ridge. (b) Non-rifted or slightly rifted spreading ridge. (c) Seamount (non-spreading volcanic structure). (d) Diapiric structure (non-spreading tectonic structure). (e) Island arc (modified from Karig, 1971; terminology from Karig, 1971, 1972).

cross-laminations. They also may contain interlayers of igneous breccias (basalt, gabbro, etc.) formed by slumping of rock exposed on adjacent fault scarps. As a consequence of repeated tectonism, unconformities may be frequent within the resulting ophiolite and sediment sequences.

A modern example of this setting is the Mid-Atlantic Ridge; ancient analogues may be the Ligurian ophiolites and associated radiolarian cherts in the Northern Apennines (Abbate *et al.*, 1972; Bonatti *et al.*, 1973) and the ophiolite complexes in the northern part of Macquarie Island southwest of New Zealand (Varne & Rubenach, 1972).

Non-rifted or slightly rifted ridges (Fig. 6b)

Because the flanks adjacent to ridges of this kind have relatively little topographic relief (a maximum of 100–500 m; van Andel & Bowin, 1968) redeposition and ponding of sediment are less common and sediment bodies less lenticular than in areas adjacent to highly rifted ridges (Menard, 1967; Bonatti, 1968; Luyendyk, 1970). Pelagic sediment will tend to lie depositionally above pillow lavas or submarine volcanic breccias, and will rarely, if ever, contain intercalated breccias of gabbro, serpentinite or other deep-seated rocks.

Modern examples of this kind of setting are the East Pacific Rise (Bonatti, 1968) and the Reykjanes Ridge (Talwani *et al.*, 1971). Possible ancient counterparts are the Vourinos ophiolite complex in Greece (Moores, 1969; Zimmerman, 1972), and some of the radiolarian chert-ophiolite associations in parts of Mesozoic eugeosynclinal rocks of the California Coast Ranges (Bailey *et al.*, 1964; Bailey *et al.*, 1970; Page, 1972).

Volcanic rocks erupted at both rifted and non-rifted ridges as well as the pelagic sequences deposited on their flanks may have properties which are common to the general ridge setting, be it in a major ocean basin or in a marginal sea behind an island arc (Karig, 1971). Tholeiitic pillow basalts erupted on ridges at average water depths of 2–3 km will be dense and have relatively small percentages of vesicles (Moore, 1965). Because rates of sea-floor spreading usually exceed rates of pelagic sedimentation by several orders of magnitude, the crestal areas of ridges will be virtually free of sediment, and significant sediment thicknesses will first occur at distances of several tens of kilometres from the ridge crest. Since ridge lavas, however, are evidently erupted in a very narrow zone near the ridge crest (about 10–20 km wide; Cann, 1970), interbedding of pelagic sediments with basaltic flow rocks should be uncommon compared to other oceanic settings like seamounts; basaltic intrusions into sediments should likewise be uncommon, and the prevailing mode of interpillow and intrapillow sediment should be of the infiltrated variety; for the same reason and also because ridge flanks normally lie below 2200 m (where hydrostatic pressure exceeds the critical pressure of water), explosive submarine eruptions will be rare and hyaloclastites interlayered with pelagic sediments are likely to be infrequent. Thin, areally extensive metalliferous sediments may lie above the basaltic rocks at the base of the pelagic sediments (Boström, 1969; Cronan & Garrett, 1973).

Finally, because the sea floor subsides during lateral movement away from the ridge crest, the stratigraphic sequence may consist of one pelagic facies overlain by a second pelagic facies deposited in deeper water (e.g. calcareous nannoplankton ooze overlain by siliceous radiolarian ooze or red clays).

Non-spreading volcanic structures (seamounts, guyots, volcanic ridges) (Fig. 6c)

Pelagic sediments associated with volcanic edifices of this kind accumulate under two sets of circumstances. The first occurs as the volcanic structure builds up from the sea floor, and consists of episodes of submarine volcanism alternating with pelagic sedimentation during quiescent periods. Consequently, previously deposited sediment frequently is intruded or eroded by basaltic magma to produce sediment lenses of variable but generally small size (e.g. a few tens of metres across), and the intrusive mode of interpillow sediment will dominate. Commonly, the pelagic sediments will be interbedded with alkali flow basalts and hyaloclastites, particularly at the margins of the structure, and the volcanic rocks may show variable vesicularity reflecting both deep- and shallow-water eruptions. The igneous rocks should consist almost entirely of effusive or shallow hypabyssal rocks (pillow basalts, diabases, etc.) rather than deep-seated ones.

The second set of circumstances prevails once the volcanic structure becomes inactive and begins to subside as lateral plate motion carries it into deeper water (Menard, 1969b). Sediment begins to form a cap on top of the structure, with deeper-water facies accumulating above relatively shallow-water facies as subsidence continues. Where the top of the volcano once stood near sea level, this facies contrast can be extreme, as in the case of some Pacific seamounts where Cretaceous reef limestones lie above basalt and in turn are overlain by Tertiary nannoplankton-foraminiferal oozes (Hamilton, 1956). Extreme *lateral* facies contrasts are also likely between the pelagic sediment perched on top of seamounts and contemporaneous sediment deposited on the adjoining deep-sea floor; the former may be calcareous ooze, the latter siliceous ooze or red clay sedimented below the CCD (Lonsdale *et al.*, 1972), or, in areas near continental margins, rapidly deposited detrital sediment (Palmer, 1964).

The caps of pelagic sediment on top of dormant volcanic structures are lenticular bodies, but they are usually very much larger than the small sediment lenses formed by ponding and igneous disruption in the underlying volcanic pile. For example, the cap of Tertiary nannoplankton ooze on top of Horizon Guyot in the Pacific is a lenticular mass more than 100 km long, 30–40 km wide, and up to 150 m thick (Lonsdale *et al.*, 1972). Strong currents periodically sweep this sediment causing erosion and producing ripples, dunes and winnowed foraminiferal sands (Dangeard & Lonsdale, 1972). Bartlett & Greggs (1970) have noted a proclivity for pelagic carbonate sediments on ridges and seamounts to become lithified and impregnated with ferromanganese crusts, and Palmer (1964) discovered glauconite and crusts of manganese oxides and phosphorite on the current-swept top of Rodriguez Seamount off California. Thus mineralized limestone hardgrounds and other kinds of non-depositional surfaces may be common in these settings.

In addition to Horizon Guyot and Rodriguez Seamount, other well documented examples of sediment-volcanic relationships on modern volcanic structures include studies of seamounts by Pratt (1963), Budinger (1967), Karig *et al.* (1970), and McGregor, Betzer & Krause (1973). Ancient examples may include interbedded pelagic limestones and submarine basaltic rocks of Eocene age on the Olympic Peninsula of Washington State (Garrison, 1972, 1973; Glassley, 1974); Miocene interstitial limestones between basalt pillows on Southern Macquarie Island (Varne & Rubenach, 1972; J. R. Cann, personal communication); and possibly also the Upper Pillow Lavas and overlying pelagic sediments of the Troodos Massif, Cyprus (Gass & Smewing, 1973; Robertson & Hudson, 1974).

Non-spreading tectonic structures (Fig. 6d)

This category includes a variety of complex situations whose common denominator is that oceanic crustal or upper mantle rocks become exposed on the sea floor by tectonism which is geographically far removed from the actively spreading ridges where they originated. Subsequently these rocks can be reburied beneath much younger pelagic sediment. Some modern examples which illustrate the variability of these settings are as follows.

(1) The submerged portions of Macquarie Ridge (Hayes & Talwani, 1972) where a variety of probable Miocene or older mafic igneous rocks are exposed on the sea floor along a complex boundary between the Pacific and Indian Plates. Most sediment along the ridge is limited to small, local basins; a maximum age discrepancy between basement rock and immediately overlying sediment of about 25×10^6 years is possible.

(2) Palmer Ridge in the NE Atlantic, where Lower Eocene oceanic crust with a capping of Eocene-Lower Miocene pelagic sediment was uplifted and diapirically intruded by serpentinite in Early Miocene time (Cann & Funnel, 1967; Ramsay, 1970; Cann, 1971). Through a combination of Eocene ridge rifting and the Miocene off-ridge tectonism, a wide variety of rocks became exposed on the sea floor (basalts, diabases, gabbros, amphibolites, serpentinites, lithified and fractured pelagic sediments, and polymictic breccias). Overlying this complicated basement of mainly Eocene rocks are patches of winnowed pelagic sediment ranging in age from Middle Miocene to Holocene. The age difference between igneous basement rock and overlying sediment can thus be as great as 60×10^6 years.

(3) The north slope of the Puerto Rico Trench where Late Tertiary deformation due to downbuckling of a plate has exposed serpentinites and altered basalts of Cretaceous (Cenomanian) age on the sea floor (Bowin, Nalwalk & Hersey, 1966; Chase & Hersey, 1968). As pointed out by Abbate *et al.* (1972, pp. 258–259) these igneous rocks may become covered again beneath much younger pelagic sediments in the trench, and seismic profiling reported by Chase & Hersey (1968) shows small sediment ponds of Holocene sediment on the north slope. Thus a maximum age difference of about $80–90 \times 10^6$ years is possible between basement rocks and overlying sediment on the north slope of the Puerto Rico Trench.

(4) The Vema Fracture Zone in the North Atlantic, which offsets the Mid-Atlantic Ridge by more than 300 km; here, a combination of lateral and vertical movements has exposed serpentinites and altered basalts on the walls of the fracture zone (van Andel, 1969; van Andel, von Herzen & Phillips, 1971; Melson & Thompson, 1971). In addition to purely tectonic processes, Thompson & Melson (1972) have proposed that new oceanic crust may be created in some fracture zones by intrusion and extrusion of mafic magmas into them.

Pelagic sedimentary sequences deposited in these settings may have a number of characteristics in common with those sedimented on the flanks of highly rifted ridges. Both may rest depositionally on a variety of igneous rocks including deep-seated plutonic ones; both may contain intercalated breccias of deep-seated rocks; both will tend to occur in lenticular bodies. The major distinguishing feature in these non-spreading tectonic structures will be the large discrepancy between the geological age of the igneous rocks and that of the overlying pelagic sediments (except in those instances of off-ridge magmatism suggested by Thompson & Melson, 1972). Determination of this in ancient, on-land ophiolite complexes thus requires a combination of radiometric and biostratigraphic age dating of high precision. Although rarely attempted, combined dating of this kind may have revealed a tectonic setting similar to those discussed above in the Jurassic Point Sal ophiolite complex of the California Coast Range; here, a well-developed ophiolite sequence is depositionally overlain by pelagic tuffaceous cherts and limestones which are some 10–20 × 10^6 years younger than the igneous rocks (Hopson *et al.*, 1973; C. A. Hopson, personal communication; Pessagno, 1973).

Island-arc settings (Fig. 6e)

Because intense volcanism and volcanogenic sedimentation act to dilute most other kinds of sedimentation (Dickinson, 1971, 1972), pelagic sediments are usually only minor components of stratigraphic assemblages formed around the frontal parts of island-arc systems. Detritus-free settings which would allow accumulation of relatively pure pelagic oozes on volcanic rocks will be most common behind the frontal arc. Using Karig's terminology (1971, 1972), such areas will be within inter-arc basins (which are a variant of the spreading ridge settings previously discussed) and especially on top of submerged remnant arc ridges which lie above levels of rapid clastic deposition. Very thin layers of pelagic ooze may lie between volcanic turbidite sands which form sediment aprons at the front and rear of the volcanically active frontal arc while basinal pelagic oozes accumulate in offshore areas.

The sedimentary bodies of pelagic deposits formed in island-arc systems will thus tend to be small, very lenticular, and scattered. They will be interbedded with thick sequences of vesicular pillow lavas and fragmental volcanic rocks of andesitic composition, as well as with redeposited shallow-water deposits such as reef limestones.

Tertiary pelagic limestones in island-arc systems of the western Pacific are discussed by Schlanger (1964), Mitchell (1970), Mitchell & Warden (1971), and by Garrison, Schlanger & Wachs (1974). Wachs (1973a, b) concluded that some of the Cretaceous nannoplankton-foraminiferal limestones in the Franciscan Formation of California were deposited in restricted, relatively shallow-water environments, possibly within an island-arc system.

CONCLUSIONS

The models proposed above must be regarded, at best, as very simplified approximations of the original depositional-tectonic settings now represented in eugeosynclinal rocks. But recognition in eugeosynclinal terrains of even these extemporized categories represents a significant advance toward the level of detailed palaeogeographical reconstruction which is routine in other, less complicated geological regions. These reconstructions, however, depend entirely on the kind of detailed stratigraphic information and mapping which is rarely available in eugeosynclinal terrains, except for a few areas like the northern Apennines. From the sedimentological point of view, the type of information required includes the size and shape of pelagic sedimentary bodies, their internal organization and their precise physical and age relationships with the adjacent igneous rocks, and the succession of sedimentary and volcanic facies. These data may then be collated with petrological and geochemical information from the associated igneous rocks to construct more comprehensive models.

Among the problems concerning pelagic sedimentary rocks in eugeosynclinal assemblages, none is more baffling nor has persisted longer than the origin and exact significance of the radiolarian cherts, especially the Mesozoic cherts in the Tethyan region.

ACKNOWLEDGMENTS

Acknowledgment is made to the Donors of the Petroleum Research Fund (PRF 5962-AC2), administered by the American Chemical Society, for the partial support of this research. This article was written while I was a J. S. Guggenheim Memorial Foundation Fellow at the Department of Geology and Mineralogy, University of Oxford; I thank Professor E. A. Vincent and his staff for many courtesies extended. And I gratefully acknowledge field excursions and/or enlightening discussions with E. Abbate, J. R. Cann, C. A. Hopson, J. C. Moore, E. Nesbit, A. H. F. Robertson, A. G. Smith, A. C. Waters, and especially E. Spooner.

REFERENCES

Abbate, E. (1969) Geologia delle cinque terre e dell'entroterra di Levanto (Liguria Orientale). *Memorie Soc. geol. ital.* **8,** 923–1014.

Abbate, E. & Sagri, M. (1970) The eugeosynclinal sequences. In: *Development of the Northern Apennines Geosyncline* (Ed. by G. Sestini). *Sedim. Geol.* **4,** 251–340.

Abbate, E., Bortolotti, V. & Passerini, P. (1972) Studies on mafic and ultramafic rocks: 2—Paleogeographic and tectonic considerations on the ultramafic belts in the Mediterranean area. *Boll. Soc. geol. ital.* **91,** 239–282.

Adelseck, C.G., Geehan, G.W. & Roth, P.H. (1973) Experimental evidence for the selective dissolution and overgrowth of calcareous nannofossils during diagenesis. *Bull. geol. Soc. Am.* **84,** 2755–2762.

Allemann, F. & Peters, T. (1972) The ophiolite-radiolarite belt of the North-Oman Mountains. *Eclog. geol. Helv.* **65,** 657–697.

Atwater, T. & Mudie, J.D. (1968) Block faulting on the Gorda Rise. *Science,* **159,** 729–731.

Aubouin, J. (1965) *Geosynclines,* pp. 335. Elsevier Publish Co., Amsterdam.

Audley-Charles, M.G. (1965) Some aspects of the chemistry of Cretaceous siliceous sedimentary rocks from eastern Timor. *Geochim. cosmochim. Acta,* **29,** 1175–1192.

Audley-Charles, M.G. (1973) Paleoenvironmental significance of chert in the Franciscan Formation of Western California: discussion concerning the significance of chert in Timor. *Bull. geol. Soc. Am.* **84,** 363–367.

Aumento, F. & Loncarevic, B.D. (1969) The Mid-Atlantic Ridge near 45° N. III. Bald Mountain. *Can. J. Earth Sci.* **6,** 11–23.

Bailey, E.H., Irwin, W.P. & Jones, D.L. (1964) Franciscan and related rocks, and their significance in the geology of Western California. *Bull. Calif. Div. Mines Geol.* **183,** pp. 177.

Bailey, E.H., Blake, M.C. & Jones, D.L. (1970) On-land Mesozoic oceanic crust in California Coast Ranges. *Prof. Pap. U.S. geol. Surv.* **700-C,** C70–C81.

Bartlett, G.A. & Greggs, R.G. (1970) The Mid-Atlantic Ridge near 45° 00′ North. VIII. Carbonate lithification on oceanic ridges and seamounts. *Can. J. Earth Sci.* **7,** 257–267.

Berger, W.H. (1970) Planktonic Foraminifera: selective solution and the lysocline. *Mar. Geol.* **8,** 111–138.

Berger, W.H. (1972) Deep-sea carbonates: dissolution facies and age-depth constancy. *Nature, Lond.* **236,** 392–395.

Berger, W.H. (1973) Cenozoic sedimentation in the eastern tropical Pacific. *Bull. geol. Soc. Am.* **84,** 1941–1954.

Bernoulli, D. (1972) North Atlantic and Mediterranean Mesozoic facies: a comparison. In: *Initial Reports of the Deep Sea Drilling Project*, Vol. XI (C. D. Hollister, J. I. Ewing *et al.*), pp. 801–871. U.S. Government Printing Office, Washington.

Bernoulli, D. & Jenkyns, H.C. (1974) Alpine, Mediterranean and central Atlantic Mesozoic facies in relation to the early evolution of the Tethys. In: *Modern and Ancient Geosynclinal Sedimentation* (Ed. by R. H. Dott, Jr and R. H. Shaver). *Spec. Publs Soc. econ. Paleont. Miner., Tulsa*, **19,** 129–160.

Blake, M.C. Jr, & Jones, D.L. (1974) Origin of Franciscan melanges in northern California. In: *Modern and Ancient Geosynclinal Sedimentation* (Ed. by R. H. Dott, Jr and R. H. Shaver) *Spec. Publs Soc. econ. Paleont. Miner., Tulsa,* **19,** 345–357.

Bonatti, E. (1966) Deep-sea authigenic calcite and dolomite. *Science,* **153,** 534–537.

Bonatti, E. (1968) Fissure basalts and ocean-floor spreading on the East Pacific Rise. *Science,* **161,** 886–888.

Bonatti, E., Elter, P., Ferrara, G. & Innocenti, F. (1973) Northern Apennine ophiolitic complex: analogies with the Mid-Atlantic Ridge peridotite-gabbro-basalt complex. *Abstracts of Papers, Int. Symposium, 'Ophiolites in the Earth's Crust'*, Moscow, 66–67.

Bonatti, E., Honnorez, J. & Gartner, S. (1973) Sedimentary serpentinites from the Mid-Atlantic Ridge. *J. sedim. Petrol.* **43,** 728–735.

Bortolotti, V., Passerini, P., Sagri, M. & Sestini, G. (1970) The miogeosynclinal sequences. In: *Development of the Northern Apennines Geosyncline* (Ed. by G. Sestini). *Sedim. Geol.* **4,** 341–444.

Boström, K. (1969) Aluminum poor ferromanganoan sediments on active ocean ridges. *J. geophys. Res.* **74,** 3261–3270.

Bowin, C.C., Nalwalk, A.J. & Hersey, J.B. (1966) Serpentinized peridotite from the north wall of the Puerto Rico Trench. *Bull. geol. Soc. Am.* **77,** 257–270.

Bramlette, M.N. (1958) Significance of coccolithophorids in calcium-carbonate deposition. *Bull. geol. Soc. Am.* **69,** 121–126.

Budinger, T.F. (1967) Cobb Seamount. *Deep Sea Res.* **14,** 191–201.

Calvert, S.E. (1966) Accumulation of diatomaceous silica in sediments of the Gulf of California *Bull. geol. Soc. Am.* **77,** 569–596.

Calvert, S.E. (1968) Silica balance in the oceans and diagenesis. *Nature, Lond.* **219,** 919–920.

Cann, J.R. (1968) Geological processes at mid-ocean ridge crests. *Geophys. J. R. astr. Soc.* **15,** 331–341.

Cann, J.R. (1970) New model for the structure of the ocean crust. *Nature, Lond.* **226,** 928–930.

Cann, J.R. (1971) Petrology of basement rocks from Palmer Ridge, N.E. Atlantic. *Phil. Trans. R. Soc.* **A268,** 605–617.

Cann, J.R. & Funnell, B.M. (1967) Palmer Ridge: a section through the upper part of the ocean crust? *Nature, Lond.* **213,** 661–664.

Carlisle, D. (1963) Pillow breccias and their aquagene tuffs, Quadra Island, British Columbia. *J. Geol.* **71,** 48–71.

Cayeux, L. (1924) La question des jaspes à radiolaires. *C. r. Séanc. Soc. géol. Fr.* **1924,** 11–12.

Chase, R.L. & Hersey, J.B. (1968) Geology of the North Slope of the Puerto Rico Trench. *Deep Sea Res.* **15,** 297–317.

Chipping, D.H. (1971) Paleoenvironmental significance of chert in the Franciscan Formation of Western California. *Bull. geol. Soc. Am.* **82,** 1707–1712.

Cooper, L.H.N. (1965) Radiolarians as possible chronometers of continental drift. *Progr. Oceanogr.* **3,** 77–82.

Conolly, J.R. & Payne, R.R. (1972) Sedimentary patterns within a continent-mid-oceanic ridge-continent profile: Indian Ocean south of Australia. In: *Antarctic Oceanology II, The Australian-New Zealand Sector* (Ed. by D. E. Hayes). *Am. Geophys. Union, Antarctic Res. Series,* **19,** 295–315.

Corliss, J.B. (1971) The origin of metal-bearing submarine hydrothermal solutions. *J. geophys. Res.* **76,** 8128–8138.

Cronan, D.S. & Garrett, D.E. (1973) Distribution of elements in metalliferous Pacific sediments collected during the Deep Sea Drilling Project. *Nature, Lond.* **242,** 88–89.

Dangeard, L. & Lonsdale, P. (1972) Phénomènes d'érosion et courants profonds sur les flancs du 'Guyot Horizon', dans le Pacifique Nord: *C. r. hebd. Séanc. Acad. Sci. Paris,* **274,** D,2752–2754.

DANNER, W.R. (1967) Organic, shallow-water origin of bedded chert in the eugeosynclinal environment. *Abstracts, Geol. Soc. Am. Annual Meeting*, New Orleans, 42.

DANNER, W.R. (1970) Cherts and jaspers of the western Cordilleran eugeosyncline of southwestern British Columbia and north-western Washington. In: *West Commemoration Volume, Trans. Min. geol. metall. Inst. India*, 533–553.

DAVIS, E.F. (1918) The radiolarian cherts of the Franciscan group. *Univ. Calif. Publs Bull. Dep. Geol.* **11**, 235–432.

DECANDIA, F.A. & ELTER, P. (1969) Riflessioni sul problema delle ofioliti nell'Appennino settentrionale (nota preliminare). *Memorie Soc. tosc. Sci. nat. A*, **76**, 1–9.

DECANDIA, F.A. & ELTER, P. (1972) La 'zona' ofiolitifera del Bracco nel settore compreso fra Levanto e la Val Graveglia (Appennino ligure). *Memorie Soc. geol. Ital.* **11**, 503–530.

DEMIRMEN, F. & HARBAUGH, J.W. (1965) Petrography and origin of Permian McCloud Limestone of Northern California. *J. sedim. Petrol.* **35**, 136–154.

DEWEY, H. & FLETT, J.S. (1911) On some British pillow-lavas and the rocks associated with them. *Geol. Mag.* **8**, 202–209 and 241–248.

DEWEY, J.F. (1971) A model for the Lower Palaeozoic evolution of the southern margin of the Early Caledonides of Scotland and Ireland. *Scott. J. Geol.* **7**, 219–240.

DICKINSON, W.R. (1971) Clastic sedimentary sequences deposited in shelf, slope, and trough settings between magmatic arcs and associated trenches. *Pacific Geol.* **3**, 1–14.

DICKINSON, W.R. (1972) Sedimentation within and beside ancient and modern magmatic arcs. *Abstracts for Wisconsin Conference on 'Modern and Ancient Geosynclinal Sedimentation'*, 40–41.

DIERSCHE, V. (1973) Paleotectonics, sedimentation and volcanics of Late Jurassic radiolarites in the Northern Calcareous Alps. *Abstracts of the First Meeting, Zürich, European Geophysical Society*, p. 27.

DIETRICH, V. (1967) Geosynklinaler Vulkanismus in den oberen penninischen Decken Graubundens (Schweiz). *Geol. Rdsch.* **57**, 246–264.

DIETZ, R.S. (1963) Alpine serpentinites as oceanic rind fragments. *Bull. geol. Soc. Am.* **74**, 947–952.

ELDERFIELD, H. (1972) Effects of volcanism on water chemistry, Deception Island, Antarctica. *Mar. Geol.* **13**, M4–M6.

ELDERFIELD, H., GASS, I.G., HAMMOND, A. & BEAR, L.M. (1972) The origin of ferromanganese sediments associated with the Troodos Massif of Cyprus. *Sedimentology*, **19**, 1–19.

EWING, M. & EWING, J. (1964) Distribution of oceanic sediments. In: *Studies in Oceanography* (Ed. by K. Yoshida), pp. 525–537. University of Washington Press, Seattle.

EWING, M., CARPENTER, G., WINDISCH, C. & EWING, J. (1973) Sediment distribution in the oceans: The Atlantic. *Bull. geol. Soc. Am.* **84**, 71–88.

FEIN, C.D. (1973) Chemical composition and submarine weathering of some oceanic tholeiites from the East Pacific Rise. *Abstracts with programs*, **5**, no. 1, *Geol. Soc. Am. Cordilleran Section Meeting*, 41.

FISCHER, A.G., HONJO, S. & GARRISON, R.E. (1967) *Electron Micrographs of Limestones and their Nannofossils*, pp. 141. Princeton University Press, Princeton, New Jersey.

FISCHER, A.G. & GEALY, E.L. (1969) Summary and comparison of lithology and sedimentary sequences in northwest Atlantic and northwest Pacific. *Abstracts, Geol. Soc. Am. Annual Meetings, Atlantic City*, 65.

FLÜGEL, E., FRANZ, H.E. & OTT, W.F. (1968) Review on electron microscope studies of limestones. In: *Carbonate Sedimentology in Central Europe* (Ed. by G. M. Friedman and G. Müller), pp. 85–97. Springer-Verlag, Berlin.

FOLK, R.L. (1973) Evidence for peritidal deposition of Devonian Caballos Novaculite, Marathon Basin, Texas. *Bull. Am. Ass. Petrol. Geol.* **57**, 702–725.

FRANZINI, M., GRATZIU, C. & SCHIAFFINO, L. (1968) Sedimenti silicei non detritici dell'Appennino centro settentrionale, 1. La Formazioni dei diaspri di Reppia (Genova). *Memorie Soc. tosc. Sci. nat. A*, **75**, 154–203.

FURNES, H. (1973) Variolitic structure in Ordovician pillow lava and its possible significance as an environmental indicator. *Geology*, **1**, 27–30.

GARRISON, R.E. (1967) Nannofossils in Eocene eugeosynclinal limestones, Olympic Peninsula, Washington. *Nature, Lond.* **215**, 1366–1367.

GARRISON, R.E. (1972) Inter- and intrapillow limestones of the Olympic Peninsula, Washington. *J. Geol.* **80**, 310–322.

GARRISON, R.E. (1973) Space-time relations of pelagic limestones and volcanic rocks, Olympic Peninsula, Washington. *Bull. geol. Soc. Am.* **84,** 583–594.

GARRISON, R.E. & BAILEY, E.H. (1967) Electron microscopy of limestones in the Franciscan Formation of California. *Prof. Pap. U.S. geol. Surv.* **575-B,** B94–B100.

GARRISON, R.E. & FISCHER, A.G. (1969) Deep-water limestones and radiolarites of the Alpine Jurassic. In: *Depositional Environments in Carbonate Rocks* (Ed. by G. M. Friedman). *Spec. Publs Soc. econ. Paleont. Miner., Tulsa,* **14,** 20–56.

GARRISON, R.E., HEIN, J.R. & ANDERSON, T.F. (1973) Lithified carbonate sediment and zeolitic tuff in basalts, Mid-Atlantic Ridge. *Sedimentology,* **20,** 399–410.

GARRISON, R.E., SCHLANGER, S.O. & WACHS, D. (1974) Petrology and paleogeographic significance of Tertiary nannoplankton-foraminiferal limestones, Guam. Manuscript submitted to *Palaeogeog., Palaeoclim., Palaeoecol.*

GARTNER, S. & GENTILE, R. (1972) Problematic Pennsylvanian coccoliths from Missouri. *Micropaleontology,* **18,** 401–404.

GASS, I.G. & SMEWING, J.D. (1973) Intrusion, extrusion and metamorphism at constructive margins: evidence from the Troodos Massif, Cyprus. *Nature, Lond.* **242,** 26–29.

GIBSON, T.G. & TOWE, K.M. (1971) Eocene volcanism and the origin of Horizon A. *Science,* **172,** 152–154.

GLASSLEY, W. (1974) Geochemistry and tectonics of the Crescent Volcanics, Olympic Peninsula, Washington. *Bull. geol. Soc. Am.* **85,** 785–794.

GLENNIE, K.W., BOEUF, M.G.A., HUGHES CLARKE, M.W., MOODY-STUART, M., PILAAR, W.F.H. & RHEINHARDT, B.M. (1973) Late Cretaceous nappes in Oman Mountains and their geologic evolution. *Bull. Am. Ass. Petrol. Geol.* **57,** 5–27.

GREENWOOD, R. (1956) Submarine volcanic mudflows and limestone dikes in the Grayson Formation (Cretaceous) of central Texas. *Trans. Gulf-Cst Ass. geol. Socs,* **6,** 167–177.

GRUNAU, H.R. (1946) Die Vergesellschaffung von Radiolariten und Ophiolithen in den Schweizer Alpen. *Eclog. geol. Helv.* **39,** 256–260.

GRUNAU, H.R. (1965) Radiolarian cherts and associated rocks in space and time. *Eclog. geol. Helv.* **58,** 157–208.

HALLAM, A. (1971) Evaluation of bathymetric criteria for the Mediterranean Jurassic. In: *Colloque du Jurassique méditerranéen* (Ed. by E. Végh-Neubrandt). *Annls Inst. geol. publ. hung.* **54/2,** 63–70.

HAMILTON, E.L. (1956) Sunken islands of the Mid-Pacific Mountains. *Mem. geol. Soc. Am.* **64,** 97 pp.

HART, R. (1970) Chemical exchange between sea water and deep ocean basalts. *Earth Planet. Sci. Letts,* **9,** 269–279.

HART, R. (1973) Geochemical and geophysical implications of the reaction between sea water and the oceanic crust. *Nature, Lond.* **243,** 76–78.

HAWKINS, J.W. (1973) Petrology of marginal basins and their possible significance in orogenic belts: The Lau Basin as an example. *Abstracts with programs,* **5,** no. 1, *Geol. Soc. Am. Cordilleran Section Meeting,* 51.

HAY, R.L. & IIJIMA, A. (1968) Nature and origin of palagonite tuffs of the Honolulu Group on Oahu, Hawaii. In: *Studies in Volcanology* (Ed. by R. R. Coats, R. C. Hay and C. A. Anderson). *Mem. geol. Soc. Am.* **116,** 331–376.

HAY, W.W. (1970) Calcium carbonate compensation. In: *Initial Reports of the Deep Sea Drilling Project,* Vol. IV (R. G. Bader *et al.*), pp. 672–673. U.S. Government Printing Office, Washington.

HAY, W.W. (1973) Significance of paleontologic results of Deep Sea Drilling Project, legs 1–9. *Bull. Am. Ass. Petrol. Geol.* **57,** 55–62.

HAYES, D.E. & TALWANI, M. (1972) Geophysical investigation of Macquarie ridge complex. In: *Antarctic Oceanology II, The Australian-New Zealand Sector* (Ed. by D. E. Hayes). *Am. Geophys. Union, Antarctic Res. Series,* **19,** 211–234.

HAYES, D.E., FRAKES, L.A., BARRETT, P., BURNS, D.A., CHEN, P.H., FORD, A.B., KANEPS, A.G., KEMP, E.M., MCCOLLUM, D.W., PIPER, D.J.W., WALL, R.E. & WEBB, P.N. (1973) Leg 28. Deep-sea drilling in the southern ocean. *Geotimes,* **18,** (6), 19–24.

HEEZEN, B.C., MACGREGOR, I.D., FOREMAN, H.P., FORRISTAL, G., HEKEL, H., HESSE, R., HOSKINS, R.H., JONES, E.J.W., KANEPS, A., KRASHENINNIKOV, V.A., OKADA, H. & RUEF, M.H. (1973) Interpretation of the post Jurassic sedimentary sequence on the Pacific Plate. *Nature, Lond.* **241,** 25–32.

HERM, D. (1967) Zur Mikrofazies kalkiger Sedimenteinschaltungen in Vulkaniten der andinen Geosynklinale Mittelchiles. *Geol. Rsch.* **56,** 657–669.

HERMAN, Y. (1972) Origin of deep sea cherts in the North Atlantic. *Nature, Lond.* **238,** 392–393.

HESS, H.H. (1965) Mid-oceanic ridges and tectonics of the sea floor. In: *Submarine Geology and Geophysics. Colston Pap.* **17,** 317–333, Butterworths, London.

HONJO, S. (1969) Study of fine grained carbonate matrix: sedimentation and diagenesis of 'micrite'. In: *Litho- and Bio-facies of Carbonate Sedimentary Rocks,* a Symposium (Ed. by T. Matsumo). *Spec. Pap. palaeont. Soc. Japan,* **14,** 67–82.

HOPSON, C.A., FRANCO, C.J., PESSAGNO, E. & MATTINSON, J.M. (1973) Late Jurassic ophiolite at Point Sal, Santa Barbara County, California. *Abstracts with programs,* **5,** no. 1, *Geol. Soc. Am. Cordilleran Section Meeting,* 58.

HOUTZ, R.E. & MARKL, R.G. (1972) Seismic profiler data between Antarctica and Australia. In: *Antarctic Oceanography II, The Australian-New Zealand Sector* (Ed. by D. E. Hayes), *Am. Geophys. Union, Antarctic Res. Series,* **19,** 147–164.

HSÜ, K.J. (1971) Franciscan melanges as a model for eugeosynclinal sedimentation and underthrusting tectonics. *J. geophys. Res.* **76,** 1162–1169.

HSÜ, K.J. (1972) The concept of the geosyncline, yesterday and today. *Trans. Leicester lit. phil. Soc.* **66,** 26–48.

JONES, J.G. (1969) Pillow lavas as depth indicators. *Am. J. Sci.* **267,** 181–195.

KANEOKA, I. (1972) Evidence of subsidence of seamounts in the northwestern Pacific. *Mar. Geophys. Res.* **1,** 412–417.

KANIA, J.E.A. (1929) Precipitation of limestone by submarine vents, fumaroles, and lava flows. *Am. J. Sci.* **218,** 347–359.

KANMERA, K. (1968) On some sedimentary rocks associated with geosynclinal volcanic rocks. *Jap. J. Geol.* **1,** 23–32 (in Japanese with English abstract).

KANMERA, K. (1972) Cherts and associated volcanic rocks in Japan. *Abstracts for Wisconsin Conference on 'Modern and Ancient Geosynclinal Sedimentation'*, 32–33.

KANMERA, K. (1974) Paleozoic and Mesozoic geosynclinal volcanism in the Japanese Islands and associated chert sedimentation. In: *Modern and Ancient Geosynclinal Sedimentation* (Ed. by R. H. Dott, Jr and R. H. Shaver). *Spec. Publs Soc. econ. Paleont. Miner., Tulsa,* **19,** 161–173.

KARIG, D.E. (1971) Structural history of the Mariana island arc system. *Bull. geol. Soc. Am.* **82,** 323–344.

KARIG, D.E. (1972) Remnant arcs. *Bull. geol. Soc. Am.* **83,** 1057–1068.

KARIG, D.E., PETERSON, M.N.A. & SHOR, G.C. (1970) Sediment-capped guyots in the Mid-Pacific mountains. *Deep Sea Res.* **17,** 373–378.

KEEN, M.J. & MANCHESTER, K.S. (1970) The Mid-Atlantic Ridge near 45°N, X: Sediment distribution and thickness from seismic reflection profiling. *Can. J. Earth Sci.* **7,** 735–747.

KOBAYASHI, T. & KIMURA, T. (1944) A study on the radiolarian rocks. *J. Fac. Sci. Tokyo Univ. Sect. II,* **VII,** Pt 2, 75–178.

LEMOINE, M., STEEN, D. & VUAGNAT, M. (1970) Sur le problème stratigraphique des ophiolites piémontaises et des roches sédimentaires associées: observations dans le massif de Chabrière en Haute-Ubaye (Basses-Alpes, France). *C. r. Séanc. Soc. Phys. Hist. nat. Genève,* **5,** 44–59.

LEWIS, D.W. (1972) Tertiary limestone dikes, Oamaru, New Zealand. *Bull. Am. Ass. Petrol. Geol.* **56,** 636–637 (abs.).

LIPPS, J. (1970) Plankton evolution. *Evolution,* **24,** 1–22.

LLOYD, R.M. & HSÜ, K.J. (1972) Stable-isotope investigations of sediments from DSDP III cruise to South Atlantic. *Sedimentology,* **19,** 45–58.

LONSDALE, P., NORMARK, W.R. & NEWMAN, W.A. (1972) Sedimentation and erosion on Horizon Guyot. *Bull. geol. Soc. Am.* **83,** 289–316.

LUYENDYK, B.P. (1970) Origin and history of abyssal hills in the northeast Pacific Ocean. *Bull. geol. Soc. Am.* **81,** 2237–2260.

MCBIRNEY, A.R. (1963) Factors governing the nature of submarine volcanism. *Bull. volcan.* **26,** 455-469.

MCBIRNEY, A.R. (1971) Oceanic volcanism: a review. *Rev. Geophys. Space Phys.* **9,** 523–556.

MCBRIDE, E.F. & THOMSON, A. (1970) The Caballos Novaculite, Marathon Region, Texas. *Spec. Pap. geol. Soc. Am.* **122,** 129.

McGregor, B.A., Betzer, P.R. & Krause, D.C. (1973) Sediments in the Atlantic Corner seamounts: control by topography, paleo-winds and geochemically-detected modern bottom currents. *Mar. Geol.* **14**, 179–190.

Matthews, V. & Wachs, D. (1973) Mixed depositional environments in the Franciscan geosynclinal assemblage. *J. sedim. Petrol.* **43**, 516–517.

Mattson, P.H., Pessagno, E.A. & Helsley, C.E. (1972) Outcropping Layer A and A^{11} correlatives in the Greater Antilles. In: *Studies in Earth and Space Sciences* (Ed. by R. Shagam *et al.*). *Mem. geol. Soc. Am.* **132**, 57–66.

Maxwell, J.C. (1969) 'Alpine' mafic and ultramafic rocks—the ophiolite suite: a contribution to the discussion of the paper 'The origin of ultramafic and ultrabasic rocks' by P. J. Wyllie. *Tectonophysics*, **7**, 489–494.

Melson, W.G., Thompson, G. & van Andel, Tj.H. (1968) Volcanism and metamorphism in the Mid-Atlantic Ridge, 22°N Latitude. *J. geophys. Res.* **73**, 5925–5941.

Melson, W.G. & Thompson, G. (1971) Petrology of a transform fault zone and adjacent ridge segments. *Phil. Trans. R. Soc.* **A268**, 423–441.

Menard, H.W. (1967) Sea floor spreading, topography, and the second layer. *Science*, **157**, 923–924.

Menard, H.W. (1969a) Elevation and subsidence of the oceanic crust. *Earth Planet. Sci. Letts*, **6**, 275–284.

Menard, H.W. (1969b) Growth of drifting volcanoes. *J. geophys. Res.* **74**, 4827–4837.

Mitchell, A.H. (1970) Facies of an Early Miocene volcanic arc, Malekula Island, New Hebrides. *Sedimentology*, **14**, 201–243.

Mitchell, A.H. & Warden, A.J. (1971) Geological evolution of the New Hebrides island arc. *J. geol. Soc. Lond.* **127**, 501–529.

Moore, J.G. (1965) Petrology of deep-sea basalt near Hawaii. *Am. J. Sci.* **263**, 40–52.

Moores, E.M. (1969) Petrology and structure of the Vourinos ophiolitic complex of northern Greece. *Spec. Pap. geol. Soc. Am.* **118**, 74 pp.

Moores, E.M. & Vine, F.J. (1971) The Troodos Massif, Cyprus and other ophiolites as oceanic crust: evaluation and implications. *Phil. Trans. R. Soc.* **A268**, 443–466.

Nayudu, Y.R. (1964) Palagonite tuffs (hyaloclastites) and the products of post-eruptive processes. *Bull. volcan.* **27**, 1–20.

Nayudu, Y.R. (1971) Geologic implications of microfossils in submarine volcanics. *Bull. volcan.* **35**, 402–423.

Nisbet, E.G. & Price, I. (1974) Siliceous turbidites: bedded cherts as redeposited, ocean ridge-derived sediments. In: *Pelagic Sediments: on Land and under the Sea* (Ed. by K. J. Hsü and H. C. Jenkyns). *Spec. Publs int. Ass. Sediment.* **1**, 351–366.

Page, B.M. (1972) Oceanic crust and mantle fragment in subduction complex near San Luis Obispo, California. *Bull. geol. Soc. Am.* **83**, 957–972.

Palmer, H.D. (1964) Marine geology of Rodriguez Seamount. *Deep Sea Res.* **11**, 737–756.

Parea, G.C. (1970) Ricerche sulla genesi delle rocce silicee non detritiche. *Memorie Soc. geol. ital.* **9**, 665–707.

Park, C.F. (1946) The spilite and manganese problems of the Olympic Peninsula, Washington. *Am. J. Sci.* **244**, 305–323.

Park, J. (1918) The geology of the Oamaru District, North Otago (Eastern Otago Division). *Bull. geol. Surv. N.Z. Survey Branch* (N.S.), **20**, 119 pp.

Pearce, J.A. & Cann, J.R. (1973) Tectonic setting of basic volcanic rocks determined using trace element analyses. *Earth Planet. Sci. Letts*, **19**, 290–300.

Pessagno, E.A. Jr (1973) Age and geologic significance of radiolarian cherts in the California Coast Ranges. *Geology*, **1**, 153–156.

Piper, D.Z. (1973) Origin of metalliferous sediments from the East Pacific Rise. *Earth Planet. Sci. Letts*, **19**, 75–82.

Pratt, R.M. (1963) Great Meteor Seamount. *Deep Sea Res.* **10**, 17–25.

Ramsay, A.T.S. (1970) The pre-Pleistocene stratigraphy and palaeontology of the Palmer Ridge area, N.E. Atlantic. *Mar. Geol.* **9**, 261–285.

Ramsay, A.T.S. (1973) A history of organic siliceous sediments in oceans. In: *Organisms and Continents through Geologic Time* (Ed. by N. F. Hughes). *Spec. Pap. Palaeont.* **12**, 199–234.

Rheinhardt, B.M. (1969) On the genesis and emplacement of ophiolites in the Oman Mountains Geosyncline. *Schweiz. miner. petrogr. Mitt.* **49**, 1–30.

RITTMAN, A. (1958) Geosynclinal volcanism, ophiolites and Baramiya rocks. *Egypt. J. Geol.* **2,** 61–65.
ROBERTSON, A.H.F. & HUDSON, J.D. (1973) Cyprus umbers: chemical precipitates on a Tethyan ocean ridge. *Earth Planet. Sci. Letts,* **18,** 93–101.
ROBERTSON, A.H.F. & HUDSON, J.D. (1974) Pelagic sediments in the Cretaceous and Tertiary history of the Troodos Massif, Cyprus. In: *Pelagic Sediments: on Land and under the Sea* (Ed. by K. J. Hsü and H. C. Jenkyns). *Spec. Publs int. Ass. Sediment.* **1,** 403–436.
SCHLAGER, W. & SCHLAGER, M. (1973) Clastic sediments associated with radiolarites (Tauglboden-Schichten, Upper Jurassic, Eastern Alps). *Sedimentology,* **20,** 65–89.
SCHLANGER, S.O. (1964) Petrology of the limestones of Guam. *Prof. Pap. U.S. geol. Surv.* **403-D,** 52 pp.
SCHNEIDER, E.D. (1972) Sedimentary evolution of rifted continental margins. In: *Studies in Earth and Space Sciences* (Ed. by R. Shagam *et al.*). *Mem. geol. Soc. Am.* **132,** 109–118.
SCLATER, J.G. & DETRICK, R. (1973) Elevation of midocean ridges and the basement ages of JOIDES deep sea drilling sites. *Bull. geol. Soc. Am.* **84,** 1547–1554.
SIEVER, R. & KASTNER, M. (1967) Mineralogy and petrology of some Mid-Atlantic Ridge sediments. *J. Mar. Res.* **25,** 263–277.
SNYDER, G.L. & FRASER, G.D. (1963) Pillowed lavas, I: Intrusive layered lava pods and pillowed lavas, Unalaska Island, Alaska. *Prof. Pap. U.S. geol. Surv.* **454-B,** 23 pp.
SPOONER, E.T.C. & FYFE, W.S. (1973) Sub-sea-floor metamorphism, heat, and mass transfer. *Contr. Miner. Petrol.* **42,** 287–304.
STEFANSSON, U. (1966) Influence of the Surtsey eruption on the nutrient content of the surrounding sea water. *J. Mar. Res.* **24,** 241–268.
STEINMANN, G. (1905) Geologische Beobachtungen in den Alpen. II. Die Schardtsche Uberfaltungs-theorie und die geologische Bedeutung der Tiefseeabsätze und der ophiolithischen Massen-gesteine. *Ber. naturf. Ges. Freiburg,* **16,** 18–67.
STEINMANN, G. (1925) Gibt es fossile Tiefseeablagerungen von erdgeschichtlicher Bedeutung? *Geol. Rdsch.* **16,** 435–468.
STEINMANN, G. (1927) Die ophiolithischen zonen in medditeranen Kettenbgebirgen. *14th Int. geol. Congr., Madrid,* **2,** 638–667.
TALWANI, M., WINDISCH, C.C. & LANGSETH, M.G. (1971) Reykjanes Ridge crest: a detailed geo-physical study. *J. geophys. Res.* **76,** 473–517.
THOMPSON, G. (1972) A geochemical study of some lithified carbonate sediments from the deep sea. *Geochim. cosmochim. Acta,* **36,** 1237–1253.
THOMPSON, G. & MELSON, W.G. (1972) The petrology of oceanic crust across fracture zones in the Atlantic Ocean: evidence of a new kind of sea-floor spreading. *J. Geol.* **80,** 526–538.
THURSTON, D.R. (1972) Studies on bedded cherts. *Contr. Miner. Petrol.* **36,** 329–334.
TRÜMPY, R. (1960) Paleotectonic evolution of the central and western Alps. *Bull. geol. Soc. Am.* **71,** 843–908.
VAN ANDEL, TJ.H. (1968) The structure and development of rifted midoceanic rises. *J. Mar. Res.* **26,** 144–161.
VAN ANDEL, TJ.H. (1969) Recent uplift of the Mid-Atlantic Ridge south of the Vema Fracture zone. *Earth Planet. Sci. Letts,* **7,** 228–230.
VAN ANDEL, TJ.H. & BOWIN, C.O. (1968) Mid-Atlantic Ridge between 22° and 23° north latitude and the tectonics of mid-ocean rises. *J. geophys. Res.* **73,** 1279–1298.
VAN ANDEL, TJ.H. & HEATH, G.R. (1970) Tectonics of the Mid-Atlantic Ridge 6–8° south latitude. *Mar. Geophys. Res.* **1,** 5–36.
VAN ANDEL, TJ.H., VON HERZEN, R.P. & PHILLIPS, J.D. (1971) The Vema Fracture Zone and the tectonics of transverse shear zones in oceanic crustal plates: *Mar. Geophys. Res.* **1,** 261–283.
VAN ANDEL, TJ.H. & KOMAR, P.D. (1969) Ponded sediments of the Mid-Atlantic Ridge between 22° and 23° north latitude. *Bull. geol. Soc. Am.* **80,** 1163–1190.
VAN ANDEL, TJ.H., REA, D.K., VON HERZEN, R.P. & HOSKINS, H. (1973) Ascension Fracture Zone, Ascension Island, and the Mid-Atlantic Ridge. *Bull. geol. Soc. Am.* **84,** 1527–1546.
VARNE, R., GEE, R.D. & QUILTY, P.G.L. (1969) Macquarie Island and the cause of oceanic linear magnetic anomalies. *Science,* **166,** 230–233.
VARNE, R. & RUBENACH, M.J. (1972) Geology of Macquarie Island and its relationship to oceanic crust. In: *Antarctic Oceanology II, The Australian-New Zealand Sector* (Ed. by D. E. Hayes). *Am. geophys. Union, Antarctic Res. Series,* **19,** 251–266.

VINE, F.J. & MOORES, E.M. (1972) A model for the gross structure, petrology, and magnetic properties of oceanic crust. In: *Studies in Earth and Space Sciences* (Ed. by R. Shagam *et al.*). *Mem. geol. Soc. Am.* **132,** 195–205.

WACHS, D. (1973a) Franciscan limestones: environment of deposition and post-depositional history. *Abstracts with programs,* 5, no. 1, Geol. Soc. Am. Cordilleran Section Meeting, 118.

WACHS, D. (1973b) *Petrology and depositional history of limestones in the Franciscan Formation of California,* 77 pp. Ph.D. Thesis, University of California.

WINTERER, E.L. (1973) Sedimentary facies and plate tectonics of equatorial Pacific. *Bull. Am. Ass. Petrol. Geol.* **57,** 265–282.

ZELENOV, K.K. (1964) Iron and manganese in exhalations of the submarine Banu Wuhu volcano (Indonesia). *Dokl. (Proc.) Acad. Sci. U.S.S.R., Earth Sciences Section,* **155,** 1317–1320 (pp. 94–96 in translation published by the American Geological Institute).

ZIMMERMAN, J. (1972) Emplacement of the Vourinos ophiolitic complex, northern Greece. In: *Studies in Earth and Space Sciences* (Ed. by R. Shagam *et al.*). *Mem. geol. Soc. Am.* **132,** 225–239.

NOTE ADDED IN PROOF

A remarkably detailed analysis of sedimentary and volcanic facies in an eugeosynclinal assemblage is given in the following, recently published paper:

CARLISLE, D. & SUSUKI, T. (1974) Emergent basalt and submergent carbonate-clastic sequences including the Upper Triassic Dilleri and Welleri Zones on Vancouver Island. *Can J. Earth Sci.* **11,** 254–279.

Spec. Publs int. Ass. Sediment. (1974) **1**, 401

Origin and fate of ferromanganoan active ridge sediments

KURT BOSTRÖM

Rosenstiel School of Marine and Atmospheric Science, University of Miami, 10 *Rickenbacker Causeway, Miami, Florida* 33149, *U.S.A.*

ABSTRACT

Major- and trace-element data, U- and Th-isotope relations, and element-accumulation rates suggest that the ferromanganoan active ridge sediments are only to a small extent due to leaching of oceanic rocks or deposition from ordinary sea water. The proportions of Fe, Mn, Ti, Al, Si, P, Ba, U and Th in such sediments are similar to those in carbonates of magmatic origin. This suggests that a CO_2-rich phase of deep-seated origin is the carrier of these elements. High spreading rates should imply intense degassing and thus high accumulation rates of active ridge sediments; low spreading rates should imply the opposite. Such correlations have been observed.

Active ridge sediments should not be confused with other Fe–Mn rich deposits on inactive ridges and on seamounts. Such deposits are often concretionary, form exceedingly slowly, and show much lower Fe/Ti and Fe/Al ratios than do the active ridge sediments.

Active ridge sediments are known since the Cretaceous, but some Fe–Mn–chert deposits on the continents that date back to the Archaean may be of a similar origin. However, many of these very old deposits rest on continental crust. Most old active ridge sediments must therefore have been subducted or metamorphosed beyond recognition; but may yet have played an important role during ore-forming processes.

The reference for the full text of this article is as follows.

BOSTRÖM, K. (1973) The origin and fate of ferromanganoan active ridge sediments. *Stockh. Contr. Geol.* **27**, 149–243.

Spec. Publs int. Ass. Sediment. (1974) **1,** 403–436

Pelagic sediments in the Cretaceous and Tertiary history of the Troodos Massif, Cyprus

A. H. F. ROBERTSON* *and* J. D. HUDSON

Department of Geology, University of Leicester, Leicester LE1 7RH

ABSTRACT

Resting on the Cretaceous pillow-lavas of the Troodos Massif, interpreted as ocean floor, there is a thick sequence of pelagic sediments ranging in age from Campanian to Mid-Tertiary.

The earliest pelagic sediments are umbers: iron- and manganese-rich mudstones comparable to the basal iron-rich sediments of the East Pacific Rise. Their origin is linked to the last stages of submarine volcanism and the activity of thermal springs. Overlying radiolarites contain well-preserved Campanian Radiolaria. Their silica diagenesis is compared to that of cherts encountered by the Deep Sea Drilling Project. In wholly autochthonous successions, Maastrichtian chalks follow, and show evidence of solution, indicating deposition near the carbonate compensation depth. All these sediments are thin (tens of metres). Palaeocene to Lower Eocene chalks are much thicker (to 300 m) but very variable. They contain abundant cherts, some of which replace foraminiferal beds of turbiditic origin. Later chalks record shallower-water conditions before a Miocene emergence.

In south-west Cyprus a thick sequence of volcanogenic clay and sandstone occurs above the umbers and radiolarites, and beneath allochthonous nappes and mélange. These include serpentinite masses, probably derived from Troodos-type oceanic basement, and the Mamonia rocks: sediments and volcanics of Triassic to Jurassic age and continental-shelf to pelagic facies. The allochthon is overlain by autochthonous Maastrichtian and Tertiary chalks.

Thus ocean-floor generation was followed by pelagic, non-calcareous sedimentation; then by intense acid volcanism, nappe and mélange emplacement, suggestive of a trench environment; then by quiet pelagic chalk deposition. All this took place during a brief period of the Late Cretaceous, a time of widespread tectonic activity from Cyprus to Oman. The geotectonic implications are briefly discussed.

INTRODUCTION

Within the last few years, the island of Cyprus has become a geological testing ground for many of the concepts of the New Global Tectonics. The Troodos Massif

*Present address: Department of Geology, University of Cambridge, Sedgwick Museum, Cambridge CB2 3EQ.

is now the most widely publicized area that has been claimed as a fragment of ocean floor, now exposed on land (Gass & Masson-Smith, 1963; Gass, 1968; Moores & Vine, 1971): a view that we think poetically appropriate as the most famous classical story associated with the island is that of Aphrodite arising from the foam on Cyprian shores (Botticelli, *c.* 1480). The interpretation of Troodos has been based almost entirely on the igneous rocks and on geophysics; relatively little has been published on the sediments overlying the lavas, although their generally pelagic nature has long been known.

In this paper, we show that some of the Upper Cretaceous and Lower Tertiary sediments which border the Troodos Massif are indeed closely comparable with modern oceanic sediments, and with older sediments found in the deep oceans by the Deep Sea Drilling Project. The Cyprus sediments can be closely examined in the field and their sedimentary structures and lateral variations assessed; thus their study may aid in the interpretation of sediments only seen in fragmentary and widely-spaced Deep Sea Drilling Project cores.

Although we believe that the 'ocean-floor' interpretation of Troodos is well founded, the Cretaceous-Tertiary history of the region is far from simple. In particular, emplacement of large-scale allochthonous nappes took place during the Maastrichtian as in many other Tethyan regions, and the mechanism and geotectonic implications of this are controversial. Indeed, the ocean-ridge interpretation of Troodos, favoured by Moores & Vine, has recently been challenged by Miyashiro (1973). It is beyond the scope of this paper, and inappropriate to this symposium, to consider these questions fully, but we do discuss in outline the implications of the sediments we have encountered for the geotectonic history of Cyprus and the Late Mesozoic and Early Tertiary plate tectonics of this area.

For the most part, in this paper, we shall concentrate on two types of pelagic sediment which are well represented both on Cyprus and in the present oceans: basal ferromanganiferous sediments, known as umbers on Cyprus; and cherts, of which we have both radiolarian and chalk-associated varieties to consider.

The allochthonous Kyrenia Range of northern Cyprus also contains Cretaceous and Tertiary pelagic sediments, but these are not dealt with here.

STRATIGRAPHY

Troodos igneous Massif

We are only concerned with the uppermost part of the now well-known igneous stratigraphy of Troodos, the Pillow Lava sequence (see Moores & Vine, 1971, for a summary and references to earlier work). The earliest substantial sediment bodies overlie the Upper Pillow Lavas (Figs 1 and 2); scattered occurrences of volcaniclastic sediments within the lava pile will not be discussed here. The famous sulphide deposits of Cyprus occur within the Pillow Lavas and their relationships are highly controversial; however, most of them seem to pre-date the Upper Pillow Lavas and are therefore not related, except indirectly, to the sediments we discuss (for an exchange of views, see Constantinou & Govett, 1972, and discussion, 1973).

We should like to stress that the Upper Pillow Lavas appear to be entirely submarine in origin; the pillows are generally beautifully formed and show all the classic

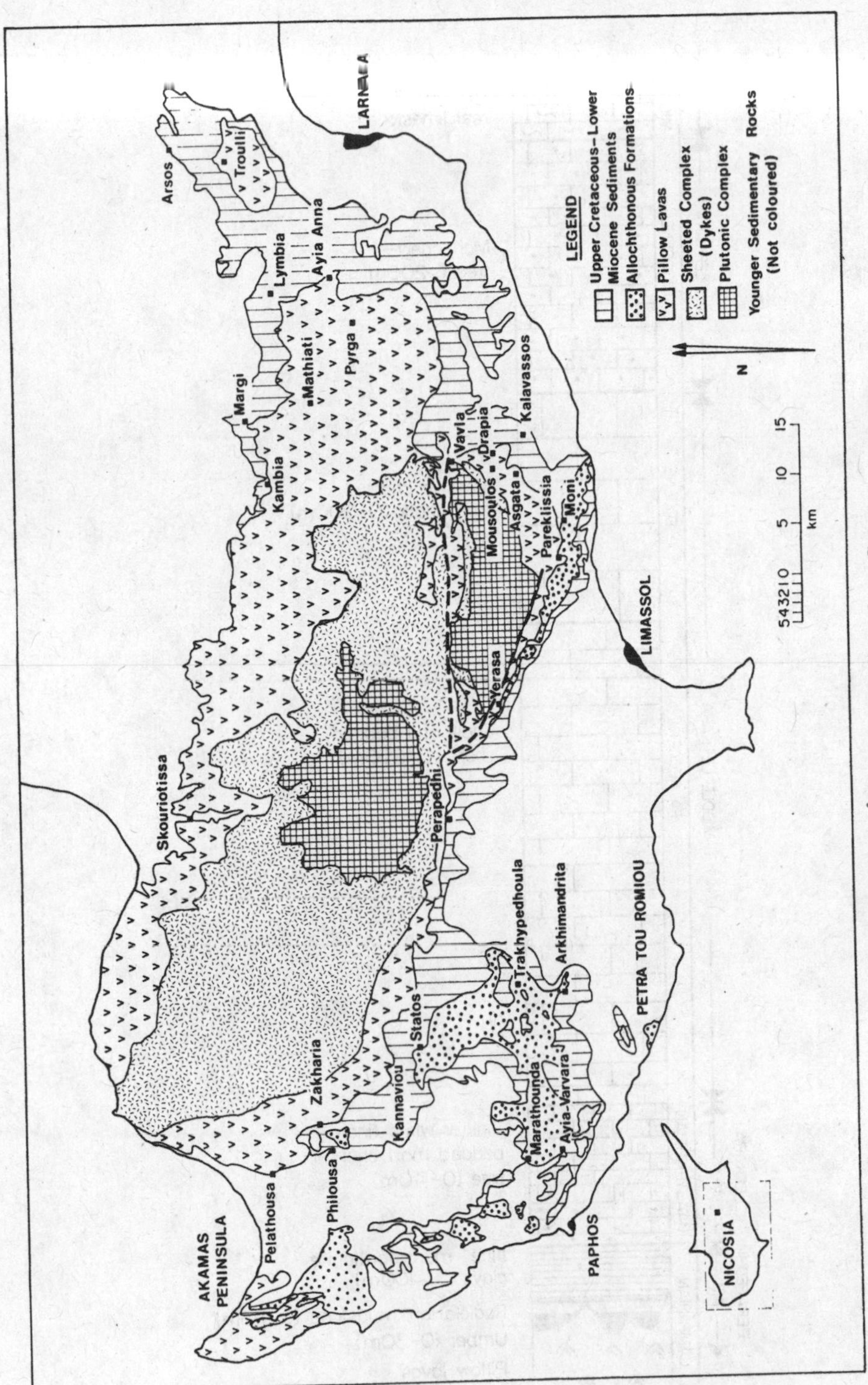

Fig. 1. Outline geology of Southern Cyprus.

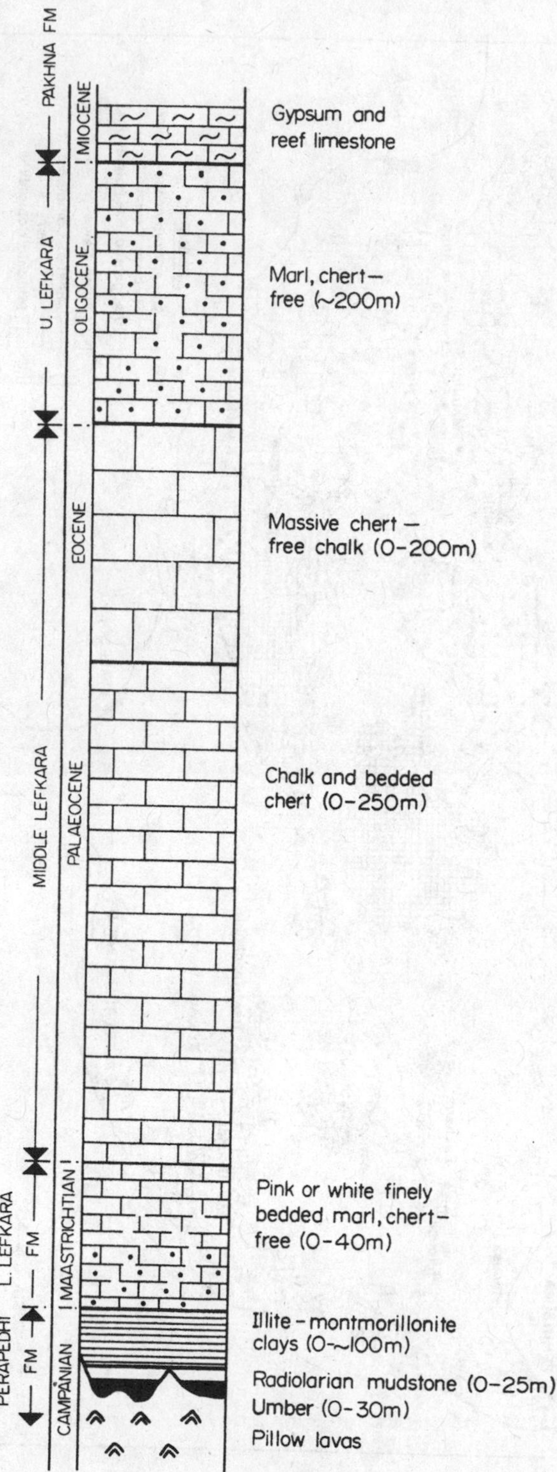

Fig. 2. Composite sequence of the Upper Cretaceous and Lower Tertiary sediments of Cyprus.

features of subaqueous flows. They give no hint of emergence, or of island-arc formation, in their structure or field relationships. Therefore we can be reasonably confident that, when deposition of sediment started, it did so in response to a cessation or slowing of volcanic activity, and under a permanent and fairly deep cover of water.

Radiometric dates now confirm the Late Cretaceous age of the Troodos Massif (I. G. Gass, personal communication; Lapierre, 1972).

Sedimentary succession

We present here a brief outline of the stratigraphy, to provide us with a basis for a more detailed account of the sediments. Regional accounts may be found in the 'Memoirs' of the Cyprus Geological Survey; this synthesis results from visits to all the main outcrops by Robertson, and we profited from discussions with, amongst others, Mr M. Mantis of the Cyprus Geological Survey, Nicosia.

The earliest sediments above the Upper Pillow Lavas are umbers (Fig. 2) which are very distinctive iron-, manganese- and trace metal-enriched fine-grained sediments. They occur only locally, restricted to hollows in the lava surface. Typically, umbers are only a few metres thick, but occasionally they reach 35 m. Upwards, but still restricted to hollows, umbers give way to radiolarites and radiolarian mudstones which are free of calcium carbonate and contain well-preserved Radiolaria. Wherever the Upper Pillow Lavas crop out, there are scattered occurrences of these sediments. In marked contrast with the overlying sediments, the umbers and radiolarian rocks show no systematic variation on a regional scale. Their age is Campanian, as discussed below.

Sediments which were deposited later vary laterally across the island. In the east, in the Nicosia, Limassol and Larnaca Districts, the radiolarian rocks pass upwards into scattered discontinuous developments of bentonitic (illite-montmorillonite) clays which tend to fill the upper parts of the hollows which are floored with umber and radiolarian rocks. Westwards along the southern margin of the volcanic massif, in the Limassol Forest area, these clays thicken, but are then concealed by younger sediments which overlap directly onto the volcanic massif. They emerge again in Paphos District where they are interbedded with fine- and medium-grained volcanogenic sandstones, the thickness of which, although difficult to estimate, may exceed 600 m (Fig. 3).

Unfortunately, published stratigraphical nomenclature concerning these sediments is contradictory. In the east, in the Nicosia, Limassol and Larnaca Districts, the umbers, radiolarites and radiolarian mudstones have been termed the Perapedhi Formation (Wilson, 1959) and the bentonitic clays have been distinguished by the term Moni Formation (Pantazis, 1967). However, in the west, in Paphos District, umbers, radiolarian rocks, bentonitic clays and volcanogenic sandstones have all been grouped as the Kannaviou Formation (Lapierre, 1968a, 1972). These contradictions are resolved if the name Perapedhi formation is restricted to the umbers and radiolarian rocks, while Kannaviou Formation is applied only to the overlying clays and volcanogenic sandstones, whether they occur in Paphos District or elsewhere (Fig. 3). Along most of the northern margin of the Troodos massif, with the exception of small pockets of umber preserved deep in hollows in the pillow lava surface, Upper Cretaceous rocks do not occur. This is almost certainly due to Miocene, or later, erosion when the Troodos Massif was at times exposed subaerially.

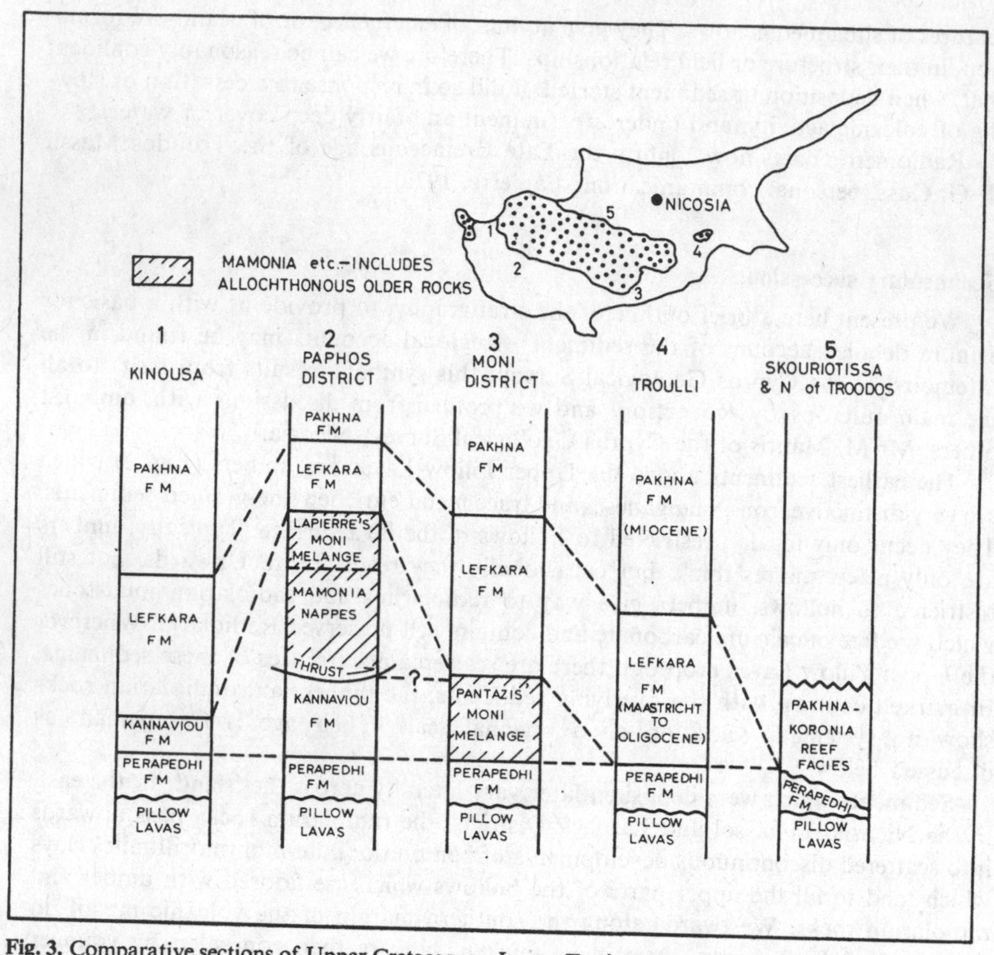

Fig. 3. Comparative sections of Upper Cretaceous—Lower Tertiary successions, Cyprus. Not to scale.

Sedimentation of bentonitic clays and volcanogenic sandstones was abruptly terminated, at least in the Paphos and Limassol Districts, by the incoming of major allochthonous masses, known in Paphos as the Mamonia Nappes and Moni Mélange (Lapierre, 1970, 1972). A smaller mass of mélange also occurs further west in the Limassol District, where the term Moni Mélange was first applied (Pantazis, 1967). It, however, differs lithologically and structurally from the formation of the same name in Paphos.

The emplacement of allochthonous units was followed by deposition of chalks. These rest with normal sedimentary contact on Kannaviou Formation, Perapedhi Formation or the allochthonous units, and in places directly on the Upper Pillow Lavas. They entirely post-date the emplacement of the nappes and mélange and are never themselves allochthonous.

The chalks were first named the Lapithos Group (Henson, Browne & McGinty, 1949; Wilson, 1959), but the name was amended to Lefkara Group (Pantazis, 1967) because the type area of the Lapithos is in the Kyrenia Range and has a totally different structural setting. The Lefkara can be divided into Lower, Middle and Upper units on combined lithological and micropalaeontological criteria (Wilson, 1959; Gass, 1960). In the following discussion the dating is taken from Henson *et al.* (1949) and Mantis (1970); it is based on Foraminifera.

The oldest chalks, the Lower Lefkara, are Maastrichtian. They rarely exceed 25 m in thickness and are generally chert-free. They are discontinuous, surrounding umber hollows or, where earlier sediments are absent, filling broad, shallow depressions in the pillow-lava surface. The thickness of the Middle Lefkara is much greater, up to 300 m, but laterally very variable. Its age is Palaeocene to Lower Eocene. It contains spectacular developments of bedded and nodular cherts. The Upper Lefkara begins with massive, generally chert-free chalks of Middle to Late Eocene age. Chalks of Oligocene to Early Miocene age are less well consolidated and lithologically highly variable. In some places, near the top of the succession, these chalks have slumped chaotically soon after deposition, attesting to rapid uplift of the Massif at this time.

During the Miocene, parts of the area were exposed subaerially, and erosion of the sediments extended in places down to the pillow-lava surface. The Middle Miocene Pakhna Formation consists of shallow-water carbonates, including reef developments, that interdigitate with gypsum deposits formed in local evaporite basins. The age of these is late Middle Miocene (Gass & Cockbain, 1961; Mantis, 1970).

Age of Perapedhi and Kannaviou Formations

There has in the past been much confusion over the age of the immediately post-Troodos succession; partly because the Triassic Mamonia rocks were at first assumed to be autochthonous, and partly because of a tentative Jurassic age assigned to Radiolaria from the Perapedhi Formation.

The correct assignment to the Late Cretaceous was first made by Allen (in Pantazis, 1967), on the basis of Radiolaria from bentonitic clays of the Limassol District. The radiolarian fauna is broadly comparable to that described by Pessagno (1963) from Puerto Rico, and suggests a Campanian age (Mantis, 1970). Most of the individual species are, however, relatively long-ranging, especially *Dictyomitra multicostata,* a very common form in these sediments.

A Campanian date for this part of the succession has now been considerably strengthened by the discovery of two diagnostic species in radiolarites of the Perapedhi Formation from Pyrga (Larnaca District). They were extracted from pure pink radiolarites using 5% hydrofluoric acid, and occur among a very diverse fauna to be described elsewhere. *Crucella espartoensis* Pessagno and *Patalibracchium lawsoni* Pessagno are regarded as diagnostic, respectively, of the Campanian Zone and the Middle Campanian subzone in the Great Valley Sequence of California (E. A. Pessagno, personal communication). It should, however, be emphasized that the use of Radiolaria for dating Mesozoic sediments is still in its earliest stages and the detailed stratigraphical ranges of most species are still unknown. In Paphos, the Kannaviou Formation can be dated as Maastrichtian, by the occurrence of diagnostic species of *Globotruncana* (M. Mantis, personal communication, 1973).

Together with the known Maastrichtian date of the Lower Lefkara Chalks, these ages date the emplacement of the nappes and mélange as intra-Maastrichtian.

We now pass on to a more detailed consideration of those parts of the succession that are critically comparable to sediments of the present oceans: the umbers, the radiolarites, and the cherts of the chalk formations.

THE UMBERS

Umbers are pale brown, chestnut-coloured or almost black, predominantly fine-grained sediments composed mostly of iron and manganese oxides. They are porous and hence have a low density. In this section pure umbers are opaque. X-ray diffraction shows them to be largely amorphous, but goethite, quartz and occasionally smectites are detectable.

General field relations

The field relations of umbers are important in that they provide clues as to the nature and origin of these most unusual sediments. Of some fifty outcrops of umber which have been examined, the vast majority are less than 4 m thick, finely laminated and undisturbed, providing few hints as to the way these sediments were formed. However, a few much larger deposits of umber also occur, reaching a thickness of 35 m, and at these localities umber often contains thin interbeds of pale tuff (Fig. 4a, b). At Troulli these tuffaceous beds are graded and micro-cross-laminated on a scale of a few centimetres. The tuffs are repetitively bedded, each one grading up from pale tuffaceous to finely laminated very fine-grained darker umber, features which point towards successive influxes of umber sediment, probably initiated by continuing instability of the underlying volcanic massif. An important implication of sediment transport is that, rather than being formed in hollows, umbers may have originated elsewhere and may only subsequently have accumulated in hollows in the pillow-lava surface. At Drapia, where one of the thickest umber deposits occurs, the lavas beneath are shattered, broken-up and physically abraded, most probably due to slumping and sliding down a steep slope, prior to umber being sedimented on top. These umbers, in addition to showing evidence of the type already outlined for sediment transport, show signs of intraformational slumping, which must have occurred soon after deposition. Angular unconformities have been observed: at Pyrga between the uppermost lava flows and overlying umber, and at Kalavassos between the umber and the overlying radiolarite. This is additional evidence for continuing disturbance and tilting of the underlying lava pile during the deposition of umbers.

Some thick deposits of umber form elongate outcrops within fault-controlled depressions, as at Arsos and Lymbia. These are likely to have been preferred sites of accumulation. At the Lymbia pit, fault movement continued after umber was deposited, which resulted in a spectacular chaotic mixture composed of derived masses of lava floating in structureless umber.

Although most of the umbers are found above the lavas, sometimes, as at Kalavassos and Zakharia, they are intercalated between the uppermost pillowed flows in which the pillows are well formed and possess good chilled margins. These pillows were extruded after deposition of the earliest umbers. Most of the umber outcrops are underlain by pale, altered-looking, fragmented and brecciated pillow lavas with interstitial umber. Elsewhere umbers overlie abnormally small pillows, 10–15 cm in diameter, which are pale and vesicular (Fig. 4d). Veins of calcite, pyrolusite and orange, iron-rich, fine-grained sediment can be traced downwards in places, as in the Mathiati area, well into the Lower Pillow Lavas. These most interesting field relations will be enlarged upon elsewhere.

Fig. 4. (a) Bedded umber with pale tuff interbeds, Troulli Umber Pit; (b) tuff interbeds in finely laminated umber, Troulli Umber Pit; (c) alternating beds of radiolarite and radiolarian mudstone, Valva; (d) small pillows with interstitial umber, Yerasa.

The field evidence, therefore, demonstrates that umbers were deposited on an uneven topography subject to continuing tectonic instability, marked by tilting and faulting of parts of the volcanic pile. This instability had ended by the time the radiolarian rocks were deposited, for these sediments show no signs of any disturbance during deposition. Another important implication of the umber field relations, especially their association with volcanism, sedimentary structures and variable thickness, is that they accumulated very rapidly indeed relative to the overlying radiolarian rocks and chalks, which are more comparable with normal pelagic sediments.

Genesis of umbers and comparison with oceanic sediments

In a previous paper (Robertson & Hudson, 1973) we showed that the umbers are closely comparable, chemically and mineralogically, with ferromanganiferous, trace element-enriched sediments currently forming along the East Pacific Rise (Boström *et al.*, 1969), and to their earlier-formed counterparts the Tertiary basal metal-rich sediments of the Atlantic (Boström *et al.*, 1972) and Pacific Oceans (Cronan *et al.*, 1972). A summary of some relevant analyses is given in Table 1. If this comparison is accepted, data from the umbers are relevant to the debate about the source of metals in the sediments of the present oceans; and with the umbers we have the great advantage of well-exposed field relations and access to the subjacent pillow lavas.

In our opinion, combined chemical and field evidence strongly favours an origin for umbers in the activity, during the last stages of volcanicity, of submarine thermal

Table 1. Averages of partial analyses of Cyprus sediments, and comparison with oceanic sediments

	Al (wt %)	Fe (wt %)	Mn (wt %)	Ti (wt %)	Ba (ppm)	Co (ppm)	Cu (ppm)	Ni (ppm)	Pb (ppm)	V (ppm)	Zn (ppm)
Umber	2·3	27·9	5·8	0·15	427	101	897	212	161	755	235
Radiolarite and radiolarian mudstone	5·0	5·1	1·05	0·27	80	40	142	71	52	98	222
Ochre (Skouriotissa)	0·91	42·65	0·58	—	—	209	11,620	40	—	—	338
Average deep-sea clay	8·4	6·5	0·67	0·46	2300	74	250	225	80	120	165
Average of metal-rich carbonate-free cores from East Pacific Rise	0·5	18·0	6·0	0·02	—	105	730	430	—	450	380

Sources of data: Umbers and radiolarian rocks, Robertson & Hudson (1973); Ochres, Constantinou & Govett (1972) (recalculated); Deep sea clays, Turekian & Wedepohl (1961); East Pacific Rise sediments, Boström & Peterson (1969).

springs. It is envisaged that sea water gained access to still-hot pillow lavas as a result of tilting and faulting of the lava pile, for which we have the field evidence discussed above. Sea water reacted with the hot lavas and in particular with their residual fluids, as postulated by Corliss (1971), leaching metals and transporting them, as chloride complexes, in reducing solution. These solutions migrated up through the lava pile and were exhaled into cold, oxygenated sea water, whereupon iron and manganese oxy-hydroxides precipitated as a fine cloud of particles in suspension. These were either transported, or settled locally to form the umbers, depending upon local topography and currents.

Further geochemical work is in progress to test this model. Field relations imply rapid sedimentation of umber, and this (together with preliminary geochemical data; Robertson & Hudson, 1973) rules out an origin by slow submarine weathering of basalt, or by slow sedimentary precipitation as is involved in the enrichment of iron, manganese and trace elements in stratigraphically condensed pelagic sediments (e.g. Elderfield, 1972). Apparently, only some kind of exhalative process can result in very rapid precipitation of iron and manganese oxyhydroxides in a pelagic environment. We hope to obtain additional evidence by studying the altered-looking, supposedly leached pillow lavas underlying the umbers (Robertson, in preparation).

Even if the iron and manganese come predominantly from the lavas, the problem of the source of the minor elements remains, and, in the case of the deep-sea sediments, is the subject of current controversy (Dasch, Dymond & Heath, 1971; Dymond *et al.*, 1973, and others). However, Corliss *et al.* (1972), in an abstract, state that the source of metals in the umbers is predominantly from lava by leaching, in agreement with our views.

We do not wish to imply that umbers are identical to the deep-ocean ferromanganese sediments which are, in any case, diverse; nor that ferromanganese sediments of this type can only form on spreading ridges. The metal-rich sediments encountered by the Deep Sea Drilling Project in the East Pacific by Leg 5 (von der Borch & Rex, 1970) and in the Atlantic by Leg 3 (Maxwell *et al.*, 1970) are, like the umbers, entirely restricted to the base of the sediment column, but elsewhere in the Pacific (Leg 8, von der Borch, Nesteroff & Galehouse, 1971) similar sediments contain interbeds of iron-free calcareous nannoplankton ooze, which we have not observed in the umbers. As regards the tectonic setting of formation, we have previously referred to the submarine ferro-manganese hydroxide precipitates of the Banu Wuhu volcano, Indonesia (Zelenov, 1964; see Robertson & Hudson, 1973). This apparently umber-like precipitate is forming on an andesite-dacite seamount in an island-arc setting. In the Mediterranean, comparable deposits are known, particularly around the volcano Stromboli (Bonatti *et al.*, 1973). It is also worth noting that the umbers are at least superficially similar in mineralogy to the sediments of the 'amorphous goethite facies' of the Red Sea brine pools (Bischoff, 1969). Scanning electron micrographs show that the umbers (and ochres), like the Red Sea goethite, occur as microspherules (Fig. 5e).

The essential requirement for umbers and analogous sediments may be that large quantities of water, generally sea water, circulate through hot, recently extruded volcanic rocks picking up readily leachable constituents, processes which may be concomitant with metamorphism of the lavas. The resulting solutions must then be released into a cool oxidizing environment, where precipitation of ferromanganese hydroxides is favoured. Once formed, these precipitates will only be preserved where

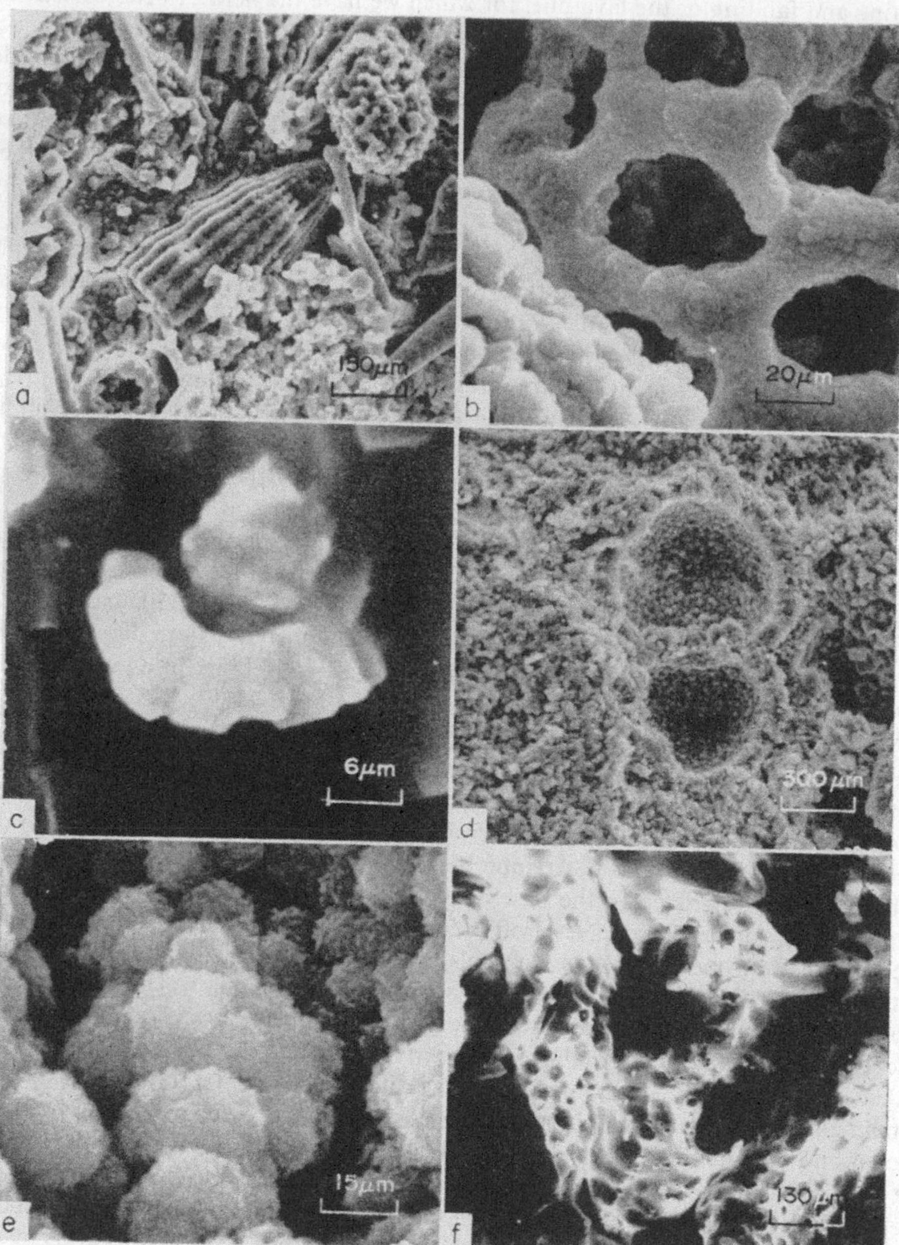

Fig. 5. Scanning electron micrographs. (a) Radiolarite etched in 5% HF, note the densely packed shells, Pyrga; (b) part of recrystallized Spumellarian lattice shell, Pyrga; (c) poorly preserved isolated coccolith shield, Maastrichtian chalk, Troulli; (d) recrystallized planktonic foraminiferal shell in recrystallized matrix, Maastrichtian chalk, Troulli; (e) microspherules which constitute umber (and ochre), Skouriotissa; (f) vesicular acid volcanic glass, Kannaviou Formation Arkhimandrita.

currents are not too strong and where dilution by clastic material is minimal. These last conditions are perhaps most often, but not uniquely, met on mid-ocean spreading ridges, the environment with which the depositional setting of the umbers can most closely be compared.

Ochres

Ochres are sediments, rich in iron and trace metals, but poor in manganese, which are closely associated with the Cyprus stratiform sulphide ore-bodies which occur within the pillow lavas (Constantinou & Govett, 1972). Here, they will not be discussed in any detail, but they are mentioned so as to clarify the distinction between ochres and umbers, a subject of considerable confusion in the past. Relative to umbers, ochres are rare heterogeneous sediments, and the strong chemical differences between them have suggested that they are totally unrelated in origin (Constantinou & Govett, 1972).

At most places where ochres occur, as at Mousoulos and Mathiati, ochre and umber are separated by unmineralized Upper Pillow Lava, proving that umbers formed later than ochres. At Skouriotissa, these upper lavas were not extruded everywhere; where they are absent umbers unconformably overlie ochres. The umbers occur in a saucer-shaped depression which was inherited from the earlier period of emplacement of massive sulphides. In these respects the situation at Skouriotissa is highly atypical and it is unfortunate that umbers from there are the most discussed in the literature (Searle, 1968; Elderfield *et al.*, 1972).

Although ochres differ chemically from most umbers, particularly in their lower manganese content (Table 1), both are oxidized sediments, composed predominantly of ferric iron oxides, mostly goethite. Both umbers and ochres are restricted to hollows and exhibit sedimentary structures including graded bedding and small-scale slumping. Scanning electron microscopy of broken rock surfaces shows that both consist of aggregates of minute microspherules (Fig. 5e).

Constantinou's data show that the trace elements of the ochres faithfully reflect variations in the original elemental compositions of each underlying ore-body, which establishes a relationship between ochres and sulphides. The dominant ore mineral is pyrite, an obvious source for much of the iron in the ochres.

Despite these differences in chemical composition, their mode of genesis may be comparable to that of umbers. We have suggested that umbers were precipitated when hydrothermal solutions were exhaled into aerobic sea water. In contrast, the sites of ochre formation, which are hollows closely related to the underlying massive sulphides, are likely to have been characterized by less oxidizing conditions, which would inhibit the precipitation of manganese, in any event negligible in the sulphide ores. Whereas we believe that the formation of umbers resulted from the widespread leaching of the underlying unmineralized Upper Pillow Lavas, the ochres relate to earlier, subaqueous oxidative weathering and local leaching of the sulphides soon after they were emplaced.

THE CHERTS

We shall now consider the siliceous rocks which occur in various parts of the *in situ* stratigraphical succession. These will be related to current concepts of chert formation, particularly those which have emerged from the findings of the Deep Sea Drilling Project.

Cherts in the umbers

Occasionally umbers contain cherts of two markedly different types. At Drapia, beds of thin, 10–15 cm thick, pale grey, loosely silicified chert occur in the upper part of the umber succession. These are radiolarian-rich and are likely to have formed during a period of reduced umber sedimentation. The other type resembled typical umber in colour, but is totally silicified to hard splintery chert. This is found where umbers have been disturbed by post-depositional slumping, where the umbers have been tectonically affected, or where they have been exposed by erosion sometime after deposition. These cherts are insignificant volumetrically and will not be further described here.

Radiolarites and related rocks

Radiolarites and radiolarian mudstones almost invariably directly overlie umbers; the transition is abrupt, over a few centimetres only. The radiolarian rocks are pink or pale grey, finely laminated and in gross aspect they appear to be regularly bedded due to variations of clay content (Fig. 4c). Occasionally, as at Vavla and in the Kambia areas, the radiolarian rocks reach thicknesses of 35 m, but typically they are very much thinner and like umbers they are restricted to hollows in the pillow-lava surface. One of the thickest exposures of radiolarian mudstones is at Vavla and there umbers are absent and the radiolarian rocks lie directly on the lavas. These sediments totally lack the tuffaceous interbeds which occur in the umbers. In hand specimen, they have a granular, porcellaneous texture and show a hackly, irregular fracture surface, which sets them aside from the waxy, vitreous flint-like conchoidally fracturing cherts, which although abundant in the chalks, are very rare in the radiolarian rocks. They also differ in this respect from the widespread Triassic and Jurassic radiolarites of the Tethyan and Alpine regions.

An unusual type of chert consisting of concentrically banded nodules, often oblate spheroids, occurs very rarely, invariably, as at Pyrga, near the transition between umber and radiolarian rocks. Bands of pink radiolarian mudstone alternate with dark grey flint-like cherts seen in thin section to be chalcedonic quartz. Concentrically banded cherts do not occur elsewhere in the stratigraphical succession, but similar cherts have been described from the Monterey Formation (Taliaferro, 1934), and from the Mishash Formation of Negev in Israel (Kolodny, 1969). For reasons of space, the origin of these most unusual cherts will not be discussed here.

Petrography and mineralogy

Thin-section examination shows that the radiolarites are densely packed with well-preserved radiolarian shells. Although most are replaced by chalcedonic quartz (Fig. 6a, b), some are preserved in part, or in total, as isotropic silica. Radiolaria which are composed of a multi-shelled dense meshwork (e.g. many Actinommids and Spongodiscoids) tend to remain as isotropic silica (Fig. 6c, d) but the Radiolaria with more open skeletal frames (many Nassellarians) are usually replaced, with the shells filled with radiating sheaf-like bundles of chalcedonic quartz (Fig. 6e, f). Radiolarian spines are numerous and almost always exist as isotropic silica. The radiolarian skeletal material is set in a brown or yellowish brown, translucent matrix of isotropic silica.

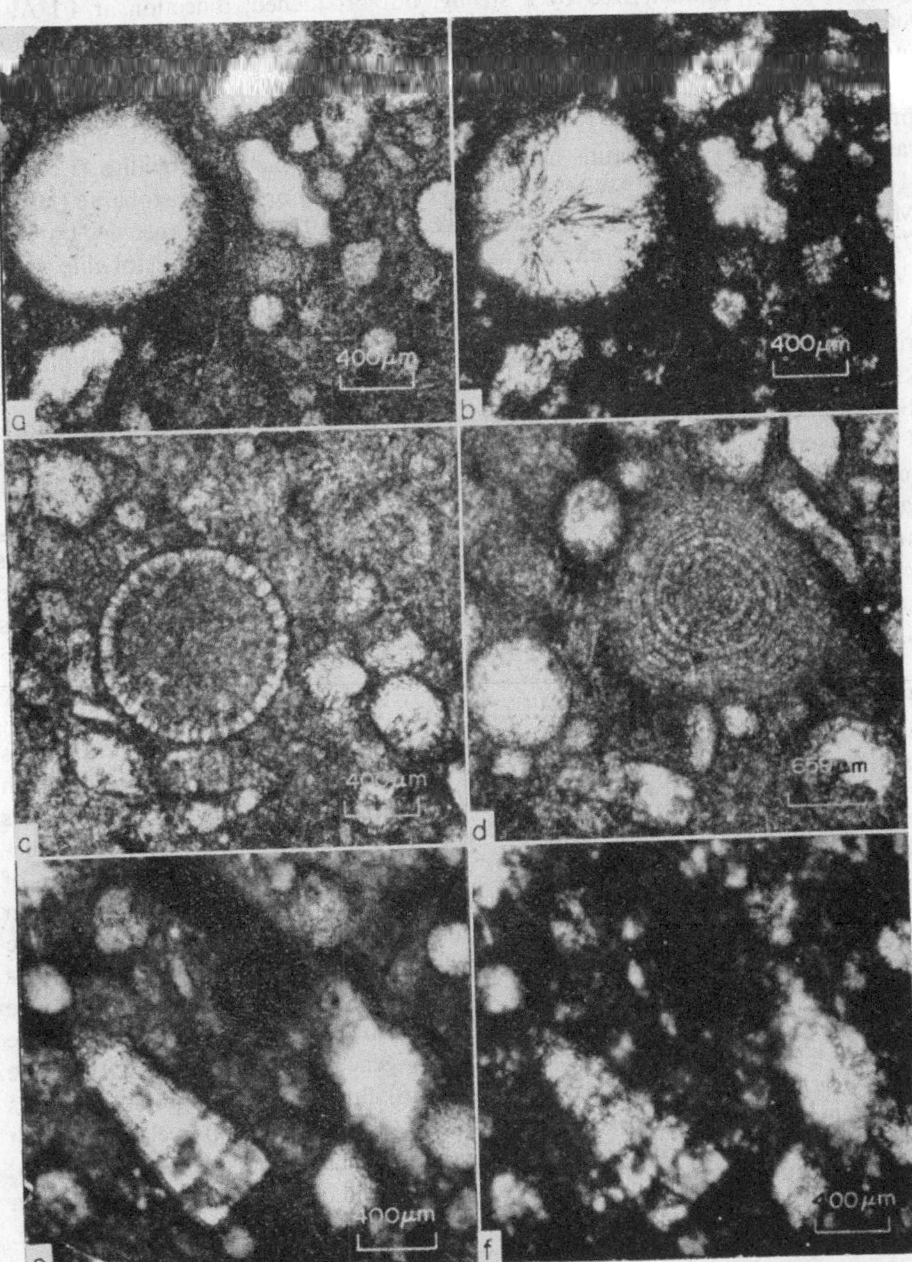

Fig. 6. Photomicrographs of radiolarites. (a, b) Radiolarian shells mostly recrystallized to sheaf-like bundles of chalcedonic quartz; (c) unrecrystallized Spumellarian shell, interior filled with minute interlocking crystals of isotropic silica; (d) unrecrystallized multi-shelled Spumellarian with surrounding shells partly or totally recrystallized to chalcedonic quartz; (e, f) partly recrystallized Dictyomitrid. (a, c, d, e) Plane polarized light, (b, f) crossed nicols. All from Campanian radiolarite, Pyrga outlier.

X-ray powder diffraction identifies lussatite, the opal C–T of Jones & Segnit (1972) (Fig. 7), which is characterized by a strong, but broadened, reflection at 4·1 Å, a weaker subordinate one from 4·25 to 4·30 Å and a much weaker diffuse reflection around 2·5 Å. This pattern has been interpreted as resulting from cristobalite with various degree of stacking disorder which produces peaks attributable to tridymite (Flörke, 1955; Jones & Segnit, 1972). The identification of lussatite was confirmed by infrared spectrometry. Lussatite has elsewhere been termed cristobalite (Calvert, 1971; Weaver & Wise, 1972; Wise, Buie & Weaver, 1972) or opal-cristobalite (Heath & Moberly, 1971a). Here, following Jones & Segnit (1972) and Berger & von Rad (1972), the term lussatite is preferred as it avoids confusion with cristobalite *sensu stricto* and emphasizes the differences between opal, almost amorphous, and lussatite, which is characteristically crystalline. Apart from these varieties of silica, X-ray diffraction has also revealed smectites and clinoptilolite in some of the radiolarian rocks.

Scanning electron microscopy of broken rock surfaces, etched in 5% hydrofluoric acid for several hours, confirms that these rocks are densely packed with well-preserved Radiolaria (Fig. 5a). Those which are lussatitic are comparatively smooth surfaced, but others, which are recrystallized to chalcedonic quartz, resemble scanning electron micrographs of etched chalcedonic Radiolaria which occur in Jurassic cherts from north-west Italy (Thurston, 1972) (Fig. 6b). Apart from occasional sponge spicules, the radiolarian rocks are free of other microfossils, including all calcareous ones.

Origin of the radiolarian rocks

Studies of cherts have demonstrated that silica exists in different structural forms of varying crystallographical maturity. The most immature of these is opal which occurs as opaline tests of siliceous microfossils, generally in post-Eocene sediments (von Rad & Rösch, 1974). Lussatite is much more abundant. This is derived diagenetically from siliceous microfossils (Calvert, 1971) or from devitrification of volcanic glass and is thought to have crystallized from an amorphous silical gel (von Rad & Rösch, 1974). Cherts of intermediate maturity are characterized by chalcedonic quartz, which may replace the calcareous sediment, or form by inversion of less mature silica initially precipitated within the shells of microfossils. The most mature form of silica, microcrystalline quartz, is extremely fine grained, composed of minute interlocking crystals which show pin-point birefringence (Folk & Weaver, 1952). The Monterey Formation of California (Bramlette, 1946) illustrates the progressive changes from biogenically precipitated opal to cristobalite (locally called porcelanite) which eventually produces chalcedonic or microcrystalline quartz (Ernst & Calvert, 1969). Most pre-Upper Cretaceous cherts exist as chalcedonic or microcrystalline quartz, a good example being the Jurassic bedded cherts of north-west Italy (Thurston, 1972). In mode of preservation and general appearance, the Cyprus radiolarian cherts resemble Cretaceous brown and reddish brown radiolarian mudstones from the central Pacific which are associated with zeolitic clays (Pimm, Garrison & Boyce, 1971). They differ by having a much more extensive matrix with only scattered radiolarian shells.

As in many other radiolarian cherts, the origin of the matrix of isotropic silica is, in the Cyprus radiolarian rocks, a difficult problem to resolve. The Radiolaria are well preserved and show no signs of dissolution during diagenesis and although

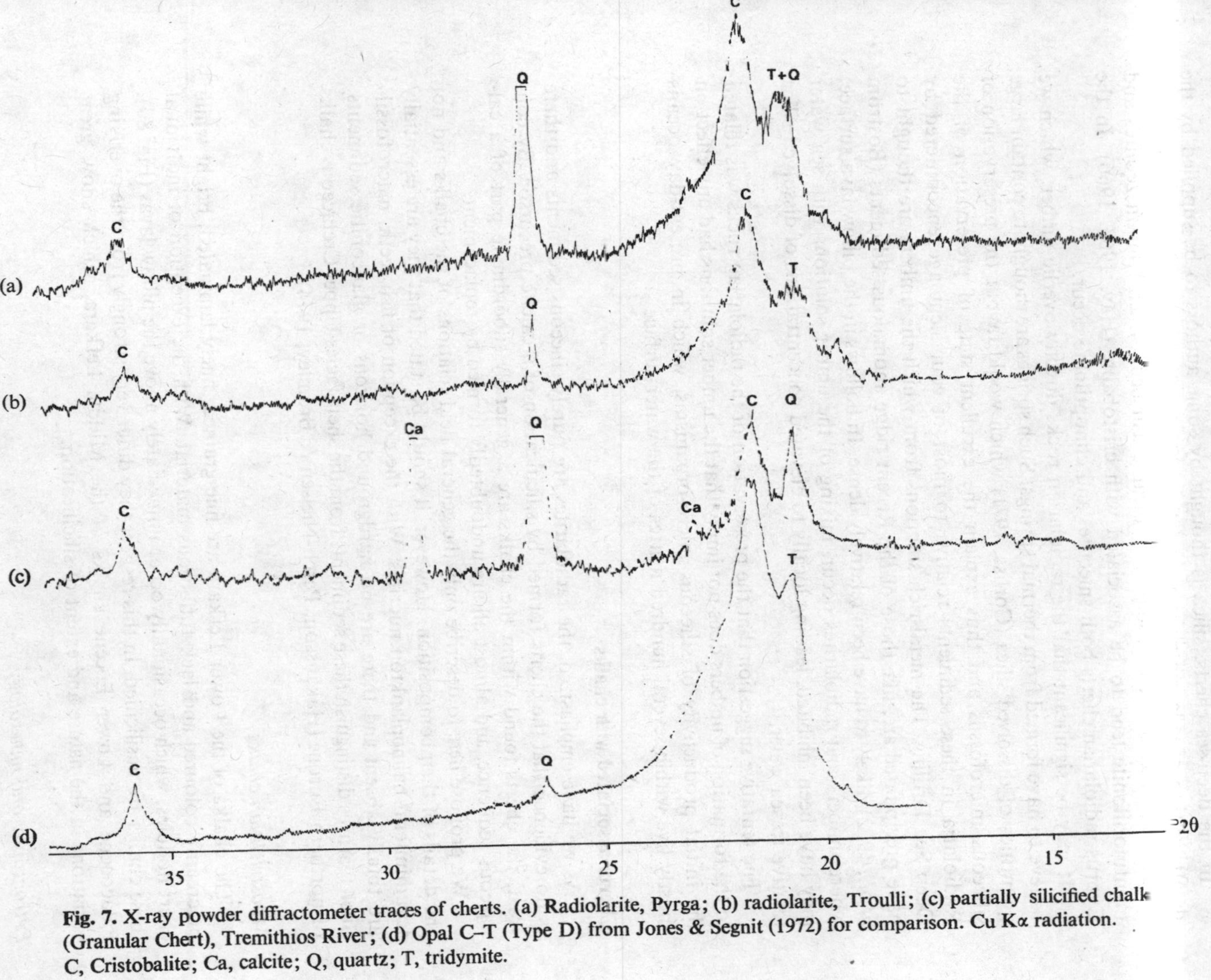

Fig. 7. X-ray powder diffractometer traces of cherts. (a) Radiolarite, Pyrga; (b) radiolarite, Troulli; (c) partially silicified chalk (Granular Chert), Tremithios River; (d) Opal C–T (Type D) from Jones & Segnit (1972) for comparison. Cu Kα radiation. C, Cristobalite; Ca, calcite; Q, quartz; T, tridymite.

diatoms are a theoretical source for silica, they have not been observed and there is no way of assessing their initial abundance.

In the deep-sea cherts, silica is thought by some authors to be supplied by the devitrification of volcanic glass, especially acid glass (Hay, 1966; Gibson & Towe, 1971; Hathaway & Sachs, 1965) and volcanic ash beds which contain zeolites and montmorillonite tend to be associated with cristobalite (Hay, 1963, 1966). In the Cyprus radiolarian cherts both smectites and clinoptilolite occur.

It may be significant that the radiolarian rocks directly overlie umber, which we believe to have formed from thermal springs. Such springs are thought to contain large quantities of dissolved silica (Corliss, 1971) which would favour the preservation of siliceous microfossils and thus explain the excellent state of preservation of the Radiolaria in these sediments relative to most of equivalent age encountered by Deep Sea Drilling. The metal-rich solutions from which the umbers are thought to have precipitated are, like those of the Recent ridge anomalous sediments (Boström *et al.*, 1972), likely to have been relatively dense. In hollows in the pillow lava surface, where umbers and radiolarites occur, mixing of the umber solutions with sea water may have been inhibited, leading locally to elevated concentrations of dissolved silica relative to sea water.

The tentative suggestion that the preservation of the radiolarian rocks was related to the formation of umbers does not imply that the umber solutions had any effect on the initial productivity of siliceous micro-organisms, which in present-day oceans mostly live within several hundred metres of the water surface.

Cherts associated with chalks

As we have emphasized, the radiolarites are purely siliceous sediments and there is no evidence that the cherts formed by silicification of a calcite precursor. By contrast, the cherts found within the chalks are a generally subordinate part of a calcareous sequence, and almost all are undoubtedly formed by replacement.

We propose here to describe only the general field relations of the chalks and not the details of their composition. However, it should be stated that they are essentially foraminiferal-nannoplankton micrites. With the exception of fish teeth, macrofossils are totally absent and there are no hardground horizons or glauconitic sediments. Such features distinguish these sediments from the shelf-facies Upper Cretaceous chalks of northern Europe (Håkansson, Perch-Nielsen & Bromley, 1974).

Maastrichtian chalks

The chalks of the Lower Lefkara are fine-grained, finely laminated, bright white or creamy coloured and lack tuffaceous material. With the exception of individual chert nodules, which occasionally occur immediately above the pillow lavas (Fig. 8d), these chalks are unsilicified. In this respect they differ very strongly from the overlying Palaeocene and Lower Eocene chalks of the Middle Lefkara, which show great variations in the nature and extent of silicification.

Petrography and mineralogy

Examination of the Maastrichtian chalks using thin sections and scanning electron microscopy shows that they consist of numerous well-preserved unrecrystallized pelagic Foraminifera and coccoliths in an unsilicified matrix of micrite. Near the base

Fig. 8. Cherts in the chalks. (a) Laterally continuous beds of partly silicified chalk (Granular Chert), Trakypedhoula (Paphos); (b) beds of chalcedonic quartz (Vitreous Chert) in partly silicified chalk, Nata (Paphos); (c) coarse, tuffaceous, cross-laminated base of calci-turbidite bed, Kalavassos; (d) individual chert nodule in unsilicified chalk, Margi.

of the succession many of the foraminiferal shells are recrystallized (Fig. 5d); most are empty, but some are filled with a drusy mosaic of sparry calcite. In thin section, numerous cavities are seen to preserve the characteristic shapes of Radiolaria (Fig. 9b). Examination of broken rock surfaces with a scanning electron microscope clearly illustrates these cavities (Fig. 10a) and shows that they are lined with microspherules, composed of minute radiating bladed crystals which look similar to those described from the deep sea (Fig. 10b) (Wise, Buie & Weaver, 1972; Weaver & Wise, 1972; Wise, 1972) where they were identified as cristobalite. The X-ray powder traces published by these authors show that strictly their 'cristobalite' is lussatite.

In addition to the lussatite microspherules, bladed crystals identical with those which constitute the microspherules occur independently in elongate solution cavities, where they form masses of bladed crystals orientated at right angles to the cavity walls (Fig. 10c, d). Apparently, the form which these crystals took depended on the morphology of the individual solution cavity. Many of the siliceous microspherules are discrete but some are coalescent, probably due to later diagenetic modification (Fig. 10e). Apart from lussatite, some of the solution cavities also contain euhedral crystals, which although not yet definitely identified, are possibly zeolites.

Palaeocene and Lower Eocene cherts

In the Palaeocene and Lower Eocene, massive or weakly bedded chalks predominate. At a distance the cherts in these look like unaltered chalks, but closer inspection

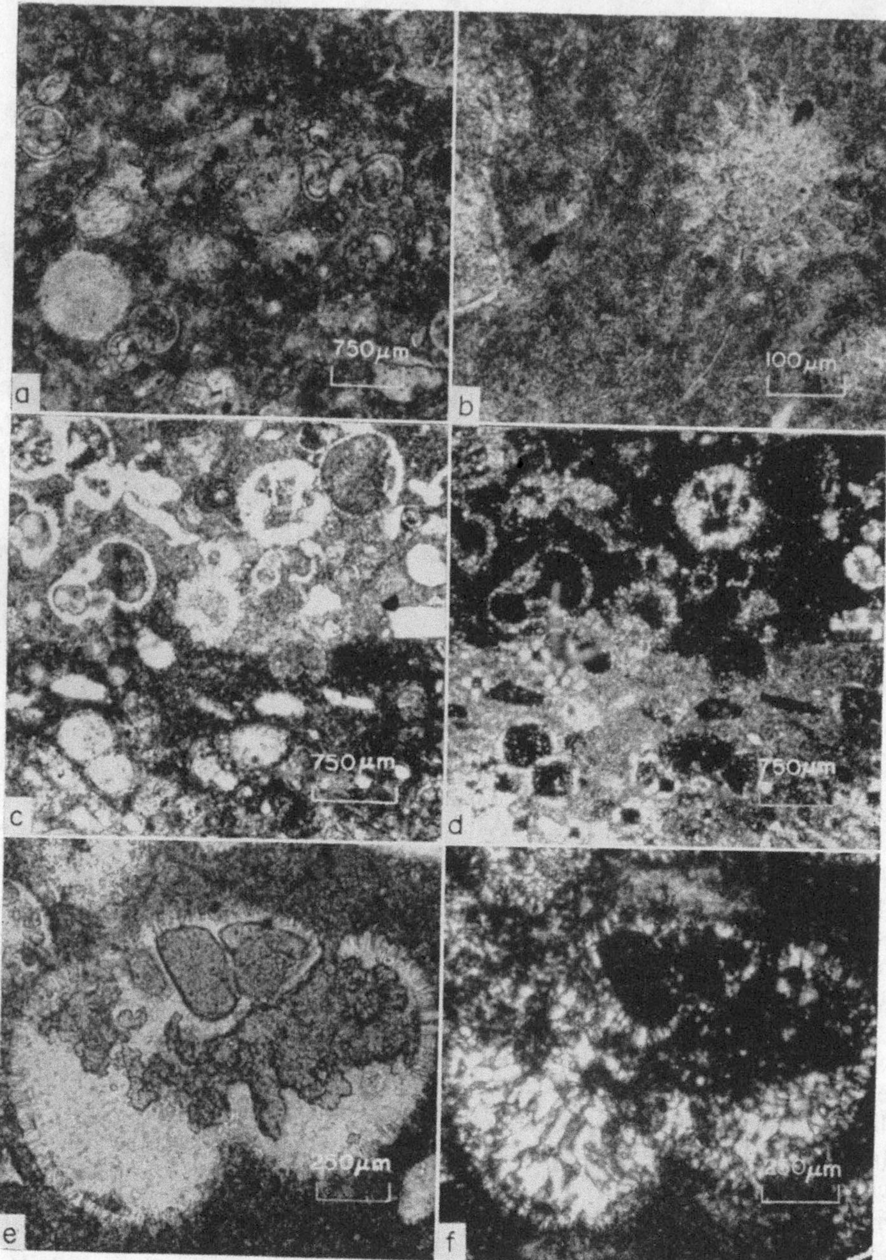

Fig. 9. Photomicrographs of cherts in the chalks. (a) Small mostly fragmentary foraminiferal shells; note the impression of a dissolved Spumellarian; (b) dissolution cavity of Spumellarian; (c, d) contact between Vitreous (chalcedonic quartz) and Granular (isotropic silica) Chert (note preferential replacement of foraminiferal shells by chalcedonic quartz where the matrix is isotropic silica); (e–f) planktonic foraminiferal shell filled with isotropic silica which has partly inverted to chalcedonic quartz. (c, e) Plane polarized light, (d, f) crossed nicols. (a, b) Maastrichtian chalk from Trouilli, (c–f) Eocene, Tremithios River.

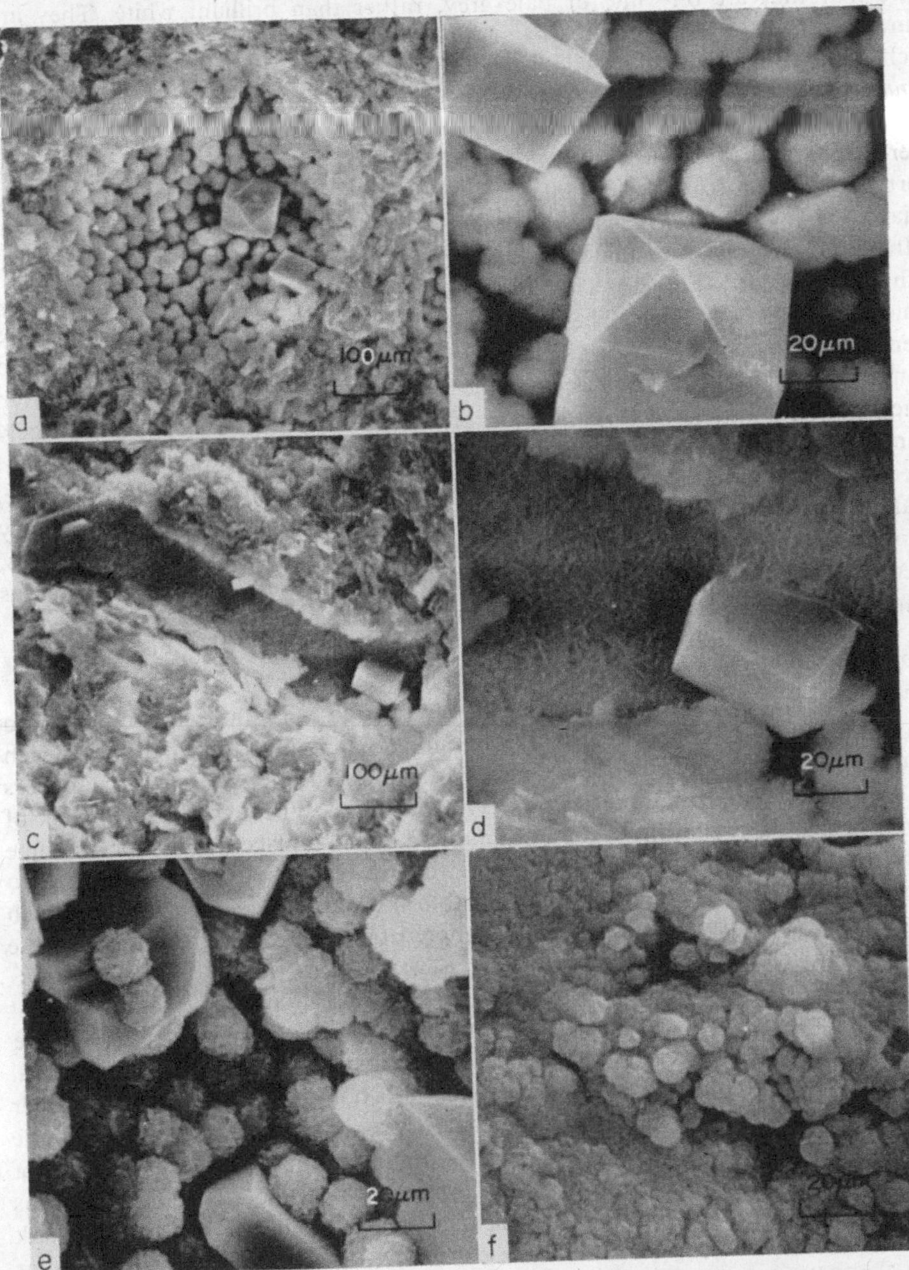

Fig. 10. Scanning electron micrographs of cherts. (a, b) Lussatite microspherules in dissolution cavity of radiolarian, the euhedral crystals are ? zeolites; (c, d) bladed crystals of lussatite lining walls of elongate solution cavity; (e) siliceous microspherules consisting of randomly orientated cigar-shaped crystals. (a–e) Maastrichtian chalk, Troulli, (f) Lower Eocene Granular Chert, Tremithios River.

shows that they are off-white or pale grey, rather than brilliant white. They are indurated, brittle and fizz only weakly in dilute hydrochloric acid. Rocks of this type, which are only partially silicified, will in the following account be referred to as *Granular Cherts*. This will distinguish them from totally silicified, conchoidally fracturing, waxy, flint-like cherts, a type which is here termed *Vitreous Chert* (Fig. 8b). Where these massive or weakly bedded chalks are partially silicified to Granular Chert they contain scattered irregular nodules of Vitreous Chert a few centimetres in diameter. These are well developed in the Limassol District.

In strong contrast with these weakly bedded chalks, there are others which are rhythmically bedded (Fig. 8a). The individual beds are laterally continuous, ranging in thickness from 10 to 60 cm. The best examples occur in the valley of the Tremethios River in the Larnaca District, where repetitively bedded chalks and cherts exceed 300 m in thickness. The base of each of the repetitive beds is relatively coarse, often tuffaceous, cross-laminated (Fig. 8c) and grades up into finer grained, more finely laminated chalk. Generally, the coarser basal beds have been totally silicified to form Vitreous Chert. Upwards, this chert has a sharp contact with Granular Chert (Fig. 9c, d), and the uppermost few centimetres of many of the repetitive beds are finely laminated and unsilicified. Thus a single hand specimen may show the three varieties, Vitreous Chert, Granular Chert and unsilicified chalk. These beds are interpreted as having originated as calci-turbidites prior to their silicification.

Petrography and mineralogy

The Granular Cherts are seen in thin section and with the scanning electron microscope to have been originally foraminiferal nannoplankton chalks in which Radiolaria are only rarely preserved. Under cross nicols, the biomicrite matrix appears dark and cloudy due to the presence of isotropic silica. Where there is a matrix of isotropic silica, the shell walls of the Foraminifera are all replaced by microcrystalline quartz. Radiolaria are rarely preserved, but where present they are recrystallized to aggregates of chalcedonic quartz crystals. Within the matrix of isotropic silica, which is often patchy and irregular, islands of unaltered biomicrite occur and in these the foraminiferal shells are not recrystallized.

The basal Vitreous Cherts of the turbiditic beds are seen in thin section to consist of densely packed Foraminifera, all with their walls replaced by microcrystalline quartz. Bundles of chalcedonic quartz fill the interiors of most of the foraminiferal shells but others are filled with isotropic silica and both types may occur together within a single shell (Fig. 9e, f). Where the Foraminifera are densely packed the matrix is minimal, but where present it exists as a mass of small interlocking grains of microcrystalline quartz. At the bottom of the graded units there is some tuffaceous material, rounded clay particles and fragments of fine-grained extrusive rocks. Chalk intraclasts are also numerous. Above these Vitreous Cherts, the zone of Granular Chert is similar to the Granular Cherts already described. Although much less numerous than in the Vitreous Cherts, the foraminiferal shells are still preferentially recrystallized to microcrystalline quartz. The strong preferential replacement of the walls of the foraminiferal shells, relative to the surrounding biomicrite matrix, is a very marked feature of these rocks (Fig. 9e, d).

X-ray powder diffraction shows that Vitreous Cherts are composed of low α-quartz with only traces of calcite. In contrast, Granular Cherts are composed of lussatite and calcite (Fig. 7). When examined with the scanning electron microscope, the surfaces of the Vitreous Cherts are seen to be smooth, structureless and non-porous. The granular, lussatitic cherts have a more irregular open and porous texture. Sometimes these cherts are made of siliceous microspherules, similar in diameter to the lussatite microspherules in the Maastrichtian chalks. However, rather than consisting of fine bladed crystals, they are formed of numerous elongate, cigar-shaped bodies which are randomly orientated. These rudely microspherular bodies coalesce and eventually become totally unrecognizable (Fig. 10f).

The uppermost portion of finely laminated chalks in the graded beds is completely unsilicified and the foraminiferal shells are often filled with sparry calcite. Radiolaria are never preserved, but radiolarian-shaped cavities have occasionally been seen.

Later Tertiary chalks

Above the chalks with abundant cherts there are Upper Eocene chalks, which are only incipiently silicified to Granular Chert, and Vitreous Cherts do not occur. The Oligocene and Lower Miocene chalks are almost totally unsilicified.

Origin of the cherts in the chalks

Maastrichtian chalks

Coccolith and foraminiferal biomicrites constitute the basal Maastrichtian chalks and these also contain cavities which preserve the characteristic shapes of Radiolaria. This shows that at least some siliceous organisms were initially deposited in the sediment and there is clear evidence that Radiolaria have been dissolved during diagenesis. The silica which was liberated was locally precipitated as lussatite microspherules within the dissolution cavities. Siliceous microspherules form only a tiny proportion of the Maastrichtian chalks which are otherwise totally unsilicified. Either siliceous micro-organisms were never very abundant in these sediments, or silica has migrated out of these beds during diagenesis, for which there is no evidence. Low productivity of siliceous micro-organisms in the surface waters at this time is a possibility, but we lack any information about this. A more simple and satisfactory explanation is that most of the tests of the siliceous micro-organisms were dissolved close to the sediment-water interface and were thus never incorporated into the sediment. The finely laminated, thin and discontinuous nature of these chalks implies lower accumulation rates than for the much thicker overlying chalks. On the present sea floor, off Portugal and Morocco, siliceous skeletons are destroyed by leaching in the upper 50 cm of the sediment within 4000 years (Schrader, quoted in Berger & von Rad, 1972). Diatoms are highly susceptible to solution, but different groups of Radiolaria also dissolve selectively (Berger, 1968; Berger & von Rad, 1972). In the Atlantic, off north-west Africa, sedimentation rates and hence rates of burial are known to be major factors which affect the preservation of siliceous microfossils. We suggest that, in the Maastrictian chalks, siliceous micro-organisms were dissolved close to the sediment-water interface and the silica released was dispersed into the water column rather than being retained within the sediment. Only a small number of Radiolaria, presumably the ones most resistant to solution, survived long enough to be incorporated in the sediment and then undergo the diagenetic processes of dissolution and subsequent precipitation of amorphous silica.

Palaeocene and Eocene cherts

Compared with the underlying Maastrichtian chalks, the Palaeocene and Lower Eocene chalks, largely deposited as calciturbidites, reach much greater thicknesses, which suggests more rapid sedimentation rates. The Granular Cherts, which are by far the most extensive volumetrically, generally lack preserved Radiolaria, and the large volumes of isotropic silica in the matrix must have been precipitated diagenetically. These cherts are free of any extensive source of volcanically derived silica and hence the dissolution of siliceous micro-organisms, possibly including diatoms, is the likely source of the matrix silica. The scanning electron microscope shows that the lussatite in Granular Cherts is either structureless or forms poorly defined coalescent microspherules composed of irregularly disposed cigar-shaped crystals. The lussatite microspherules made of bladed crystals, which occur in the Maastrichtian chalks, have not been seen in the Tertiary cherts.

Thin-section evidence showed that the graded beds which were preferentially silicified to Vitreous Cherts are largely organogenic, as they consist of pelagic foraminiferal shells and Radiolaria with some tuffaceous debris, sponge spicules and fish teeth. They resemble organogenic turbidites described from deep-sea sediments. In the Atlantic off the south-east coast of U.S.A., such turbidites represent a mixture of shelf-derived and pelagic sediment (Beall & Fisher, 1969). Other organogenic turbidites, in the West Pacific, consist entirely of reworked pelagic sediment derived from local topographical elevations (Heath & Moberly, 1971b). Beall & Fisher (1969) noticed that some of the organogenic turbidites had been silicified preferentially, as have those which occur in Cyprus. In the Cyprus turbidites shelf-derived fossils do not occur, showing that these sediments are entirely pelagic in origin and that they were derived from within the basin of deposition, probably from local topographical highs.

DISSOLUTION FACIES

A number of the observations already made on the nature of the umbers, radiolarian rocks and chalks carry with them implications about the conditions in which these sediments accumulated and we will briefly discuss these here.

The umbers and radiolarian rocks are invariably calcite-free. The calcareous microfossils must have been lost by dissolution before or very soon after these sediments were deposited. Also, the basal beds of the Maastrichtian chalks contain large numbers of small poorly preserved, often fragmentary planktonic Foraminifera (Fig. 9a). Coccoliths are very numerous, much more so than in the overlying chalks. Berger & von Rad (1972) have shown that in deep-sea sediments close to the carbonate compensation depth, fragmentary Foraminifera and large numbers of coccoliths are two factors which point towards extensive dissolution of calcareous micro-organisms. That the basal chalks have suffered extensive dissolution is supported by the poor preservation of the coccoliths as demonstrated by the scanning electron microscope (Fig. 5c); they exist as isolated shields which have been extensively corroded. The chalks higher in the succession show fewer signs of selective dissolution.

The evidence for differential dissolution of the radiolarian sediments and chalks at once suggests analogy with the oceanic carbonate compensation depth, the depth at which calcite is totally dissolved. In the world's present oceans this depth is not fixed, nor constant between different ocean basins, but is in a state of dynamic equilibrium between rates of sediment supply and dissolution (Heath & Culberson, 1970). The findings of the Deep Sea Drilling Project suggest that the carbonate compensation depth fluctuated considerably during the Tertiary (Hsü & Andrews, 1970; Hay, 1970; Berger & von Rad, 1972; Berger & Winterer, 1974) and that with the exception of a possible hiatus at the Mesozoic-Tertiary boundary when the compensation depth may have been very shallow (Worsley, 1971), levels were relatively low during much of the Mesozoic (Berger & von Rad, 1972). Recent sediments close to ocean ridges are accumulating above the compensation depths in both the Atlantic and Pacific Oceans, and are thus calcareous; siliceous sediments only accumulate further from the ridges in deeper water. Our earliest sediments, the umbers and radiolarian rocks which overlie the Upper Pillow Lavas, are siliceous, which makes them anomalous relative to Recent ridge sediments and their upward transition into chalks hard to explain. However, the palaeotectonic position of Cyprus at the time of accumulation of these sediments is still speculative and the depth of the carbonate compensation depth in this part of Tethys in the Campanian and Maastrichtian is by no means certain. As has already been discussed, hydrothermal solutions from which the umbers were formed may have selectively preserved some of the siliceous sediments and promoted the solution of calcareous ones. The upward transition from radiolarian sediments to chalks may relate to rapid uplift of the whole area immediately following the emplacement of tectonic mélanges (see later discussion), themselves invariably overlain by calcareous sediments. Until it becomes possible to disentangle these variables, the depths of water in which these sediments accumulated will remain uncertain.

THE KANNAVIOU AND THE ALLOCHTHONOUS FORMATIONS

In the foregoing account we have concentrated on demonstrating the pelagic nature of many of the Cyprus sediments, and if we are correct this applies to Campanian umbers and radiolarites as well as to Maastrichtian to Tertiary chalks. Yet we know that, in the south-west at least, Cyprus was far from normally pelagic during a part of the Maastrichtian, because a great thickness of Kannaviou volcanogenic sediment accumulated at that time, and major nappes, including continental material, were emplaced on top of it. How this can have happened in what was before and afterwards a realm of pelagic sedimentation is one of the greatest enigmas of Cyprus geology, and its solution must hold the key to understanding the geotectonic evolution of the Troodos massif.

It would be inappropriate to discuss these matters fully here, and we merely intend to draw attention to the main issues. Although we do not agree with her on all points, we should like to acknowledge stimulating discussions with Mlle H. Lapierre before and after publication of her thesis (Lapierre, 1972).

Kannaviou Formation

Probably the greatest contribution to the sedimentary geology of Cyprus in recent years has been the recognition that in the Paphos District there are two major groups of rocks which were previously confused under such terms as Trypa Group and Mamonia Complex. These are the Kannaviou Formation, essentially autochthonous and of Late Cretaceous (Maastrichtian) age, and the Mamonia proper, allochthonous and of Triassic-Jurassic age (Lapierre, 1968b, 1972; Kluyver, 1969; Turner, 1971).

The Kannaviou Formation reaches its greatest thickness in the type area in the Paphos District, where it consists of over 600 m of loosely consolidated, predominantly fine- and medium-grained sandstones interbedded with bentonitic clays. The upper parts of the succession are frequently disturbed and intercalated with the overlying allochthonous units, making the original thickness indeterminate, but in some localities at least, as in the Akamas Peninsula and at Kinousa, the base is gradational into umbers and radiolarites of the Perapedhi Formation, leaving no doubt that the formation is autochthonous. Locally, near Statos, the base is an angular unconformity on Troodos volcanics.

Individual beds of sandstone vary up to 30 m in thickness and are generally lenticular (Fig. 11). Many are massive and structureless, others exhibit parallel lamination and convolute lamination is common. To the north, in Akamas and at Kinousa, sandstones appear only a few tens of metres above the base of the succession, but farther south at Arkimandrita several hundred metres of clays precede the incoming of sandstones. The easterly equivalents of the Kannaviou in the Limassol District (Moni Clays) are thin and do not contain sandstones; they also lack calcareous microfossils. All indications are that the Kannaviou accumulated very rapidly indeed.

The volcanogenic nature of the Kannaviou is evident from clay mineral studies of the bentonitic clays (Desprairies, in Lapierre, 1972), but additional highly significant information is provided by the sandstones. Their detailed petrography will be described elsewhere, but briefly, in general, the lower ones contain abundant fragments

Fig. 11. Sedimentary structures in Kannaviou Sandstone. (a) Parallel lamination, near Kannaviou; (b) lenticular basal clay-rich sandstone, near Pelathousa (Paphos); (c) rare cross-lamination near base of the succession, Akamas Peninsula; (d) convolute lamination, near Philousa (Paphos).

of undevitrified, vesicular acid glass ('pumice') (Fig. 5f), as well as volcanic quartz, alkali feldspar (including sanidine) and plagioclase. Foraminifera and Radiolaria also occur in the rock, which is cemented by sparry calcite. Other sandstones, generally higher in the succession, contain polycrystalline, mica-bearing quartz, rock fragments of phyllite and chert and generally lack pumice.

Petrographically these rocks are predominantly lithic arkoses or feldspathic litharenites in the sense of Folk (1968). They are not turbidites; their sedimentation may be compared with parallel-laminated volcanogenic sandstones found by the Deep Sea Drilling Project north-west of Hawaii. These sands are non-turbiditic, but apparently originated in slumping and bottom transport by currents (Moberly & Heath, 1971). The structureless, highly immature 'pumice' beds are easily interpretable as ash-fall deposits.

Two important conclusions can be drawn. First, the Kannaviou records an episode of intense acid volcanism soon after the cessation of the Troodos volcanic activity (Campanian *v.* Maastrichtian), and these acid volcanics could not be derived from the potassium-poor igneous rocks of Troodos. Although provenance must be from outside the present limits of Cyprus, the freshness of the detritus implies derivation from nearby, newly erupted volcanics. Measurements of poorly developed cross-lamination in some of the coarser volcanogenic sandstones indicates that the source lay to the (present) south-west, west or north-west of the island. Secondly, this volcanism terminated as abruptly as it began because the upper part of the Kannaviou contains detritus most probably derived from the advancing Mamonia Nappes (see below), and we have no evidence of acid volcanism after their emplacement.

This acid volcanism may suggest initiation of an island arc system, but it would have been short lived as well as very localized. The succession in Hatay (Lapierre & Parrot, 1972), otherwise similar to that of Paphos, contains no trace of the Kannaviou.

Allochthonous formations

In the Paphos District, Lapierre (1968b, 1972) recognizes two main allochthonous units overlying the Kannaviou, the Mamonia Nappes and Moni Mélange (Fig. 3).

Mamonia Nappes

By piecing together fragmentary sections, two nappes could be reconstructed (Lapierre, 1972). Both contain Triassic-Jurassic sedimentary rocks of a variety of facies, including sandstone with fossil plant remains and radiolarian cherts; and the lower nappe, only, contains volcanics of predominantly alkaline affinities. Rocks of undoubted continental origin are represented by a mass of multi-folded amphibolitic gneiss and quartzite which occurs near Ayia Varvara (Paphos District). Numerous substantial thrust masses of serpentinized harzbergite occur within, and generally near the base of, the Mamonia Nappes. In the Akamas Peninsula, serpentinite is thrust over a large inlier of Troodos igneous rocks. Near Marathouda faulted and sheared masses of Troodos pillow lavas, diabase and gabbro occur enclosed within the serpentinite. Genetically these serpentinites appear to belong to Troodos rather than to the Mamonia Nappes.

'Moni Mélange' of Lapierre

Resting on the Mamonia Nappes, which are themselves often strongly fragmented, is a mélange consisting of a chaotic mixture of all Mamonia lithologies, as blocks mainly less than a few metres in diameter, in a matrix of bright red 'rock flour'. This matrix consists mainly of rock fragments, but also contains abundant kaolinite as well as other clay minerals (Desprairies in Lapierre, 1972). Thrust masses of serpentinite do not occur in this mélange. Near the margins of the Troodos Massif, the Mamonia Nappes are absent and the Mélange rests directly on the Kannaviou Formation, for instance near Kannaviou itself.

Moni Mélange of Pantazis

In the Limassol District, near Moni, there is a smaller mass of mélange described by Pantazis (1967) (Figs 1 and 2). It also consists of fragments of Mamonia lithologies, and thrust masses of serpentinite, but in a matrix of grey bentonitic clay (Moni Clay) identical to the clays of the Kannaviou Formation. The 'Petra tou Romiou Formation', 'Pareklissha Sandstone' and 'Upper Tuffaceous Member', all regarded by Pantazis as essentially *in situ*, are interpreted by us as blocks of varying size in the mélange. The Pareklissha Sandstone is lithologically identical to the Akamas Sandstone of the Mamonia Nappes. The term Moni Mélange belongs by priority to this unit, and for the sake of clarity Lapierre's 'Moni Mélange' should receive a new name.

Emplacement of the allochthonous units

Interpretations of Mamonia tectonics are severely complicated by discontinuous exposures, later erosion and particularly by the prevalence of recent landslipping facilitated by the substrate of Kannaviou bentonitic clay. Even allowing for this, there seems little doubt that the Mamonia nappes were severely disrupted at the time of their emplacement, and they may be regarded as transitional between coherent thrust sheets and tectonic mélange (cf. Hsü, 1968). The Paphos (Lapierre) Moni Mélange could be an ideal 'tectonic mélange' lithologically, except that it occurs above, not below, the nappes. We do not yet fully understand this apparent anomaly, but we cannot support the suggestion of Lapierre (1972) that the mélange formed by rapid erosional destruction of the underlying nappes soon after their emplacement.

The Pantazis Moni Mélange seems readily interpreted as a deposit in which dislodged blocks from the Mamonia Nappes were rapidly transported by gravity sliding down a slope, ending up embedded in the soft bentonitic clays of the Kannaviou Formation. However, interpretation is also complicated by later folding and landslipping.

The direction of emplacement of the Mamonia thrust sheets is even more controversial. On the basis of facing directions of the thrusts and orientation of minor folds, Lapierre (1972) suggested that transport was from the north-east. This interpretation is indeed supported by the predominantly north- to north-east-facing orientation of most of the serpentinite thrusts in both the Paphos and Limassol Districts. Against this, no Mamonia rocks have been found along the northern and eastern margins of the massif, and the critical parts of the Upper Cretaceous sedimentary succession—thin Kannaviou Formation passing up into Maastrichtian chalks—appears in most places perfectly gradational and undisturbed, as seen for instance at Vavla and Troulli. However, at a few localities around the northern margin of Troodos, as above Kokkinoyia Mine, umbers are deformed into small, plicate folds; their axial planes all dip in a northerly direction. Significantly the overlying Maastrichtian sediments are undeformed.

A resolution of these apparent contradictions lies in the suggestion that discontinuous serpentinite masses of Troodos ocean-floor origin were thrust southwards over the Troodos Massif and its sedimentary cover; they then collided with a Triassic to Jurassic continental margin represented by the Mamonia Nappes.

THE LATE CRETACEOUS AND EARLY TERTIARY GEOTECTONICS OF CYPRUS

The comparisons we have made between the Cyprus sediments and the sediments of the present oceans support the view that the Troodos Massif represents a fragment of Mesozoic ocean floor (Gass & Masson-Smith, 1963; Gass & Smewing, 1973). They cannot, by themselves, settle the disputed question as to whether the ocean-ridge model of Moores & Vine (1971) is correct, but they are consistent with it. Any geotectonic theory must take account of the fact that pelagic sediments, chalks and associated cherts, accumulated in Cyprus after, as well as before, the emplacement of the Mamonia Nappes in the Late Cretaceous and rest indifferently on the allochthon and Troodos pillow lavas alike.

A major difference between earlier views and present ones is that Troodos, then thought to be pre-Triassic, was regarded as part of the floor of a wide ocean (the traditional Tethys: see Sylvester-Bradley, 1968) situated between Africa and Eurasia (Gass & Masson-Smith, 1963). It now appears that Troodos, if it is indeed an ocean ridge, is part of a short-lived and rather small ocean basin, formed and perhaps also largely destroyed during the Cretaceous. Evidence for the crust of the supposed older ocean, pre- or early Mesozoic, is not preserved in Cyprus.

Structurally, Cyprus is a continuation of Turkey, and a solution to these problems will clearly lie in synthesis of a wider region than that of Cyprus alone. In the nappes of Antalya in S.W. Turkey, ophiolites, including fragments of plutonic and dyke complexes, have been discovered. These show some petrological affinities with Troodos, but are much older, probably Triassic (Juteau, 1970). In Anatolia, Brinkmann (1972) recognized that several radiolarite-ophiolite belts occur, separated by older sialic crust. This suggests that Turkey was at various times in the Mesozoic fragmented into continental slivers separated by small impermanent ocean basins; the Troodos could be the most southerly of these, and one of the latest.

Recent palaeogeographical reconstructions suggest that during the late Mesozoic Tethys was narrow in the area of the present East Mediterranean, but that eastwards the ocean was much wider (Smith, 1971; Hsü, 1972). For example, in Oman there is evidence that only half an ocean, which was formed between Middle Permian and Middle Cretaceous, exceeded 1200 km in width (Glennie *et al.*, 1973). In contrast, the sediments of the Mamonia nappes, deposited between Late Triassic and Late Jurassic, are composed of predominantly continental-margin facies, with some indisputably continentally derived metamorphic rocks. Tholeiitic sea-floor lavas have not been found.

It is probable, as briefly discussed above, that the Mamonia nappes were thrust on to Upper Cretaceous ocean floor of the Troodos Massif as the result of collision with a continental margin, represented by the Mamonia sediments. This emplacement was immediately preceded by the Kannaviou acid volcanism, suggesting subduction and embryonic island arc development. Why subduction did not continue, leading to complete destruction of Troodos, is not clear. It must be significant, though, that the catastrophic emplacement of the nappes took place during the Maastrichtian, a time of widespread tectonic activity in the ophiolite belt which stretches eastwards from Cyprus through Hattay in Syria (Lapierre & Parrot, 1972), Kurdestan and Neyriz, in Iran (Ricou, 1971) to Oman (Glennie *et al.*, 1973). The driving force for these widespread synchronous tectonic movements was probably the reversal of movement of Africa relative to Eurasia, consequent on a phase of spreading of the Mid-Atlantic Ridge (Smith, 1971). The result was that, relative to Eurasia, Africa moved westwards giving an anti-clockwise sense of rotation (Dewey *et al.*, 1973). The details of the correlation between this major regional event and the emplacement of the Mamonia nappes will not be understood until other, more local, problems are fully resolved: the direction of transport of the nappes, and the post-emplacement rotation of Cyprus relative to Turkey. Turkey, as well as Cyprus, has evidently suffered some rotation (van der Voo, 1968).

After the nappes were emplaced, Maastrichtian pelagic chalks were deposited over the area, apparently as if nothing had happened. Subsequently, during the Early Tertiary, differential movements in the underlying massif were reflected in the deposition of organogenic turbidites. By the Early Miocene rapid uplift was occurring as is shown by chaotically slumped beds of Lower Miocene chalk which occur in the Larnaca District (Bagnall, 1960). Thin black shales were deposited in restricted basins until continued uplift resulted in emergence and the establishment of reefs separated by local evaporite basins. The Cyprus gypsum may be a part of the circum-Mediterranean desiccation which is thought to have occurred during the Late Miocene (Hsü, Ryan & Cita, 1973) but the precise age and stratigraphical relations are still controversial. This rapid uplift may be attributed to a switch from lateral to convergent movement between Africa and Europe which is thought to have occurred in the east Mediterranean at this time (Dewey *et al.*, 1973). This convergent motion may have been a factor in the emplacement of the Kyrenia Range in the north of Cyprus.

ACKNOWLEDGMENTS

Of the many who have helped with this project, we would like to thank especially Professor I. G. Gass, Mlle H. Lapierre, Mr M. Mantis, Dr A. G. Smith and Dr F. J. Vine. Robertson also profited from discussions with, in particular, Dr R. E. Garrison, Mr J. A. Pearce and Mr I. Price. He also acknowledges a Natural Environment Research Council Studentship. Hudson's fieldwork was aided by the research board of Leicester University.

REFERENCES

Bagnall, P.S. (1960) The geology and mineral resources of the Pano Lefkara-Larnaca Area. *Mem. geol. Surv. Cyprus*, **4**, 116 pp.

Beall, A.O. Jr & Fischer, A.G. (1969) Sedimentology. In: *Initial Reports of the Deep Sea Drilling Project*, Vol. I (M. Ewing *et al*), pp. 521–594. U.S. Government Printing Office, Washington.

Berger, W.H. (1968) Radiolarian skeletons: solution at depth. *Science*, **159**, 1237–1238.

Berger, W.H. & von Rad, U. (1972) Cretaceous and Cenozoic Sediments from the Atlantic Ocean, Leg 14, Deep Sea Drilling Project. In: *Initial Report of the Deep Sea Drilling Project*, Vol. XIV (D. E. Hayes, A. C. Pimm *et al.*), pp. 787–954. U.S. Government Printing Office, Washington.

Berger, W.H. & Winterer, E.L. (1974) Plate stratigraphy and the fluctuating carbonate line. In: *Pelagic Sediments: on Land and under the Sea* (Ed. by K. J. Hsü and H. C. Jenkyns). *Spec. Publs int. Ass. Sediment.* **1,** 11–48.

Bischoff, J.L., (1969) Red Sea geothermal brine deposits: their mineralogy, chemistry and genesis. In: *Hot Brines and Recent Heavy Metal Deposits in the Red Sea* (Ed. by E. T. Degens and D. A. Ross), pp. 368–401. Springer-Verlag, Berlin.

Bonatti, E., Honnorez, J., Joensuu, O. & Rydell, H. (1973) Submarine iron deposits from the Mediterranean Sea. In: *The Mediterranean Sea, A Natural Sedimentation Laboratory* (Ed. by D. J. Stanley), pp. 701–710. Dowden, Hutchinson and Ross, Stroudsburg, Pennsylvania.

Boström, K. & Peterson, M.N.A. (1969) The origin of aluminium-poor ferromanganoan sediments in areas of high heat flow in the East Pacific Rise. *Mar. Geol.* **7,** 427–447.

Boström, K., Peterson, M.N.A., Joensuu, O. & Fisher, D.E. (1969) Aluminium-poor ferromanganoan sediments on active oceanic ridges. *J. geophys. Res.* **74,** 3261–3270.

Boström, K., Joensuu, S., Valdes, S. & Riera, M. (1972) Geochemical history of S. Atlantic Ocean sediments since Late Cretaceous. *Mar. Geol.* **12,** 85–123.

Bramlette, M.N. (1946) The Monterey Formation of California and the origin of its siliceous rocks. *Prof. Pap. U.S. geol. Surv.* **212,** 57 pp.

Brinkmann, R. (1972) Mesozoic troughs and crustal structure in Anatolia. *Bull. geol. Soc. Am.* **83,** 819–826.

Calvert, S.E. (1971) Composition and origin of North Atlantic deep sea cherts. *Contr. Miner. Petrol.* **32,** 273–288.

Constantinou, G. & Govett, J.G.S. (1972) Genesis of sulphide deposits, ochre and umber of Cyprus. *Trans. Am. Inst. Min. Metall. Engrs,* **81,** B34–B46. Discussion in **82,** Nos 798, 801, 1973.

Corliss, J.B. (1971) The origin of metal-bearing hydrothermal solutions. *J. geophys. Res.* **76,** 8128–8138.

Corliss, J.B., Graf, J.L., Skinner, B.J. & Hutchinson, R.W. (1972) Rare earth data for iron and manganese sediments associated with the sulfide ore bodies of the Troodos Massif, Cyprus. *Prog. Abstr. geol. Soc. Am.* **4** (7).

Cronan, D.S., van Andel, T.H., Heath, G.R., Dinkleman, M.G., Bennett, R.H., Bukry, D., Charleston, S., Kaneps, A., Rodolfo, K.S. & Yeats, R.S. (1972) Iron-rich basal sediments from the eastern Equatorial Pacific; Leg 16, Deep Sea Drilling Project. *Science,* **175,** 61–63.

Dasch, E.J., Dymond, J.R. & Heath, G.R. (1971) Isotopic analysis of metalliferous sediments from the East Pacific Rise. *Earth Planet. Sci. Letts,* **13,** 175–180.

Dewey, J.F., Pitman, W.C. III, Ryan, W.B.F. & Bonnin, J. (1973) Plate tectonics and evolution of the Alpine System. *Bull. geol. Soc. Am.* **84,** 3137–3180.

Dymond, J., Corliss, J.B., Heath, G.R., Field, C.W., Dasch, E.J., Veeh, H.H. (1973) Origin of metalliferous sediments from the Pacific Ocean. *Bull. geol. Soc. Am.* **84,** 3355–3372.

Elderfield, H. (1972) Compositional variations in the manganese oxide component of marine sediments. *Nature Phys. Sci.* **237,** 110–112.

Elderfield, H., Gass, I.G., Hammond, A. & Bear, L.M. (1972) The origin of ferromanganese sediments associated with the Troodos Massif of Cyprus. *Sedimentology,* **19,** 1–19.

Ernst, W.G. & Calvert, S.E. (1969) An experimental study of the recrystallisation of porcelanite and its bearing on the origin of some bedded cherts. *Am. J. Sci.* (Schairer Vol.), **267-A,** 114–133.

Folk, R.L. (1968) *Petrology of Sedimentary Rocks,* pp. 170. Hemphills, Austin, Texas.

Folk, R.L. & Weaver, C.E. (1952) A study of the texture and composition of chert. *Am. J. Sci.* **150,** 498–510.

Flörke, O.W. (1955) Zur Frage des 'Hoch'—Cristobalit in Opalen, Bentoniten und Glasern. *Neues Jb. Miner. Mh.* **1955,** 217–233.

Gass, I.G. (1960) The geology and mineral resources of the Dhali Area. *Mem. geol. Surv. Cyprus,* **4,** 116 pp.

Gass, I.G. (1968) Is the Troodos Massif of Cyprus a fragment of Mesozoic Ocean Floor? *Nature, Lond.* **220,** 39–42.

Gass, I.G. & Cockbain, A.E. (1961) Notes on the occurrence of Gypsum in Cyprus. *Overseas Geol. Miner. Resour.* **VIII,** 279–287.

GASS, I.G. & MASSON-SMITH, E.M. (1963) The geology and gravity anomalies of the Troodos Massif, Cyprus. *Phil. Trans. R. Soc. Lond.* **A255**, 417–467.

GASS, I.G. & SMEWING, J.D. (1973) Intrusion, extrusion and metamorphism at constructive margins: evidence from the Troodos Massif, Cyprus. *Nature, Lond.* **242**, 26–29.

GIBSON, T.H. & TOWE, K.M. (1971) Eocene volcanism and the origin of horizon A. *Science,* **172,** 152–154.

GLENNIE, K.W., BOEF, M.G.A., HUGHES CLARKE, M.W., MOODY-STEWART, M., PILLAR, W.F.H. & REINHARDT, B.M. (1973) Late Cretaceous nappes in Oman mountains and their geologic evolution. *Bull. Am. Ass. Petrol. Geol.* **57,** 5–27.

HÅKANNSON, E., PERCH-NIELSEN, K., BROMLEY, R.G. (1974) Maastrichtian chalk of north-west Europe—a pelagic shelf sediment. In: *Pelagic Sediments: on Land and under the Sea* (Ed. by K. J. Hsü and H. C. Jenkyns). *Spec. Publs int. Ass. Sediment.* **1,** 211–234.

HATHAWAY, J.C. & SACHS, P.L. (1965) Sepiolite and clinoptilolite from the Mid-Atlantic Ridge. *Am. Miner.* **50,** 852–867.

HAY, R.L. (1963) Stratigraphy and zeolitic diagenesis of the John Day Formation of Oregon. *Univ. Calif. Publs. geol. Sci.* **42,** 199–262.

HAY, R.L. (1966) Zeolites and zeolitic reactions in sedimentary rocks. *Spec. Pap. geol. Soc. Am.* **85,** 130 pp.

HAY, W.W. (1970) Calcium carbonate compensation. In: *Initial Reports of the Deep Sea Drilling Project,* Vol. IV (R. G. Bader *et al.*), pp. 669–672. U.S. Government Printing Office, Washington.

HEATH, G.R. & CULBERSON, C. (1970) Calcite: degree of saturation, rate of dissolution, and the compensation depth in the deep oceans. *Bull. geol. Soc. Am.* **81,** 3157–3160.

HEATH, G.R. & MOBERLY, R. JR (1971a) Cherts from the Western Pacific, Leg 7, Deep Sea Drilling Project. In: *Initial Reports of the Deep Sea Drilling Project,* Vol. VII (E. L. Winterer *et al.*), pp. 991–1007. U.S. Government Printing Office, Washington.

HEATH, G.R. & MOBERLY, R. JR (1971b) Deep-sea turbidites from the Western Pacific, Leg 7, Deep Sea Drilling Project. In: *Initial Reports of the Deep Sea Drilling Project,* Vol. VII (E.L. Winterer *et al.*), pp. 1009–1010. U.S. Government Printing Office, Washington.

HENSON, F.R.S., BROWNE, R.V. & MCGINTY, J. (1949) A synopsis of the stratigraphy and geological history of Cyprus. *Q. Jl geol. Soc. Lond.* **105,** 1–41.

HSÜ, K.J. (1968) Principles of mélanges and their bearing on the Franciscan-Knoxville paradox. *Bull. geol. Soc. Am.* **79,** 1063–1074.

HSÜ, K.J. (1972) The concept of the geosyncline, yesterday and today. *Trans. Leicester lit. Phil. Soc.* **66,** 26–48.

HSÜ, K.J. & ANDREWS, J.E. (1970) History of South Atlantic basin, in, Summary and Conclusions. In: *Initial Reports of the Deep Sea Drilling Project,* Vol. III (A. E. Maxwell *et al.*), pp. 464–467. U.S. Government Printing Office, Washington.

HSÜ, K.J., RYAN, W.B.F. & CITA, M.B. (1973) Late Miocene desiccation of the Mediterranean. *Nature, Lond.* **242,** 240–244.

JONES, J.B. & SEGNIT, E.R. (1972) The nature of Opal L. nomenclature and constituent phases. *J. geol. Soc. Aust.* **18,** 57–68.

JUTEAU, TH. (1970) Pétrogenèse des ophiolites des nappes d'Antalya (Taurus Lycien oriental, Turquie), leur liaison avec une phase d'expansion océanique active au Trias supérieur. *Sciences Terre,* **15,** 265–288.

KOLODNY, Y (1969) Petrology of siliceous rocks in the Mishash Formation (Negev, Israel) *J. sedim. Petrol.* **39,** 165–175.

KLUYVER, H.M. (1969) Report on a regional geological mapping in Paphos District. *Bull geol. Surv. Cyprus,* **4,** 21–36.

LAPIERRE, H. (1968a) Découverte d'une série volcano-sédimentaire probablement d'age Crétacé supérieure au S.W. de l'île de Chypre. *C.r. hebd. séanc. Acad. Sci., Paris,* **D,266,** 1817–1820.

LAPIERRE, H. (1968b) Nouvelles observations sur la série sédimentaire de Mamonia (Chypre), *C.r. hebd. séanc. Acad. Sci., Paris,* **D,267,** 32–35.

LAPIERRE, H. (1970) Découverte de plusieurs phases orogénetiques Mésozoiques au Sud de Chypre, *C.r. hebd. séanc. Acad. Sci., Paris,* **D,270,** 1876–1878.

LAPIERRE, H. (1972) *Les Formations sédimentaires et éruptives des nappes de Mamonia et leurs relations avec le massif de Troodos* (*Chypre*), pp. 420. Doctoral Thesis, University of Nancy.

LAPIERRE, H. & PARROT, J.F. (1972) Identité géologique des régions de Paphos (Chypre) et du Baër-Bassit (Syrie). *C.r. hebd. séanc. Acad. Sci., Paris,* **D,274,** 1999–2002.

MANTIS, M. (1970) Upper Cretaceous-Tertiary Foraminiferal Zones in Cyprus. *Scientific Research Centre of Cyprus, Epithris,* **3,** 227–241.

MAXWELL, A.E., VON HERZEN, R.P., ANDREWS, J.E. *et al.* (1970) Site 15. In: *Initial Reports of the Deep Sea Drilling Project,* Vol. III (A. E. Maxwell *et al.*), pp. 113–132. U.S. Government Printing Office, Washington.

MIYASHIRO, A. (1973) The Troodos ophiolitic complex was probably formed in an island arc. *Earth Planet. Sci. Letts,* **19,** 218–224.

MOBERLY, R. JR & HEATH, G.R. (1971) Volcanic rocks from the West and Central Pacific: Leg 7, Deep Sea Drilling Project. In: *Initial Reports of the Deep Sea Drilling Project,* Vol. VII (E. L. Winterer *et al.*), pp. 1011–1025. U.S. Government Printing Office, Washington.

MOORES, E.M. & VINE, F.J. (1971) The Troodos Massif, Cyprus and other ophiolites as oceanic crust: evaluation and implications. *Phil. Trans. R. Soc.* **A268,** 433–466.

PANTAZIS, TH. M. (1967) The geology and mineral resources of the Phamakas-Kalavassos area. *Mem. geol. Surv. Cyprus,* **8,** 120 pp.

PESSAGNO, E.A. JR (1963) Upper Cretaceous Radiolaria from Puerto Rico. *Micropaleontology,* **9,** 197–214.

PIMM, A.C., GARRISON, R.E. & BOYCE, R.E. (1971) Sedimentology synthesis: lithology, chemistry and physical properties of sediments in the north-west Pacific Ocean. In: *Initial Reports of the Deep Sea Drilling Project,* Vol. VI (A. G. Fischer *et al.*), pp. 1131–1253. U.S. Government Printing Office, Washington.

RICOU, L.E. (1971) Le croissant ophiolitique peri-arabe, une ceinture de nappes mises en place au Crétacé supérieur. *Revue Géogr. Phys. Géol. dyn.* **13,** 327–349.

ROBERTSON, A.H.F. & HUDSON, J.D. (1973) Cyprus umbers: chemical precipitates on a Tethyan ocean ridge. *Earth Planet. Sci. Letts,* **18,** 93–101.

SEARLE, D.L. (1968) *Summary of the geology of the Cyprus cupriferous sulphide deposits and notes on their mineralogy and origin,* pp. 32. Unpublished report of the U.N. Bureau of Technical Assistance Operations in Cyprus.

SMITH, A.G. (1971) Alpine deformation and oceanic area of Tethys, Mediterranean and Atlantic. *Bull. geol. Soc. Am.* **82,** 2039–2070.

SYLVESTER-BRADLEY, P.C. (1968) Tethys: the lost ocean. *Science J.* **4,** 47–52.

TALIAFERRO, N.L. (1934) Contraction phenomena in cherts. *Bull. geol. Soc. Am.* **45,** 189–232.

THURSTON, D.R. (1972) Studies on bedded cherts. *Contr. Miner. Petrol.* **36,** 329–334.

TUREKIAN, K.K. & WEDEPHOL, K.H. (1961) Distribution of the elements in some major units of the Earth's crust. *Bull. geol. Soc. Am.* **72,** 175–192.

TURNER, W.M. (1971) A progress report on the geology of western Cyprus including the Akamas Peninsula; the general stratigraphy of western Cyprus including the Akamas Peninsula. *Publ. University of New Mexico, Albuquerque, New Mexico,* **1,** 1–141.

VAN DER VOO, R. (1968) Jurassic, Cretaceous and Eocene pole positions from northeastern Turkey. *Tectonophysics,* **6,** 251–269.

VON DER BORCH, C.C., NESTEROFF, W.D. & GALEHOUSE, J.S. (1971) Iron-rich sediments cored during Leg 8 of the Deep Sea Drilling Project. In: *Initial Reports of the Deep Sea Drilling Project,* Vol. VIII (J. I. Tracey, Jr *et al.*), pp. 829–835. U.S. Government Printing Office, Washington.

VON DER BORCH, C.C. & REX, R.W. (1970) Amorphous iron oxide precipitates in sediments cored during Leg 5, Deep Sea Drilling Project. In: *Initial Reports of the Deep Sea Drilling Project,* Vol. V (D. A. McManus *et al.*), pp. 541–544. U.S. Government Printing Office, Washington.

VON RAD, U. & RÖSCH, H. (1974) Petrography and diagenesis of deep-sea cherts from the centra Atlantic. In: *Pelagic Sediments: on Land and under the Sea* (Ed. by K. J. Hsü & H. C. Jenkyns). *Spec. Publs int. Ass. Sediment.* **1,** 327–347.

WEAVER, F.M. & WISE, S.W., JR (1972) Ultramorphology of deep sea chert. *Nature Phys. Sci.* **237,** 56–57.

WILSON, R.A.M. (1959) The geology and mineral resources of the Xeros-Troodos Area. *Mem. geol. Surv. Cyprus,* **1,** 135 pp.

WISE, S.W. JR, BUIE, B.F. & WEAVER, F.M. (1972) Chemically precipitated sedimentary cristobalite and the origin of chert. *Eclog. geol. Helv.* **65,** 157–163.

Wise, S.W., Jr (1972) Inferred diagenetic history of a weakly silicified deep sea chalk. *Trans. Gulf-Cst Ass. geol. Socs*, **22**, 177–204.

Worsley, T.R. (1971) Terminal Cretaceous events. *Nature, Lond.* **230**, 318–320.

Zelenov, K.K. (1964) Iron and manganese in exhalations of the submarine Banu Wahu volcano (Indonesia). *Dokl. (Proc.) Acad. Sci. U.S.S.R., Earth Sciences Section*, **155**, 1317-1320 (pp. 94–96 in translation published by the American Geological Institute).

Spec. Publs int. Ass. Sediment. (1974) **1,** 437–447

Encrusting organisms in deep-sea manganese nodules*

J. WENDT

Geologisch-Paläontologisches Institut der Universität Tübingen, Sigwartstrasse 10, D-74 Tübingen, West Germany

ABSTRACT

Sessile arenaceous Foraminifera are intimately associated with Recent manganese nodules and crusts. This association is independent of water depth and submarine topography. Where present these Foraminifera may be used to estimate rates of accretionary growth. Serpulids, corals, bryozoans and sponges have only been observed as surficial encrustations on samples from seamounts.

INTRODUCTION

Since manganese nodules have attracted increasing interest with a view to future economic exploitation, the literature on this topic has rapidly increased during the last decade. The majority of these contributions deal with the geochemical, mineralogical and sedimentological aspects of nodule composition and formation; but some of the problems of their origin are still far from solved. Apart from the evidence of bacterial contributions to manganese accretion (Ehrlich, 1963, 1966, 1968, 1971; Trimble & Ehrlich, 1968, 1970), the role of organisms and biological environment in nodule formation and construction has been almost neglected until now. The only report of organic constituents in manganese nodules seems to be the minute observations made by Murray & Renard (1891, p. 341 ff., pl. 9) who described several groups of benthonic animals attached to these nodules. (The influence of Foraminifera in mineral accretion on the sea-floor, stressed by Graham & Cooper (1959), has never been taken seriously into account by subsequent workers.)

Prior to their discovery on the ocean floor, ferromanganese concretions had been described from the geological record (e.g. Gümbel, 1861, p. 436, 447). Later on, the same author (Gümbel, 1878, p. 209) remarked upon the close similarity of these occurrences in the Alpine Lower Jurassic to the Recent forms discovered subsequently on the ocean floors. Fossil manganese nodules occur as early as the Devonian (Tucker, 1973) and more frequently in the Triassic and Jurassic (Jenkyns, 1970; Wendt, 1970),

* Publication No. 21 of the Research Project 'Føssil Assemblages' (Fossil-Vergesellschaftungen) supported by the Special Research Programme (Sonderforschungsbereich) 53-Palökologie, at the University of Tübingen.

and are almost restricted to off-shore platforms and ridges developed during an intermediate stage of geosyncline formation. Such topographic features were characterized by limited depth and highly reduced sedimentation rates. A peculiar feature of many of these fossil ferromanganese nodules is that they are intimately associated with encrusting Foraminifera (Wendt, 1969, pl. 25; Tucker, 1973, Fig. 7) which seem to indicate a different biological and bathymetrical environment when compared with their presumed Recent counterparts. It has been the aim of the present study to examine the occurrence and distribution of similar organisms in Recent ferromanganese nodules and crusts in relation to water depth and sedimentary environment.

MATERIAL

The material has been personally selected among thousands of deep-sea samples (dredge-hauls, cores) stored in the Woods Hole Oceanographic Institution (Woods Hole, Mass.), Lamont Geological Observatory (Palisades, N.Y.), Institute of Marine Science (Miami, Florida) and Scripps Institution of Oceanography (La Jolla, Calif.) and compared with all available data on submarine topography, surrounding sediment and biological description in order to cover a wide range of environments where manganese nodules are actually being formed. Special attention has been given to seamounts and similar topographic highs which are presumed to be the best topographic counterparts to similar mineral enrichments in the fossil record (Jenkyns, 1970). A total of forty-four manganese nodules, seventy-six ferromanganese crusts and forty-one accompanying unconsolidated sediment samples and rocks have been thin-sectioned and examined.

Before sectioning, the material was dried and embedded in epoxy (Araldite) in a vacuum. Grinding of the slides had to be especially precise because only at about 0·01 mm thickness do the sections become completely transparent. Element distribution has been checked by microprobe analysis. In order to identify the Foraminifera, fracture surfaces of crusts and nodules were photographed using a scanning electron microscope; in addition a few samples were treated with 30% H_2O_2 to obtain better isolation of the included organisms.

ATTACHED BENTHONIC ORGANISMS IN AND ON MANGANESE NODULES AND CRUSTS

Foraminifera

It has been a surprising observation that in about half of the examined material micro-organisms, almost exclusively sessile Foraminifera, have contributed to the accretion of ferromanganese nodules and crusts. Most abundant are several types of arenaceous Foraminifera; these, however, may represent only a few different genera of the family Ammodiscidae, most probably *Tolypammina* and *Glomospira.*

Three types of arenaceous Foraminifera differing in size, growth form and wall thickness have been encountered: very thick-shelled coarsely agglutinated wormlike or very loosely coiled forms with tube-diameters of up to 0·6 mm (probably *Tolypammina,* Fig. 1); finely agglutinated forms of medium wall thickness which are rather regularly coiled or exhibit an irregular tight globular pattern of coiling, with tube-diameters of up to 0·1 mm (probably *Glomospira*, Figs 2–5); and extremely small thin-shelled forms (tube-diameter up to 0·03 mm) which commonly fill the tiny open spaces between the arborescent ferromanganese segregations (colloform structures) (Fig. 6). The particles of the smallest forms are so fine that they cannot be clearly detected under polarized light; microprobe analysis, however, reveals that Si is highly concentrated in the shell wall with respect to the inorganic mineral environment (Fig. 7), a fact which can only be interpreted as being the result of incorporation of fine Si-rich volcanic detritus in the wall of the Foraminifera. These tiny forms seem to be the same as those in ferromanganese concretions of the Upper Lias in the Northern Alps (Wendt, 1969, pl. 25).

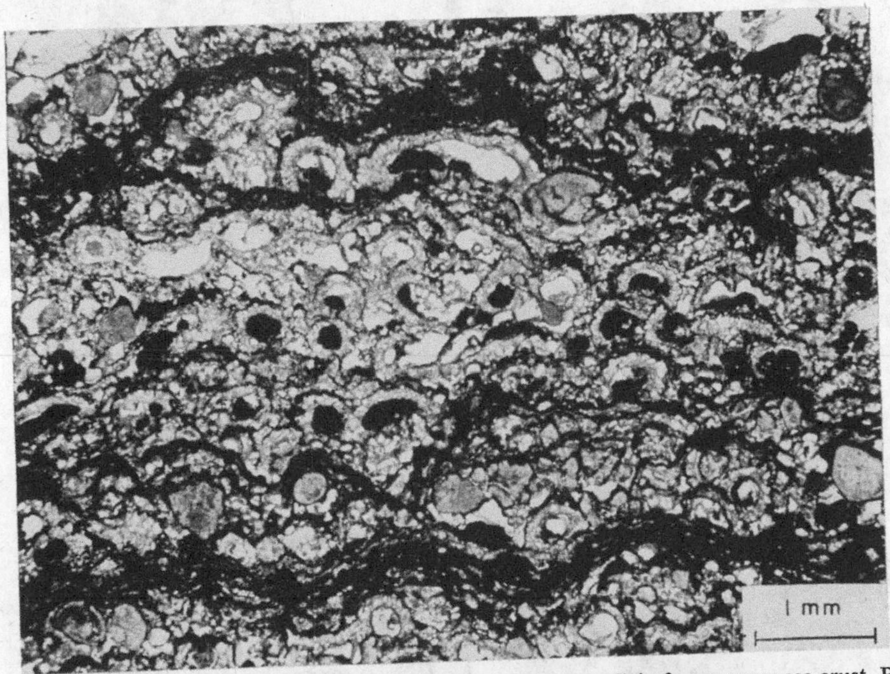

Fig. 1. Layer of coarsely agglutinated Foraminifera (*Tolypammina*?) in ferromanganese crust. Baca-Seamount: 22° 50′ N, 109° 15′ W, 1250 m (uncorr.). Scripps Inst. Oceanogr., BAC 17. Thin section.

Calcareous multilocular Foraminifera with a radial fibrous wall structure and with indistinct or non-existent perforations (*Gypsina*? and *Cymbalopora*-like forms) occur only in a few samples. In one case (Fig. 8) a cluster of large individuals with a distinct wall layering had settled on an internal surface of a nodule, the rest of which was virtually devoid of other Foraminifera. In the deepest sample in which calcareous attached Foraminifera have been encountered (3329 m), the fibrous wall structure is less evident due to incipient recrystallization.

Other organisms

The only other trace of organic activity within ferromanganese nodules and crusts is represented by thin borings (φ about 0·01 mm), probably of fungal origin, which

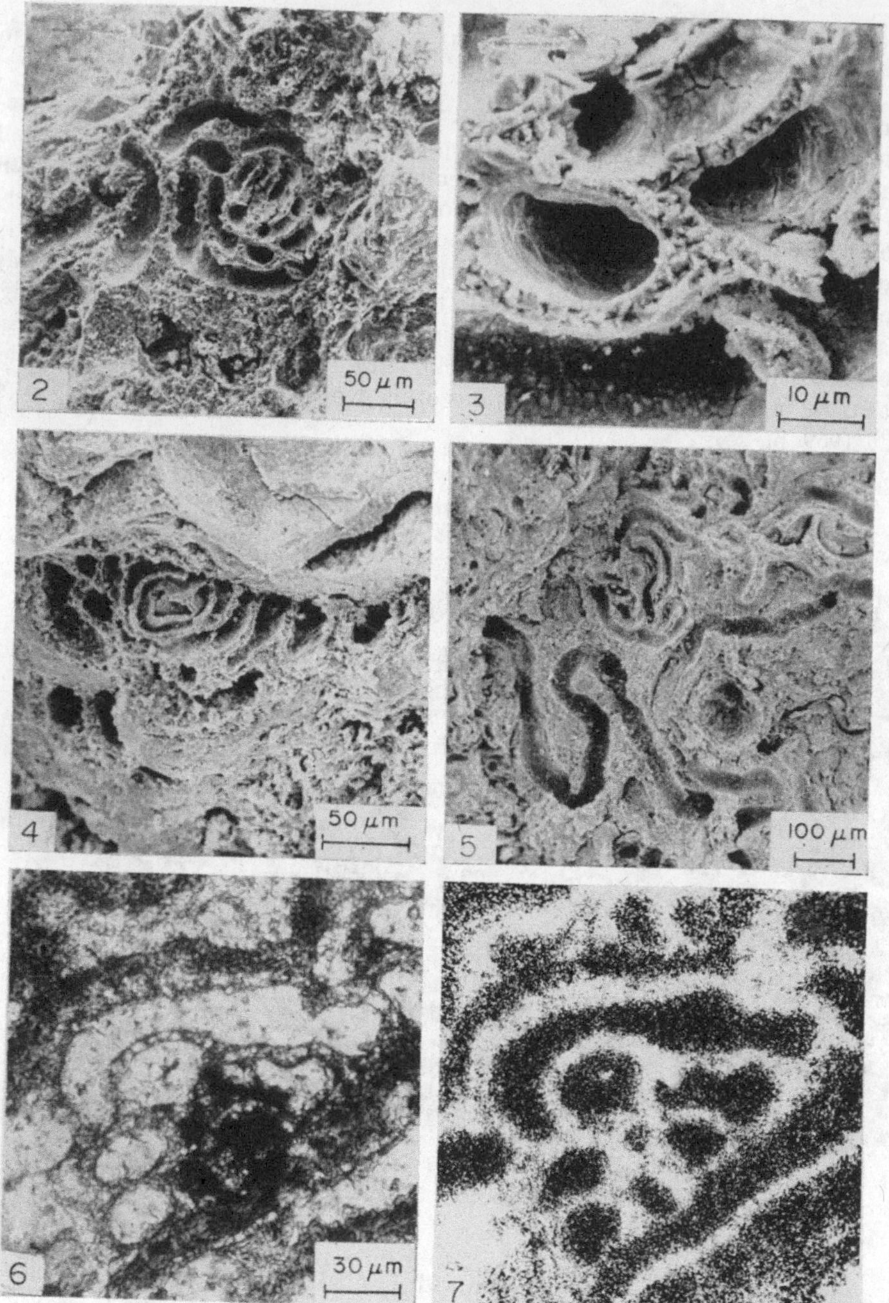

Fig. 2. Finely agglutinated Foraminifera (*Glomospira*?) attached on internal surface of manganese nodule. Caryn Seamount: 36° 41,9′ N, 67° 54,7′ W, 4162–3707 m. Lamont Geol. Observatory, RC 8–RD 12. Scanning electron micrograph.

Fig. 3. Close-up of Fig. 2 showing fine agglutination of foraminiferal walls.

Figs 4 and 5. Finely agglutinated Foraminifera (*Glomospira*? *Tolypammina*?) attached on internal surface of manganese nodule. Blake Escarpment: 27° 14′ N, 76° 26,3′ W, 4710 m (uncorr.). Lamont Geol. Observatory, EW 50-11068. Scanning electron micrograph.

Fig. 6. Arenaceous Foraminifera with indistinct agglutination occurring between colloform structure of manganese nodule. Caryn Seamount: 36° 43,7′ N, 67° 56′ W, 3329 m. Lamont Geol. Observatory, RC 18-148, 20 cm. Thin section.

Fig. 7. Same as Fig. 6. Electron scanning X-ray micrograph showing distribution of Si.

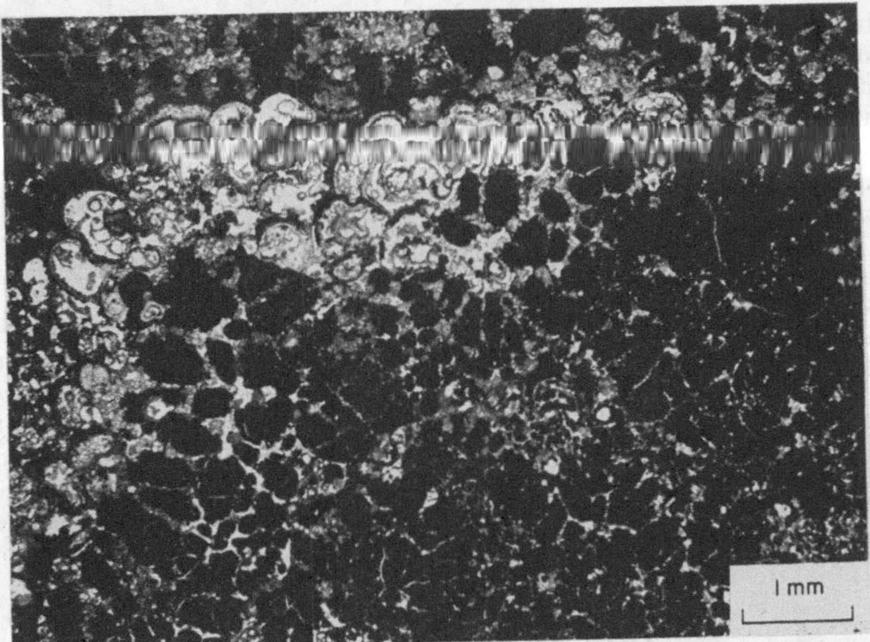

Fig. 8. Cluster of calcareous Foraminifera (*Gypsina*?) settled on internal surface of arborescent ferromanganese segregations in manganese nodule. Top of Crown Princess Cecilia Seamount, S. Pacific: 17° 49′ S, 154° 01′ W, 1027–1452 m. Lamont Geol. Observatory, V 18-BBD 4. Thin section.

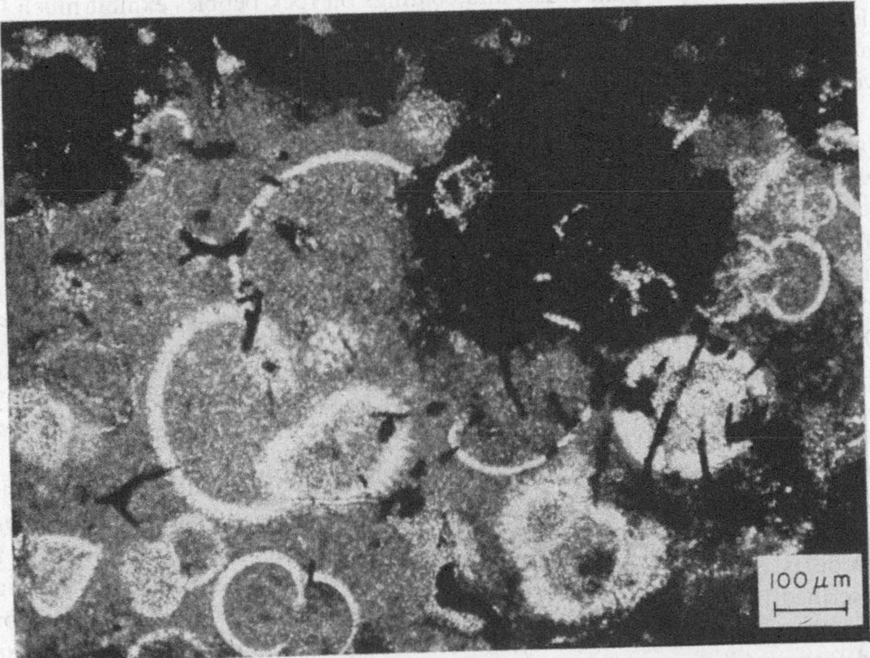

Fig. 9. Fungal (?) borings penetrating from mineralized zones of manganese crust into sediment-filled interstices with globigerinids. San Pablo Seamount: 38° 54′ N, 61° 01′ W, 1207–1262 m (uncorr.). Woods Hole Oceanogr. Inst., A 280–3. Thin section.

penetrate from the dark mineralized zones into sediment-filled interstices (Fig. 9) and into attached Foraminifera. Similar borings are rather common in sunken globigerinid tests of seamount samples and in manganese-stained deep-sea coral debris from the Mid-Atlantic Ridge and the Blake Plateau. These simple or branched perforations are generally filled with dark ferromanganese material indicating that they originated in the same environment and water depth as the surrounding mineral encrustations.

In contrast to this rather uniform and species-poor faunal composition within the nodules and crusts, seamount and near-shore samples have a much higher diversity of organisms on the *outer* surface. The best examples are offered by manganese nodules from the Blake Plateau which show encrustation with corals (probably of the same species as those described by Stetson, Squires & Pratt (1962) and Pratt (1971), bryozoans, serpulids and glass sponges. Flat manganese slabs show epifaunal overgrowth on either side—additional evidence for the strong bottom currents in that region (Pratt, 1963; Pratt & McFarlin, 1966). Manganese-stained and encrusted sedimentary rocks from here and other topographic highs are often heavily bored by cirripedes or burrowed by organisms of uncertain origin.

FAUNAL DISTRIBUTION

Sessile Foraminifera are common in about 30% of the examined manganese nodules, slabs and thick encrustations. A similar percentage of the material studied revealed only scattered individuals, while in the rest these organic constituents are extremely rare or absent. Thin crusts and coatings on rock pebbles exhibit much less organic overgrowth, which leads to the assumption that these attached organisms have a certain affinity for zones of accretion of iron and manganese. In contrast to the average distribution of these two elements in manganese nodules (Mero, 1965, Table 28; Price & Calvert, 1970, Table 3), the Foraminifera-rich zones generally contain much more iron than manganese (average of twelve microprobe analyses).

Foraminifera generally settled on the plane surfaces of the close-spaced mineral layers and in the interstices between the arborescent colloform structures. The latter pattern of foraminiferal intergrowth leaves no doubt that the characteristic columnar segregations in manganese nodules are a primary feature of the mineral accretion and not a post-depositional phenomenon as has been stressed by Cronan & Tooms (1968, p. 221). Indeed, they are reminiscent of the well-known algal stromatolites reduced to a microscopic scale. But how can one imagine a seasonal influence on bacteria-induced manganese accretion in water depths of thousands of metres, as has recently been visualized by Monty (1973)?

Foraminifera are not always uniformly distributed within nodules, but are sometimes concentrated in individual layers or clusters (Figs 1, 8, 10). This is most evident in those cases where mineral accretion has been distinctly interrupted, such as in hiatus-nodules. Such a restricted occurrence of organisms in a manganese nodule has been described by Sorem (1967); however, from his Fig 2, it is not clear if the Foraminifera that occur locally in small pockets of an erosional internal surface are attached forms or if they are free-living ones accidentally trapped by the subsequent mineral growth. In Fig. 11 of this paper, Foraminifera are abundant only in the centre of the nodule, which is a fragment of an older one; they are completely lacking in the

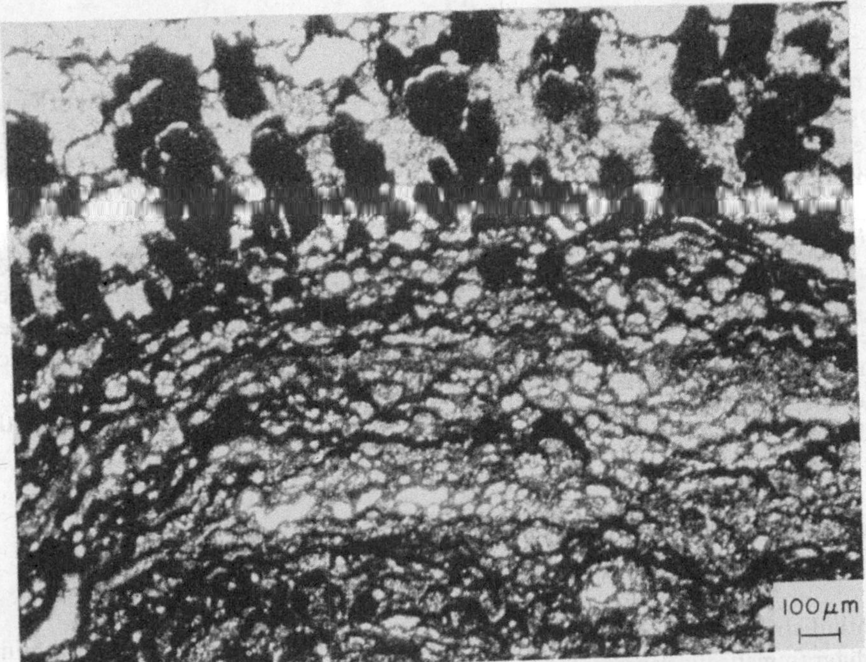

Fig. 10. Changes in the intensity of organic intergrowth in a manganese nodule: layer below almost entirely composed of arenaceous Foraminifera, overlain by rim of colloform structures without Foraminifera. Caryn Seamount: 36° 41,9′ N, 67° 54,7′ W, 4162–3707 m. Lamont Geol. Observatory, RC 8-RD 12. Thin section.

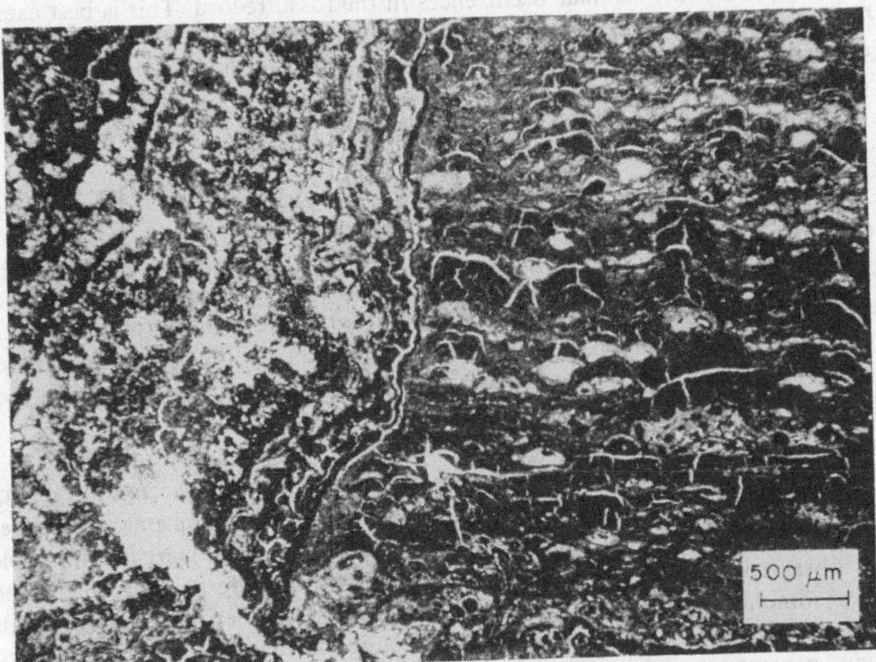

Fig. 11. Hiatus-nodule: In the centre (right) a fragment of a manganese nodule with abundant foraminiferal intergrowth and fine open shrinkage cracks (white), overlain, with a sharp contact, by a subsequent accretionary phase without Foraminifera. 140 miles NE of Burdwood Bank: 53° 00′ S, 52° 54′ W, 3131–3074 m. Lamont Geol. Observatory, V 18-RD 11. Thin section.

outer rim that represents a later accretionary phase. A similar configuration of faunal distribution is represented in septarian manganese nodules which are very common on the Blake Plateau (McFarlin, 1967; Pratt, 1971; Morgenstein, 1971): sessile Foraminifera are virtually absent between the concentric layers of these nodules, but appear rather suddenly on the walls of the cracks with the onset of a new phase of manganese growth. The numerous fine shrinkage cracks (Fig. 11; same as Monty's (1973) 'laminae claires'?), are, however, always devoid of attached organisms. These rather abrupt changes of foraminiferal content in the construction of manganese nodules must reflect changes in the microbiological, sedimentary and/or geochemical environment. A similar interpretation must be given to the obvious differences in organic incrustation on the outer and inner surfaces of nodules, especially those of the Blake Plateau.

REGIONAL DISTRIBUTION OF FORAMINIFERA-RICH MANGANESE NODULES AND CRUSTS

Depth distribution of the examined samples in relation to frequency of Foraminifera is diagrammatically represented in Fig. 12. Though the bulk of the material comes from relatively shallow depths of up to 2000 m, it is evident that attached Foraminifera occur at all depths up to more than 5000 m. They do not even show any obvious preference for special features of submarine morphology, such as seamounts or other topographic highs (e.g. Blake Plateau, Mid-Atlantic Ridge), as has been presumed from the comparison with similar occurrences in the fossil record. This is best exemplified by the twenty-one examined samples from the Blake Plateau (depth range 500–1400 m) in which foraminiferal intergrowth is absent or very rare (Fig. 12). For these reasons, attached Foraminifera in manganese nodules cannot be used either as depth indicators or as criteria for a certain submarine topography.

CONCLUSIONS

Over large areas of the ocean floor, including abyssal plains, manganese nodules are presumably the only suitable substrate for a bottom-dwelling hardground fauna. A more detailed examination of these micro-environments by biologists and micropalaeontologists should yield new information about depth-range and ecology of benthonic Foraminifera.

From the above discussion it is evident that these organisms may contribute to a noticeable extent in the growth of manganese nodules and crusts. Accretion rates of 1–6 mm/10^6 years, as have been calculated on the basis of radiochemical analyses (Ku & Broecker, 1969), or even of only 1 mm/10^3 years (Price, 1967), are incompatible with the observed intensity of organic intergrowth. Foraminifera in organism-rich layers are sometimes so close-spaced that boundaries between single individuals cannot be clearly discerned (Figs 1, 8, 10). This suggests vegetative reproduction of subsequent generations rather than an occasional settling of free-living gamonts (or agamonts). On the basis of an estimated average life time of about half a year for one generation, the accretion rate of these layers composed exclusively of sessile Foraminifera should be in the order from 1 mm/year to 1 mm/25 years. Future calculations

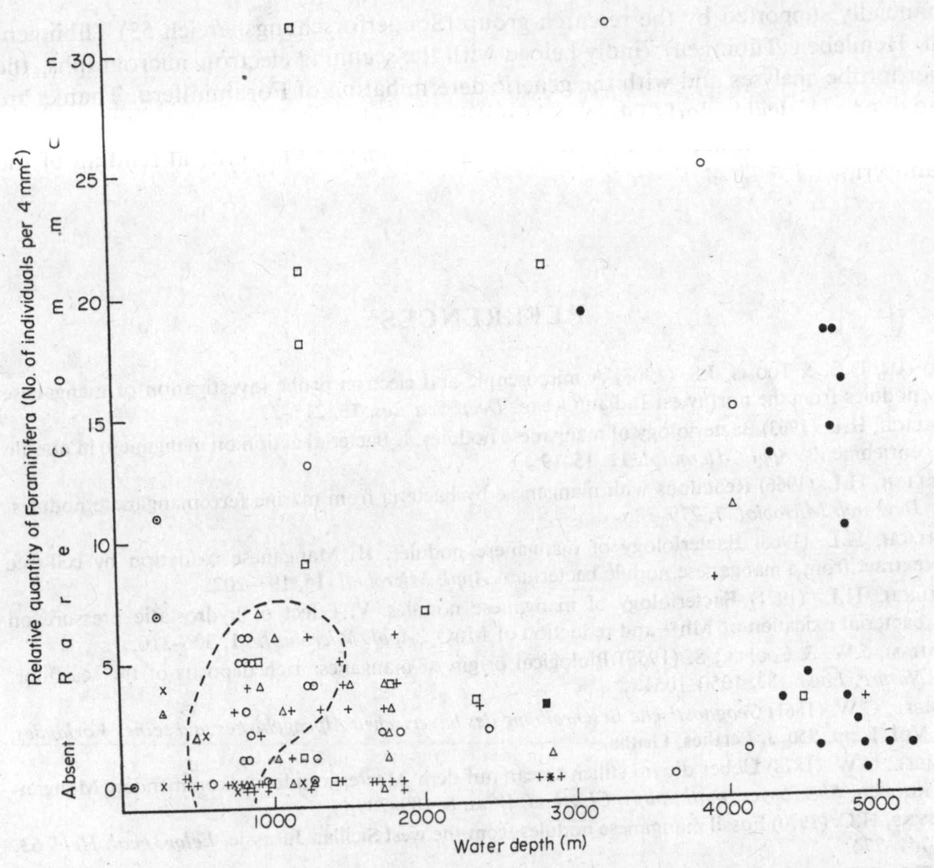

Fig. 12. Relative quantity of attached Foraminifera in manganese nodules and crusts in relation to water depth and submarine topography. 1–4, Samples from seamounts and other topographic highs; 5–8, samples from ocean floors; 9–12, near-shore samples. 1, 5, 9=nodules, 2, 6, 10=thick crusts, 3, 7, 11=thin crusts, 4, 8, 12=thin coatings. The broken line encircles twenty-one samples from the Blake Plateau. ⊙, 1; □, 2; △, 3; +, 4; ●, 5; ■, 6; ▲, 7; *, 8; ⊙, 9; ⊡, 10; △, 11; ×, 12.

of the rates of manganese nodule accretion should therefore take into account the presence and density of their autochthonous faunal constituents.

ACKNOWLEDGMENTS

During his visit to several oceanographic institutions in the United States, the author received generous assistance with the preparation and loan of deep-sea samples from F. T. Manheim (Woods Hole Oceanographic Institution, Woods Hole, Mass.), R. Capo (Lamont-Doherty Geological Observatory of Columbia University, Palisaides, N.Y.), R. J. Hurley (Institute of Marine Science, University of Miami, Florida), W. R. Riedel, T. Walsh, E. L. Winterer (Scripps Institution of Oceanography, La Jolla, California). Acquisition of the Lamont samples, used in the present study, has been made possible by support of the National Science Foundation (NSF-GA-29460) and the Office of Naval Research ONR (N00014-67-A-0108-0004). The study has been financially supported by the research-group (Sonderforschungsbereich 53) Tübingen. Ch. Hemleben (Tübingen) kindly helped with the scanning electron micrographs, the microprobe analyses and with the generic determination of Foraminifera. Thanks are also due to U. Heidersdorf (Tübingen) for a numerical examination of the thin sections, to W. Berger (La Jolla, Calif.) and H. Jenkyns (Durham) for critical reading of the manuscript.

REFERENCES

Cronan, D.S. & Tooms, J.S. (1968) A microscopic and electron probe investigation of manganese nodules from the northwest Indian Ocean. *Deep Sea Res.* **15,** 215–223.

Ehrlich, H.L. (1963) Bacteriology of manganese nodules. I. Bacterial action on manganese in nodule enrichments. *Appl. Microbiol.* **11,** 15–19.

Ehrlich, H.L. (1966) Reactions with manganese by bacteria from marine ferromanganese nodules. *Devs ind. Microbiol.* **7,** 279–286.

Ehrlich, H.L. (1968) Bacteriology of manganese nodules. II. Manganese oxidation by cell-free extract from a manganese nodule bacterium. *Appl. Microbiol.* **16,** 197–202.

Ehrlich, H.L. (1971) Bacteriology of manganese nodules. V. Effect of hydrostatic pressure on bacterial oxidation of Mn^{II} and reduction of MnO_2. *Appl. Microbiol.* **21,** 306–310.

Graham, J.W. & Cooper, S. (1959) Biological origin of manganese rich deposits of the sea floor. *Nature, Lond.* **183,** 1050–1051.

Gümbel, C.W. (1861) *Geognostische Beschreibung des bayerischen Alpengebirges und seines Vorlandes,* Vol. I, pp. 950. J. Perthes, Gotha.

Gümbel, C.W. (1878) Ueber die im stillen Ocean auf dem Meeresgrunde vorkommenden Mangan-knollen. *Sber. bayer. math.-phys. Cl. Akad. Wiss.* **8,** 189–209.

Jenkyns, H.C. (1970) Fossil manganese nodules from the west Sicilian Jurassic. *Eclog. geol. Helv.* **63,** 741–774.

Ku, T.L. & Broecker, W.S. (1969) Radiochemical studies on manganese nodules of deep-sea origin. *Deep Sea Res.* **16,** 625–637.

McFarlin, P.F. (1967) Aragonitic vein fillings in marine manganese nodules. *J. sedim Petrol.* **37,** 68–72.

Mero, J.L. (1965) *The Mineral Resources of the Sea,* pp. 312. Elsevier, Amsterdam.

Monty, C. (1973) Les nodules de manganèse sont des stromatolithes océaniques. *C. r. hebd. séanc. Acad. Sci., Paris,* D, **276,** 3285–3288.

Morgenstein, M. (1971) A study of the growth morphologies of two deep-sea manganese mega-nodules. *Pacif. Sci.* **25,** 308–312.

Murray, J. & Renard, A.F. (1891) Report on deep-sea deposits based on the specimens collected during the voyage of H.M.S. *Challenger* in the years 1872–1876. In: '*Challenger Reports*', pp. 525. H.M.S.O., Edinburgh.

Pratt, R.M. (1963) Bottom currents on the Blake Plateau. *Deep Sea Res.* **10,** 245–249.

Pratt, R.M. (1971) Lithology of rocks dredged from the Blake Plateau. *S.-East. Geol.* **13,** 19–38.

Pratt, R.M. & McFarlin, P.F. (1966) Manganese pavements on the Blake Plateau. *Science,* **151,** 1080–1082.

Price, N.B. (1967) Some geochemical observations on manganese-iron oxide nodules from different depth environments. *Mar. Geol.* **5,** 511–538.

Price, N.B. & Calvert, S.E. (1970) Compositional variation in Pacific Ocean ferromanganese nodules and its relationship to sediment accumulation rates. *Mar. Geol.* **9,** 145–171.

Sorem, R.K. (1967) Manganese nodules: nature and significance of internal structure. *Econ. Geol.* **62,** 141–147.

Stetson, T.R., Squires, D.F. & Pratt, R.M. (1962) Coral banks occurring in deep water on the Blake Plateau. *Am. Mus. Novit.* **2114,** 1–39.

968) Bacteriology of manganese nodules. III. Reduction of MnO_2
ria. *Appl. Microbiol.* **16,** 695–702.
1970) Bacteriology of manganese nodules. IV. Induction of an
arine bacillus. *Appl. Microbiol.* **19,** 966–972.
se nodules from the Devonian of the Montagne Noire (S. France)
ch. **62,** 137–153.
d Paläogeographie des Roten Jurakalks im Sonnwendgebirge
eol. Paläont. Abh. **132,** 219–238.
Kondensation in triadischen und jurassischen Cephalopoden-
eol. Paläont. Mh. **1970,** 433–448.